THE PHYSIOLOGY COLORING BOOK

THE PHYSIOLOGY COLORING BOOK

WYNN KAPIT   **ROBERT I. MACEY**   **ESMAIL MEISAMI**

1817

HARPER & ROW, PUBLISHERS, NEW YORK

Cambridge   Philadelphia   San Francisco
London   Mexico City   São Paulo   Singapore   Sydney

Sponsoring Editor: Claudia M. Wilson
Production: Kewal K. Sharma
Printer and Binder: The Murray Printing Company

The Physiology Coloring Book

ISBN: 0–06–043479–1

87 88 89 90 9 8 7 6 5 4 3 2 1

CONTENTS

CONTENTS

CONTENTS

In this book, we provide a self-contained synopsis of modern human physiology. The material begins at the beginning; it is developed from the ground up and is suitable for both college and health professional students, as well as for self-study by educated laypersons. To accomplish this within the confines of 153 plates, we utilize the unique pedagogical features made available by the active process of coloring. The result is a non-conventional book providing an alternative or supplement to commonly used texts.

What are these features, and how do they apply to physiology? In the case of anatomy, the virtues of coloring are unmistakable. Classical anatomy is a visual science concerned with well-defined physical structures. Drawing these structures is a time-honored process that works because it cannot be done without personal attention to detail. In many ways, coloring structures is similar to drawing. It develops appreciation for shapes and relative proportions, and, perhaps more important, it introduces a kinesthetic sense into the learning process as hand motion is integrated with visual stimuli. Further, the use of color coding provides a simplification and awareness of relationships in complex drawings that are hardly attainable by other means.

To the extent that physiology depends on structure, the same benefits accrue to it. However, descriptions of static anatomical structures are merely the starting point. Physiology's most distinctive feature is that it deals with dynamic processes. This is reflected in the prevailing and effective use of flow diagrams to describe forces, flows, chemical reactions, steady states, signals, feedbacks, etc. These concepts are necessarily more abstract, they have not been standardized into any universally accepted symbolic representation, and they introduce significant difficulties for beginning students. This book addresses these difficulties in several ways. In the first place, the liberal use of illustrations and even cartoons adds "flesh and blood" to flow diagrams, allowing students to associate process with locale or other more familiar ideas. Further, the use of color codes makes it easy to follow common elements (e.g., H^+ ions in acid-base balance) in complex diagrams. But most importantly, the coloring process provides an urgently needed focus for first encounters with complex phenomena. In these instances, beginners commonly quit in despair while the more experienced will break the problem into smaller, manageable parts and gradually piece the entire problem together. The act of coloring forces the student to confront a complex diagram one part at a time, making the novice feel more secure in a state of "not knowing" for longer periods so that learning has a greater chance. Finally, the individual choices of colors make the project both personal and fun — a welcome diversion from stereotype studies, where many long hours are often spent going through the motions of soaking up information. We have had fun producing this volume; we hope you will have good times too.

Although the chapters are placed in linear sequence it is not always necessary to follow them in the order of presentation. Some readers may find the beginning plates too abstract on first encounter; they may profit by starting with one of the organ systems in the latter sections of the book, returning to earlier plates as needed. In any case, it is highly desirable to read (and refer back to) the introduction on the following page, which explains a number of codes and symbols that are utilized consistently throughout the book.

Our attempts to depict physiological phenomena in semi-literal caricatures inevitably involves compromise. Furthermore, some topics have been developed more intensely, at the expense of others. We are interested in readers' opinions on these issues, and we will especially appreciate responses that point out any inaccuracies. We are grateful for the expert critical advice of a number of our colleagues who have reviewed sections of the book. These include Kenneth Andersen, Mary Banich, Jasper Brahm, Debra Draves, Albert Feng, John Forte, Martha Gillette, Jon Goerke, Janice Juraska, Terry Machen, Curtis Okimoto, Suzanne Palmer, Alex Quintanilha, Lawson Rosenberg, Sharon Russell, Paola Timiras, Tony Waldrop, and Francis White. In addition, we were fortunate to enlist the copy editing talents of Sylvia E. Stein. We also thank Deborah Kamali for her critical review of the illustrated material and supplying us with a most artistic color rendition of each plate. At Delphey & Chase in Berkeley, we wish to thank Barbara Martisch for her typesetting efforts and David Gartland, proofreader. Finally, we thank Claudia M. Wilson, our sponsoring editor at Harper & Row, for her encouragement and support through this project.

Wynn Kapit
Robert I Macey
Esmail Meisami

Berkeley, California

INTRODUCTION
IT IS VITAL THAT YOU READ THIS!

HOW TO USE THIS BOOK

The book is arranged in sections. It isn't necessary to follow these sections in the order they appear, provided that you begin with the first plate of whichever section you choose.

There are 153 plates, each dealing with a specific topic. A plate consists of a text page and a coloring page. The text provides an introduction and overview of the subject under consideration. Within the text, the use of *italics* occurs when important points are presented, or when a *title* (a word that also appears in outlined letters on the illustration page) appears for the first time. At the bottom of the text page, you will find color notes (CN) that recommend specific colors (where they are needed) and tell what order to follow when coloring the page. On the opposite page, waiting to be colored, are the illustrations and titles. Captions are also present to explain the meaning of the illustrations. Within the caption, you may come across capital letters. These are the abbreviations of words that maybe repeated in captions (e.g., M for membrane, C for cytoplasm, etc.). The word in question will be spelled out first and abbreviated afterward. Do not confuse these abbreviations with the letter labels that follow the titles and are used to identify the parts of the illustration that have to be colored. These letter labels do not appear in the captions but may be found in the color notes.

You will soon discover whether you prefer to read the text first and then color, or color the titles and illustrations first, read the relevant caption, and consult the text last. Whatever you do, read the color notes before doing any coloring. Above all, it is essential that you color this material because much of the information contained in these plates becomes apparent only when color is applied. Only by coloring is it possible to achieve a unique level of awareness and concentration. And without color, the book's usefulness as a reviewing tool would be severely reduced.

HOW TO COLOR THIS BOOK

The following instructions may seem overwhelming when read all at once, but we promise that in a very short time you will become quite at home with the coloring format and these instructions will be easily remembered.

Use felt-tipped pens with fine points (first choice) or colored pencils. A minimum number to have is twelve (including a medium gray color). Twenty colors would be ideal, allowing for more variety and enjoyment. If possible, buy mostly light colors. These pens or pencils may be purchased individually at art supply or stationery stores. You would also be able to replace individual colors when necessary. If all you can afford is a small, prepackaged set of colors, do not despair; you can still do the job with less than twenty.

Light colors are desirable because they are not as likely to conceal the surface detail of the illustration or, when applied to a large area, overpower everything else on the page. The color notes (CN) will often recommend the use of a specific color, the repetition of colors used on the preceding page, or the use of a dark color where emphasis is needed. Examples of specific colors might be red for arterial (oxygenated) blood, purple for the capillaries (where blood is giving off oxygen), and medium blue for veins (deoxygenated blood). A lighter blue is usually recommended for the color of water. Where certain colors are specified, you may discover that a useful procedure to follow is to first find all the titles that have these colors assigned to them and color them in before coloring anything else on the page (not even coloring the structures to which the titles refer). Then start at the beginning of the page and begin coloring in the prescribed order, while using your remaining colors for any of the remaining titles.

The basic principle of the coloring format is first to color a title and then to use that same color for any structure on

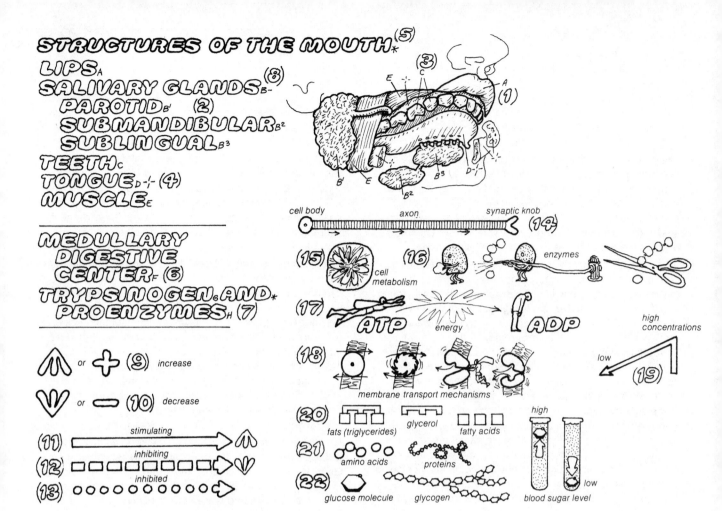

STRUCTURES OF THE MOUTH* (5)

LIPS_A

SALIVARY GLANDS_B- (8)

   PAROTID_B¹ (2)

   SUBMANDIBULAR_B²

   SUBLINGUAL_B³

TEETH_C

TONGUE_D -/- (4)

MUSCLE_E

MEDULLARY
DIGESTIVE
CENTER_F (6)
TRYPSINOGEN_G AND*
PROENZYMES_H (7)

(9) increase

(10) decrease

(11) stimulating

(12) inhibiting

(13) inhibited

cell body — axon — synaptic knob (14)

(15) cell metabolism

(16) enzymes

(17) ATP — energy — ADP

(18) membrane transport mechanisms

(19) high concentrations / low

(20) fats (triglycerides) / glycerol / fatty acids

(21) amino acids / proteins

(22) glucose molecule / glycogen / blood sugar level (high / low)

---

the page identified with the letter label that follows the title (see above, 1). Occasionally, the same letter may be followed by different superscripts, identifying different but related structures. In that case, because the same letter is used, this suggests that the structures are sufficiently related to warrant the use of the same color (2). Where an identical structure, or part of a diagram, is repeated many times but is labeled only once or twice, it is still necessary to color all such repeated structures, even if unlabeled. Structures that are meant to be colored will usually be drawn with a darker outline (in addition to being labeled) (3). If the title or its structure are not meant to be colored, they will be so identified with a -/- sign following the letter label (4). Where the color gray is requested, an asterisk ✗ will follow the title or structure. Gray is commonly used in the case of titles representing subject headings (5).

Color all titles on the illustrated page. If a single title consists of words lettered in two or three lines, with a letter label appearing only after the last word, color all the preceding words in the title with that color (6). If a single title should contain words requiring different colors, those words will be so identified with different labels (7). Sometimes a title will not refer to any particular structure but will still need to be colored. In that case, the letter label will be followed by a hyphen (8). Many of the technical words in the titles have been lettered with a slight space between syllables. This is designed to aid you in the recognition and pronunciation of unfamiliar terms.

We are using the following symbols throughout the book as a space-saving, memorable way of conveying as much information as possible within the space limitations of the illustrations and diagrams:

Any structure or substance associated with these symbols

is increasing, rising, growing, activating or stimulating (9).

These symbols suggest a decrease, decline or reduction in size (10).

A solid line signifies a direct connection, stimulation, excitation, or secretion from one structure or substance to another (11).

A broken line indicates a reducing or inhibiting effect (12).

A dotted line shows a cessation of activity (the state of being inhibited or turned off) (13).

A generalized nerve. The pattern of lines along the length of the axon suggests nerve impulses, though these lines may not always be present when the symbol appears (14).

Metabolism occurring within the cell (burning or rapid consumption) (15).

Various enzyme symbols representing a devouring, breaking down, or cutting apart action. (16)

The ATP and ADP molecules. ATP turns into ADP in the process of releasing its energy (17).

Symbols for cell membrane transport mechanisms. The third one is being driven by ATP (18).

A gradient from high to low concentrations (19).

Fats or triglycerides, glycerol, and fatty acids (20).

Amino acids, proteins (long chains of amino acids) (21).

Glucose molecule, and glycogen (long chain of glucose molecules) blood sugar levels (22).

Once again let us assure you that the coloring format is quite logical and fun to do, that the preceding rules and symbols don't all appear all at once on the same page, and that, in a very short time, you will be perfectly at home with the ground rules.

CELL STRUCTURE

"All living things consist of one or more cells."
"Each cell can live independently of the rest."
"Cells can arise only from other cells."

These three statements express the "cell doctrine," an insight that has been accepted for over 100 years. It implies that those parts of our body that live — that eat, breathe, move about, and reproduce — do so only through the cells that make up about two-thirds of our body weight. If physiology seeks to discover how living things work, it must ultimately express the explanations in terms of cellular activities.

In the past, a good deal of physiology was developed by concentrating on cell environments: the fluid that makes up blood plasma and the fluid that surrounds cells. It was found that human cells survive only in highly specialized environments where the relative proportions of minerals, water, nutrients, and other constituents remain within narrow limits. As a result, interpretations often focused on how the body's organs protect the cellular environment from change. This concept has been extraordinarily fruitful for both physiologists and clinicians, and we shall apply it repeatedly in this book. However, physiology is in transition. Modern technology has given us access to the interior of living cells allowing measurements and experiments that could hardly be dreamt of a generation ago. Today research is moving away from its former preoccupation with cellular environments and beginning a concentrated assault on processes occurring within the cell itself. In a way, we are no longer on the outside looking in.

Cells come in different sizes, shapes, and internal structures. Liver cells differ from brain cells, which differ from blood cells. All cells contain "miniorgans" called *organelles*, each specialized to perform a function. Although the cell portrayed in the plate cannot represent all cells of the body, it does contain the following structures and organelles that commonly occur in most.

Cell (plasma) membrane. This outer boundary of the cell consists of a thin (4-5 nm), continuous sheet of fatty (lipid) molecules in which protein is embedded. Some of these proteins provide pathways for transport and regulate the flow of materials into and out of the cell. Other proteins serve as receptors for chemical signals coming from other cells.

Nucleus. The most prominent cellular organelle, the nucleus contains genetic material: genes, DNA, and chromosomes. By expressing information stored in genes, it directs everyday cell life and reproduction. The nucleus contains a smaller body, the *nucleolus*, that consists of densely packed chromosome regions together with some protein and some RNA strands. The nucleolus initiates the formation of *ribosomes*, structures that are required for protein synthesis. The nucleus is surrounded by a double membrane that is riddled with pores involved in transporting materials between the nucleus and the rest of the cell.

Cytoplasm. Occupying the space between the nucleus and the plasma membrane, the cytoplasm contains membrane-bound organelles, ribosomes for synthesizing cytoplasmic proteins, and a complex network of filaments and tubules called the cytoskeleton (see below). The fluid portion of the cytoplasm in between these structures, the *cytosol*, contains many protein enzymes (catalysts used in cellular chemistry).

Mitochondria. These "power houses" of the cell are the sites where chemical energy contained within nutrients is trapped and stored through the formation of ATP molecules. ATP, in turn, serves as an energy "currency" to carry out cellular work, supplying the energy required for movement, secretion, and synthesis of complex structures.

Endoplasmic reticulum. The endoplasmic reticulum (ER) is a network of tubes and flattened sacs, formed by membranes, that is distributed throughout the cytoplasm. Some ER (*rough ER*) has a granular appearance because of attached ribosome particles. These are sites for synthesis of proteins destined for organelles, for cell membrane components or for secretion to the cell exterior (e.g., hormones). *Smooth ER* lacks attached ribosomes. It is commonly involved in lipid metabolism, but it can also serve in detoxification of drugs and deactivation of steroid hormones. In muscle cells, smooth ER (called sarcoplasmic reticulum) sequesters large amounts of calcium, which are used to trigger muscular contraction.

Golgi apparatus. Sets of smooth membranes that form flattened, fluid-filled sacs that are stacked like pancakes, the Golgi apparatus is involved in modifying, sorting, and packaging proteins for delivery to other organelles or for secretion out of the cell. Numerous membrane-bound vesicles are frequently found around the Golgi apparatus. They probably carry material between the Golgi and other organelles of the cell (e.g., receiving protein-laden vesicles from the rough ER or delivering other vesicles to the plasma membrane).

Endo- and Exocytotic vesicles. These membrane-enclosed vesicles traveling from (and to) the plasma membrane are important carriers for protein delivery into (or out of) the cell. *Exocytosis* (secretion) involves an actual fusion of the vesicular membrane with the plasma membrane, enabling vesicle contents to be expelled (secreted) outside the cell. In *endocytosis* (*pinocytosis, phagocytosis*) the reverse occurs: the plasma membrane infolds and engulfs extracellular material; then a membrane-bound vesicle (containing the material and surrounding fluid) buds off and is incorporated into the cell.

Lysosomes. These membrane-bound vesicles contain enzymes capable of digesting natural particles, damaged organelles, and bacteria brought into the cell via endocytosis.

Cytoskeleton. The cytoskeleton consists of arrays of protein filaments that form networks within the cytosol, giving the cell its shape. These filaments also provide a basis for movement of both the entire cell and its components (e.g., organelles). They are the "bones and muscles" of the cell. The cytoskeleton appears to be organized from a region near the nucleus containing a pair of *centrioles* (which are particularly important during cell division). The three major type of cytoskeleton filaments are microtubules (25 nm diameter), actin filaments (7 nm — shown on next plate), and intermediate filaments (10 nm — shown next plate).

CN: *Use your lightest colors for A and G.*
1. Begin in the upper left corner by coloring the title, structural example, and the related structure in the central illustration of the entire cell. Do this for each structure as you work clockwise around the page. Note that the space between membranes of the rough endoplasmic reticulum (I) is not colored in the example on the right but is colored in the central drawing for reasons of identification.

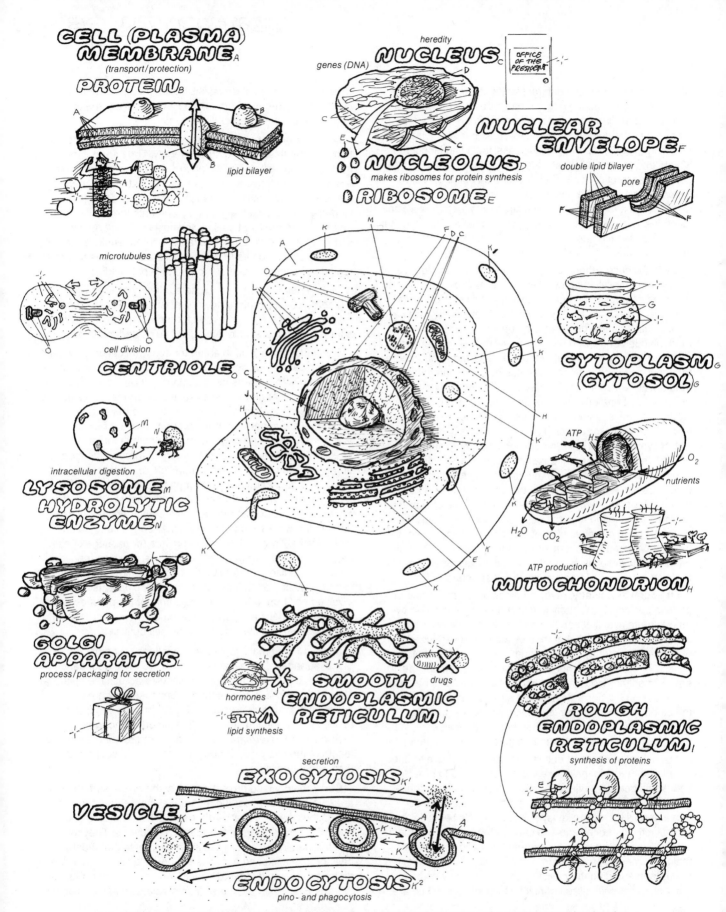

CELL (PLASMA) MEMBRANE A
(transport/protection)

PROTEIN B

lipid bilayer

heredity
NUCLEUS C

genes (DNA)

OFFICE OF THE PRESIDENT

NUCLEAR ENVELOPE F

NUCLEOLUS D
makes ribosomes for protein synthesis

RIBOSOME E

double lipid bilayer
pore

microtubules

cell division

CENTRIOLE O

CYTOPLASM (CYTOSOL) G

intracellular digestion

LYSOSOME M
HYDROLYTIC ENZYME N

ATP
O₂
nutrients
H₂O
CO₂
ATP production

MITOCHONDRION H

GOLGI APPARATUS L
process/packaging for secretion

hormones
SMOOTH ENDOPLASMIC RETICULUM J
drugs
lipid synthesis

ROUGH ENDOPLASMIC RETICULUM I
synthesis of proteins

secretion
EXOCYTOSIS K'

VESICLE

ENDOCYTOSIS K²
pino- and phagocytosis

Cells in different organs of the body are highly specialized, and this specialization is often reflected in structural variations. Although the generalized cell depicted above cannot represent any particular cell, it does contain structures and organelles that commonly occur in most. All cells are bounded by a plasma membrane, a continuous double-layered sheet of lipid molecules containing embedded proteins. Similar membranes form a number of structures within the cell. Essentially, all cells have a membrane-bound nucleus that contains the genetic instructions (genes). By expressing information stored in the genes, the nucleus directs everyday cell life and reproduction. The space between the plasma membrane and the nucleus is called the cytoplasm. Membrane-bound organelles, as well as the filaments and microtubules that compose the cytoskeleton, are suspended within the cytoplasmic fluid, the cytosol.

Although there are many different kinds of human cells, they can be classified into four broad types: (1) muscle cells, specialized for generation of mechanical force and movement; (2) nerve cells, specialized for rapid communication; (3) connecting and supporting tissue cells, including blood and lymph; and (4) epithelial cells for protection, selective secretion, and absorption. This plate focuses on epithelial cells to illustrate how groups of these cells adhere to one another to form tissues and how specialized structures (in this case, cell junctions, microvilli, and cilia) support special functions. Other cell types are taken up in more detail in the context of specific organs.

Epithelial cells adhere to one another, often forming layered sheets with very little space between cells. They are found at surfaces that cover the body or that line the walls of tubular or hollow structures. Thus, epithelial cells are found in the skin, kidney, glands, and linings of the lungs, gastrointestinal tract, bladder, and blood vessels. Sheets of them often form the boundaries between different body compartments, where they regulate exchange of molecules between compartments. Virtually all substances that enter or leave the body must cross at least one epithelial layer. For example, the small intestine forms a hollow cylinder whose interior lining is populated by several types of epithelial cells. Some secrete digestive enzymes, others absorb nutrients, still others secrete a protective mucus. In each case, the epithelial cells are called upon to transport materials in one direction only: either from blood vessels (embedded within the intestinal walls) to the hollow interior (lumen) of the cylinder in the case of secretion, or from lumen to blood in the case of absorption. Thus, the cell must have a "sense of direction"; it must "know" the difference between the lumen side and the blood side. The cell cannot be completely symmetrical, and its asymmetry in function is reflected in an asymmetrical structure.

Structural asymmetry, revealed by both cell shape and organelle position, is probably established and maintained by an elaborate cytoskeleton. In addition, there are striking differences in the plasma membranes located at various sides of the cell. We identify three different surfaces of epithelial cells: (1) The apical or mucosal surface faces the outside environment or the lumen of a particular organ. (2) The basal surface is on the opposite side, the side that lies closest to the blood vessels. (3) The lateral sides face neighboring epithelial cells. Each of these membrane surfaces contains different proteins and structures required for normal function.

The lateral surfaces of epithelial cells must adhere to one another to maintain their sheetlike structure and to provide tight seals between adjacent cells so that fluids and other substances cannot leak between them. If substances do move across the epithelial layer, it is generally because they are selectively recognized and transported by the cells themselves. Discrete structures called desmosomes provide a major source of this adhesion. They lie close to, or within, the membrane and bind the cells together where they come in contact. Other specialized contact sites (tight junctions) are used to plug potential leaks; still others (gap junctions) are used for cell-to-cell communication. Collectively, these contact sites are called cell junctions.

Desmosomes are regions of tight adhesion between cells that give the tissue a structural integrity. They are concentrated in tissues like skin, which are subjected to mechanical stress. At a desmosome, there is a small extracellular space between the two cell membranes that is filled with a fine filamentous material that probably cements the two cells together. There are two types of desmosomes: belt desmosomes (continuous zones of attachment that encircle the cell) and spot desmosomes (more localized attachments to small regions of contact, often compared to "spot welds").

Tight junctions form very close contacts between neighboring cells, leaving virtually no space between. These junctions extend around the entire circumference of the cell, providing a tight seal that prevents leakage of fluids and materials.

Gap junctions are specialized for communication between adjacent cells. They consist of an array of six cylindrical protein subunits that spans the plasma membrane and reaches out a short distance into the extracellular space. The subunits are bunched together with their long axes parallel to one another in a manner that forms an open space or channel about 1.5 nm wide running the entire length of the array. These channels act as pores that tunnel through the membrane, but the tunnels do not empty into the extracellular space. Instead, each array attaches to a similar array in an adjacent cell, forming a tunnel of double the length with the entrance in one cell and the exit in the adjacent cell. These tunnels are wide enough to allow small solutes and common ions to pass. Thus, the junctions provide for passage of electrical and chemical signals between cells, allowing them to function in unison. Under certain circumstances (e.g., a rise in intracellular Ca^{++}), the central channel closes, isolating the involved cell from others. The most common type of cell junction, gap junctions are particularly important in coordinating heart, smooth muscle, and epithelial cell activities.

Microvilli are small, fingerlike projections found on the apical surface of epithelial cells. They are most abundant in tissues that primarily transport molecules across the epithelial sheet. Microvilli are advantageous because they greatly increase the surface area available for transport (e.g., by a factor of 25 in the intestine). Actin filaments, anchored at their base in the terminal web of fibers and running the entire length of the microvilli, are believed to provide support for their upright position.

Cilia are very long projections from the apical surface that are involved in transporting material along (i.e., tangential to) the epithelial surface rather than through it. They are abundant in the respiratory tract, oviducts, and uterus. They function by "beating" (i.e., by whiplike movements that mechanically propel fluids and particles on the cellular surface in the direction of a rapid forward stroke). An array of microtubules that runs the length of each cilium mediates these motions.

CN: Use the same color for the plasma membrane (F) as was used on plate 1.
1. Begin with the 3-dimensional drawing of epithelial cells on the right. As you color each structure, complete its corresponding structure in the cross-sectional diagram on the left. The latter contains additional structures which should be colored as well. Note that A, D, and L are all parts of the plasma membrane (F) but receive different colors.
2. Among the list of titles, notice that the functions of structures H-N are placed in parentheses and are colored gray.

APICAL SURFACE*
CILIA_A
MICROTUBULE_B
BASAL BODY_C
MICROVILLI_D
MICROFILAMENT_E (ACTIN)

LATERAL SURFACE*
LATERAL PLASMA MEMBRANE_F
TIGHT JUNCTION_G (IMPERMEABLE)*
DESMOSOME: BELT_H (CELL-TO-CELL
DESMOSOME: SPOT_I (ADHERENCE)
GAP JUNCTION_J (INTERCELLULAR
COMMUNICATION)*

BASAL SURFACE*
HEMIDESMOSOME_K (ADHERENCE)*
BASAL PLASMA MEMBRANE_L

TERMINAL WEB_M
CYTOSKELETON_N (INTERMEDIATE FILAMENTS)
EXTRACELLULAR SPACE_O
NUTRIENTS & METABOLITES_P

Epithelial cells adhere to one another, often forming layered sheets that cover the body and organs, or line the walls of tubular or hollow structures (skin, kidney, glands, and linings of the lungs, gastrointestinal tract, bladder, and blood vessels). They have 3 distinct surfaces: (1) the apical surface facing the outside environment or the lumen of an organ, (2) the basal surface, on the opposite side, facing blood vessels, and (3) the lateral sides facing neighboring epithelial cells.

THE APICAL SURFACE sometimes contains microvilli and cilia. Microvilli increase the area of the apical surface several fold. Actin filaments, anchored in the terminal web and running the length of each microvillus, are believed to support their upright position. Cilia are involved in transporting material tangential to the epithelial surface by whiplike movements that propel fluids and particles on the cellular surface in the direction of a rapid forward stroke. These motions are mediated by microtubules that run the length of each cilium in a characteristic 9 + 2 array (9 pairs of microtubules forming a ring around a central pair). Each cilium is anchored in a basal body. The cilium is bent as the pairs of microtubules slide past one another.

THE LATERAL SURFACE contains 3 types of junctions: (1) desmosomes, for adherence of neighboring cells, (2) tight junctions, for sealing leaks between cells and (3) gap junctions which provide open channels be-

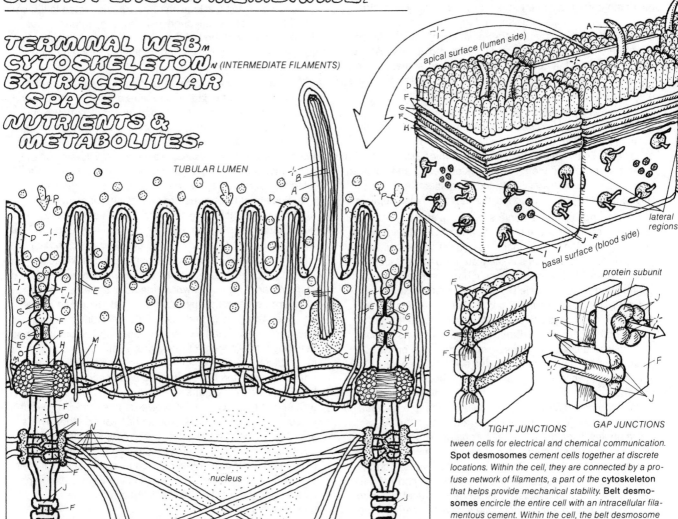

TUBULAR LUMEN

apical surface (lumen side)

lateral regions

basal surface (blood side)

protein subunit

TIGHT JUNCTIONS

GAP JUNCTIONS

nucleus

basement membrane (basal lamina)

blood capillary

nerve

tween cells for electrical and chemical communication. Spot desmosomes cement cells together at discrete locations. Within the cell, they are connected by a profuse network of filaments, a part of the cytoskeleton that helps provide mechanical stability. Belt desmosomes encircle the entire cell with an intracellular filamentous cement. Within the cell, the belt desmosome has a band of cylindrical actin filaments (shown in cross-section) just adjacent to the inner part of the cell membrane.

THE BASAL SURFACE'S plasma membrane is attached to the basement membrane (basal lamina), a porous structure containing collagen and glycoproteins that separates the epithelial cells from underlying connective tissues, nerves, and blood vessels. The attachment is strengthened by hemidesmosomes (half-spot desmosomes).

DNA REPLICATION & CELL DIVISION

No cell lives forever. With a few exceptions (notably nerve and muscle cells), the cells of your body are not the same ones that were present just a few years ago. "Old" cells apparently wear out, die, and are continually replaced by new ones. On average, intestinal cells live for only 36 hours, white blood cells for 2 days, and red blood cells for 4 months; brain cells may live for 60 years or more. Growth also requires the production of new cells. As cell size increases, cells become less efficient because distances from the plasma membrane to the more central portions of the cell also increase, making the transport of such essentials as O_2 into and CO_2 out of the cell more difficult. These difficulties do not arise because growth occurs primarily by increasing the number of cells rather than increasing the mass of individual cells.

CELL DIVISION. In cell division one parent cell divides into two daughter cells to create new cells. Although some characteristics (e.g., weight) of the daughters may be different from the parent, they are identical in the most important way: they both carry the same fundamental set of genetic instructions that govern their activities and reproduction. This instruction set, the *genetic code*, is provided by the detailed structure of *DNA* (deoxyribonucleic acid) molecules that are packaged within the cell nucleus. Replication of these molecules and their distribution to each daughter cell ensures the continuity of cell characteristics with each division.

Processes involved in cell division take place in three phases. 1. During *interphase*, the cell increases in mass by synthesizing a diversity of molecules, including an exact copy of its DNA. That portion of interphase in which DNA synthesis takes place is called the S period; it is preceded and followed by two "gap" periods called G_1 and G_2 respectively (see illustration). During the S period, the centrioles also duplicate.

2. Following G_2, the cell enters *mitosis*, a stage in which the replicate sets of DNA are bundled off to opposite ends of the cell in preparation for the final stages in which the cell splits in two (follow the diagrams in the plate for details). Mitosis begins when DNA molecules, which had been unwound during interphase, become highly coiled and condense into rod-shaped bodies known as *chromosomes*. At this stage, each chromosome is split longitudinally into two identical halves called chromatids. Each *chromatid* contains a copy of the duplicated DNA along with some protein that provides a scaffold for the long DNA molecules and helps regulate DNA activity. Meanwhile, the nuclear envelope begins to degenerate, and, outside the nucleus, *centrioles* migrate to opposite ends of the cell to form an elaborate structure of *microtubules* called a *spindle*. Each chromosome, attached to these microtubules, lines up at the cell's equator in such a way that its two chromatids are attached to microtubules leading to opposite ends of the cell. The microtubules then pull on the chromatids, moving a complete set to opposite parts of the cell. Finally, the chromatids at both ends of the cell begin to unwind and become indistinct while a new nuclear envelope forms around each of the two sets of chromatids. 3. *Cytokinesis* is the final stage. Cytoplasm division takes place as a

furrow develops, becoming deeper and deeper until the original cell is pinched in two, and the daughter nuclei, formed during mitosis, are enclosed in separate cells. At this point, the daughter cells enter the G_1 stage of interphase, completing the cycle.

DNA REPLICATION. If DNA is the heredity material, two important questions arise. First, how is DNA replicated so that it can be passed undiluted from generation to generation? Second, how does DNA carry the information needed for directing cellular activities? Answers to both questions require information about the chemical structure of DNA.

A DNA molecule contains 2 extremely long "backbone" chains made of many 5-carbon sugars (*deoxyriboses*) connected, end on end, via a phosphate linkage (i.e., . . . sugar-phosphate-sugar-phosphate . . .). Like the legs of a ladder, these backbone chains run parallel to one another. They are connected at regular intervals by *nitrogenous bases*, which form the "rungs" of the ladder. It takes two bases to span the distance between the legs; the two are connected in the center of the span by weak chemical bonds, *hydrogen bonds*. Finally, the legs of the ladder are twisted into a helical structure, making one complete turn of the helix for every ten "rungs" of the ladder.

The particular bases that form the rungs and their relative placement within the ladder structure are the key to our problems. Only four different base species form DNA; *adenine* (abbreviated as A), *guanine* (G), *cytosine* (C), and *thymine* (T). Formation of each ladder rung requires two of these, but not any two. The two bases, like pieces in a jig-saw puzzle, must have the proper size and shape and must be able to interlock (form hydrogen bonds) within a given constellation. Examination of DNA structure shows that rungs can be formed by a combination of A with T (A-T) or G with C (G-C), but all other possible combinations, like A-A, A-C, or G-T, will not work. A-T and G-C are called *complementary base pairs*.

Imagine that you and another person eack take hold of one leg of the ladder and pull. It will come apart at the seams (i.e., at the ctner of the rungs where the complementary base pairs are held together by relatively weak hydrogen bonds). You each take one strand (half of the structure) consisting of one long leg with single bases attached and, separately, you both begin to reconstruct the missing half. The missing leg is no problem; it is always the same string of deoxyribose and phosphate. But the bases are also prescribed: to every A on the single strand, you attach a T, to every T, an A, to every G, a C, and to every C, a G. You have reconstructed an exact replica of the original DNA, and so has your partner. There are now copies of the original; precise replication has been accomplished. A similar process takes place within the cell; only here the strands are separated bit by bit, and synthesis of new DNA follows closely behind in the wake of the separation, aided by the action of special enzymes, *DNA polymerases*.

A discussion of our second problem, how DNA carries the hereditary material, is taken up in plate 4.

CN: Use a dark color for D.
1. Begin by coloring the cell at the top of the page, and then color the circular diagram of the cell cycle immediately below it.
2. Color the stages of the cell cycle, beginning with interphase near the top left-hand side and continu-

ing progressively through the stages of mitosis and cytokinesis.
3. Color the schematic respresentation of DNA structure replication along the right-hand side. Among the bases note that guanine (I) and cytosine (I') are cross-hatched).

PLASMA MEMBRANE A
NUCLEAR ENVELOPE B
CHROMATIN C
CHROMOSOME (4-6) D
CENTROMERE E
CENTRIOLE F
SPINDLE FIBERS G

Reproduction of somatic cells is accomplished during three general stages of the cell cycle: interphase, mitosis, and cytokinesis. Interphase is further divided into three substages, a period preceding DNA synthesis (G_1), the period during DNA synthesis (S), and the period following DNA synthesis (G_2).

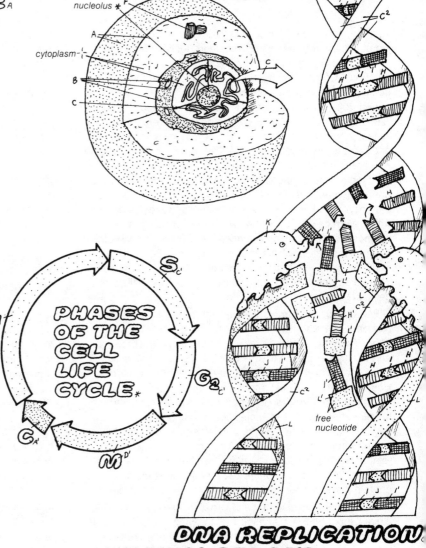

nucleolus ← F
cytoplasm
nucleolus ← F

INTERPHASE C'

During interphase, 1. uncoiled DNA (contained in chromatin) replicates. 2. Following replication, DNA becomes active in directing the RNA and protein synthesis required for cell division. 3. The centrioles duplicate.

MITOSIS D'
PROPHASE D'

During prophase, 1. the nuclear envelope begins to break down. 2. The two copies of DNA begin to coil and supercoil, forming chromosomes. Each chromosome consists of two sister chromatids attached at the middle by a structure called the centromere. 3. Centrioles separate, migrating toward the cell poles (located at the extreme right and left in the illustration). 4. The centrioles organize microtubules, which form the mitotic spindle apparatus.

PHASES OF THE CELL LIFE CYCLE *

G_1 C'
S C'
G_2 C'
M D'
C A'

METAPHASE D'

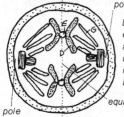

pole
pole
equator

During metaphase, 1. the nuclear envelope and the nucleolus disappear. 2. The chromosomes line up around the cell equator (imaginary line connecting the top and bottom in the illustration).

ANAPHASE D'

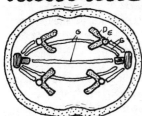

During anaphase, 1. the centromeres divide, separating the sister chromatid pairs. 2. In each chromosome, one of the two sister chromatids migrates to one pole and the other to the opposite pole. Migration is brought about by the action of the microtubules as they pull on the chromosomes.

TELOPHASE D'

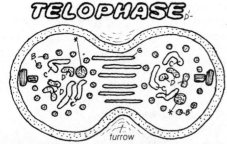

furrow

During telophase, 1. a new nuclear envelope forms around the chromosomes near each of the two poles as two nuclei and two nucleoli begin to appear. 2. The chromosomes uncoil, forming chromatin. 3. The spindle fibers disappear.

DNA REPLICATION
ORIGINAL STRAND C²
BASES: *
ADENINE H, THYMINE H'
GUANINE I, CYTOSINE I'
HYDROGEN BOND J
DNA POLYMERASE K
NEW STRAND L
BACKBONE L'

free nucleotide

Replication is the process in which an exact copy of the DNA molecule is made. The double stranded DNA in the chromatin unfolds, separating at sites where the two strands are attached (hydrogen bonding sites between the complementary bases adenine, thymine, guanine, and cytosine). Each strand consists of a backbone (leg of the ladder) plus attached bases. Aided by enzymes such as DNA polymerase and using the original separated strands as a template, two new strands are synthesized as nucleotide building blocks (molecules containing the base together with the backbone materials, sugar and phosphate) are attached to the template. An exact copy is always obtained because complementary bases are the only ones that are attached. An adenine will attach only to a thymine (and vice versa); a guanine only attaches to a cytosine (and vice versa).

CYTOKINESIS A'

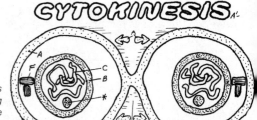

furrow

During cytokinesis, the two daughter cells separate. A furrow (narrowing) forms along the equator and progressively constricts the cell until it separates in two. First signs of a furrow can be seen as early as anaphase.

DNA EXPRESSION & PROTEIN SYNTHESIS

To understand how DNA directs the cell, we begin by observing that such activities as growth, reproduction, secretion, and motility are all derived in the final analysis from chemical reactions. Of the large numbers of products that theoretically could be formed from chemicals used by the cell, only a few appear to be produced within the cell. These products are "selected" by the action of substances called *enzymes*, catalysts that speed specific reactions. Left by themselves, most of the plausible reactions proceed too slowly to be significant. The presence of a specific enzyme "turns on" a specific reaction simply by speeding it. In this way, enzymes control chemical reactions and cellular activities. But what controls the enzymes? They are made of protein and are synthesized within each cell. It follows that whatever controls protein synthesis controls which enzymes are present and therefore controls the cell. DNA plays its dominant role because it contains detailed plans for each protein that is synthesized. This determines the growth and development of individual cells, of tissues, and of the entire organism.

Proteins are giant molecules constructed by linking large numbers of amino acids, end to end, by special chemical bonds (*peptide bonds*) so that they form a chain. There are only 20 different kinds of amino acids in proteins, and because proteins often contain hundreds of them, the same kind of amino acid must appear in more than one position along the chain. We can compare amino acids with letters of the alphabet and protein molecules with huge words. Just as the word is determined by the precise sequence of letters, so is the protein (and its properties) determined by the sequence (placement) of amino acids along the chain. It follows that if DNA contains the "blueprints" for protein construction it must contain the amino acid sequence of that protein. But how?

DNA (plate 3) is also made of large numbers of building blocks, the *nitrogenous bases*, and the properties of the DNA molecule are determined by the sequential placement of these bases as "rungs" in the ladderlike chain structure. Each DNA is also like a huge word with the bases representing letters of the alphabet. However, although proteins are based on a 20 letter "alphabet" (20 amino acids), DNA has only 4 bases: *adenine* (A), *guanine* (G), *cytosine* (C), and *thymine* (T). Somehow the sequence of just 4 different kinds of bases along the DNA ladder provides a code for the placement of 20 different kinds of amino acids in a protein chain. There cannot be a one-to-one correspondence of the letters in the two alphabets, for if each base corresponded to a single amino acid, then DNA would be able to code only for proteins containing at most 4 different amino acids. Instead, a sequence of 3 bases is used to code for each amino acid. For example, when the bases C, C, G occur one right after the other in the DNA ladder, it is a code for the amino acid glycine; the sequence A, G, T, codes for the amino acid serine. The sequence C, C, G, A, G, T is a signal for part of a protein where serine follows glycine. By using bases 3 at a time, it is possible to form 64 unique combinations (e.g., AAA, AAG, . . . CCA, CTC, . . . GGA, . . . TTC, . . . etc.), far more than necessary to code for 20 amino acids.

How do cells actually translate the code and build proteins? DNA always remains within the nucleus, yet proteins are synthesized in the cytoplasm. A first step is to make a copy of the "blueprints" and transport it into the cytoplasm, a process called *transcription*. The transcript (copy) of this genetic code is a molecule called *messenger ribonucleic acid (mRNA)*, which moves to the cytoplasm, where it associates with particles called *ribosomes*, the assembly sites for new proteins. Meanwhile other RNA molecules, *tRNA (transfer ribonucleic acid)*, pick up loose amino acids in the cytoplasm that have been activated (energized) in preparation for use. Each tRNA molecule, with a single specific amino acid attached, migrates to the ribosomes, where its amino acid will be utilized at the appropriate position as it detaches from the tRNA and links to the emerging protein chain.

Given this scenario, two problems arise. The first is *transcription*: how are DNA blueprints copied onto RNA? The second is *translation*: how is the code utilized so that amino acids are always linked to the protein in the proper sequence? Answers to both questions are based on RNA's close resemblance to DNA. They differ in that 1. they have slightly different sugars (deoxyribose and ribose); 2. RNA is usually single-stranded, containing only one leg of the ladder together with nitrogenous bases forming half "rungs" along its length; and 3. like DNA, RNA contains A, G, and C, but T is replaced by a very similar molecule, *uracil* (U). Thus, RNA is a similar "4 alphabet" molecule with letters A, G, C, and U. All RNA, but in particular mRNA, is formed from DNA in the same way that DNA makes more DNA. The double-stranded DNA "unzips" a bit, and one of the legs serves as a template for RNA construction. As in DNA synthesis, the sequence of bases in RNA is complementary to the sequence in the DNA template that formed it. A piece of DNA with sequence AGATCTTGT, for example, will make a piece of RNA with sequence UCUAGAACA. Each base triple (3 letters) in mRNA is called a *codon*. The transcription problem is solved by constructing a strand of RNA, which does not duplicate the base sequence of the original DNA, but rather contains the complementary base sequence as a codon.

tRNA molecules are shaped something like a cloverleaf. The stem contains the attachment site for the amino acid, and the loop contains a specific set of three bases (called an *anticodon*), which are the code for the amino acid that will become attached. Because the mRNA codons contain the complementary bases to the DNA and hence to the amino acid code, it follows the mRNA and tRNA have complementary sets of bases and that they will easily form loose H bonds. The tRNA simply lines up along the mRNA sites as illustrated so that the amino acids are now in proper sequence and can be linked by peptide bonds. Actually, the ribosome moves along the mRNA strand and, as illustrated, handles only 2 amino acids at a time. After the peptide bond is formed between the 2 amino acids, the tRNA that has resided longest on the ribosome detaches. The ribosome then moves toward a new codon (to the right in the illustration), leaving a vacant position for the next tRNA (and amino acid) with the complementary anticodon to attach. In this way, the protein chain grows until the final 1 or 2 codons on the mRNA signal the end. Following this *translation process*, proteins are often modified by folding, shortening, or adding carbohydrates, a process called *postranslational modification*.

CN: Use a dark color for E and a very light color for I. Use the same colors as on the previous plate for nuclear envelope (A), chromatin (B), and cell membrane (L). Note that codon triplets (E) and anticodon triplets (H) are actually complementary bases, but receive different colors for identification purposes.
1. Begin by coloring the cell at the top of the page.
2. Color the events taking place within the cell nucleus shown along the right-hand side.
3. Color the scheme for protein synthesis in the bottom third of the plate.

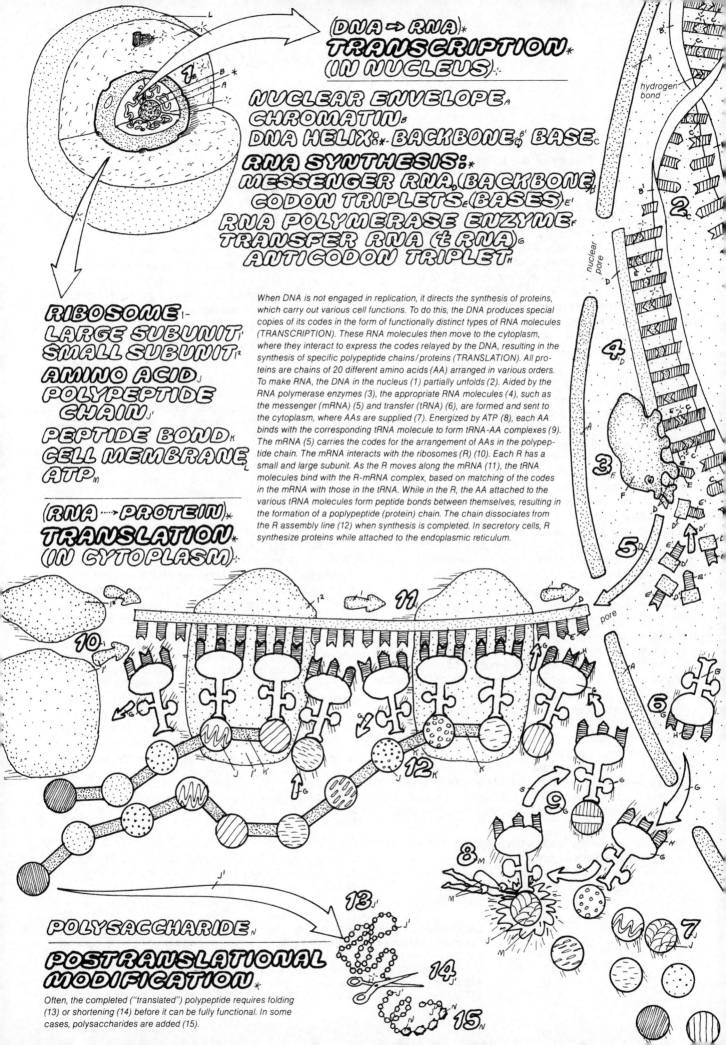

(DNA → RNA) TRANSCRIPTION (IN NUCLEUS)

NUCLEAR ENVELOPE A
CHROMATIN B
DNA HELIX: * BACKBONE B' BASE C

RNA SYNTHESIS: *
MESSENGER RNA (BACKBONE D
CODON TRIPLETS (BASES) E'
RNA POLYMERASE ENZYME F
TRANSFER RNA (t RNA) G
ANTICODON TRIPLET H

RIBOSOME I
LARGE SUBUNIT I'
SMALL SUBUNIT I²
AMINO ACID J
POLYPEPTIDE CHAIN J'
PEPTIDE BOND K
CELL MEMBRANE L
ATP M

(RNA ⟶ PROTEIN) TRANSLATION (IN CYTOPLASM)

When DNA is not engaged in replication, it directs the synthesis of proteins, which carry out various cell functions. To do this, the DNA produces special copies of its codes in the form of functionally distinct types of RNA molecules (TRANSCRIPTION). These RNA molecules then move to the cytoplasm, where they interact to express the codes relayed by the DNA, resulting in the synthesis of specific polypeptide chains/proteins (TRANSLATION). All proteins are chains of 20 different amino acids (AA) arranged in various orders. To make RNA, the DNA in the nucleus (1) partially unfolds (2). Aided by the RNA polymerase enzymes (3), the appropriate RNA molecules (4), such as the messenger (mRNA) (5) and transfer (tRNA) (6), are formed and sent to the cytoplasm, where AAs are supplied (7). Energized by ATP (8), each AA binds with the corresponding tRNA molecule to form tRNA-AA complexes (9). The mRNA (5) carries the codes for the arrangement of AAs in the polypeptide chain. The mRNA interacts with the ribosomes (R) (10). Each R has a small and large subunit. As the R moves along the mRNA (11), the tRNA molecules bind with the R-mRNA complex, based on matching of the codes in the mRNA with those in the tRNA. While in the R, the AA attached to the various tRNA molecules form peptide bonds between themselves, resulting in the formation of a poplypeptide (protein) chain. The chain dissociates from the R assembly line (12) when synthesis is completed. In secretory cells, R synthesize proteins while attached to the endoplasmic reticulum.

hydrogen bond

nuclear pore

pore

POLYSACCHARIDE N

POSTRANSLATIONAL MODIFICATION

Often, the completed ("translated") polypeptide requires folding (13) or shortening (14) before it can be fully functional. In some cases, polysaccharides are added (15).

METABOLISM: ROLE & PRODUCTION OF ATP

Moving about, pumping blood, producing complex cellular structures, transporting molecules - these and other everyday activities that we normally take for granted all extract a price: they require energy. That energy is supplied by food. On the one hand, we have the machines that do the work (muscles, for example); on the other hand, we have the food as an energy source. Somehow they must be linked; energy has to be extracted from the food and stored in a form that is directly utilizable by the machine. The primary storage form living organisms use is the molecule ATP (adenosine triphosphate). ATP contains three phosphate groups joined in tandem. When the terminal phosphate is split off, it becomes ADP (adenosine diphosphate), and considerable energy is released. If the proper machinery is present, most of this energy can be captured and used for work. The ADP is not a simple waste product; it is recycled and utilized to synthesize new ATP.

$$\text{ATP} \underset{\text{energy trapped from food*}}{\overset{\text{energy source for work*}}{\rightleftarrows}} \text{ADP} + \text{P} + \text{energy*}$$

The reaction goes to the right to power cellular machinery for contraction, transport, and synthesis. But if the split phosphate group is simply transferred to water, this energy is wasted; it is given off as heat. However, if the phosphate is transferred to the machine, the energy goes with it, and the machine becomes energized. (The molecular part of the machine that receives the phosphate now has a higher energy content, which allows it to enter reactions it otherwise could not have entered. The finer details of how the actual machinery works are not understood.) ATP is the universal energy currency because of its ability to phosphorylate (transfer the phosphate to) cellular machines and boost them into a higher energy state.

The reaction goes to the left as carbohydrates, fats, and proteins are broken down by chemical reactions occurring within the cell (metabolism). In this plate, we focus on ATP formation via carbohydrate metabolism. Glucose contains large quantities of energy that can be released when the chemical bonds holding its atoms together are broken. For example, if 1 mole (180 grams) of glucose is oxidized, forming CO_2 and water, 686,000 calories of energy are liberated. We can imagine many different ways of splitting the glucose to arrive at the same products, but in each case the same energy would be released. The cell must take the glucose apart in small controlled steps and capture most of this energy in the form of ATP before it is dissipated as heat. The cell accomplishes this in part because it contains a number of specific enzymes that speed the reaction along a specific path (i.e., by their presence, they single out the path of "least resistance").

Energy release from glucose or from glycogen (the storage form of glucose) always begins with a sequence of reactions called *glycolysis* that converts glucose into pyruvate with the concomitant production of ATP. Beginning with the 6-carbon glucose, the reaction sequence is primed by *investing* 2 molecules of ATP to phosphorylate the molecule before it is broken into two 3-carbon fragments. These are processed further to yield 4 new ATP, a net profit of 2 (4 − 2 [priming ATP] = 2). The entire sequence involves 10 reactions, each catalyzed by a specific enzyme, ending in the production of 2 molecules of pyruvate (a 3-carbon structure).

The presence of O_2 is not required for any of these steps, and, although only a small fraction (about 2%) of the available energy in the original glucose has been trapped as ATP, the cell apparently can generate ATP anaerobically (in the absence of air or free oxygen). However, this glycolytic process of breaking down glucose works only if H atoms are stripped off the carbon skeletons and transferred to other molecules called NAD^+.

$$\text{2H (from carbohydrate)} + NAD^+ \longrightarrow NADH + H^+$$

For every glucose, 4 H are transferred to 2 NAD^+. But the total amount of NAD^+ is very small (it is built from the vitamin niacin), and the reaction will stop if we run out of NAD^+. NADH needs to dump its H somewhere so it can return for more. Normally, O_2 serves as the final resting place for H, and H_2O forms. In the absence of O_2, pyruvate itself serves as a dumping ground for H, and lactic acid forms. NAD^+ circulates, carrying H from high up in the glycolytic scheme to pyruvate and back (see plate).

When O_2 is present, glycolysis proceeds as before, but now the role of NAD^+ (and a similar H carrier, FAD) becomes more apparent. They have succeeded in trapping a good portion of the energy in the orginal glucose, and the presence of O_2 allows this energy to be utilized to form ATP. Now, instead of using pyruvate, the H carriers transfer their H and energy to the respiratory chain, a system of carriers that reside within the mitochondrial membranes. In turn, the energized membranes of the mitochondria are able to produce 3 ATP for each NADH passed (only 2 ATP if the H donor is FAD).

Moreover, the availability of the respiratory chain allows energy contained in pyruvate to be tapped. Instead of absorbing H and forming lactate, pyruvate splits off a CO_2, and the remaining 2-C (acetate) portion is transferred via acetyl-CoA to the Krebs cycle, where it is further degraded into 2 molecules of CO_2 (see plate 6). Again H are stripped off the carbon skeletons by the H carriers, which deliver them to the respiratory chain and return for more. The final bookkeeping record for cellular combustion of 1 molecule of glucose is

glycolysis:	2 ATP +	2 NADH + 0 FADH$_2$
2 pyruvate ⟶ acetyl-CoA:	0 ATP +	2 NADH + 0 FADH$_2$
2 turns of Krebs cycle:	2 ATP +	6 NADH + 2 FADH$_2$
Total:	4 ATP +	10 NADH + 2 FADH$_2$

Total ATP *(after cashing H carriers in at resp. chain)*   4   + (10×3) + (2×2) = 38 ATP!

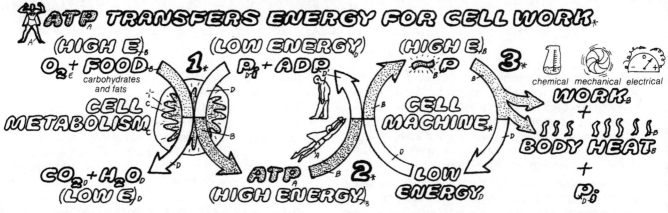

ATP TRANSFERS ENERGY FOR CELL WORK.

(HIGH E.) O_2 + FOOD carbohydrates and fats — 1 — (LOW ENERGY) P_i + ADP — (HIGH E.) ~P — 3

CELL METABOLISM

CELL MACHINE

chemical mechanical electrical

WORK + BODY HEAT + P_i

CO_2 + H_2O (LOW E.) — ATP (HIGH ENERGY) — 2 — LOW ENERGY

Cellular machines do work. Some transport materials, others lift weights (muscle cells), and still others build complex molecules and structures out of simple raw materials. Food supplies the energy, but an intermediary substance, ATP (adenosine triphosphate), transfers the energy from food to the cellular machine. When food is burned in a furnance, energy is liberated as heat and light. When the same food is "burned" in the metabolic reactions of the cell (1), a good portion of this energy is trapped through the synthesis of ATP from ADP and inorganic phosphate (P_i). ATP, in turn, can energize the cell machine. It does this (2) by transferring its terminal phosphate (~P) to it, raising the machine to a higher energy state where it can participate in more reactions and perform cellular work (3).

HOW ATP IS MADE.

ANAEROBIC (NO O_2)
1 MOL. GLUCOSE

+1   +1

NAD+ HYDROGEN NADH + H+ (hydrogen carrier)

FAD FADH$_2$ (hydrogen carrier)

LAC. PYRUVATE LAC.

ROAD BLOCK — NO O_2

AEROBIC (O_2)
1 MOL. GLUCOSE

+2

GLYCOLYSIS IN CYTOPLASM

RESPIRATION IN MITOCHONDRIA

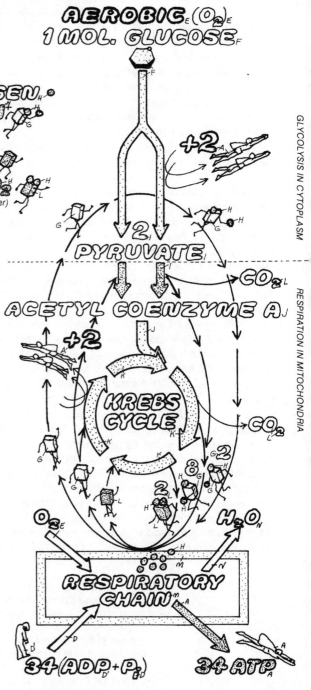

2 PYRUVATE CO_2

ACETYL COENZYME A

+2   KREBS CYCLE CO_2

8   2   2

O_2 H_2O

RESPIRATORY CHAIN

34 (ADP + P_i)   34 ATP

GLYCOLYSIS

ATP formation can occur in the absence of O_2 by glycolysis. Beginning with the 6-carbon (6-C) sugar glucose, the reaction sequence is primed by investing 2 molecules of ATP to phosphorylate the molecule before it is broken into 2 3-carbon fragments. These are processed further to yield 4 new ATP, a net profit of 2 (4 - 2 [priming ATP] = 2). The process works only if H atoms are stripped off the carbon skeletons and transferred to the H carriers, NAD+. For every glucose, 4 H are transferred to 2 NAD+. But the total amount of NAD+ is very small, and the reaction will stop if we run out of NAD+. NAD+ must dump its H somewhere before it can return for more. Normally, O_2 serves as the final resting place for H, and H_2O forms. In the absence of O_2, pyruvate, a product of glycolysis itself, serves as a dumping ground for H, and lactic acid forms. NAD+ circulates, carrying H from high up in the glycolytic scheme to pyruvate and back.

GLYCOLYSIS + RESPIRATION

When O_2 is present, glycolysis proceeds as before, but now the role of NAD+ (and a similar H carrier, FAD) becomes more apparent. When they transfer their H to the respiratory system (instead of pyruvate), they energize the membranes of mitochondria, which are able to produce 3 ATP for each NADH processed (only 2 ATP if the H donor is FAD). Moreover, the availability of the respiratory system allows the energy contained in pyruvate to be tapped as it is broken down to acetyl-CoA and then, via the Krebs cycle, to CO_2. Again H is stripped off the carbon skeletons and transferred to the H carriers, which deliver them to the respiratory chain and return for more. The net result, from 2 turns of the cycle, is the production of 10 NADH and 2 FADH$_2$. From these, the respiratory system produces $10 \times 3 + 2 \times 2 = 34$ ATP. Add the 2 ATP produced during glycolysis + an additional 2 produced in the Krebs cycle, and we have a net of 38 ATP! Compare this to the 2 formed when O_2 was absent and the path took the detour to lactate.

METABOLISM: RESPIRATION AND THE KREBS CYCLE

Plate 5 focused on the degradation of carbohydrates, in particular glucose, to form ATP. Fats and proteins are also used for these purposes, but the final common pathway is the same — through the Krebs cycle and the respiratory chain, as outlined below. Taking an overview of the processes involved in ATP generation, we can conveniently divide the oxidation of foodstuffs into three stages:

Stage I. Large molecules in food are broken down into simpler forms. *Proteins* are broken down into amino acids, *fats* are broken down into *glycerol* and *fatty acids*, and large *carbohydrates* (e.g., starch, glycogen, sucrose) are broken down into *simple 6-carbon sugars* such as glucose.

Stage II. The large number of stage I products are broken down into a few simple units that play a central role in metabolism. Most of these products, including simple sugars, fatty acids, glycerol, and several amino acids, are broken down into the same 2-carbon fragment called *acetate* that attaches to the same pivotal molecule, *coenzyme A* (abbreviated CoA), and enters the Krebs cycle as the compound *acetyl-CoA*.

Stage III. This final stage consists of the Krebs cycle and the respiratory chain (also called the electron transport chain) together with the ensuing synthesis of ATP. From acetyl-CoA onward, the metabolic path is the same for all foodstuffs. This plate focuses on this final common pathway, stage III, which occurs only in the presence of O_2.

Returning to our example of glucose metabolism, recall that 1 molecule of *glucose* produces a net gain of *2 ATP, 2 NADH and 2 pyruvate.* In the presence of O_2, pyruvate is not utilized to form lactic acid because the NADH deposits its H on O_2 (via the respiratory chain — see below), freeing pyruvate to enter into further reactions. The 2 pyruvate move into the *mitochondria* in preparation for their entrance into the *Krebs cycle.* A precycle step breaks them into 2-C fragments (acetate), which are attached to a coenzyme A (CoA), forming acetyl-CoA. In the process, *energy is recovered by transfer of H to NAD^+*, and CO_2 is formed. Acetyl-CoA is also formed during the combustion of fats and proteins, and it plays a key role in feeding the 2-C acetate to the Krebs cycle. The acetate is split from CoA and combines with a 4-C structure, forming the 6-C citrate molecule, and the cycle begins. As illustrated, *each turn of the cycle produces 3 NADH, 1 $FADH_2$, and 1 ATP*, with the carbon remains of the original acetate finally discarded as 2 CO_2.

Like glycolysis, the Krebs cycle will come to grinding halt as soon as all the H carriers NADH and $FADH_2$ are loaded with H. However, H is unloaded into the respiratory chain at the mitochondrial membrane, regenerating NAD^+ and FAD to participate in further metabolism. The *respiratory chain* is a system of *electron and H carriers* embedded in the inner member of the *double membrane* that surrounds the mitochondria. In addition to regenerating NAD^+ and FAD, the respiratory chain also pumps H^+ ions into the space between the two mitochondrial membranes. These ions will be used in the final synthesis of ATP.

To understand the respiratory chain, recall that a neutral H atom consists of 1 electron and 1 H^+ (i.e., $H = H^+ + 1$ electron). When NADH arrives at the first carrier of the respiratory chain at the inner face of the inner mitochondrial membrane, it transfers 2 electrons, and 1 H^+. Another H^+ is picked up from the surrounding solution, and the electrons and H^+ (= 2H) are carried from the inner to the outer face. At this point, the H components, H^+ and electrons, part company. The H^+ are deposited in the small space between the 2 mitochondrial membranes, and the electrons return to the inner face to pick up another pair of H^+ from the surrounding solution. This trip across the inner mitochondrial membrane is repeated twice for a total of 3 round-trips. After the third trip, the electrons are picked up by O_2 and, together with H^+ from the surrounding fluids, form water.

At each of the 3 trips, 2 H^+ are deposited in the intermembrane space so that the concentration of H^+ builds up. These H^+s leak back into the matrix of the mitochondria through special protein complexes that form channels through the membrane. The energy dissipated by the H^+ moving through these channels from high to low concentrations is somehow used to *synthesize ATP from ADP and phosphate (Pi).* Just how the respiratory chain pumps H^+ into the intermembrane spaces and just how the back leakage of H^+ is used to synthesize ATP are not known. It is as though we were dealing with darkened tunnels; we can see what goes in one side and what comes out the other. From this, we can only guess what happens inside.

$FADH_2$ differs from NADH; it transfers its 2H to the respiratory chain downstream from the transfer point of NADH, where only 2 round-trips across the membrane are available. As a result, it transfers only 4H^+ across the mitochondrial membrane, and this accounts for the fact that it provides energy for synthesis of 2 (rather than 3) ATP.

The Krebs cycle and ATP synthesis take place within the mitochondria; other functions (e.g., glycolysis) occur in the cytoplasm. Special transport systems within the mitochondrial membrane that move materials in and out overcome these restrictions. The transport system for newly synthesized ATP moves ATP out in exchange (countertransport, see plate 9) for ADP, which will be used for further ATP production. Other specialized transport systems are available for pyruvate and the NADH that arises from glycolysis. NADH itself does not cross the membrane; instead, it transfers its H at the outer face to H carriers that transport the H to the inner surfaces. Here the H is picked up by NAD^+ that is trapped inside the mitochondria. It becomes NADH, which now has access to the respiratory chain. In some mitochondria, the H are picked up by FAD rather than NAD^+, resulting in some energy loss. In these mitochondria, the net ATP production arising from combustion of 1 glucose molecule will be 36 rather than 38.

CN: *Use the same color for the following titles as was used on the preceding page (note that the letter labels are different, so check carefully): A C, D, E, F, G, I, J, P, Q, R, and S. Use dark colors for B and K.*
1. *Begin at the top of the page with the entry, from the cytoplasm, of pyruvate into the Krebs cycle. It isn't necessary to color the titles of the various acids in this cycle.*
2. *Begin the lower panel with the small cutaway drawing of an entire mitochondrion. Color the enlarged rectangular por-* *tion, noting the arrows on the left representing the entry of pyruvate (F) and hydrogen ions (D).*
3. *Color the next enlargement. Note that the inner matrix (N) and intermembrane space (L) are left uncolored. Begin the coloring on the upper left side with the passage of H^+ through the membrane via the electron carrier (P). Note that the electrons that are carried along the path of this system are not shown. Follow the buildup of H^+ in the intermembrane space and its passage back into the matrix.*

THE KREBS CYCLE

CARBON C

NAD+

HYDROGEN

NADH + H+

FAD

FADH₂

During glycolysis, each glucose forms 2 pyruvates, which move into the mitochondria in preparation for their entrance into the Krebs cycle. A precycle step breaks them into 2-C fragments (acetate), which are attached to a coenzyme A (CoA), forming a compound called acetyl-CoA. In the process, energy is recovered by transfer of H to NAD+, and CO_2 is formed. Acetyl-CoA is also formed during the combustion of fats and proteins. The acetate combines with a 4-C structure, forming the 6-C citrate molecule, and the cycle begins. As illustrated, each turn of the cycle produces 3 NADH (+3H+), 1 FADH₂, and 1 ATP, with the C remains of the original acetate finally discarded as 2 CO_2.

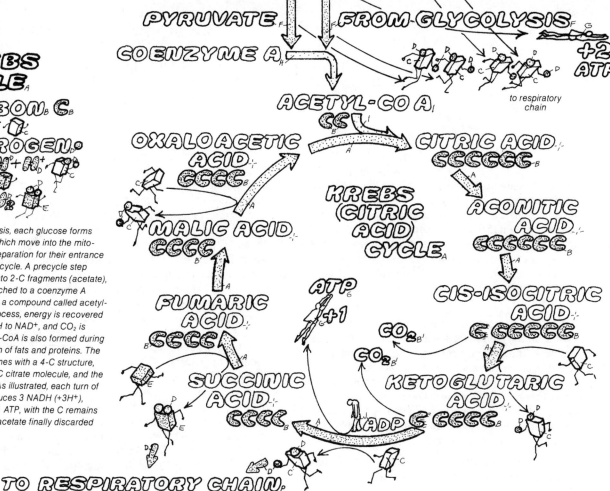

PYRUVATE   FROM GLYCOLYSIS

COENZYME A

+2 ATP

to respiratory chain

ACETYL-CO A

OXALOACETIC ACID   CITRIC ACID

KREBS (CITRIC ACID CYCLE)

ACONITIC ACID

MALIC ACID

CIS-ISOCITRIC ACID

FUMARIC ACID

ATP +1

CO_2

CO_2

KETOGLUTARIC ACID

SUCCINIC ACID

ADP

TO RESPIRATORY CHAIN

THE MITOCHONDRION

OUTER MEMBRANE
INTER MEM. SPACE
INNER MEMBRANE
MATRIX

CYTOPLASM

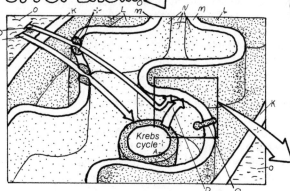

Krebs cycle

Mitochondria have a double membrane. The respiratory chain is a system of electron and H+ carriers imbedded in the inner membrane. NADH arrives at the first carrier and transfers 2 electrons and one H+. Another H+ is picked up from the surrounding solution, and the 2 electrons and 2 H+ are carried from the inner to outer face, where the H+ are deposited in the intermembrane space. The electrons return, pick up another pair of H+, and repeat

the trip 2 more times for a total of 3 round-trips. Each time 2 H+ are deposited in the intermembrane space so that the concentration of H+ builds up. These H+ leak back into the matrix of the mitochondria through special protein complexes. The energy dissipated by the H+ moving through these channels from high to low concentrations is somehow used to synthesize ATP from ADP and phosphate (Pᵢ). The newly synthesized ATP moves out of the mitochondrial matrix via a special transport system that exchanges it for ADP to be used in further ATP production. Note that after the third round-trip, the electrons are no longer "useful"; their energy has been tapped, and if they pile up, the respiratory process will stop. The role of O_2 is simply to prevent this pileup. O_2 picks up the electrons and, together with H+ from the surrounding fluids, forms water.

THE RESPIRATORY CHAIN

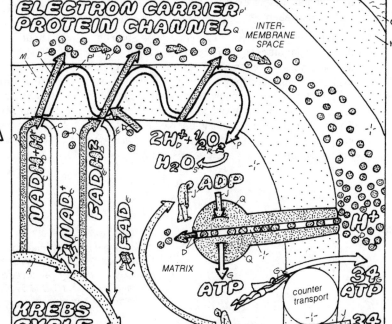

ELECTRON CARRIER
PROTEIN CHANNEL

INTER-MEMBRANE SPACE

NADH+H+

NAD+

FADH₂

FAD

$2H^+ + \frac{1}{2}O_2$

H_2O

ADP

H+

MATRIX

ATP

counter transport

34 ATP

34 ADP

KREBS CYCLE

THE STRUCTURE OF CELL MEMBRANES

Membranes are ubiquitous! Not only do they define the boundaries of cells and subcellular organelles, they transmit signals across these boundaries, contain cascades of enzymes that are essential for metabolism, and regulate which substances enter and which leave. In many ways, the cell membrane resembles the wall around an ancient city; by regulating traffic in and out, the membrane is a major determinant of cellular economy.

Although different membranes have different compositions, all cell membranes share a common primitive structure. They are composed of proteins floating in a fluid *bilayer* (two layers back to back) of *lipid*. How do they form? What holds them together? Although membrane molecules interact with one another, the primitive forces holding the membrane together appear to arise from interactions of membrane with water and from interactions of water with water.

Water is not a symmetrical molecule; the two hydrogens are at one side of the molecule, and the oxygen is at the other. Moreover, the electron clouds that make up the molecule tend to hover closer to oxygen than to hydrogen. As a result, because water bears no net charge, it will have a negative pole near the oxygen and a positive pole near the hydrogens. We classify water as *polar* to contrast it with electrically symmetric molecules like *hydrocarbons*, which are *nonpolar*. In ordinary water, not only is the positive hydrogen end of one water attracted to the negative oxygen end of its neighbor, but it may even form a weak chemical bond (called an H bond) with it. Thus, liquid water has a structure. Although it is not a rigid crystal like ice, it does contain numerous aggregates of up to eight to ten molecules weakly held together by H bonds.

Just as water molecules interact with one another through polar attractions or H bond formation, so do they interact and form microstructures with other polar molecules. However, they do not interact with nonpolar molecules. As a result, nonpolar molecules are excluded from the water phase; they are insoluble. The water molecules, "seeking each other," eject the nonpolar solute much like the ejection of a bar of soap from a clenched fist. The forces arising from this phenomenon are called hydrophobic forces, and nonpolar solutes are called hydrophobic solutes.

Most membrane lipids are called *phospholipids*. Phospholipids have a dual structure. One end, the "*head*," is polar (*hydrophilic*), and it is attracted to water. The remaining "*tail*" is nonpolar (*hydrophobic*), and it is ejected from water. This incompatibility of the two ends is the key to membrane structure. When phospholipids are mixed with water, forces that tend to incorporate the head within the water and eject the tail are set up. The illustrations show how both of these forces can be accommodated through formation of *micelles* or *bilayers*.

The same principles apply to those *proteins* that are part of the membrane. Proteins consist of long chains of amino acids, some of which are more polar and some of which are more hydrophobic. Although proteins do not have an obvious head and tail, they are folded so that their hydrophobic portions are out of the water phase, imbedded in the hydrophobic body of the membrane, while their polar portions are anchored in the water.

Specific proteins imbedded in the membrane perform specific functions. Some are involved in transport of materials into or out of the cell, others serve as receptors for hormones, others catalyze specific chemical reactions, and still others act as links between cells or as "anchors" for structural elements inside the cell.

Some proteins traverse the entire membrane; others are confined to a single side. Proteins exposed to the external surface of the cell membrane frequently have *carbohydrate* side chains attached; they are called glycoproteins. Those that span the membrane are exposed to both surfaces. They are often involved in transport of materials into and out of the cell. Some appear to form clusters of perhaps two or four molecules in the membrane, and some are believed to fold in ways that confine polar parts to an interior core that runs through the center of the molecule or through the center of a cluster. This would provide a *channel* for small polar molecules, particularly water and ions, to move through the membrane. Some of these channels may contain "gates," "filters," or other devices that regulate traffic. These are taken up in plates that follow.

CN: Use light blue for B and a dark color for H.
1. Begin with the polar portion of the phospholipid molecule, noting the heavy outline dividing it into three areas of color (C, D, & E). These areas make up the head (A). In the symbolic representation to the right.
2. Color the nonpolar portion, and continue with the central cartoon. Then color the two forms the result can take: micelle or lipid bilayer.
3. Color the material describing the cell membrane, including the small drawing of a cell. Note that the areas of water bordering both sides of the membrane are left uncolored except for the titles describing their location.

THE PHOSPHOLIPID MOLECULE.

POLAR PORTION. (HYDROPHILIC).
CHARGED GROUP:*
ALCOHOLS.
PHOSPHATE.
GLYCEROL.

WATER.

carbon
nitrogen
hydrogen
oxygen
phosphorus

HEAD

NONPOLAR PORTION (HYDROPHOBIC).
FATTY ACID CHAIN.

Mixing phospholipids with water yields structures where polar head groups remain in the water while nonpolar tails are excluded. This can result in different structures: micelle spheres are simplest; their tails avoid contact with water by pointing toward the interior of the sphere where water is absent. Bilayer vesicles are more like cells. Their tails avoid water by contacting one another at their tips so that only head groups are exposed to water at the exterior or interior of the sphere.

TAIL

SYMBOL

STRUCTURAL MODEL

MICELLE.

OR

THE CELL MEMBRANE.
LIPID BILAYER:* POLAR. NONPOLAR.
PROTEIN.
CHANNEL.
CARBOHYDRATE.

Membrane proteins are folded so that their polar parts are exposed to water, but their nonpolar parts are imbedded more deeply into hydrophobic portions of the bilayer. Some proteins traverse the entire membrane; they are often involved in transport of materials into and out of the cell. Other proteins are confined to a single side of the bilayer. Specific proteins imbedded in the membrane perform specific functions. Some are involved in transport, others serve as receptors for hormones, others catalyze specific chemical reactions, and still others act as links between cells.

sugar molecule

glycoprotein

LIPID BILAYER.

EXTRACELLULAR FLUID.

CYTOPLASM OF CELL.

non-polar
polar

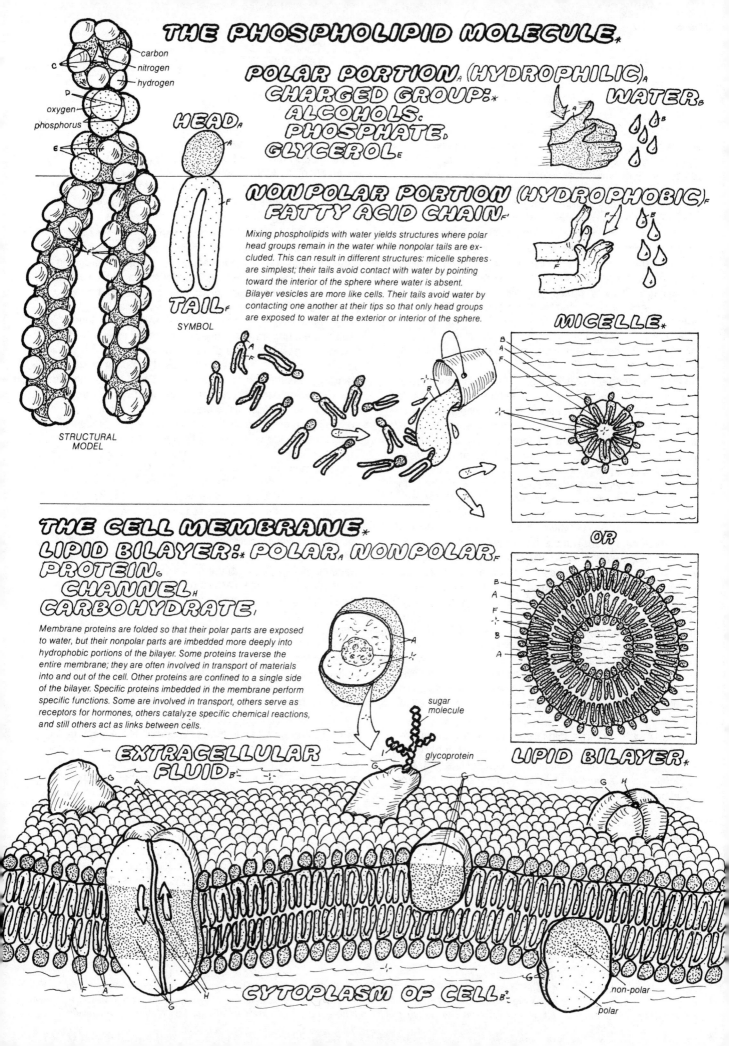

SOLUTE AND WATER MOVEMENTS

Forces generate movements. There are several different types of force; balls roll downhill because of gravitational force, and electrons (negative charge) flow toward protons (positive charge) because of electrical forces. Movements through membranes are driven by forces that arise from differences in concentration, pressure, and electrical charge on the two sides of the membrane. These differences are often called gradients. All else being equal, the larger the gradient (i.e., the larger the difference), the greater the flow through the membrane. *Concentration gradients* give rise to movements called *diffusion*, *pressure gradients* cause *bulk flow*, and *electrical gradients*, more commonly called *voltage gradients* or *membrane potentials*, drive the *flow of ions* (*ionic current*). In addition, *water* flows to the side of the membrane that has the most concentrated solute, a process known as *osmosis*.

DIFFUSION. Whenever there is a difference of concentration of a solute between two regions, the solute tends to move (diffuse) from the highly concentrated region to the less concentrated region. Solutes diffuse down their concentration gradient. This net movement arises because molecules are always moving about at random. At first thought, it seems surprising that an orderly net movement arises from molecular chaos, but the simple example described in the legend to the diffusion figure illustrates how this comes about. Net movement stops when concentrations are equal. Under most circumstances, diffusions of different substances are independent of each other. For example, the diffusion of sugar down its concentration gradient would take place at the same speed even if another diffusing substance, say urea, were present.

The time required for diffusion over small distances, such as the size of a cell, is only a fraction of a second, but larger distances take surprisingly longer times. To diffuse 10 cm takes about 53 days, and it would take years for O_2 to diffuse from your lungs to your toes. But O_2 does not travel from lungs to body tissues by diffusion. Instead, it is transported by bulk flow through the circulating blood. Once the O_2 reaches the tissues, it diffuses the short distance through the wall of the blood vessel (capillary) into the tissue and to any cell in the near vicinity. The process is over in seconds.

BULK FLOW. In contrast to diffusion, in bulk flow the whole mass (fluid plus any dissolved solutes) moves. For example, when you push the plunger on the top of a syringe, fluid flows out the needle (by bulk flow). In the diagram, the man on the left is pushing harder than the one on the right, so fluid flows from left to right. We call the "push" *pressure*. (More precisely, the pressure is the force he exerts on each square centimeter of plunger.) Fluid flows down pressure gradients, from high to low pressures, by bulk flow.

The diagram to the right shows how pressure can be measured. A light (ideally weightless) moveable partition is placed on top of the fluid, and mercury is poured on top until the weight of the mercury is just sufficient to stop the flow. At this point, the pressure gradient has fallen to zero; pressures to

the left and right are equal. But the pressure on the right is determined by the weight of the mercury (divided by the area of the partition), and this is determined by the height of the mercury column. We measure pressure in millimeters of mercury (mm Hg); it is the height of the column of mercury required to balance the pressure so that no movement occurs.

OSMOSIS. The diagram (next to bottom) shows a membrane separating two solutions. The membrane is permeable to water but impermeable to the solute. Water will flow from left to right, from the region where the concentration of dissolved solute is low (actually zero in our example) toward the region where it is high. The flow is called osmosis. The forces that cause osmosis can be measured by the same technique used to measure pressure (see illustration on the right). Use the "weightless" moveable piston to cover the solution, and pour on mercury until movement stops. The height of the mercury measures the pressure required to stop the osmotic flow; it is called *osmotic pressure*. The osmotic pressure will depend on the solution used on the right-hand side; the more concentrated the solution, the higher the osmotic pressure. As a result, we identify the osmotic pressure with the solution. The osmotic pressure of a solution is simply the pressure that would be measured if the solution were placed on the right-hand side of our device. Note that according to this convention, water flows *up* the osmotic pressure gradient. To a good approximation, all molecules and ions make equal contributions to osmotic pressures.

Osmosis involves bulk flow. Suppose, in our example, that the solute on the right were protein and that we dissolve equal concentrations of sugar on both sides of the membrane. Because proteins are so much larger than sugar, it is easy to find a porous membrane that will allow water and sugar to pass but will restrain the protein. What happens in this case? As before, the water flows from left to right, but now it carries the sugar with it as if the water were being driven by the same type of pressure gradient described in the bulk flow panel. Osmotic flow takes place in bulk; the solvent (water) drags all solutes along with it except those that are restrained by the membrane.

IONIC CURRENT. The bottom diagram shows ionic movements (current) that arise from tiny differences in electrical charge on the two membrane surfaces. Ions of like sign repel; ions of unlike sign attract. When positive and negative ions are separated, they tend to move back together. Energy associated with this attraction (or repulsion) is easy to measure. It is called a voltage. Positive ions move down voltage gradients; negative ions move up voltage gradients.

In each of our examples, materials moved from regions of high energy to regions of lower energy. Concentration gradients, pressure gradients, osmotic gradients, and voltage gradients are all exmaples of *energy gradients* (more precisely, free energy gradients - see plate 9). Our discussion can be generalized: *energy gradients are forces that generate movements.*

CN: Use light blue for D.
1. Begin at the top of the page and complete each panel before proceeding to the next.
2. Flow, flux, diffusion, osmosis, and ionic currents are all examples of movement (A).
3. Energy gradients, pressure gradients, concentration gradients, osmotic gradients, osmotic pressure and voltage gradients are all examples of forces that cause flow (B).

FLOW (FLUX) ∝ FREE ENERGY GRADIENT

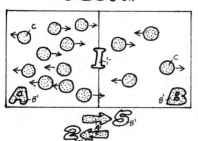 **HIGH ENERGY** **SOLUTE** **FLOW GRADIENT** **LOW ENERGY STATE**

Substances flow down free energy gradients (i.e., from regions of high free energy toward regions with low free energy). Free energy gradients are the forces that propel the flow. The steeper the free energy gradient (i.e., the greater the difference in energies), the faster the flow (flux).

DIFFUSION: CONCENTRATION GRADIENT

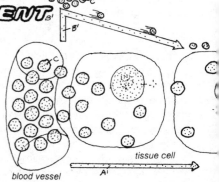

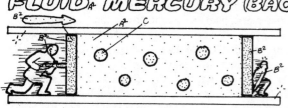 **FLUX**

The higher the concentration, the higher the free energy. Solutes flow (diffuse) down concentration gradients. This is a result of random motion; on average, equal numbers of molecules move in all directions. In compartment A, 5 move to the right (5 also move to the left). In B, with a smaller concentration, only 2 move to the left. The net flow across the interface at I is 5 - 2 = 3 moving to the right. Diffusion is from A (high concentration) to B (low concentration). It stops when the concentrations in the two compartments are equal. Diffusion is the process for transporting O_2 and nutrients from capillary blood vessels (high concentration) to tissue cells (low concentration).

blood vessel   tissue cell

BULK FLOW: PRESSURE GRADIENT
FLUID MERCURY (BACK PRESSURE)

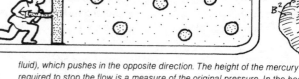

heart

Pressure (mechanical "push") also contributes to free energy. Fluid flows down pressure gradients, from regions where the pressure is high to where it is low. On the right, the pressure is balanced by a column of mercury (heavy fluid), which pushes in the opposite direction. The height of the mercury required to stop the flow is a measure of the original pressure. In the body, contractions of the heart supply the pressure gradients to propel blood.

OSMOSIS: OSMOTIC GRADIENT
WATER (SOLVENT), MEMBRANE

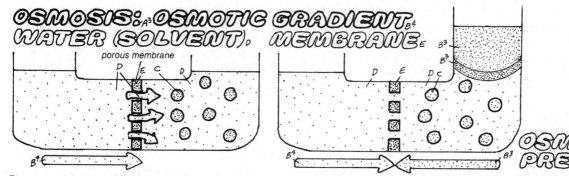

porous membrane

OSMOTIC PRESSURE

The presence of solutes lowers the free energy of water. When a membrane separating two solutions allows water to pass through but restrains the solute, water flows down its free energy gradient toward the solute by a process called osmosis. This osmotic flow can be prevented by applying a pressure (column of mercury) to push in the opposite direction. This pressure measures the osmotic pressure. Note that by this definition, osmotic water flow goes from regions of LOW to HIGH osmotic pressure. Osmotic flow is responsible for swelling and shrinking of tissue.

IONIC CURRENT: VOLTAGE GRADIENT
POSITIVE IONS
NEGATIVE IONS   ELECTRIC FORCE

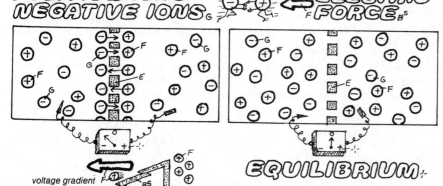

Solutes that carry an electrical charge (ions) also move under the influence of special (electrical) forces that arise because ions of like sign repel, and ions of unlike sign attract. When positive and negative ions are separated, they tend to move back together. Energy associated with this attraction (or repulsion) is easy to measure. It is called a voltage; positive ions move down voltage gradients. Negative ions move in the opposite direction (up voltage gradients).

voltage gradient (voltage)

EQUILIBRIUM

VOLTAGE

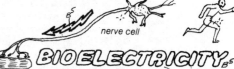

nerve cell

muscle cell   BIOELECTRICITY

PATHWAYS FOR MEMBRANE TRANSPORT

To deal with movements through membranes, we require a "common denominator" that allows us to compare magnitudes of forces and predict motions. Free energy provides that concept. Free energy is the amount of energy that can be "set free" to do work. When substances move from regions where their free energy is high to regions where it is low, down the free energy gradient, we call the movement passive because it can occur without any "aid" or work done by an external agency. The substance simply loses some of its energy to the environment. However, substances cannot move in the opposite direction (from low to high free energy) without obtaining energy (work) from the environment. When substances move uphill, from low to high free energy, we call the process active. One of the major problems of membrane physiology is to identify the source of energy supplied by the environment and to describe in detail how it is utilized.

Favorable free energy gradients by themselves are not sufficient to ensure transport. It doesn't matter how large a gradient is if the membrane does not allow the substance to pass through. In addition to a favorable gradient, there must also be a pathway. The common pathways we describe in this plate have not been fully identified; our understanding is incomplete, and our descriptions of mechanisms are oversimplified.

PASSIVE PATHWAYS. Some solutes, particulary steroid hormones, fat soluble vitamins, oxygen, and carbon dioxide, are lipid soluble. They simply dissolve in the *lipid bilayer* portions of the membrane and diffuse to the other side (1). Many other important solutes, including ions, glucose, and amino acids, are more polar; they are soluble in water, but not in lipids. These substances move through special pathways provided by *proteins* that span the membrane. Small solutes like Na^+ pass through *channels* (2). Larger ones like glucose enter the cell by *facilitated diffusion* (3). They bind to a protein carrier that "rocks" back and forth or moves in some other way, exposing the binding site first to one side, then to the other side of the membrane. The solute hops on or off the site, depending on the concentration. If there is a higher concentration outside the cell, then the binding site will have a greater chance of picking up a solute on the outside, and more solutes will move in than out. This will continue until the concentrations on both sides are equal. At this point, movement in one direction is just balanced by movement in the opposite direction; net movement ceases. It is a purely passive transport because any glucose movement is always down its *concentration gradient*. Similar facilitated diffusion systems exist for many other substances.

TRANSPORT AGAINST GRADIENTS. Proteins also provide pathways for solute movements against concentration gradients (uphill). *Primary active transport* (4) is probably similar to facilitated diffusion. The transported molecule binds to a site on a protein that can "rock" or otherwise expose the binding site first to one side then to the other side of the

membrane. Now, in contrast to the passive facilitated diffusion described above, suppose the binding site properties change and depend on which side of the membrane it faces. If the solute can bind on only one side of the membrane, say on the surface facing the inside of the cell, then transport is in only one direction, from inside to out, but never the reverse. Now if the concentration is less inside than out, our protein will transport against a gradient; it will be an active transport system. Energy for the transport will have to be supplied in order to change the binding site properties each time it cycles back and forth. This energy is generally derived from the splitting of ATP.

Solutes can also move uphill by co- and countertransport. Both utilize the passive transport of one solute to transport a different solute. Our example of co-transport (5) is similar to facilitated transport, but now the protein carrier has binding sites for two different solutes, Na^+ (represented by circles) and glucose (triangles). The carrier will not "rock" if only one of the sites is occupied. In order to "rock," both sites have to be empty or both sites occupied (both a Na^+ and a glucose have to be bound). Outside the cell, Na^+ is much more concentrated than glucose, but inside the cell, the concentration of Na^+ is very low because it is continually pumped out by an active transport process operating elsewhere in the membrane. Both Na^+ and glucose will move into the cell, but few molecules will come back out because the low concentration of intracellular Na^+ makes it difficult for glucose to find a Na^+ partner to ride the co-transport system in the reverse direction. By this mechanism, glucose can be pulled into the cell even against its concentration gradient. The energy for transporting glucose uphill against its concentration gradient comes from the energy dissipated by Na^+ as it moves down its concentration gradient. The concentration gradient for Na^+ is maintained by a primary active transport pump, which is driven by energy released by the splitting of ATP, so that ATP is indirectly involved in this co-transport example. Similar co-transport systems exist for other solutes.

Countertransport (6) is similar to co-transport, but now the two solutes move in opposite directions. In our example, there are binding sites for two different solutes, say Na^+ (circles) and Ca^{++} (triangles). Again the carrier will not "rock" if only one of the sites is occupied. In order to "rock," both sites have to be occupied (both Na^+ and Ca^{++} have to be bound). Because the Na^+ concentration is much higher than Ca^{++}, it tends to dominate and keeps the countertransporter moving in a direction that allows Na^+ to flow down its gradient (into the cell). It follows that Ca^{++} will flow out of the cell, even though the Ca^{++} concentration is higher outside the cell than in. Once again the energy dissipated by Na^+ moving down its gradient is coupled to the uphill transport of another solute.

For simplicity, we have neglected the influence of electrical forces on the ions. The combination of electrical and concentration gradients is covered in plate 11.

CN: Use a dark color for F.
1. Begin with example number one, lipid solubility, and note that the arrows representing flow (F) receive a different color. Color the summary of this process in the central cell diagram. The various transport mechanisms in these summaries are given a circular shape for diagrammatic purposes.

2. Color the other five methods of transport. The flow arrows (F) which go against the gradient (uphill) are drawn with a bolder outline. Note that in 5 and 6 the wider gradient arrow (B¹) is the dominant gradient. A gradient of outside to inside the cell is used in most of these examples. If it were reversed, the flow would be in the opposite direction.

LIPID BILAYER$_A$ SOLUTE$_B$ CHANNEL$_C$ PROTEIN$_D$ ATP$_E$ FLOW$_F$ GRADIENT$_{B'}$

PASSIVE TRANSPORT (DOWNHILL)*

Solutes move passively (down concentration gradients) through cell membranes, provided there is a pathway. Some solutes such as steroids are lipid soluble (1). They dissolve in the membrane and diffuse to the other side. Most solutes are not lipid soluble; they move through special pathways provided by proteins that span the membrane (2). Small solutes like Na$^+$ pass through channels. Larger ones like glucose enter the cell by facilitated diffusion (3). They bind to a protein carrier that "rocks" back and forth exposing the binding site first to one side, then to the other side of the membrane. The solute hops on or off the site depending on the solute concentration. Solutes move through the membrane until the concentrations on both sides are equal. At this point, movement in one direction is just balanced by movement in the opposite direction; net movement ceases.

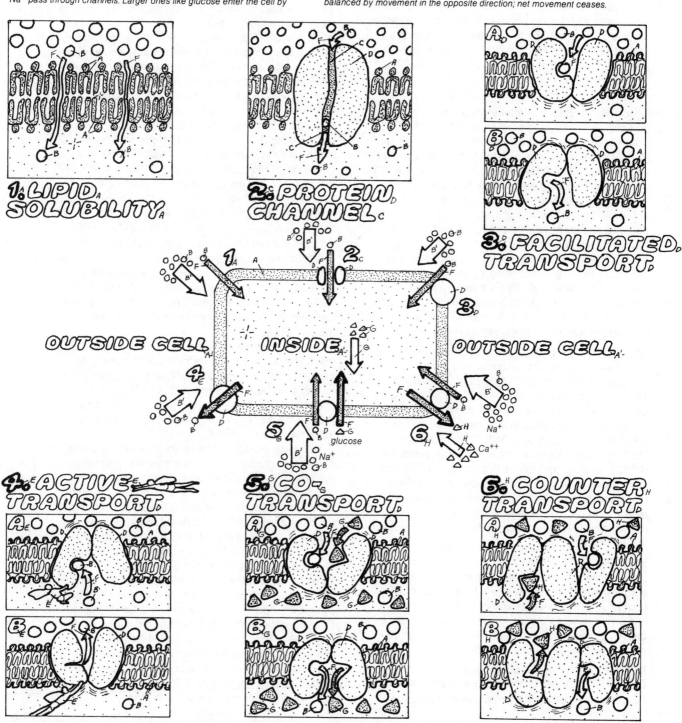

TRANSPORT AGAINST GRADIENTS (UPHILL)*

Proteins also provide pathways for solute movements against concentration gradients (uphill). Primary active transport is probably similar to facilitated diffusion except that the binding site properties change and depend on which side of the membrane the site faces. If the site can bind only on one side of the membrane, transport is only in one direction (uphill in the illustration). Energy for changing site properties and for uphill movement is derived from splitting of ATP (4). Solutes can also move uphill by co- and countertransport. Both utilize the passive transport of one solute to transport a different solute. In co-transport, one solute (triangle) rides piggyback on the carrier of another (circle) that is moving down its gradient; the triangles may move uphill (5). Countertransport is similar, but now the two solutes move in opposite directions (6).

The Na^+-K^+ pump refers to an active transport system that continually pumps Na^+ out of and K^+ into cells. Generally, three Na^+ are pumped out for every two K^+ pumped in. The pump, sometimes called the Na^+ pump for short, is found in the plasma membranes of all body cells, and it is one of the major energy-consuming processes of the body. The pump may account for more than a third of the resting energy consumption of the entire body. What functions does the pump perform to warrant this huge investment of energy?

OSMOTIC STABILITY. Proteins and many other smaller intracellular substances exert an osmotic pressure that is balanced by extracellular solutes, of which Na^+ and Cl^- are the most abundant. But both Na^+ and Cl^- leak into cells. If nothing intervened, this leak would create a continuous osmotic gradient, drawing water into the cell, which would swell and burst. In plant cells, this is prevented by a stiff wall. Animal cells are more flexible and mobile; they have no wall. Instead animal cells have a Na^+-K^+ pump that pumps Na^+ out, keeping internal solute concentrations low enough to prevent cells from swelling and bursting. (Some Cl^- follows to maintain electrical neutrality.) A steady low level of internal Na^+ is maintained because it is pumped out as fast as it leaks in. If the Na^+-K^+ pump is poisoned, animal cells swell and eventually burst.

CO- AND COUNTERTRANSPORT. (See plate 9.) The Na^+ gradient generated by the pump is used to drive transport of other solutes. Transport of glucose and amino acids by cells of the intestine and kidney are good examples of co-transport. In these cells, the solute (glucose or amino acid) enters or leaves the cell only when accompanied by Na^+. Outside the cell (high Na^+), the solute easily finds a Na^+ partner; inside it is difficult because Na^+ has been pumped out. The result: more solute enters than leaves even when solute is higher inside than out. Countertransport is exemplified by Na^+ and Ca^{++} exchange in the heart. In this case, the energy of Na^+ moving down its gradient is coupled to Ca^{++} moving out.

BIOELECTRICITY. The K^+ gradient generated by the pump creates a voltage gradient (negative inside) across the membrane because K^+ leaks out of cells (carrying positive charges out) much faster than Na^+ leaks in. The pump also contributes to this voltage because it pumps three Na^+ out for each two K^+ in. (See plate 11.)

METABOLISM. The pump creates an intracellular environment that is rich in K^+ and poor in Na^+. These conditions are optimal for the operation of a variety of cellular processes including those involved in protein synthesis and in the activation of some enzymes.

PROPERTIES OF THE PUMP. The Na^+-K^+ pump transports Na^+ out of and K^+ into the cell. Na^+ is more than ten times more concentrated in the plasma than inside the cell. The reverse is true for K^+. Thus, Na^+ leaks in, and K^+ leaks out. Nevertheless, a steady level of these two ions is achieved because as fast as they leak in (or out), they are pumped back out (or in). Under normal conditions, both ions are pumped against a concentration gradient, and the energy for this active transport is from degradation of ATP. This latter fact is easily demonstrated in isolated cells whose metabolic machinery has been artificially inactivated or removed. These cells fail to actively pump Na^+ or K^+ unless ATP is introduced into the cytoplasm; other substrates will not work. Detailed studies show that for each ATP split, three Na^+ are pumped out and two K^+ are pumped in:

$$3Na^+_{in} + 2K^+_{out} + ATP_{in} \longrightarrow 3Na^+_{out} + 2K^+_{in} + ADP_{in} + P_{in}$$

Given this scheme, we can think of the pump not only as an ion-pumping machine; it is also an ATP-splitting machine. That is, it behaves like an enzyme that splits ATP; for this reason, we call it an *ATPase*. Further, the reaction written above will not proceed unless both Na^+ and K^+ are present; for this reason, we call it a Na^+-K^+ ATPase.

The virtue of regarding the pump as an enzyme is that it immediately suggests methods to study it. We can grind up the membrane, extract proteins, and look for those proteins that show Na^+-K^+ ATPase activity. Presumably, these proteins contain the pump, and these are the ones to study. Exploiting this strategy has led to the following view:

The pump consists of two pairs of protein subunits, and all four members of the complex are required for transport. Three Na^+ ions arising from inside the cell bind to sites on the inner surface of the pump, which show a strong preference for Na^+ over K^+. This triggers the breakdown of ATP to ADP, and in the process, a high-energy phosphate is transferred to the pump protein (i.e., the pump is phosphorylated). Next the pump changes shape (conformation); the bound Na^+ now face outside, and the binding sites are altered. They no longer favor Na^+ over K^+, but just the reverse is true. Accordingly, they release the three Na^+ and bind two K^+ instead. Bound K^+ favors dephosphorylation; the proteins revert to their original shape with the binding sites back on the inside releasing K^+ because they have been transformed back to the Na^+-preferring state. The cycle repeats itself up to 100 times per second under optimal pumping conditions.

CN: Use red for C and dark colors for F and G.
1. Begin with the upper 4 panels, completing each one before going on to the rest in clockwise order. Then color the summary diagram between the panels.
2. Color three functions of the pump, beginning with osmotic stability.

CELL MEMBRANE BILAYER_A SODIUM PUMP_B ATP_C /ADP_D* PHOSPHATE_E SODIUM (Na⁺)_F BINDING SITE_F' POTASSIUM (K⁺)_G BINDING SITE_G'

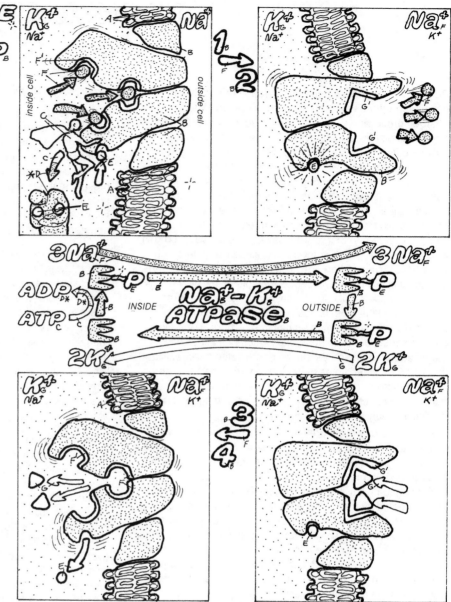

The Na⁺-K⁺ pump transports Na⁺ out of and K⁺ into the cell. Both are transported against a concentration gradient: the energy is derived from degradation of ATP. Beginning in the upper left panel, we see one version of the pump. It consists of 2 pairs of protein subunits. Three Na⁺ ions arising from inside the cell bind to sites on the inner surface of the pump. This triggers the breakdown of ATP to ADP: in the process, a high-energy phosphate is transferred to the pump protein (i.e., the pump is phosphorylated). Next the pump changes shape (conformation): the bound Na⁺ now face outside, and the binding sites are altered. They release the 3 Na⁺ and bind 2K⁺ instead. Bound K⁺ favors dephosphorylation: the proteins revert to their original shape, with the binding sites back on the inside releasing K⁺ because they have been transformed back to the Na⁺- preferring state. The pump behaves like an enzyme that splits ATP, but only in the presence of Na⁺ and K⁺. Hence the name Na⁺-K⁺ ATPase.

The net result: $3Na^+_{in} + 2K^+_{out} + ATP_{in} \longrightarrow$
$3Na^+_{out} + 2K^+_{in} + ADP_{in} + P_{in}$

FUNCTIONS OF THE SODIUM PUMP*

OSMOTIC STABILITY*

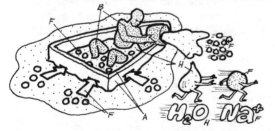

Osmotic pressure exerted by solutes trapped in the cell (like proteins) draws water in. The Na⁺-K⁺ pump compensates by pumping Na⁺ out so that internal solute concentrations are kept low enough to prevent cells from swelling and bursting. A steady low level of internal Na⁺ is maintained because it is pumped out as fast as it leaks in.

GRADIENT FOR CO-TRANSPORT*

The Na⁺ gradient generated by the pump is used to drive transport of other solutes (e.g., in intestinal cells, glucose enters or leaves the cell only when accompanied by Na⁺). Outside the cell (high Na⁺), it easily finds a Na⁺ partner: inside it is difficult because Na⁺ has been pumped out. The result: more glucose enters than leaves when glucose is higher inside than out.

GLUCOSE_I CARRIER*

BIOELECTRICITY*

POSITIVE CHARGE_J
K⁺ CHANNEL_G₂
NEGATIVE CHARGE_K

The K⁺ gradient generated by the pump creates a voltage gradient (negative inside) across the membrane because K⁺ leaks out of cells (carrying positive charges out) much faster than Na⁺ leaks in. The pump also contributes to this voltage because it pumps 3 K⁺ out for each 2 K⁺ in (see next plate).

MEMBRANE POTENTIALS

Plates 9 and 10 focused on concentration gradients as the driving force for transport through membranes. When charged particles (i.e., ions) are involved, *electrical driving forces* become equally effective. These forces appear as voltage differences (voltage gradients) across cell membranes. The forces arise from separation of *positive* and *negative charge* by the membrane. The inside surface of a cell membrane generally has slightly more negative charge in its immediate vicinity, and the outside surface has slightly more positive charge. This creates an electrical force (voltage difference) that may be as large as 0.1 volt (100 mv), attracting positive charge inward and negative charge outward. This voltage difference is called the *membrane potential*, and it is always expressed as the voltage inside the cell minus the voltage outside. Because the inside is generally negative (and the outside positive), the normal membrane potential will be negative. It ranges from -10 mv in red blood cells to about -90 mv in heart and skeletal muscle. The magnitude of these forces can be appreciated if we realize that when K^+ is ten times more concentrated on one side of the membrane, its diffusion to the less concentrated side can be completely stopped by opposing it with a membrane potential of only 60 mv.

Membrane potentials are measured by connecting two electrodes (properly conditioned metallic wires) to each side of the membrane and connecting these electrodes to each other through a meter. Electrons will flow through the wire from the negative to the positive side. The magnitude of the flow, detected by the meter, is proportional to the voltage difference. In practice, special precautions are taken to ensure that the drainage of charge through the electrodes is so small that it does not disturb the original voltage. (In the diagrams, we follow the convention that the needle of the meter points in the direction that *positive* charge would flow — i.e., it is opposite to the direction of electron flow.)

How does the charge separation responsible for membrane potentials arise? Consider the impermeable membrane shown in panel A of the middle diagram. KCl is more concentrated on the left than on the right. Because both sides are electrically neutral ($[K^+] = [Cl^-]$ on each side), the meter shows no voltage difference between the two sides. In panel B, patch *K^+ channels* into the membrane, which allow K^+ but not Cl^- to go through. K^+ begins to diffuse to the right, building up positive charge on the right and abandoning negative charge on the left; a *voltage gradient* is created across the membrane, which tends to move K^+ in the opposite direction (from left to right). Each time a K^+ moves, the charge separation becomes larger so that the voltage opposing diffusion becomes larger. Finally (actually after a very short time), the voltage is just able to balance the concentration gradient, and net K^+ movement ceases (panel C). At this point, the system is in equilibrium; the voltage gradient required to stop diffusion of the K^+ ion is called the K^+ *equilibrium potential*. The larger the concentration gradient, the larger the equilibrium potential.

Notice in panel C that the excess ions charging up the membrane hover close to the membrane; the excess positive and negative ions attract each other. These excess ions are confined to a very thin layer adjacent to the membrane, and their numbers are very small compared to the number of ions present in the remaining bulk solution. Nevertheless, they produce significant electrical forces. Within the bulk of the solution (away from the membrane), the numbers of positive and negative charges are equal.

Instead of using K^+ channels, we could use Cl^- channels. Our analysis would be similar, only now the diffusing ion is negatively charged; it leaves a positive ion behind on the left, and the right-hand side becomes negatively charged. Equilibrium ensues when the membrane potential has the same magnitude as before, but is oriented in the opposite direction (negative on the right). In this case, the Cl^- equilibrium potential is equal and opposite to the K^+ equilibrium potential. When we are dealing with more complex mixtures of ions, their concentration gradients may be virtually independent of one another, so that each ion would have its own equilibrium potential.

Similar effects occur in cell membranes, but now we deal with more ions, and the system does not settle down to equilibrium. *The Na^+-K^+ pump* establishes a K^+ gradient: high K^+ on the inside, low on the outside. (It also establishes an oppositely directed Na^+ gradient, but this is not as important because there are many more operative K^+ channels than Na^+ channels.) K^+ diffuses through K^+ channels to establish a voltage gradient (membrane potential) with the inside negative. The magnitude of the membrane potential is not quite equal to the K^+ equilibrium potential for two reasons: (1) other ions besides K^+ (e.g., Na^+ and Cl^-) can also permeate the membrane, and (2) the Na^+-K^+ pump may also make a direct contribution by pumping charge. Recall that each time the pump cycles, three Na^+ move out, but only two K^+ move in; this results in a net movement of one positive charge out. Thus, the pump not only plays an indirect role by setting up the original concentration gradient of K^+ so that it can diffuse out through K^+ channels, but it also pumps positive charge out. The relative size of this latter (direct) contribution varies with circumstances. It is often small.

The concentrations of Na^+ and K^+ are fairly constant, but they are not in equilibrium. They settle down to steady values because as fast as they leak in (or out), they get pumped back out (or in). If the pump is poisoned, the concentrations of Na^+ and K^+ will change as they drift toward equilibrium, and the membrane potential diminishes. Final equilibrium is then determined primarily by negatively charged ions inside the cell (e.g., protein anions) that cannot permeate the membrane. These are represented by A^- in the large diagram.

CN: Use very light colors for A and B.
1. Begin with the upper panel.
2. Color the stages in the development of a membrane potential. Include the gradient symbols shown below each stage.
3. Color the large panel in the lower right corner illustrating membrane potentials in cells.

ELECTRICAL FORCES *
POSITIVE CHARGE A
NEGATIVE CHARGE B

Some atoms or molecules carry a net electrical charge, [+] or (-). They are called ions. [+] charged ions repel each other; (-) ions also repel each other, but [+] ions attract (-) ones and vice versa. Voltages are high around [+] ions and low around (-) ions. [+] ions flow from high to low voltages. (-) ions flow in the opposite direction.

FLOW B

FLOW A'

LOW VOLTAGE B   **HIGH VOLTAGE** A

MEMBRANE POTENTIAL *
MEMBRANE BILAYER C
POTASSIUM (K⁺) ION A'
K⁺ CHANNEL PROTEIN A²
CONCENTRATION GRADIENT A³
CHLORIDE (Cl⁻) ION B'
VOLTAGE GRADIENT D

A. *An impermeable membrane separates two solutions of K⁺, Cl⁻. The left side is more concentrated than the right. Nothing happens because the ions cannot permeate the membrane.* **B.** *Now K⁺ channels are introduced; K⁺ can get through, but Cl⁻ cannot. K⁺ begins to diffuse to the right, building up [+] charge on the right and abandoning (-) charge on the left. A voltage gradient is created across the membrane, which tends to move K⁺ in the opposite direction (from left to right).* **C.** *As more K⁺ diffuses, the voltage gradient grows larger until it is just able to balance the concentration gradient. At this point, net K⁺ movement ceases; the system is in equilibrium. The voltage gradient required to stop the diffusion of any ion is called its equilibrium potential.*

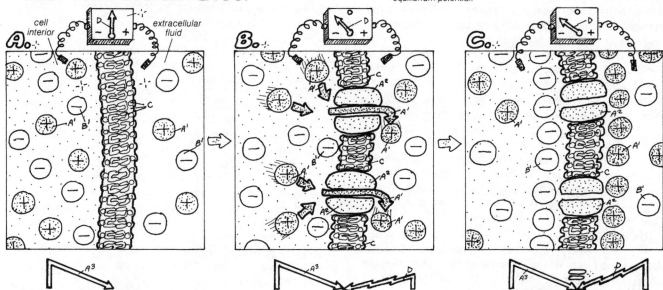

cell interior   extracellular fluid

A.   B.   C.

EQUILIBRIUM POTENTIAL D

THE LIVING CELL *
SODIUM (Na⁺) ION E
Na⁺ CHANNEL E'
SODIUM-POTASSIUM ATPase PUMP F ATP G
ANION (A⁻) H

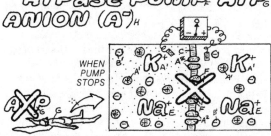

WHEN PUMP STOPS

Similar events occur across cell membranes. The Na⁺ K⁺ pump establishes a K⁺ gradient, high K⁺ on the inside, low outside. (It also establishes an oppositely directed Na⁺ gradient, but this is not as important because of the paucity of operative Na⁺ channels.) K⁺ diffuses through K⁺ channels to establish a voltage gradient (also called membrane potential), with the inside negative (outside positive). The magnitude of the membrane potential is not quite equal to the K⁺ equilibrium potential because other ions (e.g., Na⁺) can also permeate the membrane. There are many (-) charged ions other than Cl⁻ inside the cell represented by A⁻. When the pump stops, the ions equilibrate, and the size of the membrane potential diminishes.

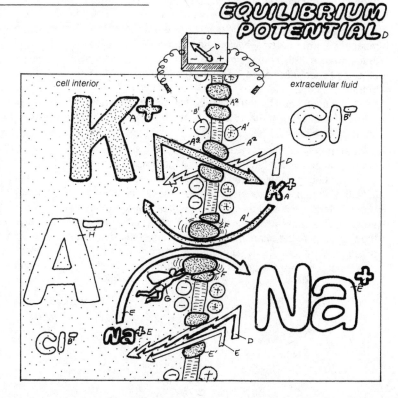

cell interior   extracellular fluid

K⁺   Cl⁻

A⁻

Cl⁻   Na⁺

THE NERVE IMPULSE: AN INTRODUCTION

Cells that transmit "messages" or impulses throughout the nervous system are called neurons. Although the size and shape of these cells show large variations from place to place, a typical neuron consists of three characteristic parts: (1) *Dendrites*, which specialize in receiving stimuli from other neurons, from sensory epithelial cells, or simply from their environment. These are narrow extensions of the cell and are often short and branched. (2) *Cell body*, which can also receive impulses. This contains the nucleus, mitochondria, and other standard equipment of cells. (3) *Axon*, a single, cylindrical extension of the cell, specialized in conducting impulses over large distances to other nerve, muscle, or gland cells. The final portion of the axon is generally branched, and each branch terminates in an *axon terminal*, a swollen bulblike structure that is important in transmitting information to the next cell. The whitish cables, easily seen in gross dissection, that connect the brain or spinal cord to various parts of the body are made up of numerous axons held together by connective tissue.

The existence of these "messages" or "signals," which are most often called *nerve impulses*, is easily demonstrated. If the nerve to a limb is cut, the limb becomes paralyzed. The muscles in the limb remain healthy, they will move if they are stimulated directly with weak electrical shocks, but otherwise the muscles never get the message to move. After some time, the paralysis subsides and the return of normal activity coincides with the regeneration of the cut axons so that the original connections are reestablished.

Using a single axon together with its associated muscle, we can study many properties of the "messages." If we stimulate the axon (e.g., with an electrical shock), the subsequent movement of the muscle will reveal whether it got the "message." This primitive strategy has been used to study how fast messages travel. Suppose we stimulate an axon at a specific point, A, and measure the time it takes before the muscle contracts. Next we stimulate at point B, 5 cm closer to the muscle. This time the muscle responds about .001 sec. (1 msec.) sooner because the message does not have as far to travel, and the difference in the two times (.001 sec.) reflects the time taken for the message to travel the 5 cm from A to B. If it takes .001 sec. to travel 5 cm, then the velocity of the message must be 5/.001 = 5000 cm/sec. = 50 m/sec.

What are these "messages"? When an axon is stimulated, many changes occur before the muscle contracts. Of these, electrical changes have been the easiest to measure and intrepret. At rest, the axon behaves like any cell; its interior bears a negative electrical charge when compared to the outside. This is easily demonstrated by placing electrodes on the two sides of the axon membrane. They register a voltage difference of about 70 mv across the membrane with the inside negative (membrane potential = -70 mv). When the axon is stimulated, this voltage reverses itself, moving momentarily to positive inside and then quickly back to the resting negative state. This sudden change in membrane potential that accompanies activity is called an *action potential*. It occurs first at the point of stimulation and moments later at each position along the axon; the farther from the stimulation point, the longer the delay before the action potential appears. In other words, the action potential begins at the stimulus and travels along the axon toward the muscle. The speed of travel is identical to the speed of the "message" (measured as described above). In fact, comparing the properties of "messages" with those of action potentials leads us to conclude that they are one and the same; "messages" are action potentials. At the height of the action potential, the outside of the membrane is negatively charged, and this negative region moves along the axon. Viewed from outside the axon, the action potential appears to be a *wave of electrical negativity* that travels down the axon.

How do action potentials arise? At rest, the membrane potential is about -70 mv. Because of this polarity (negative inside, positive outside), the membrane is said to be *polarized*. To understand the genesis of this resting potential as well as the action potential, review plate 11. Remember that the intracellular fluid in neurons (i.e., in axons) has high K^+ and low Na^+ concentrations; the reverse is true for extracellular fluid which is rich in Na^+ and poor in K^+. At rest, most operative channels allow passage of K^+ but not of Na^+. K^+ diffuses down its concentration gradient, delivering a positive charge on the outside surface while leaving negatively charged "partners" behind on the inner surface. The membrane becomes polarized, with the inside negative with respect to the outside.

A stimulus causes a brief increase in the number of open Na^+ channels. If the stimulus is weak, only a few channels open, and the membrane potential is hardly perturbed. However, if the stimulus is sufficiently strong (i.e., if it is stronger than a critical level called the *threshold*), then the number of open Na^+ channels becomes very substantial. Na^+ ions, poised at high concentration outside the axon, leave their negatively charged "partners" behind on the outside and rush in fast enough to overwhelm the K^+ moving out. The inside of the cell is inundated with positive charge so that the polarity is reversed; now the inside is positive and the outside is negative. A moment later the Na^+ channels close and extra K^+ channels open. The membrane becomes very permeable to K^+. K^+ moves out, making the membrane potential even more negative than it was at rest, driving it very close to the K^+ equilibrium potential. Finally (after several milliseconds), the extra K^+ channels close, and the membrane permeability returns to its resting condition.

CN: Use the same colors as on the previous page for K^+ (G) and Na^+ (L).
1. Begin with elements of the nerve cell.
2. Color each membrane state, including the chart to its right, before going on the next.

THE NERVE CELL *

DENDRITE A
CELL BODY B
AXON: -
 POLARIZED MEMBRANE C
 DEPOLARIZING MEMBRANE D
 (NERVE IMPULSE) D'
 REPOLARIZING MEMBRANE E
AXON TERMINAL F

A nerve cell has several short processes called dendrites, which extend from the cell body, and branch extensively. It also has a long cylindrical axon, which transmits signals called nerve impulses from one nerve cell to the next, or to muscles or glands. Electrical measurements on the cell surface show that the nerve impulse consists of a wave of electrical negativity that moves along the axon with freeway speeds.

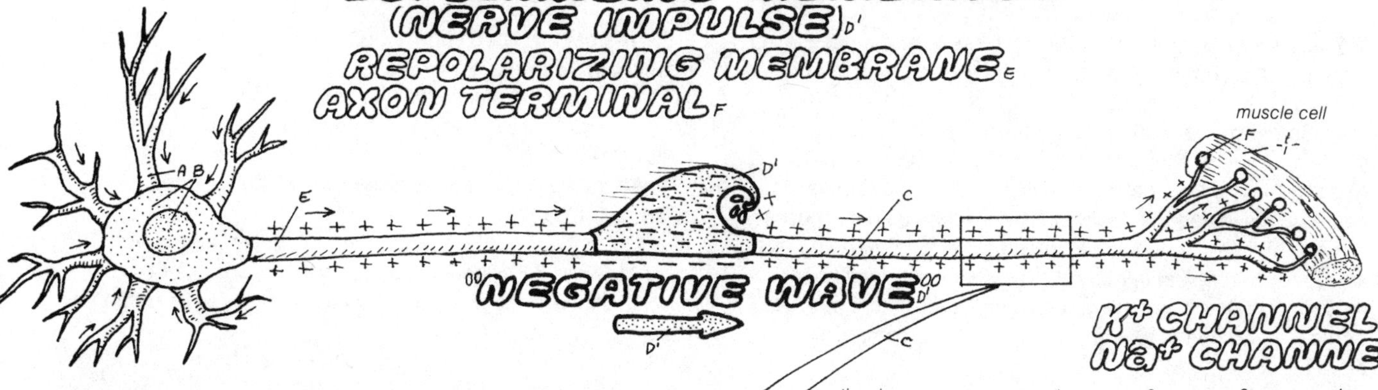

muscle cell

"NEGATIVE WAVE"

K^+ CHANNEL G'
Na^+ CHANNEL H'

POLARIZED (RESTING): HIGH K^+ PERMEABILITY

At rest, intracellular electrodes show a membrane potential, V_m, of about -70 mv. V_m results from the high K^+ permeability; very little K^+ leaks out because V_m is close to the equilibrium potential for K^+.

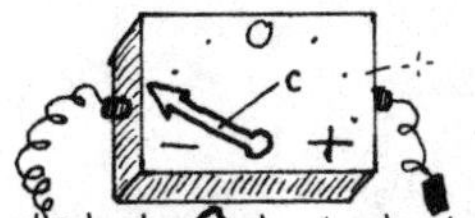

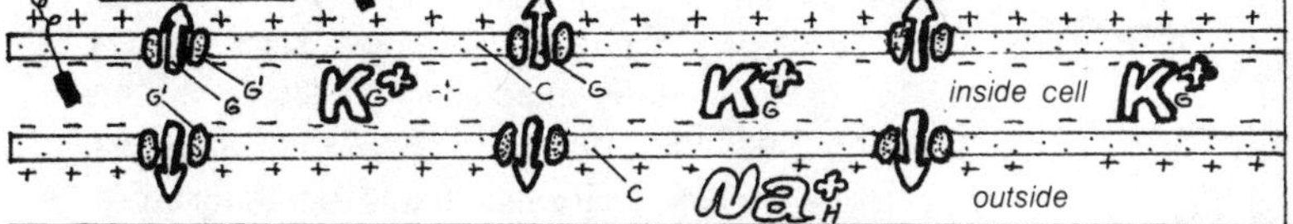

inside cell
outside

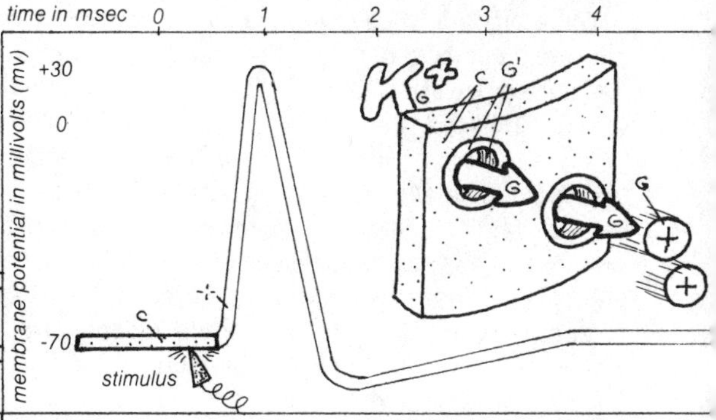

time in msec
membrane potential in millivolts (mv)
+30
0
-70
stimulus

REVERSED POLARITY: VERY HIGH Na^+ PERMEABILITY

A stimulus causes a brief increase in Na^+ permeability. If the stimulus is above threshold, Na^+ ions, poised at high concentrations outside the axon, rush in fast enough to make the V_m positive.

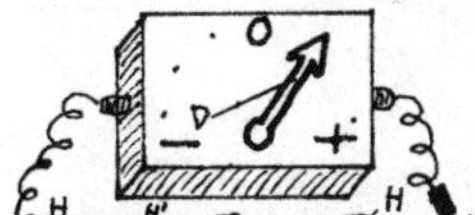

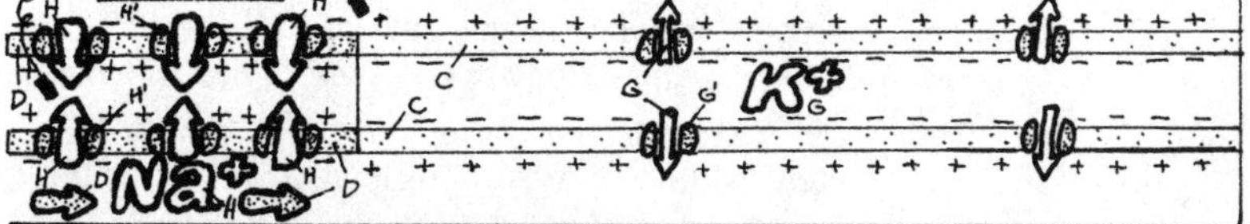

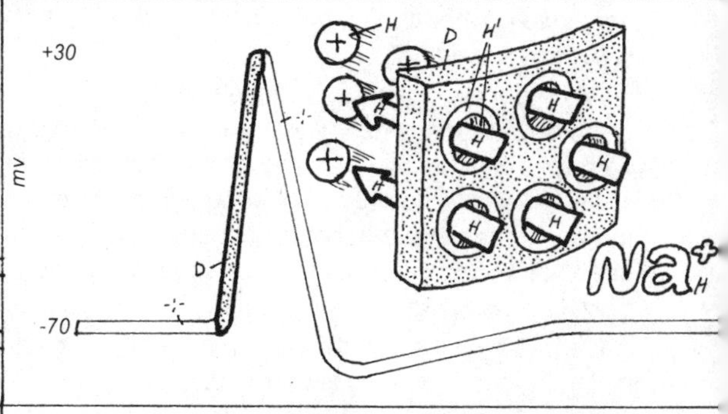

+30
mv
-70
Na^+

REPOLARIZED: VERY HIGH K^+ PERMEABILITY

A moment later, the increased Na^+ permeability subsides, and the membrane becomes very permeable to K^+. K^+ moves out, making V_m negative and driving it even closer to the K^+ equilibrium potential.

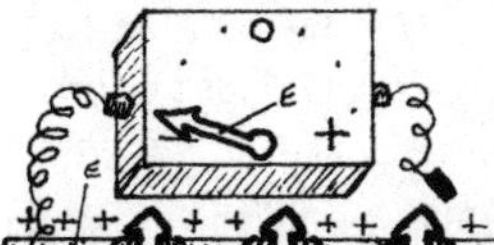

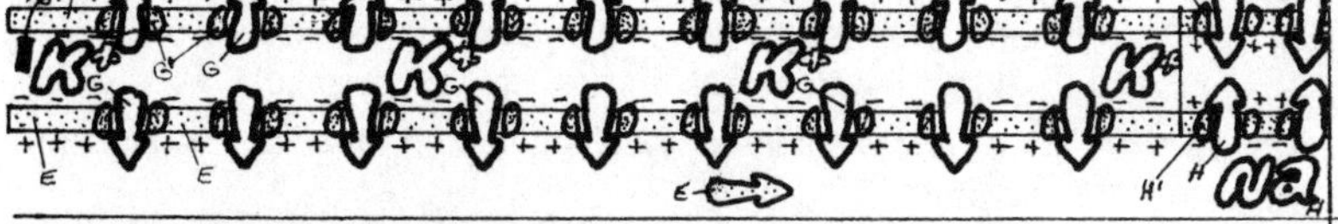

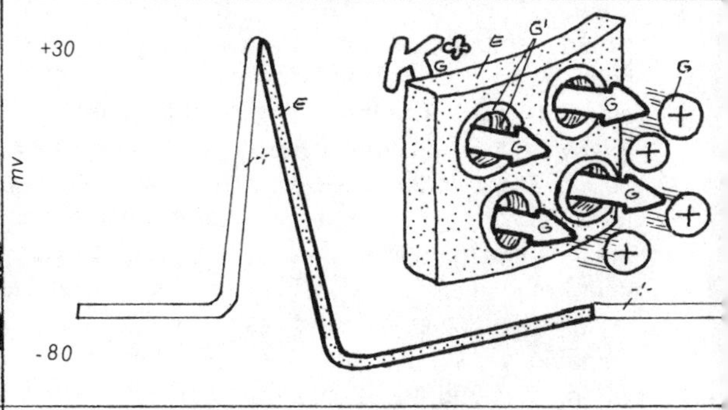

+30
mv
-80

POLARIZED: HIGH K^+ PERMEABILITY

Finally (after several msec), the membrane returns to its resting condition.

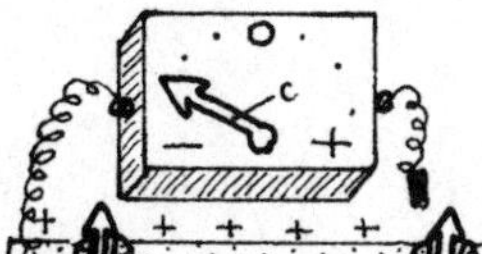

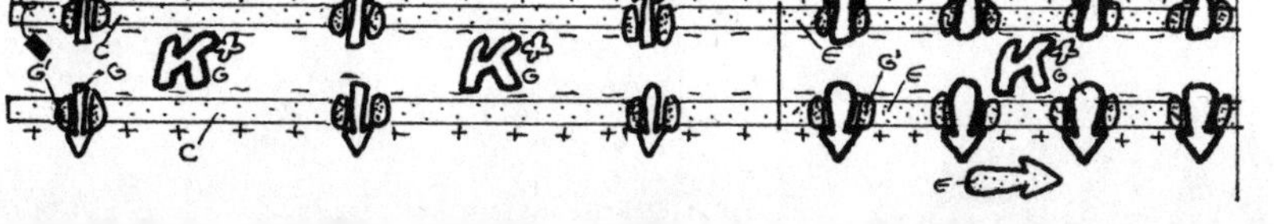

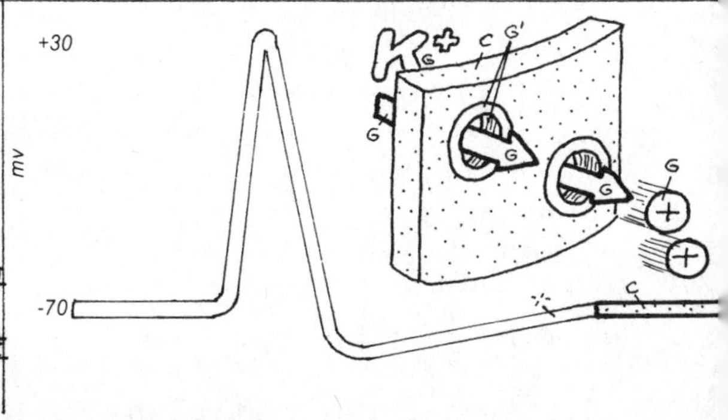

+30
mv
-70

CONTROL OF ION CHANNELS BY MEMBRANE POTENTIAL

Modern interpretations of excitation in nerve and muscle are based on the sequential opening and closing of *ion channels* in the membrane. There are separate channels for different ions. Somehow each channel "recognizes" the appropriate ion (e.g., K^+) and allows it to pass while restraining others (e.g., Na^+). The channel mechanism responsible for this selectivity is called a *selectivity filter*. The status (i.e., open or closed) of many channels depends on the *membrane potential*. To form a mental image of these channels, imagine that they are regulated by *voltage activated "gates"* that open or close in response to changes in membrane potential or voltage.

There are two types of K^+ channels. One type is voltage activated, but most of these are closed during rest, when the membrane potential is about -70 mv. The other type is not voltage activated; it is always open and provides the pathway for the small but continuous K^+ leakage that creates the resting potential. A resting membrane potential of -70 mv implies that the inside is negative while the outside is positive. Because of this electrical distinction between the membrane's inner and outer surfaces, the membrane is said to be *polarized*. By definition, the membrane is *depolarized* whenever the magnitude of this membrane potential becomes smaller than the resting potential (i.e., close to zero); conversely, when the magnitude is increased, the membrane is *hyperpolarized*.

When nerves are excited with electrical shocks, the impulse always arises at the negatively charged electrode (the cathode). By attracting positive ions and repelling negative ones, the electrode reduces the membrane potential so that the nerve membrane becomes depolarized. This simple observation can be generalized: *A stimulus will be effective only if it depolarizes the membrane.*

Depolarizing the membrane works because the permeability of nerve membranes to ions is very sensitive to the voltage gradient (membrane potential). The crucial relationship between membrane potential and ion flow has been studied in detail by an ingenious electrical method called a voltage clamp. Using this method, we can set the membrane potential to any desired value and keep it at that value for a prolonged period. At the same time, we can estimate the amounts of Na^+ and K^+ that flow through the membrane in response to the imposed membrane potential. These results are then interpreted in terms of opening and closing of Na^+ and K^+ channels, and they form the basis of our current understanding. In particular, they allow us to ask what happens when the membrane is stimulated (i.e., when it is depolarized).

If the membrane is depolarized (e.g., the membrane potential is changed from the resting value of -70 mv to a new value of -50 mv and *maintained* at this new potential), then the response of the ion channels can be arbitrarily divided into two phases:

1. An early response (<1 msec.) when the Na^+ channels open.
2. A late response (>1 msec.) when the Na^+ channels close and the K^+ channels open. During this late period, the Na^+ channels appear to be inactivated; they will not respond to further depolarization.

We can interpret these changes in terms of our hypothetical gates as follows:

1. *Early response.* The Na^+ channel contains two gates, a slow one and a fast one. At rest (polarized membrane), the slow gate is open, but the fast gate is closed so that the channel is closed. Upon depolarization, the fast gate opens quickly; now both gates are open so that the channel is freely permeable to Na^+, and Na^+ rushes into the axon.
2. *Late response.* A moment later the slow Na^+ gate closes. The membrane is no longer highly permeable to Na^+; the rapid inflow of Na^+ ceases. In addition, the slowly responding gates in the K^+ channel open and K^+ flows out of the axon.

Thus, a sustained depolarization leads to a transient increase in Na^+ permeability followed by a sustained increase in K^+ permeability. The increase in Na^+ permeability is attributed to the presence of two gates that give opposite responses to the depolarization. The fast gate opens, but the slow gate closes. The time between the opening of the fast gate and the closing of the slow gate corresponds to the period of increased Na^+ permeability. In contrast, a K^+ channel has only one voltage activated gate that opens (slowly). Once open, it will stay open as long as the depolarization is sustained.

Immediately following the depolarization, even though the membrane potential has been returned to resting level (back to -70 mv), the axon is not fully recovered. This is because the slow gates require a millisecond or two to respond to the newly established resting potential. If, during this brief period, a rapid second stimulus (depolarization) is delivered, the Na^+ channels will fail to open. The fast gates respond and open, but the slow gates are still closed as a result of the original depolarization. Only after a recovery period of one or two msec. will the slow gates open and allow a second stimulus to trigger the transient increase in Na^+ permeability.

Application of these results to action potentials is detailed on plate 14.

STIMULUS: DEPOLARIZES MEMBRANE.

NEGATIVE ELECTRODE_A
MEMBRANE_B
POSITIVE CHARGE_C
NEGATIVE CHARGE_A'

When nerves are excited with electrical shocks, the impulse always arises at the negatively charged electrode (the cathode). This is where the nerve membrane becomes depolarized. By attracting positive ions and repelling negative ones, the electrode reduces the membrane potential. This is what is required for excitation. A stimulus will be effective only if it depolarizes the membrane.

extracellular fluid

cell interior

EFFECTS OF DEPOLARIZATION ON ION CHANNELS.

SODIUM (Na⁺) IONS_D
CHANNEL_D'
SELECTIVITY FILTER_E
FAST GATE_F
SLOW GATE_G
DEPOLARIZING Na⁺ FLUX_D²
(CURRENT)
POTASSIUM (K⁺) IONS_H
CHANNEL_H'
SELECTIVITY FILTER_E'
SLOW GATE_G
POLARIZING K⁺ FLUX_H²
LEAK CHANNEL_H³

Nerve cell membranes contain separate channels for different ions. Each channel "recognizes" the appropriate ion (e.g., K^+) and allows it to pass while restraining others (e.g., Na^+ ions). The channel mechanism responsible is called a selectivity filter. In addition, many channels contain voltage activated "gates" that open or close the channel, depending on the membrane polarization. The response of an excitable membrane to a **sustained depolarization** (stimulus) can be arbitrarily broken into two phases: (1) an early response (<1 msec.) when the Na^+ channels open and (2) a late response (>1 msec.) when the Na^+ channels close and the K^+ channels open. During this period, the Na^+ channels will not respond to further depolarization.

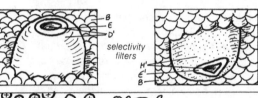

selectivity filters

NORMAL (RESTING POTENTIAL).

There are two types of K^+ channels. One is always open; it is the pathway for the small K^+ leakage that creates the resting potential. The other is voltage activated; it is mostly closed when the membrane is highly polarized. Voltage activated Na^+ channels are also closed in the highly polarized state.

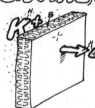

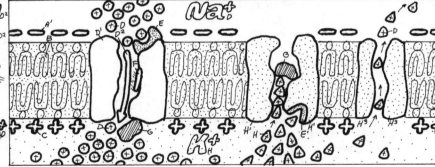

extracellular fluid

cell interior

DEPOLARIZATION EARLY (<1msec)_D²

The Na^+ channel contains two gates, a slow one and a fast one. At rest (polarized membrane), the slow gate is open, but the fast gate is closed so that the channel is closed. Upon depolarization, the fast gate opens quickly, making the membrane permeable to Na^+. This depolarized state can be artificially sustained.

SUSTAINED DEPOLARIZATION LATE (>1msec)_H²

A moment later the slow Na^+ gate closes, and the membrane is no longer highly permeable to Na^+. In addition, a slow K^+ opens, making the membrane more permeable to K^+ than it was at rest.

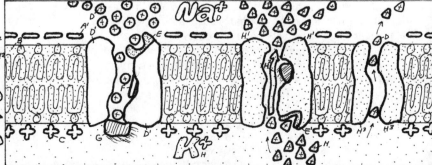

IONIC BASIS FOR THRESHOLD ALL-OR-NONE RESPONSE, AND REFRACTORY PERIOD

In this plate, we define and interpret three of the most important characteristics of excitation: a threshold stimulus, an all-or-none response, and a refractory period.

THRESHOLD. If a nerve axon is stimulated with weak electrical shocks, nothing seems to happen. When the stimulus is repeated many times with each stimulus a little stronger than the last, eventually a point will be reached where the nerve responds by transmitting an action potential. The strength of the stimulus just barely able to excite is called the threshold. Stimuli below threshold do not work; stimuli above threshold produce action potentials.

ALL-OR-NONE. A stimulus above threshold excites the nerve, but the size of the response is independent of the strength of the stimulus. All action potentials are the same, no matter how large the stimulus; the response is all-or-none. This behavior is similar to a fuse; once lit, the size of the spark that travels along is independent of the size of the match that initiated it.

To interpret these properties, recall that the inside of the axon has high K^+ and the outside, high Na^+. Further, the membrane potential is simply a measure of the electrical force on a positive charge. Sometimes the terms "membrane potential," "voltage gradient," and "difference in voltage" are used interchangeably, and we will abbreviate these terms as V_m. Finally, recall from plate 11 that the amount of charge movement necessary to make substantial changes in V_m is very small. During the short time of a single action potential, the actual amounts of Na^+ and K^+ that move into or out of the axon are very small; they have significant effects on V_m, but the change in concentration of Na^+ or K^+ is so small that it cannot be detected by chemical means.

At rest, the axon is permeable mostly to K^+, but not much K^+ leaks out because the opposing membrane potential, V_m, is close to the K^+-equilibrium potential (i.e., the concentration gradient of K^+ is almost balanced by V_m pushing in the opposite direction).

Now the nerve is stimulated. Depolarization (stimulation) has two effects:

1. Early on, voltage activated Na^+ channels open.
2. Later (delayed for about 1 msec.) Na^+ channels close, and K^+ channels open.

With a weak, subthreshold stimulus (panels 1 and 2), not enough Na^+ flows in to overcome the outflow of K^+, and the axon repolarizes.

With a stronger, suprathreshold stimulus (panels 3 and 5), more Na^+ channels open so that Na^+ inflow exceeds K^+ outflow, the net flow of charge is now positive inward, and the axon is depolarized even further. But this opens even more Na^+ channels, which causes more depolarization. A vicious cycle ensues; the membrane potential takes off in the positive direction with an explosive velocity as the interior of the axon becomes more and more positive. But this rapid upward movement of the membrane potential does not persist. Soon (panel 5) V_m becomes positive and large enough to oppose Na^+ entry despite the open channels (i.e., V_m approaches the

Na^+ equilibrium potential, where the concentration gradient moving Na^+ inward is just balanced by V_m pushing Na^+ out). At the same time, the delayed effects begin to appear (panel 6). Na^+ channels close, and voltage activated K^+ channels open, K^+ outflow exceeds Na^+ inflow, and the net flow or charge is now positive outward. V_m plummets toward its resting value, overshoots momentarily, and comes very close to the K^+ equilibrium potential because the voltage activated K^+ channels are still open, making the membrane even more K^+ permeable than it was at rest. Finally (panel 7), the repolarized membrane closes the voltage activated K^+ channels, and V_m returns to its resting value.

From this description, we see that the threshold is determined by the stimulus strength that is able to cause an inward Na^+ flow to just barely exceed the outward K^+ flow. From that point onward, the stimulus plays no further role because the seeds of the positive feedback (vicious cycle) reside in the axon itself. The all-or-none response arises naturally out of this positive feedback; once the response is triggered, the positive feedback drives the membrane potential to its maximum value (given by the Na^+ equilibrium potential). The size of the action potential is determined by the concentration gradients of Na^+ and K^+ because the concentration gradient of K^+ determines the resting potential (K^+ equilibrium potential), and the concentration gradient of Na^+ determines the height of the action potential (Na^+ equilibrium potential). Just as a stick of dynamite contains its own explosive energy, the axon membrane is "loaded" with "explosive" energy in the form of ion gradients.

REFRACTORY PERIOD. For a brief millisecond or two following excitation, the axon is no longer excitable. This recovery phase, called the refractory period, can be divided into two phases. The earlier phase is the absolute refractory period, where the threshold appears to be infinite, and no stimulus will suffice. In the later phase, the relative refractory period, the threshold returns to normal. The basis for the refractory period is found in the "delayed effects." After the first millisecond of excitation, the slow Na^+ gates close and remain closed for a brief time despite the fact that V_m is near rest. These gates are slow to respond to the initial depolarization, and they are equally slow in responding to the repolarized membrane. In addition, the voltage activated K^+ gates are still open. With the slow Na^+ gates closed and the K^+ gates open, it is difficult if not impossible for Na^+ inflow to exceed K^+ outflow (i.e., to reach threshold).

How do the activities of the Na^+-K^+ pump influence the action potential? They don't, at least not directly. Any contributions by the pump to V_m are overpowered by the more massive movements of the ions through the channels. The pump does not cycle often enough to make a difference during activity. However, action potentials are very brief, and the axon is at rest most of the time. During rest, there is ample time for the slow cycling of the pump to restore the small amounts of Na^+ and K^+ that have leaked through channels activated during the action potential.

CN: Use the same colors as on the preceding page for cell membrane (B), K^+ (C), Na^+ (D), positive charge (E), and negative charge (F).
1. Color each of the panels completely before going on.
2. Color the Na^+ (vicious) cycle.
3. Color the graph in the lower left corner; color the dark numerals gray, and note their relationship to the panels on the right.

STIMULUS (DEPOLARIZATION)ₐ
CELL MEMBRANE_B
POTASSIUM (K⁺)_C
CONCENTRATION GRADIENT_{C¹}
LEAKAGE CHANNEL_{C²}
VOLTAGE GATED CHANNEL_{C³}
SODIUM (Na⁺)_D
CONCENTRATION GRADIENT_{D¹}
VOLTAGE GATED CHANNEL_{D²}
POSITIVE CHARGE_E
ELECTRICAL GRADIENT_{E¹}
NEGATIVE CHARGE_F

At rest, the axon is permeable mostly to K^+, but not much K^+ leaks out because the opposing electrical gradient V_m is close to the K^+ equilibrium potential. Depolarization (stimulation) has two effects: A. Early on, voltage activated Na^+ channels open. B. Later (after 1 msec.) Na^+ channels close, and K^+ channels open. **With a weak stimulus** *(1)-(2): not enough Na^+ flows in to overcome the increased outflow of K^+ resulting from the reduced V_m caused by the stimulus; the axon repolarizes.* **With a stronger stimulus** *(3)-(5): more Na^+ channels open so that Na^+ inflow exceeds K^+ outflow, and the axon is depolarized even further. But this opens even more Na^+ channels, which causes more depolarization. A vicious cycle ensues; the interior of the axon becomes more and more positive until (5) V_m is large enough to oppose Na^+ equilibrium potential. Now (6) the Na^+ channels close and voltage activated K^+ channels open. V_m plummets toward its resting value, overshoots momentarily, and comes very close to the K^+ equilibrium potential before (7) the repolarized membrane closes the voltage activated K^+ channels, and V_m returns to its resting value.*

Identify portions of the action potential (below) with relevant panels on the right. The action potential is all-or-none because its size is limited by the K^+ equilibrium potential (bottom) and the Na^+ equilibrium potential (top). Note the refractory period that follows the action potential where the threshold becomes infinite and slowly returns to normal. This arises partly because the voltage activated K^+ channels are still open, but more importantly because once voltage activated Na^+ channels open, they are momentarily inactivated and cannot respond to a second stimulus unless allowed a short time to recover.

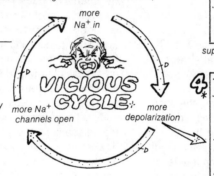

more Na⁺ in

VICIOUS CYCLE

more Na⁺ channels open

more depolarization

THE ACTION POTENTIAL_*
RESTING POTENTIAL_G
THRESHOLD POTENTIAL_H
Na⁺ EQUILIBRIUM POTENTIAL_{D³}
K⁺ EQUILIBRIUM POTENTIAL_{C⁴}

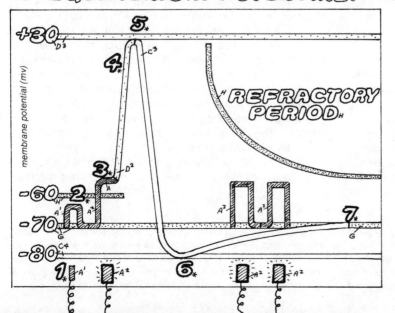

WEAK STIMULUSₐ'

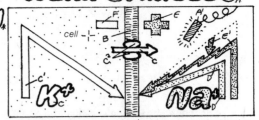

subthreshold depolarization — K⁺ leakage increases causing

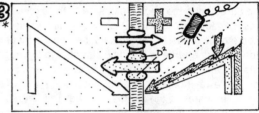

repolarization — return to resting potential

STRONG STIMULUS_{A²}

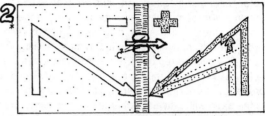

suprathreshold depolarization — more Na⁺ channels open causing

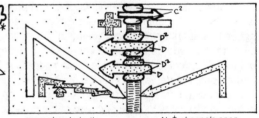

more depolarization — even more Na⁺ channels open

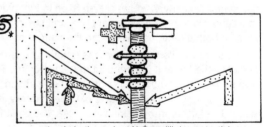

depolarization ends at Na⁺ equilibrium potential.

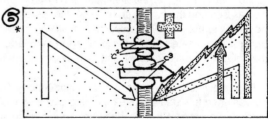

repolarization — K⁺ channels open

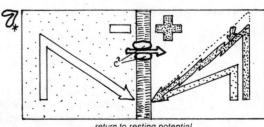

return to resting potential

TRANSMISSION OF NERVE IMPULSES

Plate 14 described how depolarization at a particular location on an axon membrane leads to the formation of an action potential *at that location*. But once an axon is excited, the action potential moves! It is propagated along the entire length of the axon. The first diagram in the plate shows an axon with an impulse located at B that is traveling from left to right. The excited region at B is at the height of the action potential, where the membrane polarity is reversed. This discrepancy in charge between excited and unexcited regions of the axon will cause the charge to move along the axon. For simplicity, we will describe only the movement of the positive charge. (If the same arguments are applied to negative charge movements, the same conclusion will be reached.)

On the external surface of the axon, the positive charge will be attracted to the negative charge of the excited region so that the adjacent regions, A and C, on either side will lose some of their positive charge. Further, on the inner surface, the positive charge at B will be attracted to the negative charge at adjacent regions. These actions take positive charge away from the external surface at A and C and add positive charge to the internal surface. The net result is to depolarize the membrane at both A and C. But depolarization stimulates! The stimulus is effective at C but not at A because A is still recovering from the impulse that just passed it (i.e., A is in a refractory period). Nerve impulse conduction is analogous to a spark traveling along a fuse. The heat of the spark ignites the powder in the region ahead of it. The spark does not travel backward because the trailing edge has been burned out. In a nerve, the impulse excites the region ahead of it, releasing some of the energy contained in ion gradients; it does not travel backward because of the refractory period. If you light a fuse in the middle, the spark will travel in both directions. Similarly, if you excite an axon in the middle, the refractory argument no longer applies, and the impulse will travel in both directions (away from the source of excitation).

Axons vary in diameter as well as length. The larger the diameter, the faster it will conduct nerve impulses. This follows because the speed of conduction depends on how far downstream the electrical effects of the excitatory impulse reach. The farther they reach, the quicker the distant regions become excited. These electrical effects are propagated by charge movement (i.e., electrical current) inside the axon as well as out, and the narrower the axon, the more resistant it becomes to these movements. As a result, the electrical disturbance created by an action potential in a narrow axon is confined to regions close by, and the velocity of conduction is small.

Rapid reflexes require fast impulse conduction along nerves. Invertebrates acquire rapid responses by using very large nerve axons. However, their behavior is uncomplicated, and they do not require very many of these nerves. But vertebrates have complex behavior and require many more axons.

If these were all large, they would be cumbersome and create a packaging problem (see below). The problem is solved by keeping the axon diameters small and by using another means, *myelin sheaths*, to achieve rapid conduction velocities.

Most axons are encased in a white, fatty, myelin sheath that is broken at intervals called *nodes of Ranvier*. These nodes are about 1 to 2 mm apart, and they are the only place that the bare axon membrane is exposed to the external solution. Myelin sheaths are formed by satellite *Schwann cells* that are wrapped spirally around the axons; the sheath is made of layers of ion-impermeable Schwann cell membranes.

Impulse conduction in myelinated and nonmyelinated axons differs because the myelin sheaths alter the charge distribution along the axon. At the nodal regions, positive and negative charges are separated by the thin plasma membrane; they are not too far apart to partially cancel each other. Between the nodes, the myelin sheath, which may contain as many as 300 tightly packed membranes, imposes a much larger distance between intra-and extracellular charge, and the partial cancellation is considerably reduced. As a result, for a given membrane potential of say -70 mv, there will be much less charge piled up at the internodal regions (along the sheath between the nodes) than at the nodes. Similarly, it will require much less charge removal to depolarize the internodes. Thus, when a node is excited, it quickly depolarizes the adjacent internodal region and reaches much farther downstream to the next node for more charge. The neighboring node becomes depolarized, and the impulse jumps from node to node. The internodal region does not become excited because the depolarization has to be "shared" by the many membranes that are stacked in series and also because there are few if any Na^+ channels in the internodal regions. These factors result in faster conduction because the impulse now jumps from node to node and does not have to wait for each (internodal) section of the membrane to become excited.

Vertebrate nerves often contain small, slow, conducting, nonmyelinated nerves mixed with faster myelinated nerves. Conduction velocities vary from 0.5 to 120 m/sec. (1 to 268 mph). The advantages obtained by use of myelin sheaths is illustrated by the following example. A typical mammalian motor nerve containing about 1000 fast conducting axons is only about 1 mm in diameter. If the only basis for achieving fast conduction is by increasing the fiber diameter, then it can be calculated that this nerve should be about 38 times larger (i.e., without myelin, the 1 mm nerve would have a diameter of 38 mm = 1.5 in.) In addition, by restricting ion movements to the nodes of Ranvier, the myelin sheath helps to prevent the dissipation of the Na^+ and K^+ gradients each time the nerve fires. Thus, less energy is required to restore these gradients.

NON MYELINATED AXON*

AXON MEMBRANE_A
Na⁺ CHANNEL,_B IONS,_B'
FAST GATE_C
SLOW GATE_D
POSITIVE CHARGE_E
NEGATIVE CHARGE_F
NERVE IMPULSE_F'

Once an axon is excited, an impulse is propagated along its entire length. Area B below shows a patch of axon that is excited and is at the height of the action potential where the membrane polarity is reversed. Follow the flow of [+] charge from, and to, the adjacent membrane area, C. Because [+] charge is attracted to [−] charge, it is removed from the external surface, and added to the internal surface at C. Both effects depolarize the membrane at C. But depolarization stimulates! An advancing impulse (B) stimulates the membrane ahead of it (C), but not the region behind (A) because that region is in a refractory period.

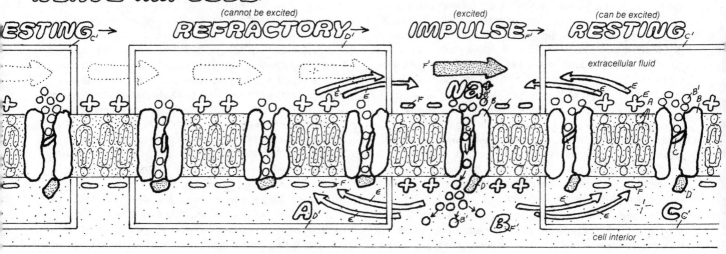

RESTING_C' →   (cannot be excited) REFRACTORY_D' →   (excited) IMPULSE_F' →   (can be excited) RESTING_C'

extracellular fluid

cell interior

MYELINATED AXON*

SCHWANN CELL_G
MYELIN SHEATH_H
NODE OF RANVIER_I

Most axons are encased in a white, fatty, myeline sheath, which is broken at intervals called nodes. Nodes are the only place that the bare axon membrane is exposed to the external solution. Myelin sheaths are formed by satellite Schwann cells that are wrapped spirally around the axon. The sheath is made of layers of ion impermeable Schwann cell membranes. Impulse conduction in myelinated and non-myelinated axons is similar, only here the intra- and extracellular charge responsible for membrane polarization tends to hover around the nodes (where [+] and [−] can get closer). Further, voltage activated channels are virtually confined to the nodal regions. These factors result in faster conduction because the impulse now jumps from node to node and does not have to wait for each (internodal) section of the membrane to become excited.

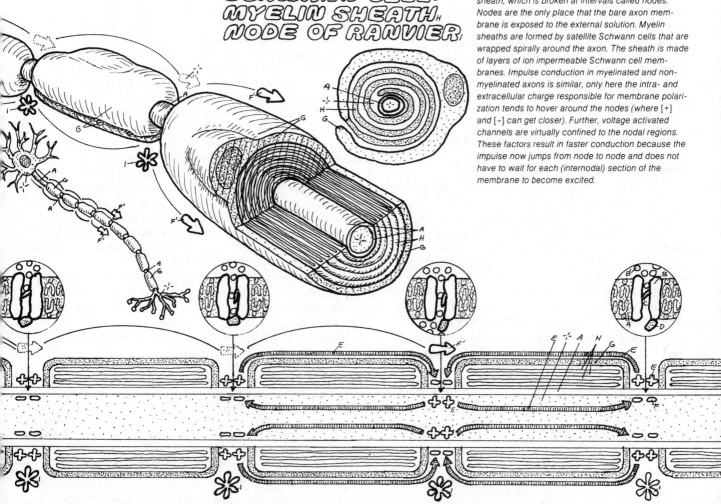
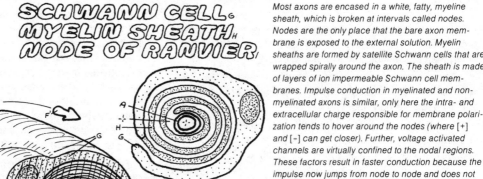

SYNAPTIC TRANSMISSION

Nerve impulses are transmitted both along axons and from cell to cell. Transmission from nerve cell to nerve cell, from nerve to muscle, or from nerve to gland is called synaptic transmission, and the sites of this transmission are called *synapses*. A typical synapse consists of a terminal branch of an incoming nerve axon, the *presynaptic* cell, in close contact with the target *postsynaptic* cell (nerve, muscle, or gland). The distance between these two cells at a synapse is only about 20 nm (1 nm = 1 millionth of a mm), and the space between them is called the *synaptic cleft*.

In some cases, transmission is electrical; the arriving impulse forces ion flow through gap junctions that connect the two cells, and the resulting electrical disturbance depolarizes and stimulates the postsynaptic cell. More often, transmission is quite different, operating by release of a chemical called a *neurotransmitter*. The sequence of events in chemical transmission is as follows:

1. The impulse arrives at the terminal branch of the incoming axon and depolarizes the *presynaptic membrane*. This depolarization opens Ca^{++} *channels* in the presynaptic membrane, and Ca^{++} flows down its gradient from outside the cell, where its concentration is high, to inside, where it is very low.

2. The raised intracellular Ca^{++} promotes the fusion of vesicles with the presynaptic membrane. This process, called *exocytosis*, releases neurotransmitters that had been stored within the *vesicles* into the synaptic cleft.

3. The neurotransmitter molecules diffuse across the synaptic cleft and bind to proteins called *receptors* on the postsynaptic membranes.

4. The transmitter-receptor complex promotes the opening of specific *postsynaptic ion channels*. (In some cases, the complex activates enzyme systems. See plate 17.)

5. Ions flow through the open channels and, if *excitatory* channels are opened, the postsynaptic membrane is *depolarized*. The resulting membrane potential generated across the postsynaptic membrane is called an *EPSP* (*excitatory postsynaptic potential*).

This depolarization (EPSP) stimulates voltage activated channels adjacent to the synaptic region. If enough of these channels are activated, the postsynaptic cell membrane becomes excited, and the impulse is propagated out from the synaptic region over the surface of the postsynaptic cell membrane by the same electrical mechanism that brought the impulse into the synapse on the presynaptic axon.

6. If the open channels are *inhibitory*, the postsynaptic membrane *hyperpolarizes*. Now the membrane potential generated across the postsynaptic membrane is called an *IPSP* (*inhibitory postsynaptic potential*) because the hyperpolarization spreads to some extent to the adjacent voltage activated channels, making it more difficult for them to respond to a stimulus (depolarization) from any other source (i.e., they are inhibited).

In either case (EPSP or IPSP), the postsynaptic channels in the synaptic cleft are different from the ordinary excitation channels that populate the other portions of nerve and muscle cell membranes. The postsynaptic channels are *not* activated by depolarization; instead, activation will occur only if a specific chemical binds to their associated receptor. Once activated chemically, they *produce* the electrical depolarization (hyperpolarization) required to excite (inhibit) ordinary voltage activated channels that lie in adjacent areas.

What distinguishes an "excitatory" from an "inhibitory" channel on the postsynaptic membrane? It all depends on which ions pass freely through the channel. In a typical excitatory synapse, the chemically activated channels are permeable to both Na^+ and K^+. More Na^+ moves into the cell than K^+ moves out because the gradient (electrical + concentration) is larger for Na^+. As a result, net positive charge moves in, and the postsynaptic membrane depolarizes. We have an EPSP.

In an inhibitory synapse, the transmitter reacts with the postsynaptic membrane and opens chemically activated channels that are permeable to K^+ and Cl^-, but not to Na^+. K^+ moves out of the cell, but movement of Cl^- is limited because its gradient is much smaller. As a result, net positive charge moves out, and the post synaptic membrane becomes even more polarized (hyperpolarized). We have an IPSP, making it more difficult for any excitatory impulse to depolarize the membrane. The postsynaptic cell is inhibited.

All synapses are not alike. Those that occur at neuromuscular junctions between nerve and skeletal muscle use *acetylcholine* as a neurotransmitter; they are always excitatory. Those that occur in visceral organs (i.e., autonomic synapses — plates 17, 25) use either norepinephrine or acetylcholine and may be either excitatory or inhibitory. Finally, the synapses that occur between neuron and neuron in the central nervous system are the most varied; they use a multitude of neurotransmitters (plate 82).

The action of a neurotransmitter does not persist for a long time because it is continuously removed from the synaptic cleft either by enzymatic attack or by re-uptake by nerve terminals. A persistent response of the postsynaptic cell can be obtained only by delivery of an equally persistent barrage of nerve impulses to the synapse.

CN: Use the same colors as on the previous page for J and K, and a dark color for C.
1. Begin with the upper diagram.
2. Color the EPSP sequence on the left. This represents an enlargement (with structures added to the action taking place in the central illustration above). Complete stage 1 before going on to 2. Color everything in all 4 examples even though only stage 1 has been completely labeled. (Labels have been added to the other three only where necessary.)
3. Do the same for the IPSP sequence.

CELL BODY & DENDRITE_A

CELL BODY & DENDRITE._{A'}
AXON_B
NERVE IMPULSE_C
SYNAPTIC TERMINAL_D
CALCIUM (Ca⁺⁺)_E
SYNAPTIC VESICLES_F
PRESYNAPTIC
    MEMBRANE_{F'}
POSTSYNAPTIC
    MEMBRANE_G
NEUROTRANS-
    MITTER_H
SYNAPTIC CLEFT_I

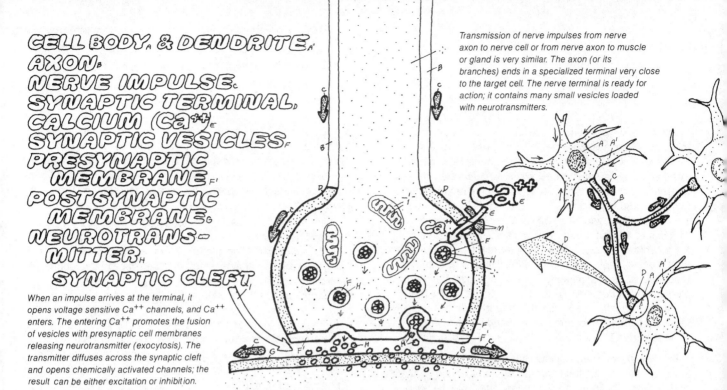

Transmission of nerve impulses from nerve axon to nerve cell or from nerve axon to muscle or gland is very similar. The axon (or its branches) ends in a specialized terminal very close to the target cell. The nerve terminal is ready for action; it contains many small vesicles loaded with neurotransmitters.

When an impulse arrives at the terminal, it opens voltage sensitive Ca⁺⁺ channels, and Ca⁺⁺ enters. The entering Ca⁺⁺ promotes the fusion of vesicles with presynaptic cell membranes releasing neurotransmitter (exocytosis). The transmitter diffuses across the synaptic cleft and opens chemically activated channels; the result can be either excitation or inhibition.

+ CHARGE_J
– CHARGE_K
RECEPTOR_O

CHEMICALLY ACTIVATED CHANNEL_*
PERMEABLE TO Na⁺ & K⁺_L
PERMEABLE TO Cl[–] & K⁺_M
VOLTAGE ACTIVATED CHANNEL_N

EXCITATORY POST SYNAPTIC POTENTIAL (EPSP)_L

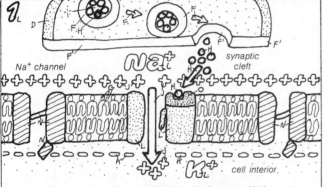

Na⁺ channel — Na⁺_L — synaptic cleft — K⁺_L — cell interior.

In an excitatory synapse, the transmitter (chemical) reacts with the postsynaptic membrane and opens chemically activated channels that are permeable to both Na⁺ and K⁺. More Na⁺ moves into the cell than K⁺ moves out because the gradient (electrical + concentration) is larger for Na⁺.

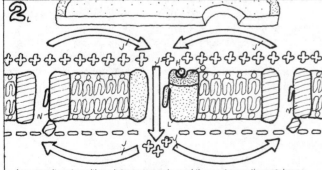

As a result, net positive charges move in, and the postsynaptic membrane depolarizes. This depolarization is strong enough to stimulate voltage activated channels on adjacent portions of the postsynaptic cell membrane. These regions become excited, and the excitation is relayed over the entire cell surface by the same electrical transmission mechanism described in plate 15.

DEPOLARIZED_L

INHIBITORY POST SYNAPTIC POTENTIAL (IPSP)_M

Cl[–]_M — K⁺_M

In an inhibitory synapse, the transmitter reacts with the postsynaptic membrane and opens chemically activated channels that are permeable to K⁺ or Cl[–], but not to Na⁺. K⁺ moves out of the cell, but movement of Cl[–] is limited because its gradient is much smaller.

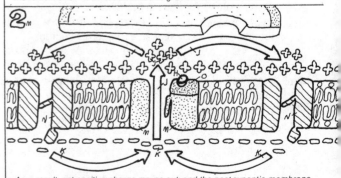

As a result, net positive charge moves out, and the post synaptic membrane becomes even more polarized (hyperpolarized). This hyperpolarization is strong enough to spread to adjacent portions of the post synaptic membrane (containing Na⁺ channels) making it more difficult for any excitatory impulse to depolarize the membrane. The post synaptic cell is inhibited.

HYPERPOLARIZED_M

NEUROTRANSMITTERS AND RECEPTORS

In this plate, we continue our discussion of synaptic transmission, paying particular attention to those synapses of nerves with muscles and glands. The chemistry of transmission follows a common pattern. *Precursors* (raw materials) for transmitter formation are taken up by the nerve terminal, *synthesis* occurs, and the *transmitters* are stored in vesicles until stimulation, when they move to the cell membrane, fuse with it, and release the transmitter into a synaptic cleft. Diffusing across the cleft, the transmitter reacts with a *receptor* on the postsynaptic membrane and initiates a characteristic response in the postsynaptic cell. Postsynaptic membranes also contain a *degrading enzyme*, which inactivates the transmitter. Finally, the transmitter, or parts of it, are actively taken up by the nerve terminal and used for re-release and resynthesis.

In *cholinergic* (acetylcholine liberating) synapses, the *acetylcholine* transmitter is formed by combination of *acetate* and *choline*. Choline is actively transported into the nerve terminal, and it is also synthesized there. The acetate is formed from ordinary metabolism and first reacts with coenzyme A (CoA) to form an activated *acetyl-CoA* (plates 5, 6), which then readily reacts with choline to form acetylcholine in the presence of a specific enzyme. The postsynaptic receptor for acetylcholine is a protein complex with a molecular weight of about 250,000 that extends through the membrane. When acetylcholine binds, small changes in shape are believed to open an ion channel running through the central core of the complex. Once open, each channel remains open for about a millisecond and then closes. As the channels close, acetylcholine molecules dissociate from the receptors and are quickly split into inactive products (acetate and choline) through the action of the degrading enzyme *cholinesterase*, which is concentrated on the postsynaptic membrane. If this enzyme is impaired by poison, the acetylcholine will persist within the synaptic cleft and reactivate channels several times before it diffuses away.

In *adrenergic* (*norepinephrine* liberating) synapses, transmitter synthesis begins with the amino acid *tyrosine*, which is actively taken up by the nerve terminal. The synthesis uses a common pathway followed in other nerve tissues for the formation of two additional transmitters, dopamine and epinephrine. (All three, *norepineprine*, *epinephrine*, and *dopamine*, are members of a chemical family called *catecholamines*.) Storage of norepinephrine in intracellular vesicles is essential to protect it from the action of an intracellular degrading enzyme, *MAO* (*monoamine oxidase*), attached to the outside surface of mitochondria. Once it is secreted into the synaptic cleft, norepinephrine continues to act until it is taken back up into the presynaptic axon terminal or diffuses away. About 70% of the liberated norepinephrine is recaptured intact by the uptake mechanism and returned to storage vesicles, where it can be reused. Although there is an extracellular degrading enzyme, *COMT* (*catechol-O-methyl transferase*), which inactivates norepinephrine, it is not concentrated in the synaptic region and appears to operate chiefly on catecholamines that have escaped or have been secreted (by the adrenal gland) into the circulation.

Responses to drugs show that all cholinergic receptors are not identical. They are classified in two groups: (1) *nicotinic* and (2) *muscarinic*. Nicotinic receptors respond to nicotine as though it were acetylcholine but are insensitive to muscarine, a poison found in some mushrooms and in rotten fish. Nicotinic receptors are all excitatory, and their response is rapid, coming to completion within milliseconds. They are found in neuromuscular junctions and in the preganglionic synapses of the autonomic nervous system. These receptors are blocked by curare, a poison used for arrowheads by South American indians.

Muscarinic receptors respond to muscarine, but not to nicotine. They can be excitatory or inhibitory, and their response is often prolonged, lasting for seconds. They are found in cardiac muscle, smooth muscle, and exocrine glands. Unlike nicotinic receptors, they are insensitive to curare, but are blocked by the drug atropine.

Adrenergic receptors are also classified into two major groups, *alpha* and *beta* receptors. Alpha receptors are more sensitive to norepinephrine than to epinephrine (a hormone of the adrenal medulla). They cause constriction of vascular smooth muscle and sphincter muscles of the gut and bladder and contraction of the spleen. They frequently operate by increasing *intracellular* Ca^{++}, which in turn causes muscular contraction or secretion.

Beta receptors are more sensitive to epinephrine than to norepinephrine. They are found in the heart, where they increase the rate and strength of contraction in the gastrointestinal tract, where they inhibit motility, and in blood vessels of skeletal muscle and heart, where they cause dilation. They operate by activating the enzyme *adenyl cyclase*, which catalyses the conversion of *ATP* to *cyclic AMP*, a second messenger that produces effects characteristic of the particular tissue involved.

The two graphs on the lower right-hand side of the plate illustrate how the existence of two receptor types can produce responses that appear contradictory. The first graph shows that injection of epinephrine increases blood pressure. This occurs because the epinephrine, acting primarily through alpha receptors, constricts small blood vessels. The second graph shows what happens when the alpha receptors are blocked with a drug. Now the action of epinephrine on beta receptors is unmasked. The same dose of epinephrine acting via beta receptors dilates blood vessels and lowers blood pressure.

CN: Use the same color as was used on the previous page for postsynaptic membrane (F¹). Use dark colors for D and G.
1. Begin with the cholinergic neuron (A). Then color the precursors (B) that form the ACh (A²), which is transported to the synaptic cleft, then to the ACh receptor (A³), degrading enzyme (G¹), and back to the precursors.
2. Do the same for the adrenergic neuron (C) on the right.
3. Color the material on cholinergic receptors. Notice that, in the diagrams that distinguish nicotinic (H) from muscarinic (I) receptors, the receptor that is blocked by contact with the opposing chemical is marked with a large X, which is colored gray.
4. Color the adrenergic receptor panel, beginning with the alpha (J) receptor and the material under it.

NEUROTRANSMITTER RELEASE

CHOLINERGIC NEURON
SYNAPTIC TERMINAL

PRECURSORS (SYNTHESIS)
NEUROTRANSMITTERS
ACETYLCHOLINE (ACh)
NOREPINEPHRINE (NE)
NERVE IMPULSE
CALCIUM CHANNEL
POSTSYNAPTIC MEM
Ach RECEPTOR
NE RECEPTOR
DEGRADING ENZYMES
AChE (ACETYLCHOLINESTERASE)
COMT (CATECHOL-O-METHYLTRANSFERASE)
MAO (MONOAMINE OXIDASE)

ADRENERGIC NEURON
SYNAPTIC TERMINAL

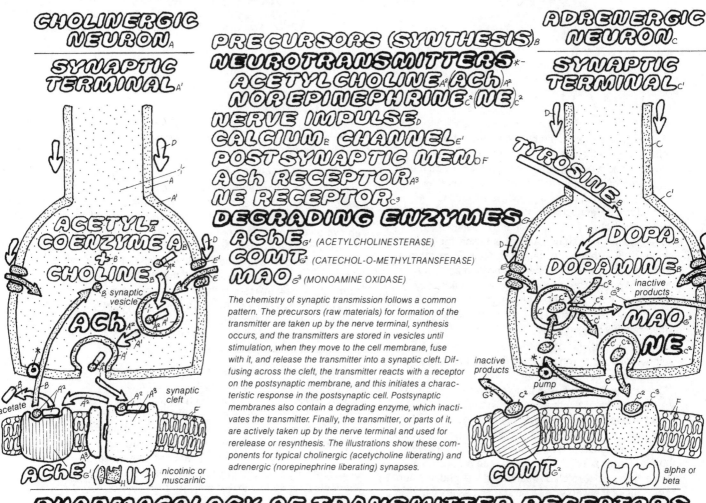

ACETYL- COENZYME A + CHOLINE

synaptic vesicle

ACh

acetate

synaptic cleft

AChE

nicotinic or muscarinic

TYROSINE

DOPA

DOPAMINE

inactive products

MAO

NE

inactive products

pump

COMT

alpha or beta

The chemistry of synaptic transmission follows a common pattern. The precursors (raw materials) for formation of the transmitter are taken up by the nerve terminal, synthesis occurs, and the transmitters are stored in vesicles until stimulation, when they move to the cell membrane, fuse with it, and release the transmitter into a synaptic cleft. Diffusing across the cleft, the transmitter reacts with a receptor on the postsynaptic membrane, and this initiates a characteristic response in the postsynaptic cell. Postsynaptic membranes also contain a degrading enzyme, which inactivates the transmitter. Finally, the transmitter, or parts of it, are actively taken up by the nerve terminal and used for rerelease or resynthesis. The illustrations show these components for typical cholinergic (acetycholine liberating) and adrenergic (norepinephrine liberating) synapses.

PHARMACOLOGY OF TRANSMITTER RECEPTORS

CHOLINERGIC RECEPTORS

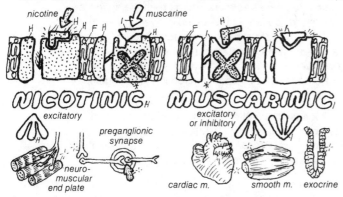

nicotine   muscarine

NICOTINIC

excitatory

preganglionic synapse

neuro-muscular end plate

MUSCARINIC

excitatory or inhibitory

cardiac m.   smooth m.   exocrine

ADRENERGIC RECEPTORS

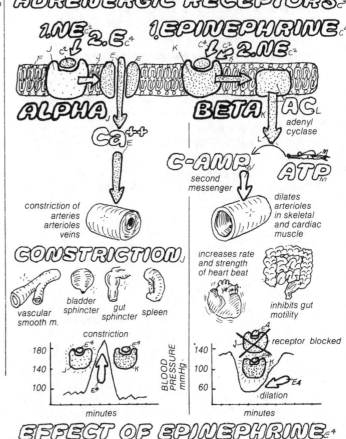

1. NE   2. E   1. EPINEPHRINE   2. NE

ALPHA   BETA   AC

Ca^{++}

adenyl cyclase

C-AMP
second messenger

ATP

constriction of arteries arterioles veins

CONSTRICTION

dilates arterioles in skeletal and cardiac muscle

increases rate and strength of heart beat

inhibits gut motility

vascular smooth m.

bladder sphincter   gut sphincter   spleen

constriction

180
140
100

BLOOD PRESSURE mmHg.

140
100
60

receptor blocked

dilation

minutes   minutes

EFFECT OF EPINEPHRINE

Cholinergic receptors: *Responses to drugs show that all cholinergic receptors are not identical. They are classified in two groups (1) nicotinic and (2) muscarinic. Nicotinic receptors respond to nicotine as though it were acetylcholine but are insensitive to muscarine. These excitatory receptors are found in neuromuscular junctions and in the preganglionic synapses of the autonomic nervous system. Muscarinic receptors respond to muscarine, but not to nicotine. They can be excitatory or inhibitory, their response is often prolonged, and they are found in cardiac muscle, smooth muscle, and exocrine glands.*
Adrenergic receptors: *These are also classified into two major groups: (1) alpha and (2) beta receptors. Alpha receptors are more sensitive to norepinephrine than to epinephrine. They cause constriction of vascular smooth muscle and sphincter muscles of gut and bladder and contraction of the spleen. They operate by increasing intracellular Ca^{++}, which in turn causes muscular contraction or secretion. Beta receptors are more sensitive to epinephrine than to norepinephrine. They are found in the heart, where they increase rate and strength of contraction, in the gastrointestinal tract, where they cause dilation. They operate by activating adenyl cyclase, which catalyses the conversion of ATP to cyclic AMP, a second messenger that produces effects chacracteristic of the particular tissue involved.*

THE STRUCTURE OF SKELETAL MUSCLE

The beat of the heart, the blink of an eye, the breath of fresh air — these obvious signs of life are all brought about by muscular contraction. How do muscles shorten? Something "inside" must move, but what? Years ago, many physiologists believed that muscles contract because the proteins of which they are made actually shorten, either by folding or by changes in the pitch or diameter of helical molecules. In the 1950s, they were startled to discover that this is not the case at all. True, the contractile machinery is made of protein, but contraction does not occur by protein folding; rather than changing their dimensions, the proteins simply slide past each other and change their relative positions.

An important clue came from early studies of the striped pattern of living skeletal muscle that could be seen under the light microscope. The stripes are localized in long fibrous cylinders called *myofibrils* that run the length of the muscle cell. The muscle cell contracts because the myofibrils contract; they contain the contractile machinery. Each myofibril is punctuated with alternating light and dark bands called *A and I bands*. These bands are "lined up" so that an A band on one myofibril is closest to an A band on its neighbor. When you look at the whole cell, you see stripes instead of a checkerboard. When a muscle contracts, the I band shortens, but the A band does not change size. The mystery of contraction seemed to reside in the I band. Soon after the electron microscope became available, however, a new picture emerged.

Examination with an electron microscope reveals that each myofibril contains many fibers called *filaments*, which run parallel to the myofibril axis. Some filaments, the thick ones, are confined to the A band; the other, thinner ones seem to arise in the middle of the I band, at the *Z line* (a structure that runs perpendicular to the myofibril through the I band, connecting neighboring myofibrils). The thin filaments run the course of the I band and partway into the A band, where they overlap (interdigitate) with the thick

filaments. The next step is to identify the filaments with the contractile machinery.

The chemical identity of the filaments can be determined by using concentrated salt solutions that selectively extract muscle proteins. When the protein called *actin* is extracted, the thin filaments disappear, and when the protein called *myosin* is extracted, the thick filaments disappear. Moreover, when the cell membrane is destroyed and substances other than these two proteins are leeched out, the thick and thin filaments remain intact, and the muscle can still contract (if it is provided with ATP as an energy source). These results imply that the thick and thin filaments are the *contractile machinery* and that the thick filaments are made of myosin, and the thin ones are actin.

Returning to interpret the muscle stripes, we now have a thick A band consisting of a lighter middle region (the *H zone*) with denser regions on each side. The denser edges are where thick myosin and thin actin filaments overlap; the middle (H zone) contains only myosin. The I bands contain only actin. Whenever a muscle or myofibril changes length, either by contracting or stretching, neither myosin nor actin filaments change length, yet they are the contractile machine! It follows that they must slide past each other, increasing their area of overlap during contraction and decreasing it during stretching. During contraction, the I band decreases as more and more of the actin filaments are "buried" in the region of overlap with myosin. The A band cannot change because it represents the length of the myosin filaments, which are invariant. However, if this picture is correct, you might expect the H zone to decrease upon contraction and lengthen on stretching. It does.

Because the motive force for contraction is provided by actin and myosin filaments sliding together, there must be some "connecting" elements that allow them to interact. These are the *cross bridges*, taken up in plate 19.

CN: Use dark colors for G and H.
1. Begin at the top with skeletal muscle (A) and work your way down the right side of the page to its molecular components. Note that the end surface of each cylindrical example receives the color of its components. In the case of the myofibril (D), only the title and the myofibrils making up the end of the cell (C) receive the color D. The length of the myofibril receives the colors of the various bands of contractile elements.
2. Color the diagrams of the contractile elements on the left side of the page. Note that the first diagram attempts to show how the two kinds of filaments actually make up the bands you previously colored. Note that the thin filaments (E) actually penetrate the A band. This wasn't shown in the drawing of the myofibril on right. Note too, that the lower two diagrams represent a vertical enlargement (in order to show cross bridge activity) of the upper diagram, but not a horizontal enlargement (the Z lines (H) still coincide with the upper diagram).

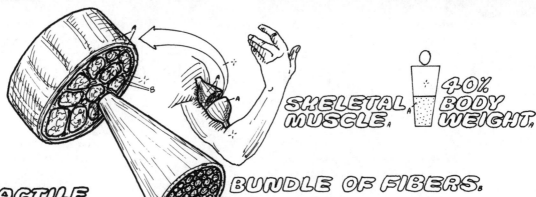

SKELETAL MUSCLE. 40% BODY WEIGHT.

CONTRACTILE ELEMENTS*
A BAND
THICK FILAMENT
CROSS BRIDGE
I BAND
THIN FILAMENT
H ZONE
Z LINE
SARCOMERE

Myofibrils are composed of repeating dark A and light I bands which are responsible for the striations (stripes). Electron microscopy shows finer detail as illustrated in the lower two diagrams; each fibril is composed of thick and thin filaments. Thick filaments run the length of the A band; thin filaments run through the I band and peripheral portions, but not the central H zone, of the A band. Thin filaments are anchored in the center of the I band by the Z line. That portion of the myofibril (2.5μ long) between the two Z lines is called a sarcomere. Thick and thin filaments interact through cross bridges which are bud-like extensions of thick filaments. The cross bridges are given a separate color for identification purposes.

When living muscle contracts, the I band shortens, the H zone shortens, but the length of the A band does not change. Thus, neither thick nor thin filaments change length; they simply slide past each other increasing the area of overlap.

BUNDLE OF FIBERS.
Whole muscles are made of bundles of cylindrical striated cells called fibers.

CELL (MUSCLE FIBER).
Cells (muscle fibers) range from 5 to 100μ in diameter but may be several thousand times longer as they extend from one bone to another.

MYOFIBRIL.
Hundreds of banded cylindrical myofibrils run the length of each cell; they are the contractile elements of the cell.

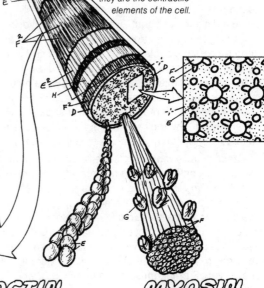

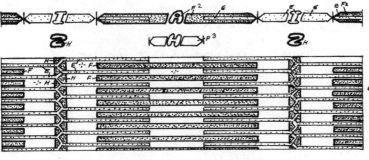

SARCOMERE
RELAXED
CONTRACTED

ACTIN FILAMENT (THIN).
Thin filaments are highly ordered assemblies of protein molecules called actin.

MYOSIN FILAMENT (THICK).
Thick filaments are highly ordered assemblies of protein molecules called myosin.

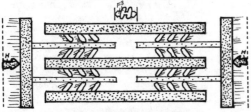

ACTIN MOLECULE.
Actin molecules are pear shaped (approx. 4nm in diameter). In thin filaments they are joined together like two strings of beads intertwined at regular intervals. (Note: Thin filaments also contain other proteins in addition to actin).

MYOSIN MOLECULE.
Myosin molecules have long (160nm) rod-shaped tails with globular heads. The heads form cross bridges between thick and thin filaments.

CROSS BRIDGES & SLIDING FILAMENTS

In relaxed muscle, the *cross bridges* are detached from *actin filaments*. During contraction, they attach and provide the contractile force. How does this come about? *Thick filaments* are ordered assemblies of *myosin* molecules; each molecule contains a long rod-shaped tail, a shorter rod-shaped neck, and two globular heads, which form the cross bridges. Only one head is shown in the drawings. (The signficance of the second head is not known.) There are two flexible *hingelike* regions. The hinge closest to the thick filament, between tail and neck, allows the cross bridges to attach and detach from the actin filament. The hinge next to the globular head allows the head to tilt. This tilt is the *power stroke*; it is responsible for propelling the actin a distance of about 75 nm relative to myosin. Following the power stroke, the bridge detaches and then repeats the cycle farther upstream. The cycles of individual bridges are not synchronized as shown. They are out of phase, some attaching while others are detaching. Thus, at each moment, some of the cross bridges are entering the "power stroke" while others leave. The movement is not jerky, and there is no tendency for the filaments to slip backward.

Gross muscle movements are brought about by a cyclic reaction of the cross bridges: *attachment* (to actin) → *tilting* (producing movement) → *release attachment* → etc. By repeating the cycle many times the small movements add up to the smooth, coordinated, macroscopic motions we all enjoy. But cyclic reactions cannot occur without an energy source (if they could, we would be able to build perpetual motion machines). Further, muscle can do physical work (i.e., lift a weight), and work requires energy. The immediate

source of this energy is *ATP*. When we incorporate ATP in our scheme the details of each cycle become more complex as we are able to distinguish more steps. These are shown in the set of diagrams at the bottom of the page. Attachment of ATP to the myosin head groups allows the myosin heads to release the actin. Further, a *"high-energy" phosphate* is transferred from the ATP to the myosin, which becomes "energized," while the original ATP, having lost a phosphate, becomes *ADP*. The energized cross bridge is now ready for action. If the muscle is stimulated, the cross bridge will attach to the actin, tilt, and move the actin along (the *power stroke*). Following the power stroke, the myosin and actin remain attached until the beginning of the next cycle, when ATP once again binds, releases the attachment, and energizes the myosin cross bridge. Note that if ATP has been used up, the myosin heads will remain locked to the actin filaments, and no sliding can take place. The muscle will become rigid, resisting both contraction and stretching. This is the condition known as *rigor mortis*, which is common after death when ATP has degenerated. Also note that ATP splitting is not directly involved in the power stroke. Its energy is used to "prime" the myosin head so that it can attach to the myosin and repeat the cycle.

If ATP is present, why doesn't the muscle continue to contract until all the ATP is used up? The answer involves an additional substance, Ca^{++}, which is required for the attachment phase of the cycle. If sufficient Ca^{++} is present, attachment can occur; at lower levels, it cannot. The action of Ca^{++} as a trigger for contraction and its removal for relaxation are taken up in plate 20.

CN: Use the same colors as on the previous page for actin (A), Z line (B), myosin (C), and cross bridge (E). Use a bright color for F and I, and a dull color for G.

1. Begin with the diagram in the upper right corner of the contractile mechanism of a myofibril. Then color the details on the left showing how cross bridges operate to cause contraction.

2. Color the diagram explaining the cross bridge orientation of the myosin filament. Color the entire myosin filament (including the bare zone).

3. Color the cycle of energy transfer and contraction after coloring the formula at the top of the panel.

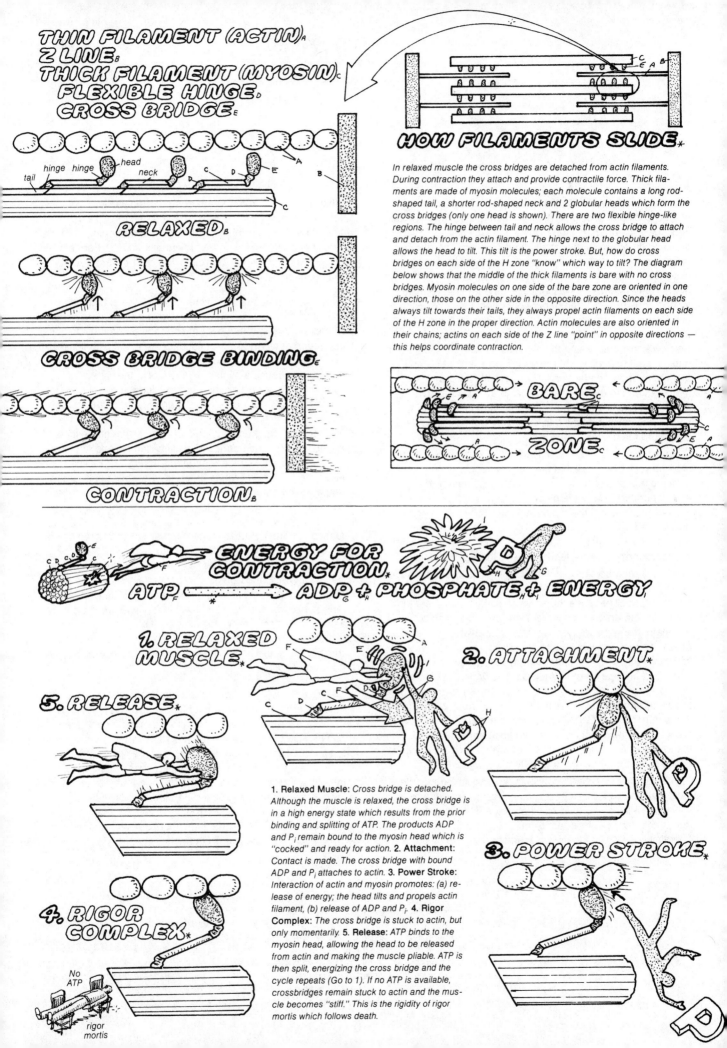

THIN FILAMENT (ACTIN)ₐ
Z LINE_B
THICK FILAMENT (MYOSIN)c
FLEXIBLE HINGE_D
CROSS BRIDGE_E

RELAXED_B

tail, hinge, hinge, head, neck

CROSS BRIDGE BINDING_E

CONTRACTION_B

HOW FILAMENTS SLIDE*

In relaxed muscle the cross bridges are detached from actin filaments. During contraction they attach and provide contractile force. Thick filaments are made of myosin molecules; each molecule contains a long rod-shaped tail, a shorter rod-shaped neck and 2 globular heads which form the cross bridges (only one head is shown). There are two flexible hinge-like regions. The hinge between tail and neck allows the cross bridge to attach and detach from the actin filament. The hinge next to the globular head allows the head to tilt. This tilt is the power stroke. But, how do cross bridges on each side of the H zone "know" which way to tilt? The diagram below shows that the middle of the thick filaments is bare with no cross bridges. Myosin molecules on one side of the bare zone are oriented in one direction, those on the other side in the opposite direction. Since the heads always tilt towards their tails, they always propel actin filaments on each side of the H zone in the proper direction. Actin molecules are also oriented in their chains; actins on each side of the Z line "point" in opposite directions — this helps coordinate contraction.

BARE ZONE_C

ENERGY FOR CONTRACTION*

ATP_F ⟶* ADP_G + PHOSPHATE_H + ENERGY

1. RELAXED MUSCLE*

2. ATTACHMENT*

3. POWER STROKE*

4. RIGOR COMPLEX*

No ATP

rigor mortis

5. RELEASE*

1. Relaxed Muscle: Cross bridge is detached. Although the muscle is relaxed, the cross bridge is in a high energy state which results from the prior binding and splitting of ATP. The products ADP and P_i remain bound to the myosin head which is "cocked" and ready for action. 2. Attachment: Contact is made. The cross bridge with bound ADP and P_i attaches to actin. 3. Power Stroke: Interaction of actin and myosin promotes: (a) release of energy; the head tilts and propels actin filament, (b) release of ADP and P_i. 4. Rigor Complex: The cross bridge is stuck to actin, but only momentarily. 5. Release: ATP binds to the myosin head, allowing the head to be released from actin and making the muscle pliable. ATP is then split, energizing the cross bridge and the cycle repeats (Go to 1). If no ATP is available, crossbridges remain stuck to actin and the muscle becomes "stiff." This is the rigidity of rigor mortis which follows death.

INTRACELLULAR CALCIUM TRIGGERS CONTRACTION

If, in the presence of ATP, the cross bridges can enter repeated cycles of attachment, propulsion (tilting), and release, how does this process stop? How do muscles relax? Two key discoveries provided important clues. One was the realization that the presence of minute quantities of *free Ca++* ions were essential for contraction. This fact had escaped detection for many years because it was virtually impossible to remove small traces of Ca++ from laboratory chemicals or even from distilled water. Apparently, these traces were sufficient for the contractile process. After learning to control traces of Ca++, we now know that raising the cytoplasmic Ca++ (inside the muscle cell) to concentrations as low as .0001 mM is sufficient to support contraction. (This is twenty thousand times more dilute than the free Ca++ level in the plasma!) When Ca++ is at this level or above, contraction ensues; when Ca++ is somewhat below this level, contraction cannot take place and the muscle relaxes.

How does Ca++ exert its influence? The other important clue was the discovery that the thin filaments contain other proteins besides actin. In particular, they contain *tropomyosin* and *troponin*. These proteins can be removed from the actin in highly purified artifical systems. When this is done, the requirement for Ca++ disappears! The system contracts in the presence of ATP and in the absence of Ca++. This, together with other observations, leads to the following description.

In order for muscle to contract, the energized cross bridges must first attach to the actin filaments. During relaxation, this does not occur because the *sites for myosin attachment* on the actin filaments are covered by tropomyosin molecules; in this state, the sites are masked and not available for the cross bridges. Another protein, troponin, is bound to and serves as a "handle" on the tropomyosin. Troponin can bind Ca++ and change shape. When Ca++ is bound, the troponin moves the tropomyosin out of the way. The sites are now exposed, attachment of cross bridges can occur, and contraction ensues. When Ca++ is absent, the tropomyosin reverts back to its original position and blocks attachment; relaxation follows. But what controls Ca++? How does its concentration rise to trigger contraction and fall to allow relaxation?

Although the free Ca++ concentration in relaxed muscle is extremely low in the cytoplasm, other vesicular structures wthin the cell may contain an abundance. This is particularly true of the *sarcoplasmic reticulum (SR)*, a compartment containing Ca++ ions that are separated from the cytoplasm by the membranes forming the compartment walls. Each myofibril is surrounded by a sheath of sarcoplasmic reticulum, which resembles a net stocking stretching from one Z line to the next. It is the movement of Ca++ from the SR interior to the cytoplasm and back that controls contraction and relaxation.

When nerve impulses activate muscles, the excitation is transmitted through the *motor endplate*, and a muscle *action potential* quickly spreads over the surface of the muscle cell. Contraction of all myofibrils, including those in the cell interior, follows within milliseconds. This all-or-none response is possible because a system of tiny tubes, the *T tubules* (transverse tubules), extends from the surface membrane deep into the interior of the muscle and encircles the perimeter of each myofibril at the level of the Z line in some muscles (frog skeletal muscle, mammalian heart) or at the level of the junction of A and I bands in mammalian skeletal muscle. The lumens of T tubules are continuous with extracellular spaces, and the membranes that form the walls conduct the surface action potential deep into the cell to each sarcomere, where the tubules come in close contact with the SR. Somehow, when the wave of excitation reaches this point, Ca++ is released into the cytoplasm; the mechanism of this release is not understood. Upon entering the cytoplasm, Ca++ reacts with troponin, tropomyosin moves (exposing actin binding sites for myosin), and contraction occurs. Following the excitatory wave, Ca++ is pumped back into the SR by an *ATP-driven active transport system for Ca++*. This lowers cytoplasmic Ca++ and, when it falls low enough, binding to troponin is no longer supported. At this point, tropomyosin returns to mask the actin binding site for myosin, and relaxation occurs.

The role of Ca++ ions in muscle contraction is an excellent example of the ubiquitous role of intracellular Ca++ as regulator of cellular processes. In addition to muscle contraction, these include ciliary activity, amoeboid movement, exocytosis, synaptic transmission and cell cleavage. In the above example, the Ca++ level was increased by releasing it from intracellular stores. Having Ca++ stored near its site of action allows the very rapid response characteristic of skeletal muscle. In some other cases, the Ca++ level is increased by simple opening of Ca++ channels in the cell membrane and allowing Ca++ to flow in from the outside.

CN: Use same colors as on previous page for action (C) and myosin (D). Use dark colors for E, F, I and J.
1. Begin with the muscle cell in the upper right corner. Note that the titles for the axon (G) and axon terminal (H) are listed below. The myofibrils within the cell are left uncolored here and below.
2. Color the enlargement of a strand of actin filament and myosin filament and then color the stages of cross bridge activation.
3. Color the anatomical description of the cell interior in the lower left corner.
4. Color the enlargement of the neuromuscular junction and the release and withdrawal of Ca++ in the cytoplasm of the cell. Follow the numbered sequence, noting that steps 5 and 6 involve contraction and relaxtion of the filament structures (only a portion of which are shown in this diagram). Color the second enlargement of step 4, following the A-D sequence of events.

FREE CALCIUM TRIGGERS CONTRACTION.

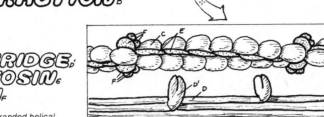

MYOFIBRIL

neuromuscular junction

CELL

ACTIN.
MYOSIN,
CROSS BRIDGE,
TROPOMYOSIN,
TROPONIN

Tropomyosin is a long, two-stranded helical protein which is aligned almost parallel to the axis of the thin actin filaments. Troponin is a protein complex of 3 globular subunits, located at regular intervals (spaced approximately 7 actin molecules apart) along the thin filament. One of the subunits attaches to tropomyosin, another to actin, and the third subunit can bind Ca++ ions.

1. Relaxation: *Myosin cross bridges cannot attach to thin filament because the site of attachment is blocked by tropomyosin.* **2. Myosin Binding:** *Ca++ ions appear on the scene. Four Ca++ bind to each troponin and the complex moves the tropomyosin away from the binding sites. Myosin can now bind to actin.* **3. Contractile Stroke:** *Once energized myosin binds to actin, the head tilts and propels the thin filament.*

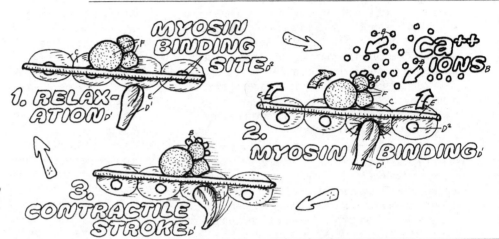

MYOSIN BINDING SITE

1. RELAXATION

Ca++ IONS

2. MYOSIN BINDING

3. CONTRACTILE STROKE

SARCOPLASMIC RETICULUM & Ca++ STORAGE.

AXON
AXON TERMINAL
ACTION POTENTIAL
ACETYLCHOLINE
MOTOR END PLATE
CELL MEMBRANE
T TUBULE
IP₃

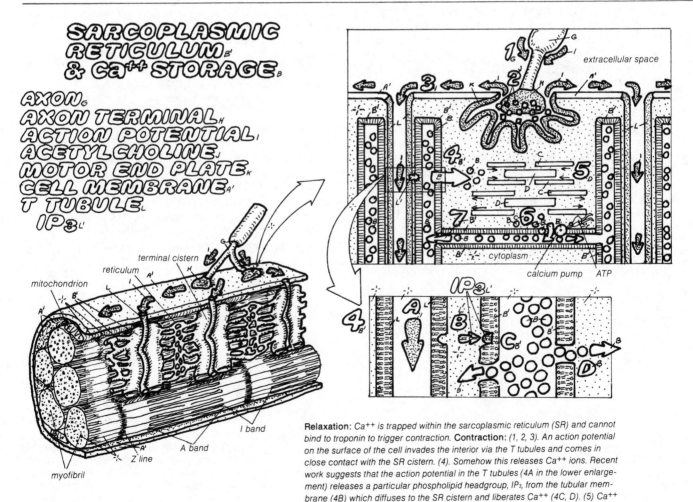

extracellular space

cytoplasm

calcium pump   ATP

IP₃

terminal cistern

reticulum

mitochondrion

I band

A band

Z line

myofibril

Relaxation: *Ca++ is trapped within the sarcoplasmic reticulum (SR) and cannot bind to troponin to trigger contraction.* **Contraction:** *(1, 2, 3). An action potential on the surface of the cell invades the interior via the T tubules and comes in close contact with the SR cistern. (4). Somehow this releases Ca++ ions. Recent work suggests that the action potential in the T tubules (4A in the lower enlargement) releases a particular phospholipid headgroup, IP₃, from the tubular membrane (4B) which diffuses to the SR cistern and liberates Ca++ (4C, D). (5) Ca++ ions bind to troponin and expose binding sites for myosin. Contraction follows.* **Relaxation:** *(6, 7) An ATP driven Ca++ pump actively transports Ca++ back into the SR. The cytoplasmic Ca++ level falls and relaxation follows.*

RELATIONSHIP OF MUSCLE TENSION TO LENGTH

Although the contraction of each muscle cell is all-or-none, it is obvious that body movements are not. Sometimes they are forceful, other times slight. This is easily accounted for by realizing that body movements are brought about by whole muscles (groups of muscle cells), not by single cells acting alone. Increasing the force of movement may simply be a matter of recruiting more and more cells into cooperative action. However, there are also more subtle means for changing the performance of individual cells.

The strength or, more precisely, the force a muscle is capable of exerting depends on its *length*. For each muscle cell, there is an optimum length or range of lengths where the contractile force is strongest. This is easily explained by the sliding filament theory. The strength of contraction depends on the number of cross bridges that can make contact with actin filaments. When the muscle is too long, few cross bridges can make contact, and contraction is weak. When the muscle is too short, cross bridge contact can be made, but the filaments begin to get in each other's way and jam up. Again, contraction is weak. Maximal force develops only at a small range of lengths where recruitment of operable cross bridges is maximal and where filaments do not interfere. For the human bicep muscle, this optimum length is attained when the forearm and upper arm are at right angles. When the arm is extended so that the angle between forearm and upper arm is 180°, the bicep is stretched, and contraction is weaker. This explains a common experience of weight lifters: when performing a "curl," it is most difficult to raise the weight from the bottom position with the arm extended. Progress is much easier once the weight has been lifted, and the fore and upper arm are at right angles.

When picking up a light weight, the muscles shorten and move the skeleton. We call this *isotonic contraction*. What happens if you attempt to pick up a weight that is too heavy? The muscle tenses but does not shorten. This is called an *isometric contraction*, a contraction with no change in length! How is this contradiction in terms resolved? Actually, when a muscle undergoes an isometric contraction, the *contractile machinery* really does shorten; the actin and myosin filaments slide past each other. But other passive parts of the cell attached to the contractile machinery, the tendon and connective tissue, are stretched, so there is no net movement. Those parts of the muscle stretched by the contractile machine are called the *series elasticity (SE)*. The exact identity of the SE is a bit vague, but it is known to include the tendons, connective tissue, and elasticity of the cross bridge hinge regions.

The SE stretches a little even when muscle undergoes isotonic contractions. This follows because at the beginning of a contraction, the SE is slack, and as the contractile machine shortens, this slack is taken up until the SE can support the load that is to be moved. From this point on, the muscle shortens.

Changing the length of a muscle is not the only way to alter the strength of contraction. If a rapid succession of stimulating impulses is delivered to a muscle, the cumulative effect will show a stronger contraction than the contraction resulting from a single impulse; the contractions summate. The contraction of a whole muscle can also be increased simply by stimulating more and more muscle cells, a process called recruitment. Summation and recruitment are described in plate 22.

CN: Use same colors as on previous pages for actin (A), and myosin (B), and Z line (C).
1. In the upper illustration, color the strength and length axis of the graph, including the percentages. Then color parts A-C of each of four muscle lengths, taking note of the position of the strength curve for the various percentages of muscle length.
2. Color the three muscle states below, noting the extreme stretching of series elasticity in isometric contraction, and the minor stretching in isotonic contraction, permitting the lighter weight to be lifted.

MUSCLE TENSION-LENGTH *

ACTIN A
MYOSIN B
 CROSS BRIDGE B'
Z LINE C
MUSCLE TENSION D
MUSCLE LENGTH E

Contractile force (strength) depends on the number of cross bridges
that can be recruited for "power strokes". This depends on the length of
muscle because cross bridges must contact actin to be effective. When
muscle is stretched, contact is poor, contraction is weak (4). When
muscle is too short, filaments jam up, interfering with each other's
movements (1). Maximal strength of contraction occurs when all cross
bridges can contact actin filaments and where there is still room for
sliding without interference by actins running into each other (2-3).

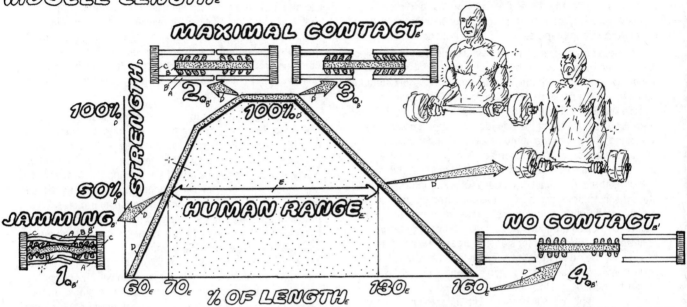

MAXIMAL CONTACT B'

JAMMING B'

NO CONTACT B'

HUMAN RANGE E

% OF LENGTH E

STRENGTH D

100%
50%

100% D

60 E 70 E   130 E   160 E

2 3   4

1

ISOMETRIC
VS.
ISOTONIC
CONTRACTION *

MUSCLE LENGTH E
LOAD F
CONTRACTILE
  ELEMENT G
SERIES
  ELASTICITY H

During contraction, the contractile
machinery stretches the series elasticity
(SE). Sufficient slack is removed until SE
can support the weight. Muscle then
shortens in an isotonic contraction.

If the weight is too large, the contractile
machinery is incapable of stretching
(removing slack) the SE sufficiently so that
SE can support weight. Contractile
machinery simply stretches SE as much
as it can, but no net movement occurs.
This is an isometric contraction: increase
in length of SE = decrease in length of
contractile machinery. (The amount of
movement of contractile elements shown
in the figure is exaggerated for purposes
of illustration.)

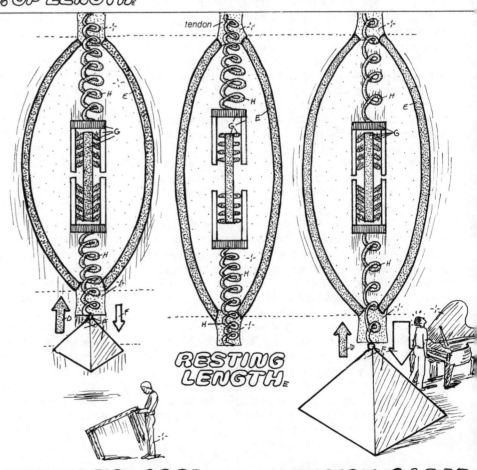

tendon

RESTING
LENGTH E

TENSION D > LOAD F
(MOVEMENT)
ISOTONIC
CONTRACTION *
(constant tension)

TENSION D < LOAD F
(NO MOVEMENT)
ISOMETRIC
CONTRACTION *
(constant length)

SUMMATION OF CONTRACTION & MUSCLE FIBER RECRUITMENT

Different tasks call for different types of motion. Sometimes our movements are rapid and vigorous, other times they may be slow, and still other times they may be fine and precise. In this plate, we explore mechanisms in muscles that are utilized to vary the strength and pattern of contractions.

A muscle can be stimulated by an electrical shock applied directly to a muscle cell or by an action potential arriving at the neuromuscular junction. When a single threshold stimulus is delivered to a muscle cell by either of these routes, the muscle responds with a *twitch* that has three phases. (1) *The latent period* consists of the brief delay of 2 or 3 msec. between the delivery of the stimulus and the first moment when some contraction can be observed. During this time, Ca^{++} is released, it activates the contractile machinery, and this stretches the series elasticity. In an isotonic contraction, changes in muscle length are measured, and no change will be observed until the developing tension matches and just begins to exceed the load (weight). In contrast, for isometric contractions, changes in tension are measured, and the change will be observable as soon as the series elasticity is stretched. It follows that the isotonic latent period will be longer than the isometric one, and the duration of the isotonic latent period will increase with increasing load. (2) During the *contraction period* in an isotonic contraction, once the tension matches the load, the continued contraction of the contractile machinery causes a net shortening or contraction of the entire muscle. In an isometric contraction, the contraction phase begins as soon as tension begins to rise. In both cases, the recorded contraction phase lasts anywhere from 5 to 50 msec., depending on the muscle. In the isotonic case, the speed of shortening decreases when the load increases. (3) The *relaxation phase* sets in when the Ca^{++} level subsides as Ca^{++} is pumped back into the sarcoplasmic reticulum (SR). Ca^{++} leaves the troponin so that attachment sites for cross bridges on actin are covered by tropomyosin, and the actin and myosin cannot interact. The filaments slide passively back to their original position.

A single twitch does not express the full potential of a muscular contraction. The twitch is brief and begins to subside before the maximum force or shortening has had a chance to develop. If several twitches are excited in rapid succession, they summate and give a combined contraction greater than a single one. In an isometric contraction, this summation occurs because, in a single twitch, contractile activity is terminated by pumping Ca^{++} back into the SR before the contractile machinery has had time to fully stretch the series elastically. If another stimulus follows on the heels of the first, it too will initiate a twitch, but this latter twitch reaps the benefit of the first. It finds the SE already partially extended, and its efforts are added to this. In an isotonic contraction, a single twitch has enough time to develop the tension (otherwise the muscle wouldn't shorten), but the extent of

shortening has been compromised by the brief twitch duration. Again, a second twitch arising before the muscle has had time to fully relax will reap benefits from the first. It will find the muscle partially shortened and will summate. In either case (isotonic or isometric), when the frequency of stimuli is sufficiently rapid, each succeeding stimulus arrives before the twitch from the preceding one has even begun to relax. The result is a smooth, sustained contraction called a tetanus.

The strength of contraction of a single muscle cell can be altered by changing its length or the frequency of stimulation (frequency of nerve impulses). Because a whole muscle is a collection of cells, it follows that the strength of contraction also can be increased simply by engaging the simultaneous contraction of more cells, a phenomenon known as *recruitment*. Each motor nerve axon that transmits impulses to a muscle branches several times before making synaptic connections with muscle cells. Branches from one axon innervate many muscle cells. Each muscle cell in a mammal receives branches from only a single axon. Thus (in the body), the muscles innervated by a single axon will all contract if and only if that axon transmits an impulse. A single motor neuron and the muscle fibers connected to it act as a unit; it is called a *motor unit*. Recruitment of motor units is the major means for varying the strength of contraction.

The number of muscle cells (fibers) in each motor unit varies in different parts of the body. Motor units involved in finely graded and skilled motions (e.g., those that move the fingers or the eye) contain a small number of muscle cells (as low as ten). This gives the nervous system the option of making very tiny adjustments in the performance of these muscles. Those units involved in more gross contractions (e.g., those that control the postural muscles in the back) have many muscle cells (perhaps over two hundred) for each nerve axon. In this instance, the nervous system makes powerful adjustments with only a few nerves.

Muscles fatigue; their activity cannot be sustained indefinitely. Yet some muscles (e.g., postural muscles) are called upon for prolonged, sustained and smooth contractions. In the laboratory, we can demonstrate smooth contractions (shown on the right side of the chart) by delivering rapid stimuli and producing a complete tetanus. In the body, motor units are rarely stimulated at those frequencies. Contractions of a motor unit often involve a train of twitches that are not completely fused; the muscle cells have moments to relax, so the motion of any individual motor unit appears jerky. Contraction of the whole muscle is smooth because many motor units are involved, and they do not fire in synchrony. When one unit is beginning to relax, another will be beginning to contract. The individual motions summate, and the net motion is smooth.

CN: Use dark colors for A, O, and P.
1. Begin with the examples of a twitch on the left, noting that this measurement of muscle shortening receives the colors of its three phases, whereas in the myogram to the right the twitch receives its single color, A (drawn on the revolving drum).
2. Color the instrument measuring the twitch

material in the upper right corner.
3. Color the chart distinguishing between various contractions.
4. Color the recruitment panel beginning with the motor unit. Only color the motor unit drawn in bold outline (O, P and G). The unit left uncolored represents inactivity.

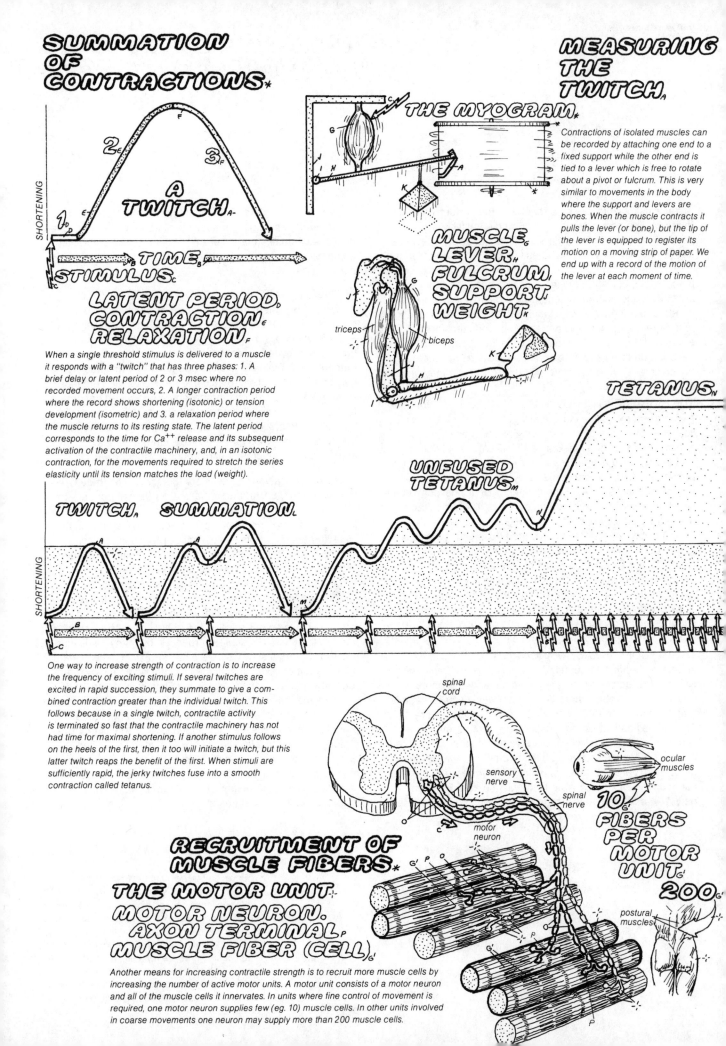

SUMMATION OF CONTRACTIONS*

A TWITCH.

SHORTENING

TIME

STIMULUS.

LATENT PERIOD, CONTRACTION, RELAXATION.

When a single threshold stimulus is delivered to a muscle it responds with a "twitch" that has three phases: 1. A brief delay or latent period of 2 or 3 msec where no recorded movement occurs, 2. A longer contraction period where the record shows shortening (isotonic) or tension development (isometric) and 3. a relaxation period where the muscle returns to its resting state. The latent period corresponds to the time for Ca^{++} release and its subsequent activation of the contractile machinery, and, in an isotonic contraction, for the movements required to stretch the series elasticity until its tension matches the load (weight).

MEASURING THE TWITCH.

THE MYOGRAM.

Contractions of isolated muscles can be recorded by attaching one end to a fixed support while the other end is tied to a lever which is free to rotate about a pivot or fulcrum. This is very similar to movements in the body where the support and levers are bones. When the muscle contracts it pulls the lever (or bone), but the tip of the lever is equipped to register its motion on a moving strip of paper. We end up with a record of the motion of the lever at each moment of time.

MUSCLE, LEVER, FULCRUM, SUPPORT, WEIGHT

triceps
biceps

TWITCH, SUMMATION.

UNFUSED TETANUS.

TETANUS.

SHORTENING

One way to increase strength of contraction is to increase the frequency of exciting stimuli. If several twitches are excited in rapid succession, they summate to give a combined contraction greater than the individual twitch. This follows because in a single twitch, contractile activity is terminated so fast that the contractile machinery has not had time for maximal shortening. If another stimulus follows on the heels of the first, then it too will initiate a twitch, but this latter twitch reaps the benefit of the first. When stimuli are sufficiently rapid, the jerky twitches fuse into a smooth contraction called tetanus.

spinal cord

sensory nerve

spinal nerve

motor neuron

ocular muscles

10 FIBERS PER MOTOR UNIT.

200

postural muscles

RECRUITMENT OF MUSCLE FIBERS*

THE MOTOR UNIT:
MOTOR NEURON.
AXON TERMINAL.
MUSCLE FIBER (CELL).

Another means for increasing contractile strength is to recruit more muscle cells by increasing the number of active motor units. A motor unit consists of a motor neuron and all of the muscle cells it innervates. In units where fine control of movement is required, one motor neuron supplies few (eg. 10) muscle cells. In other units involved in coarse movements one neuron may supply more than 200 muscle cells.

SOURCES OF ENERGY FOR MUSCLE CONTRACTION

All cells use *ATP* to fuel their reactions and perform work (plate 5). The concentration of ATP within most cells is generally around 5 mM; it is kept at this steady state level because new ATP is synthesized as fast as it is utilized. Muscle cells present a special case because they are called upon for both sudden bursts and long, sustained periods of intense activity. During endurance exercise, a muscle may utilize a hundred to a thousand times as much ATP as it does during rest. Somehow the supply has to adjust and meet these enormous demands. ATP (as shown in the upper panel) is supplied via three separate sources: *creatine phosphate* (2), *the glycolysis-lactic acid system* (4), and *aerobic metabolism* or *oxidative phosphorylation* (3).

THE HIGH-ENERGY PHOSPHATE SYSTEM. The amount of ATP present in muscle cells at any given moment is small. By itself, it is barely enough to sustain 5-6 seconds of intense activity, say a 50-m dash. But as ATP is utilized, it is quickly replenished by the small reserve of energy stored as creatine phosphate. Creatine phosphate very rapidly donates its high-energy phosphate to *ADP* the moment ADP forms, converting it back to ATP. This extra source of ATP is easily mobilized and is very effective as long as it lasts. Unfortunately, this is limited because the store of creatine phosphate is small, only about four to five times larger than the original store of ATP. Normally, the supply of creatine phosphate is replenished by oxidative metabolism via the ATP produced by the Krebs cycle (plate 6). But during sustained, intense exercise, there is not enough time for this to occur. Thus, after some 20-25 seconds of intense activity, we are back in the same place — no ATP. We require additional sources.

THE GLYCOLYSIS-LACTIC ACID SYSTEM. ATP can be supplied in a hurry through the *anaerobic breakdown of glucose* (or stored *glycogen*). Each time a glucose is chopped up by this anaerobic path, 2 ATP are formed. Its advantage is that it produces the ATP without O_2, and it produces it fast. Though half as fast as the creatine phosphate system, it is two to three times faster than aerobic metabolism. It is limited, however, because on this path the hydrogens stripped off glucose that are normally bound for O_2 to form water are taken up instead by pyruvate to form lactic acid. For each new ATP, a lactic acid is also formed. Energy production via

this pathway is limited by this accumulation of lactic acid, which produces fatigue. In addition, anaerobic glycolysis produces very small amounts of ATP, 2 per glucose consumed, compared to oxidative phosphorylation, which yields 36 ATP per glucose.

AEROBIC METABOLISM — OXIDATIVE PHOSPHORYL-ATION. This system utilizes *fats* as well as glucose and glycogen. In contrast to creatine phosphate or glycolysis. *Aerobic metabolism* is fairly slow, but, it is efficient and can provide energy for almost unlimited durations, as long as the nutrients last. Typically, it takes about 0.5 to 2 minutes for aerobic metabolism to adjust to the increased demands of exercise. Thus, anaerobic processes are required not only for brief peak physical exertion, but also to supply energy at the beginning of long-term muscular activity before aerobic metabolism becomes fully mobilized. Once this has occurred, an exhausted runner may experience a "second wind."

Not all skeletal muscle cells are the same. The three types, *red/slow*, *red/fast*, and *white/fast*, differ in their capacity to generate ATP, their speed of contraction, and their resistance to fatigue. These and related properties are illustrated in the plate. In general, whole skeletal muscles in humans contain all three types, but in different proportions. Postural muscles of the back, for example, are continually active and have a high proportion of red/slow fibers. These fibers are specialized for aerobic metabolism. They contain the red respiratory pigment *myoglobin*, which stores O_2 and facilitates the diffusion of O_2 within the muscle to mitochondria. Further, the fibers are small, surrounded by many capillaries, and they contract slowly so the blood supply of O_2 can keep up with demand. Red/fast fibers are intermediate between red/slow and white/fast. White/fast fibers are abundant in muscles that have rapid, intense bursts of activity. Myoglobin is absent, mitochondria are sparse, and capillaries are less profuse. Glycolysis is well developed so that ATP is produced rapidly, but the muscle fatigues quickly when the limited glycogen stores are depleted. Muscles of the arms, which may be called upon to produce strong contractions over short periods of time (e.g., weight lifting), have a relatively large proportion of white/fast fibers.

CN: Use purple for A, red for J, and a dark or bright color for E.
1. Begin by first coloring the boundaries of the capillary (A) and muscle fiber cell (B). Then color each title in the column on the right, beginning with number one, and coloring each process in the cell that the title refers to. Note that the group of sports figures along the right margin is not to be colored. The upper figures are representative of an exercise in which ATP is formed aerobically.
2. Color the characteristics of the three skeletal muscle fibers, coloring each line before proceeding to the next.

SOURCES OF ENERGY IN MUSCLE FIBER*

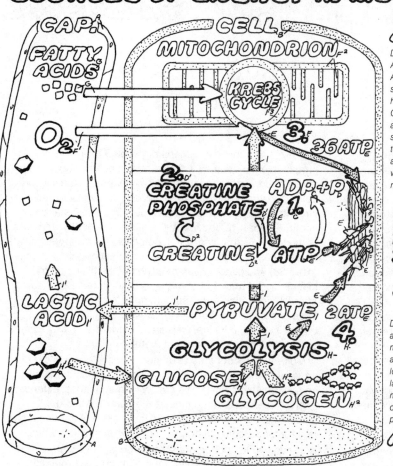

CAP. FATTY ACIDS

CELL MITOCHONDRION

KREBS CYCLE

O₂

36 ATP

2. CREATINE PHOSPHATE — CREATINE

ADP+P (1.)

ATP

LACTIC ACID

PYRUVATE, 2 ATP

4.

GLYCOLYSIS

GLUCOSE, GLYCOGEN

AEROBIC

During exercise, muscles require rapid supplies of ATP. The ATP in the cells is used quickly (1), forming ADP and inorganic phosphate. The next available source is creatine phosphate (2). It really donates its high energy phosphate to ADP to form ATP + creatine. Creatine phosphate is also easily exhausted, and fatty acids and glucose begin to be utilized. At low levels of sustained activity the blood supply of O_2 is adequate to meet demands so that utilization of these fuels is aerobic (3); it terminates in oxidative phosphorylation where O_2 is the final acceptor of the H stripped off fuel molecules, and many ATP are produced.

1. ADP + PHOSPHATE,
2. CREATINE PHOS.,
3. OXIDATIVE PHOSPHORYLATION
4. GLYCOLYSIS

During strenuous exercise, involving bursts of intense activity, blood cannot supply O_2 fast enough. The muscle cells rely on anaerobic metabolism of glucose and glycogen (4) to rapidly supply ATP. Pyruvate is no longer metabolized by mitochondria, it is converted to lactic acid which escapes in the blood. Anaerobic metabolism is very rapid, but inefficient because compared to aerobic metabolism the amount of ATP produced by each fuel molecule is small.

ANAEROBIC

---

3 KINDS OF SKELETAL MUSCLE FIBER*

	1. RED/SLOW	2. RED/FAST	3. WHITE/FAST
COLOR (MYOGLOBIN)			
SPEED OF TWITCH			
ATPase ACTIVITY	cross bridge / myosin → ADP / ATP		
TYPE OF ATP PRODUCTION	OXIDATIVE PHOSPHORYLATION		ANAEROBIC GLYCOLYSIS
NUMBER OF CAPILLARIES			
RESISTANCE TO FATIGUE			
DIAMETER OF FIBER			

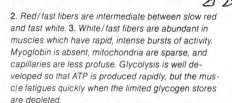

Skeletal muscles generally contain mixtures of 3 type of fibers: 1. Red/slow fibers are specialized for slow sustained activity and resistance to fatigue. They are red because they contain the respiratory pigment myoglobin which, similar to the hemoglobin of blood cells, stores O_2 by loosely binding it. They are small, surrounded by many capillaries, and contract slowly so that blood supply of O_2 can keep up with demand and their metabolism is essentially aerobic. Postural muscles of the back contain high proportions of these.

2. Red/fast fibers are intermediate between slow red and fast white. 3. White/fast fibers are abundant in muscles which have rapid, intense bursts of activity. Myoglobin is absent, mitochondria are sparse, and capillaries are less profuse. Glycolysis is well developed so that ATP is produced rapidly, but the muscle fatigues quickly when the limited glycogen stores are depleted.

SMOOTH MUSCLE

Smooth muscles are responsible for movements of the viscera and blood vessels. Unlike skeletal muscle, they are involuntary and are adapted for long, sustained contraction. Although these contractions are slower, they can generate forces of the same magnitude as skeletal muscle without fatigue and with little energy consumption. The structure of the two types also differs. Smooth muscles are smaller (about 50-400 µm long and 2-10 µm thick), spindle shaped, contain a single nucleus and a poorly developed *sarcoplasmic reticulum*, and have no obvious motor endplate. Autonomic nerve axons innervating smooth muscle have numerous swollen *varicosities* containing *neurotransmitters*. Although smooth muscle contains *actin* and *myosin*, these filaments are not held in register, so they do not show cross-striations. There are no Z lines; instead, actin filaments appear to be anchored to small *dense bodies* that are scattered throughout the cytoplasm. This lack of rigid organization probably accounts for the ability of smooth muscle to be stretched four to five times its length and still contract.

Smooth muscle is divided into two classes, single- and multi-unit. *Single-unit* muscles act together in groups because they are interconnected by *gap junctions* that are capable of transmitting excitation from one cell to the next at speeds of about 5-10 cm/second. These muscles often show spontaneous activity with slowly rising resting potentials (*pacemakers*) that culminate in *action potentials* that are entirely independent of the nerve supply. The contractions that result are slow and prolonged; a single twitch elicited by an action potential can last several seconds. If stimuli occur at a frequency of one per second, the individual twitches fuse into a sustained tetanic contraction. This differs from skeletal muscle only in the remarkably low frequency of stimuli that produce continuous tension. However, as a result, smooth muscles are usually in a state of partial contraction or tension, which is called *tone* or *tonus*. The nerve supply does not initiate this activity; it simply augments or inhibits it. When acetylcholine (transmitter for parasympathetic nerves) is applied to smooth muscle in the large intestine, the pacemaker cells are depolarized to near threshold levels so that the frequency of action potentials increases, and individual twitches fuse and summate. The greater the frequency, the stronger the net contraction. If norepinephrine (transmitter for sympathetic nerves) is applied, pacemaker cells hyperpolarize and this lowers the frequency of action potentials and the tonus (tension generated).

The smooth muscle response to stretch is not always predictable. Sometimes it shows plastic behavior; when it is stretched, it releases tension. In other cases, the stretch acts as a stimulus for contraction; when the muscles are stretched, they produce more tension. In these instances, stretch appears to depolarize the pacemaker cells, and they respond by discharging action potentials at an increased rate. This response is implicated in autoregulation of blood vessels (plate 58) and in automatic emptying of the filled urinary bladder in the absence of neural regulation (e.g., after spinal cord injuries). Single-unit smooth muscle often occurs in large sheets and is found in the walls of hollow visceral organs like the intestine, uterus and bladder. It is also found in some small blood vessels and ureters.

Multi-unit smooth muscle is more like skeletal muscle because it shows no inherent activity but depends on its nerve supply. However, this nerve supply is more diffuse, extending over a larger area of the muscle membrane. Multi-unit smooth muscles are found in the large airways to the lungs, large arteries, seminal ducts, irises, and some sphincters.

In addition to spontaneous action potentials, contraction of smooth muscle can be initiated or modified by nerve excitation, stretch, hormones, or direct electrical stimulation. In each case, the stimulation results in an increase in intracellular Ca^{++}, either from extracellular sources via Ca^{++} channels, or from the sarcoplasmic reticulum. Like skeletal muscle, intracellular Ca^{++} is the trigger for smooth muscle contraction. However, the primary mechanism of Ca^{++} action appears to be different in smooth muscle. In this case, the rising Ca^{++} (step 5 in bottom diagram) reacts with an intracellular protein called *calmodulin*, forming a complex that in turn activates an inactive form of an enzyme, *myosin light chain kinase (MLCK)*. The active MLCK catalyses the *phosphorylation* (transfer of a phosphate group to an organic molecule) of specific small (light) chains of amino acids contained in the myosin head groups. In this process, the phosphate is donated by ATP. Without this phosphorylation, the myosin heads are incapable of forming cross bridges with actin. Somehow, this phosphorylation activates the myosin ATPase, which promotes further utilization of ATP in the formation of active cross bridges so that contraction occurs. As long as intracellular Ca^{++} concentrations remain above threshold, myosin remains phosphorylated, and tension is maintained. When Ca^{++} is reduced, MLCK is inactivated, myosin is dephosphorylated, and the muscle relaxes.

CN: Use dark colors of D, J, and L.
1. Begin with the upper panel; note that varicosity (C) is left uncolored.
2. Color the comparison between single-unit and multi-unit smooth muscle.
3. Color the lower diagram of how Ca^{++} triggers contraction, following the numbered sequence. Note that the size of the myosin filaments (F) has been reduced in comparison to the MLCK (O).

THE SMOOTH MUSCLE CELL_A
ACTIN_E
MYOSIN_F
DENSE BODY_G
SARCOPLASMIC RETICULUM_H

Smooth muscles are responsible for movements of the viscera. Unlike skeletal muscle, they are involuntary and are adapted for long sustained contraction without fatigue and with little energy consumption. They have a poorly developed sarcoplasmic reticulum and have no obvious motor endplate. Autonomic nerve axons innervating smooth muscle have numerous bulb-like varicosities containing neurotransmitters. Although smooth muscle contains actin and myosin, these filaments are not held in register so they do not show cross striations. There are no Z lines, instead actin filaments appear to be anchored to small dense bodies that are scattered throughout the cell. This lack of rigid organization probably accounts for the ability of smooth muscle to be stretched 4 to 5 times its length and still contract.

AUTONOMIC NERVE AXON_B VARICOSITY_C NEUROTRANSMITTER_D

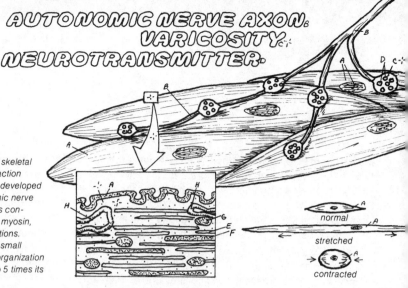

normal

stretched

contracted

SINGLE-UNIT SMOOTH MUSCLE (VISCERAL)

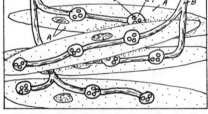

ACTION POTENTIAL_I GAP JUNCTION_J PACEMAKER POT_K

+30

0

threshold potential

−60

bladder

small blood vessel

intestine

longitudinal circular muscle

Smooth muscle is divided into two classes, single- and multi-unit.
1. **Single-unit** smooth muscles act together in groups because they are interconnected by gap junctions. They show spontaneous activity with slowly rising resting potentials (pacemakers) which culminate in action potentials that are independent of the nerve supply. The contractions that follow are slow and prolonged; as a result, the muscles are usually in a state of partial contraction or tension which is called tone, or tonus. The nerve supply does not initiate this activity, it augments or inhibits it. The response of smooth muscle to stretch is not always predictable. Sometimes it shows plastic behavior; when it is stretched it releases tension. In other cases the stretch acts as a stimulus for contraction. Single-unit muscle often occurs in large sheets and is found in the walls of hollow visceral organs like intestine, uterus, and bladder. It is also found in small blood vessels and ureters. 2. **Multi-unit** smooth muscle is more like skeletal muscle because it shows no inherent activity, but depends on its nerve supply. However this nerve supply is more diffuse, extending over a larger area of the muscle membrane.

MULTI-UNIT SMOOTH MUS._A

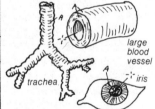

large blood vessel

trachea

iris

Ca⁺⁺_L & MUSCLE_A CONTRACTION_A
MEMBRANE RECEPTOR_M
CALCIUM CHANNEL_L'
CALMODULIN_N
MYOSIN LIGHT CHAIN KINASE (MLCK)_O

In addition to spontaneous action potentials, contraction of smooth muscle can be initiated or modified by nerve excitation, stretch, hormones, or electrical stimulation (1). In each case, stimulation results in increased intracellular Ca^{++}, either from extracellular sources via Ca^{++} channels (2) or from the sarcoplasmic reticulum (3). The rising Ca^{++} (4) reacts with an intracellular protein, calmodulin (5), forming a complex (6) which in turn activates an inactive form of an enzyme, myosin light chain kinase (MLCK) (7). The active MLCK catalyzes the transfer of phosphates to specific small (light) chains of amino acids contained in the myosin head groups. In this process the phosphate is donated by the reaction $ATP \rightarrow ADP + Pi$. Somehow this phosphorylation allows actin to activate myosin to further utilize ATP in the formation of active cross bridges, and contraction occurs. As long as the intracellular Ca^{++} concentration remains above threshold, myosin remains phosphorylated, and tension is maintained. When Ca^{++} is reduced, MLCK is inactivated, myosin is dephosphorylated, and the muscle relaxes.

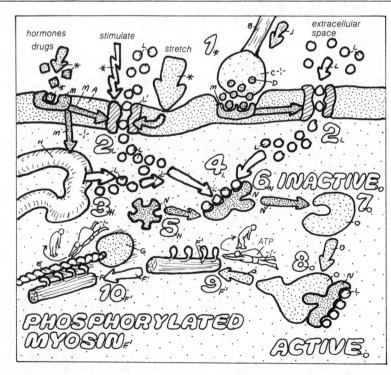

hormones drugs

stimulate

stretch

extracellular space

1

2

2

4

6. INACTIVE_N

7

3

5

ATP

8

10

9

PHOSPHORYLATED MYOSIN_F

ACTIVE_N

THE AUTONOMIC NERVOUS SYSTEM

The *autonomic nervous system (ANS)*, together with the endocrine (hormone) system, controls the body's internal organs. It innervates smooth muscle, cardiac muscle, and glands, controlling the circulation of blood, the activity of the gastrointestinal tract, body temperature, and a number of other functions. Most of this control is not conscious.

The ANS is divided into two parts, the *sympathetic* and the *parasympathetic nervous systems*, whose actions are mostly antagonistic. Many organs are supplied by nerves from each division, but some are not. The following table summarizes some of these actions.

AUTONOMIC EFFECTS OF SELECTED ORGANS .

ORGAN	EFFECT OF SYMPATHETIC STIMULATION	EFFECT OF PARASYMPATHETIC STIMULATION
Eye:		
Pupil	Dilated	Constricted
Ciliary muscle	Slight relaxation (far vision)	Constricted (near vision)
Heart:		
Muscle	Increased rate	Slowed rate
	Increased force of contraction	Decreased force of contraction (especially of atrium)
Coronaries	Dilated (β); constricted ($\propto$)	Dilated
Systemic arterioles:		
Abdominal	Constricted	None
Muscle	Constricted (adrenergic $\propto$)	None
	Dilated (adrenergic β)	
	Constricted	
Skin		None
Lungs:		
Bronchi	Dilated	Constricted
Blood vessels	Mildly constricted	? Dilated
Adrenal medullary secretion	Increased	None
Liver	Glucose released	Slight glycogen synthesis
Sweat glands	Copious sweating	None
Glands:		
Nasal, lacrimal, salivary, gastric	Vasoconstriction and slight secretion	Stimulation of copious secretion (except pancreas)
Gut:		
Lumen	Decreased peristalsis and tone	Increased peristalsis and tone
Sphincter	Increased tone	Relaxed (most times)
Gallbladder and bile ducts	Relaxed	Contracted
Kidney	Decreased output and renin secretion	None
Bladder:		
Detrusor	Relaxed (slight)	Excited
Trigone	Excited	Relaxed
Penis	Ejaculation	Erection
Basal metabolism	Increased	None

*modified from A.C. Guyton, Textbook of Medical Physiology,
7th Edition, 1986, W.B. Saunders

If we examine the effects of sympathetic stimulation, a useful pattern emerges. In many instances, sympathetic stimulation appears to prepare the animal for emergencies, for running or fighting. For example, air passages to the lungs (bronchi) dilate, making rapid breathing easier, the heart beats faster and stronger, and the liver releases glucose into the bloodstream. In addition, although it is not evident from the table, the constriction of blood vessels is most prominent in the intestinal tract and least in skeletal and heart muscle; so blood is shifted to the heart and skeletal muscle, where it is most needed. Viewed from this perspective, the often antagonistic parasympathetic nervous system appears to serve a vegetative function. However, the generalization that the sympathetic nervous system prepares the animal for emergencies has several important exceptions; for example, the sympathetic control of blood vessels to the skin is primarily responsive to changes in body temperature. Nevertheless, the generalization serves as a useful aid for remembering the diverse functions of the two divisions of the ANS.

The diagrams at the bottom of the plate illustrate that autonomic nerves differ from those going to skeletal muscle. Instead of going directly to their target, the autonomic nerves first make synaptic connections with other neurons, which then relay the impulses to the organs. These synaptic relay stations are called ganglia. Nerves conveying impulses into the ganglia are called *preganglionic fibers*; those that relay impulses to the organs are called *postganglionic fibers*. Both divisions of the ANS use the same neurotransmitter, acetylcholine, to transmit impulses over synaptic connections, from pre- to postganglionic fibers, within the ANS ganglia. However, the two divisions liberate different chemical transmitters at their postganglionic terminals making connection with the organs. Parasympathetic postganglionic transmission, again, use acetylcholine; sympathetic postganglionic transmission employs norepinephrine (plate 17).

The adrenal medulla gland resembles a sympathetic ganglion. It is stimulated by the acetylcholine liberated by sympathetic preganglionic nerves, which make direct synaptic connections with the gland. However, no postganglionic fibers arise from the ganglionlike gland. Instead, the activated cells secrete mixtures of norepinephrine and the closely related epinephrine directly into the bloodstream. More details about the adrenal medulla are given in plate 119; the brain centers controlling the ANS are covered in plates 101 and 102.

CN: Use a dark color for D.
1. Begin in the upper panel and work your way down the central portion, coloring the plus or minus symbols representing effects of both sympathetic and parasympathetic nerves on the glands and muscles of the body. Color the tiny circles that represent parasympathetic ganglia (B¹) in these various organs. The bottom diagram (under the genitals) shows a general rule for the location of the ganglia of each system (titles are in the lower left corner). Note that the upper three ganglia of the parasympathetic system are an exception; they lie outside the effector organ.
2. The lower panel expands the upper diagram by introducing preganglionic and postganglionic neurotransmitters. Note that the titles for the various effector cells are colored, but not the cells themselves.

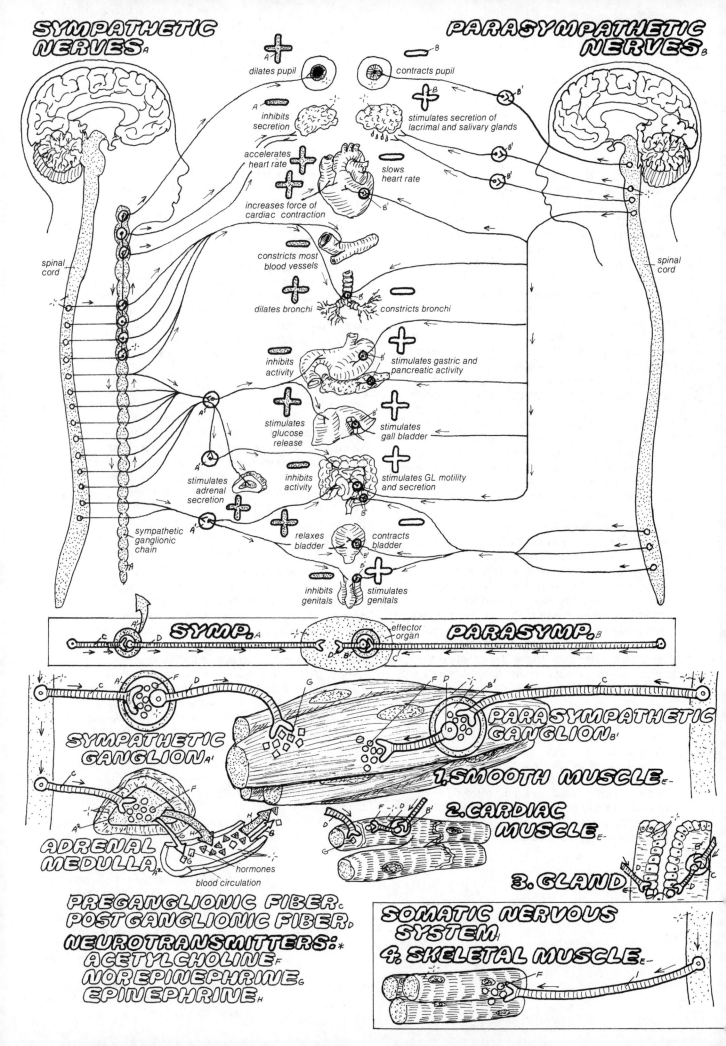

INTRODUCTION TO THE CARDIOVASCULAR SYSTEM

BLOOD FLOWS IN A CIRCLE. At first glance, the anatomy of the circulation is a mess! Its basic functional simplicity is obscured by the fact that the heart appears to be a single anatomical organ, when it is actually composed of two separate, functionally distinct, pumps. The right heart pumps blood to the lungs, where blood takes up oxygen and gives up carbon dioxide; the left heart pumps blood to tissues, where just the reverse happens. Further, although the lungs appear to be two organs (right and left), both lungs do exactly the same thing; they are really just one organ. We can use these ideas to untangle the circulation. Begin by separating the heart into its two functional units by an imaginary slice through the thick septum (wall) that divides the heart into right and left pumps. Pull these pumps apart, and place all vessels entering or leaving each heart in neat parallel arrays. This can be accomplished without really compromising the functional pathway taken by the blood. Finally, represent the lungs as a single organ, and you will arrive at the simple circle shown in the illustration in the upper right corner. Here, all *pulmonary arteries* (i.e., arteries that leave the right heart and go to the lungs) are collected into a single functional path, as are all *pulmonary veins* (veins that leave the lungs and enter the left heart). Similarly, all *systemic arteries* (i.e., those that leave the left heart bound for all non-lung tissues of the body) are collected, as are all *systemic veins* (veins leaving the non-lung tissues and emptying into the right heart). As illustrated in the functional diagram, the northern hemicircle (between right and left heart) supplying the lungs is called the *pulmonary circulation*. The southern hemicircle (between left and right heart) supplies the rest of the body tissues; it is called the *systemic circulation*. The eastern hemicircle contains oxygen-rich blood. In the western hemicircle the blood is oxygen poor.

Follow the diagrams on the opposite page that show important properties of blood flow through this circular path:

1. *Steady state blood flow is the same through any total cross-section of the circulation*. During each minute, the amount of blood flowing out of the right heart equals the amount flowing into the left heart; if this were not true, blood would be piling up continuously in the lungs. Momentarily, some fluid shifts could occur, but on average, over a substantial period of time our conclusion is correct, and the same argument can be applied to any section of the diagram. When the average adult is at rest, this flow amounts to about 5000 mL per min.

2. *Blood flow = blood velocity × cross-sectional area*. Blood *velocity* represents the speed of a blood "particle" in the stream, i.e., how far the particle moves in one minute. Blood *flow* represents how many particles (more precisely, the volume of these "particles") pass a given cross-section in one minute; it is measured in mL/min. The diagram illustrates the relationship between these two quantities.

3. *Total cross-sectional area of the vascular tree is greatest in the capillaries*. By total cross-sectional area, we mean the sum of the cross-sections of all branches of a similar type, e.g., major arteries, minor arteries, capillaries, major veins, etc. Beginning in the aorta and progressing toward the tissue, the total cross-section of the vascular tree gets larger and larger until it becomes maximal in the capillaries. Although each branch is smaller than its parent, the number of branches increases so rapidly that it more than compensates for the reduction in size of any individual branch. Progressing from capillaries to venules to veins and back to the heart, the reverse occurs.

4. *Blood velocity is slowest in the capillaries*. This follows because the blood flow is constant throughout the vascular tree, and the total cross-sectional area is largest in the capillaries. The equation *total blood flow = blood velocity × total cross-sectional area* shows that as cross-sectional area increases, blood velocity decreases, so that the product of the two does not change. (The same argument applies to a river, where a widening of the river bed is accompanied by a slowing of the stream.) It follows that the velocity will be smallest where the area is largest (in the capillaries). This is important because capillaries are very short (approximately 0.1 cm), and if the blood didn't slow down, there wouldn't be enough time for exchange (e.g., of O_2) between blood and tissues to occur. For example, blood normally spends about 1 sec. in a capillary; if it traveled at the same speed as it does in the aorta, this time would be reduced to only 0.001 sec., a 1000-fold reduction.

CN: Use blue for B, purple for C, and red for D.
1. Begin with the anatomical description of blood flow. Follow the numbered sequence starting with 1 in the right atrium (marked by an asterisk).
2. Color the functional description of blood flow, starting again with the right heart.
3. Color the diagrams describing the relationship between flow, area and velocity.
4. Color the chart in the lower right corner demonstrating these physical principles as they occur in the body.

MYOCARDIUM_A
DEOXYGENATED BLOOD_B
SYSTEMIC VEINS_B'
PULMONARY ARTERIES_B²
CAPILLARIES_C
OXYGENATED BLOOD_D
SYSTEMIC ARTERIES_D'
PULMONARY VEINS_D²

FLOW: ANATOMICAL *

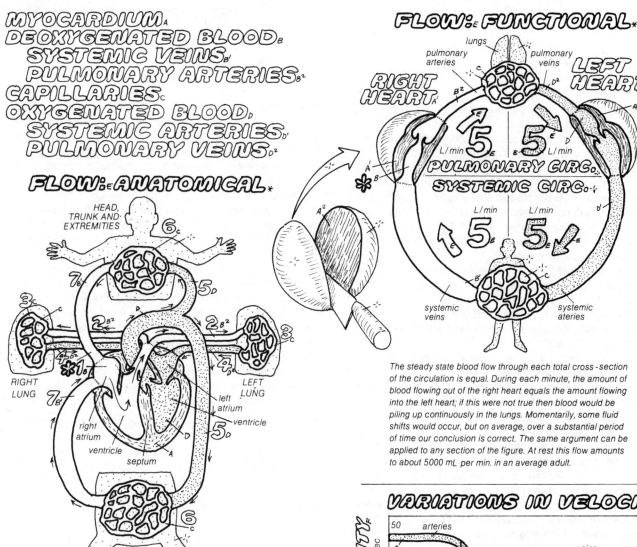

HEAD, TRUNK AND EXTREMITIES

RIGHT LUNG

LEFT LUNG

left atrium

ventricle

right atrium

ventricle

septum

LOWER TRUNK & EXTREMITIES

FLOW: FUNCTIONAL *

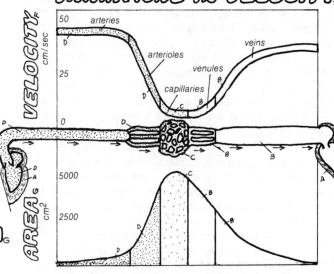

lungs

pulmonary arteries

pulmonary veins

RIGHT HEART_A'

LEFT HEART_A²

5 L/min_E

5 L/min_E

PULMONARY CIRC.

SYSTEMIC CIRC.

5 L/min_E

5 L/min_E

systemic veins

systemic ateries

The steady state blood flow through each total cross-section of the circulation is equal. During each minute, the amount of blood flowing out of the right heart equals the amount flowing into the left heart; if this were not true then blood would be piling up continuously in the lungs. Momentarily, some fluid shifts would occur, but on average, over a substantial period of time our conclusion is correct. The same argument can be applied to any section of the figure. At rest this flow amounts to about 5000 mL per min. in an average adult.

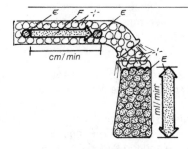

cm/min

ml/min

Blood velocity represents the speed of a blood "particle" in the stream, i.e. how far the particle moves in one min. Blood flow represents how many particles (more precisely the volume of these "particles") pass a given cross section in one minute; it is measured in ml/min.

FLOW = VELOCITY X AREA

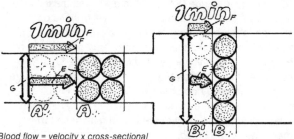

1 min

1 min

Blood flow = velocity x cross-sectional area. You can verify this in the special case shown above. The total cross section are of the tubing at B is twice that at A, so that a given length of tubing near B holds twice as many cubic elements as a similar length near A. With constant flow, the number of cubic fluid elements passing through the plane at A in one min.

must equal the number passing through plane at B in one min. (= 4 in our example). For this to occur, the elements at A must have passed the point A' one minute ago while those at B passed the point B'. During the same minute, those at A have traveled twice the distance as those at B; their velocity is twice as great.

VARIATIONS IN VELOCITY *

VELOCITY_F cm/sec

50

arteries

25

arterioles

venules

capillaries

veins

0

5000

AREA_G cm²

2500

Beginning in the aorta and progressing toward the tissues the total cross section of the vascular tree gets larger and larger until it becomes maximal in the capillaries. Although each branch is smaller than its parent, the number of branches increases so rapidly that it more than compensates for the reduction in size of any individual branch. Progressing from capillaries to venules to veins and back to the heart, the reverse occurs. Because area is largest in the capillaries, and blood flow is the same in all sections, it follows that blood velocity must be slowest in the capillaries. This is important because capillaries are very short, (approx. 0.1 cm) and if the blood didn't slow down, there wouldn't be enough time for exchange (e.g. of O_2) between blood and tissues to occur. For example, blood normally spends about 1 sec. in a capillary. If it traveled at the same speed as it does in the aorta, this time would be reduced to only 0.001 sec., a 1000 fold reduction.

ACTION POTENTIALS OF THE HEART

The heart is a hollow organ with walls made from specialized muscle called *cardiac muscle*. When excited, these muscles shorten, thicken, and squeeze on the hollow cavities of the heart, forcing blood to flow in directions permitted by the heart valves. Cardiac and skeletal muscles are similar in many ways: both contain actin and myosin filaments (see plate 18), which interdigitate and slide closer together during contraction; both can be electrically excited; and both show action potentials that propagate along the surface membrane, carrying excitation to all parts of the muscle. However, there are also significant differences:

1. The *duration of the action potential* is very brief in skeletal muscle; in cardiac muscle, it is 100 times longer lasting throughout the contraction of the muscle.

2. The *long refractory period* associated with the prolonged action potential also lasts throughout the contraction. This implies that:

3. *Cardiac muscle contractions are always brief twitches.* In skeletal muscle, contractions resulting from rapid repetitive stimulation can summate or "fuse" to provide smooth, sustained contractions. This cannot happen in cardiac muscle because the long refractory period "cancels" any stimulus that occurs before the heart has a chance to relax. Relaxation between each beat is essential for the heart to fill with blood to be pumped at the next beat.

4. *Cardiac muscles are interconnected by gap junctions (nexus).* These are channels that allow action potentials to pass from one cell to the next and ensure that the entire heart participates in each contraction. The heartbeat is *all or none.* In contrast, skeletal muscle cells are electrically isolated; one cell may contract while its neighbor remains quiescent.

5. Normally, skeletal muscle will contract only if it receives a nerve impulse. *Cardiac muscle excites itself.* Nerves that carry impulses to the heart influence the rate and strength of contraction, but they do not initiate the primitive heartbeat. When these nerves are destroyed, the heart continues to beat without any external prompt. In contrast, when nerves to skeletal muscle are destroyed, the muscle is paralyzed.

The shape of the action potential varies in different parts of the heart. The top figure shows an intracellular recording from a *Purkinje fiber*. These cardiac muscle fibers are particularly adept at conducting impulses. They also can excite themselves; when a Purkinje fiber is isolated, it continues to beat at its own rhythm. Notice that the resting potential (often called *diastolic potential*) is not level; it slowly rises to a threshold and initiates an action potential. The initial spike (very rapid rise in potential) is similar to those observed in nerve and skeletal muscle. In each case, the rise is due to the opening of Na^+ channels, which allow positively charged Na^+ ions to rush into the cell from outside, where they are highly concentrated. In all three cases, the opening of the Na^+ channels is caused by membrane depolarization so that a *positive feedback* (depolarization$\rightarrow$opening of $Na^+\rightarrow$ channels$\rightarrow$depolarization) is activated. In nerve and skeletal

muscle, this is followed by an inactivation of the Na^+ channels together with an opening of the K^+ channels, which repolarizes the membrane very quickly. Cardiac muscle is different; its Na^+ channels inactivate, but the opening of its K^+ channels is delayed. Meanwhile, the membrane potential is held in a suspended plateau by small amounts of Ca^{++} flowing through Ca^{++} channels that have opened in response to the depolarization. The small amounts of Ca^{++} that enter just balance the small amounts of K^+ that are leaking out. Finally, after 0.2 to 0.3 sec., K^+ channels open, the Ca^{++} channels close, and the membrane is rapidly depolarized. The potential falls to a minimum and then begins to slowly rise toward threshold as the cycle repeats. This slowly rising diastolic potential is due to closing of the K^+ channels so that the small resting flow of positive charge carried by Na^+ leaking inward becomes more and more effective in counterbalancing K^+ outflow. This drives the potential toward depolarization. Action potentials recorded from other areas of the ventricle are similar, except that the resting potential remains level. These cells do not show the same spontaneous activity as Purkinje fibers.

Action potentials recorded from the *SA or AV nodes* are different. Instead of Na^+ channels, Ca^{++} channels are activated by membrane depolarization, and the inward flow of Ca^{++} is responsible for the rising phase of the action potential. Further, the rise in diastolic potential is rapid and reaches threshold quickly; when SA node cells are isolated, they beat at fast rates. Isolated cells from the SA node beat faster than those from the AV node, and these beat faster than Purkinje fibers. In the intact heart, cells of the SA node set the rhythm for the entire heart; the SA node is the *pacemaker*. These rapidly beating cells become excited first and transmit their excitation to all others. Although many cells are capable of beating at their own (slower) rate, they never do because they are driven at a faster rate by impulses originating in the SA node.

In order for the pacemaker to initiate a coordinated beat, there has to be a mechanism for rapid impulse conduction to all parts of the heart. This is particularly important because the atria and ventricles are separated by a band of connective tissue that does not conduct impulses. The required pathway is provided by the AV node and the *Purkinje system*, shown in the bottom figure. The AV node supplies the only normal conductive bridge betwen atria and ventricles. It takes only about 0.04 sec. for the impulse to travel from its origin in the SA node to the beginning of the AV node, but by the time the impulse finally leaves the AV node to emerge in the bundle, there is an additional delay of about 0.11 sec. This AV delay provides time for the atria to complete their beat before the ventricles begin. Once past the AV node, the impulse is rapidly conveyed via the Purkinje network to all parts of the ventricle, ensuring that all parts beat in unison to impart maximal thrust to the blood.

CN: *Use dark colors for J and K structures and very light colors for H and I.*
1. Begin in the upper right corner. Color the membrane diagram at the top as you color each phase of the action potential. Color the smaller chart.

2. Color the two graphs representing excitation from other sites in the heart.
3. Color the large heart drawing, following the title sequence in the lower left corner. Color the diagram to the right.

DIASTOLIC POTENTIAL A
ACTION POTENTIALS: *
HIGH Na⁺⁺ PERMEABILITY B
HIGH Ca⁺⁺ PERMEABILITY C
HIGH K⁺ PERMEABILITY D
REFRACTORY PERIOD E

CONTRACTILE RESPONSE F
ELECTRODE G

An action potential in a Purkinje heart cell begins with a resting potential which drifts upward until it reaches threshold when there is a rapid rise in Na^+ permeability and Na^+ entry into the cell. The high Na^+ permeability is quickly inactivated and a prolonged plateau period follows where the stagnant potential is produced by a slow Ca^{++} entry and slow K^+ exit that almost balance. Finally K^+ permeability increases; K^+ exit dominates and the potential rapidly returns to rest levels. The action potential lasts throughout contraction, providing a long refractory period which prevents tetanic contractions and ensures relaxation for filling of the heart.

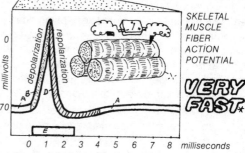

OUTSIDE CELL

Na^+_B Na^+_B Ca^{++}_C

membrane

INSIDE CELL

K^+_D K^+_D K^+_D

CARDIAC MUSCLE FIBER ACTION POTENTIAL (PURKINJE)

+20
0
−90

millivolts

depolarization   plateau   repolarization

SLOW *

milliseconds   0   100   200   300

SKELETAL MUSCLE FIBER ACTION POTENTIAL

0
−70

millivolts

depolarization   repolarization

VERY FAST *

0 1 2 3 4 5 6 7 8 milliseconds

The duration of the action potential is very brief in skeletal muscle. In cardiac muscle it is some 100 times more prolonged.

VENTRICLE I

Action potential from ordinary ventricle muscle cells are similar to those from Purkinje cells, with the exception that their resting potentials do not show any slow depolarizing drifts. These cells do not excite themselves.

NODE J K

In the SA or AV nodes, Na^+ channels no longer play a role. Instead Ca^{++} channels are activated and inward flow of Ca^{++} is responsible for the rising phase of the action potential. Further, the rise in diastolic potential is well developed. Cells of the SA node are the heart's pacemaker.

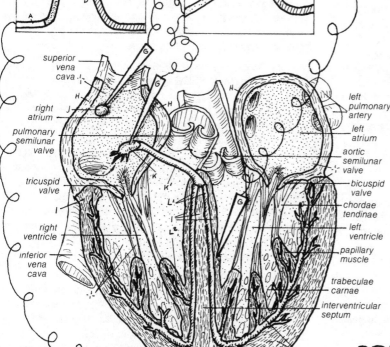

superior vena cava
right atrium
pulmonary semilunar valve
tricuspid valve
right ventricle
inferior vena cava

left pulmonary artery
left atrium
aortic semilunar valve
bicuspid valve
chordae tendinae
left ventricle
papillary muscle
trabeculae carnae
interventricular septum

IMPULSE CONDUCTION M

Although most of the atria and ventricles are separated by a band of connective tissue which does not conduct impulses, a conductive path is provided by the AV node and the Purkinje system. Purkinje fibers leading from the AV node into the ventricles conduct impulses very fast so that all parts of the heart beat in unison to impart maximal thrust to the blood.

connective tissue

ATRIUM WALL H
VENTRICLE WALL I
SINOATRIAL NODE, (SA NODE) (PACEMAKER) J
ATRIOVENTRICULAR NODE (AV NODE) K
AV BUNDLE K'
PURKINJE SYSTEM: L
RIGHT & LEFT BUNDLE BRANCHES L'
PURKINJE FIBERS L²

ARTIFICIAL PACEMAKER J'

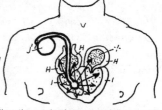

When the conduction system fails, an artifical pacemaker can be used. the ventricle is stimulated directly via leads which convey electrical impulses from a battery implanted under the skin.

THE ECG AND IMPULSE CONDUCTION IN THE HEART

The accurate recording of cardiac membrane potentials described in the previous plate requires the insertion of a microelectrode into the cytoplasm of the cell, a procedure that cannot be undertaken in a human. However, there are other, noninvasive methods for assessing the electrical activity of the heart. Although these methods are less precise, they have the advantage of providing global information about how the activities of various portions of the heart are integrated into a coherent beat.

To understand these measurements, first consider the simple case illustrated in the top figure, where two electrodes are placed near the heart's surface. The cells on the left are active and their extracellular surfaces are negatively charged while surfaces of resting cells on the right are positive. This difference is picked up by surface electrodes. The *electrode* on the left, close to the negative charge, is more negative than the electrode on the right (close to the positive charge). The meter detects only the *difference* between the electrodes. When the circumstances are reversed, with cells on the left at rest while cells on the right are active, then the right-hand electrode is close to the negative charge, and it will be negative with respect to the one on the left.

The key words are "close to." How close do the electrodes have to be to make the measurement? Fortunately, the heart is large, and body fluids contain ions that conduct electricity so that electrodes can be placed at some distance from the heart, anywhere on the body surface, as long as they are in good electrical contact with body fluids. The figure at the top right shows conventional locations for electrode placement. Measurements called *electrocardiograms* (ECGs) are taken as the difference between any two of the three electrodes. The legs and arms act as simple extensions of the electrodes; measurements from the leg approximate electrical variations occurring in the groin; measurements from the arms approximate those from the corresponding shoulder. Electrodes are placed on wrists and ankles merely for convenience.

The middle figure shows a typical ECG. Although it is not obvious, this recording represents the sum of all action potentials of all cardiac muscle cells during one beat. Remember that the recording is made some distance from the heart, that various heart cells are oriented in different directions, and that they are excited at different times and recover at others. As "seen" by an electrode on the body surface, the electrical signal from one cell may easily augment or detract from the signal of another. No wonder the composite ECG bears no obvious resemblance to the action potential of a single cell. Nevertheless, years of careful observations and correlations have established a basis for interpreting ECGs. The landmarks on a typical record are designated by the letters P, QRS, and T. Their physiological correlates are:

P WAVE. The P wave signals the beginning of the heartbeat. It corresponds to the spread of excitation over both atria.

P-R INTERVAL. The time from the beginning of the P wave to the beginning of the R wave measures the time for impulse conduction from atria to ventricles. Although the heart appears to be "electrically silent" during this time, a wave of electrical depolarization is propagated; the time includes passage of the impulse to the AV node, the delay imposed by the AV node, passage through the AV bundle, the bundle branches, and the Purkinje network. Disturbances of AV conduction induced by inflammation, poor circulation, drugs, or nervous mechanisms are often revealed by an abnormal prolongation of the P-R interval.

QRS COMPLEX. This corresponds to the invasion of the ventricular musculature by excitatory impulses. It is higher than the P because the ventricular mass is much larger than the atria. The duration of the QRS complex is shorter than the P wave because impulse conduction through the ventricles (partly via the Purkinje network) is very rapid.

S-T SEGMENT. During the interval between S and T, the ECG registers zero. All of the ventricular muscle is in the same depolarized state (recall the long plateau of the action potential of ventricular fibers), and there are no differences to record.

T WAVE. The T wave results from ventricular repolarization as different parts of the ventricle repolarize at different times.

These are only the bare rudiments of information buried in an ECG. By examination of these records, a cardiologist learns about the anatomical orientation of the heart, disturbances of heart rate and impulse conduction, the extent and location of damaged tissue, and the effect of disturbances in plasma electrolytes.

Heart block and *fibrillation* are pathological conditions that are easy to detect in ECG recordings. In *heart block*, impulse propagation through the AV node is impeded. In first degree block, the impulse is merely slowed so that there is an abnormally long P-R interval. In one form of second degree block, the AV node fails to pass every impulse. Only one out of two or one out of three impulses passes, and the ECG contains two or three P waves for every QRS. In more severe cases (third degree or complete block), the AV node fails completely, no impulses get through, and the atria and ventricles are electrically isolated. Ventricular pacemakers then take over, and the atria and ventricles beat independently of one another. The ECG in this case shows no correlation between the appearance of P waves and QRS complexes.

In *ventricular fibrillation*, individual portions of the heart beat independently, without coordination. The heart is reduced to a quivering mass with no obvious excitation period and no obvious resting period. Blood is no longer pumped. The cause of ventricular fibrillation is not completely understood, but it appears to result from rapid and chaotic pacemaker activities that develop in different locations, together with long, circuitous conduction pathways. Fibrillation confined to the atria can be tolerated because, at rest, the atrial contribution to filling of the ventricle is small. In contrast, ventricular fibrillation is always fatal unless it can be immediately arrested.

CN: 1. Begin with the upper 1/3 of the page showing the electrodes.
2. Color the heart and ECG. As you color each title, complete corresponding structures in the heart diagram and in the ECG. Note that in this particular illustration the colors used in the heart refer to the phase of ECG and not necessarily structures of the heart.
3. Color the lower material on heart block and fibrillation.

MEASURING THE HEART'S ELECTRICAL ACTIVITY*
CELL ACTIVITY.

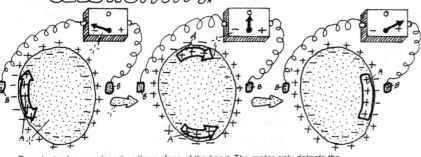

Two electrodes are placed on the surface of the heart. The meter only detects the **difference** between the two electrodes (left side-right). If cells on the left are active while those on the right are at rest, the electrode on the left will be negative with respect to the electrode on the right. When cells on the left rest while cells on the right are active, then the right-hand electrode will be negative with respect to the one on the left, or in other words, the left electrode will be positive with respect to the right.

TAKING THE ECG.
ELECTRODE.B

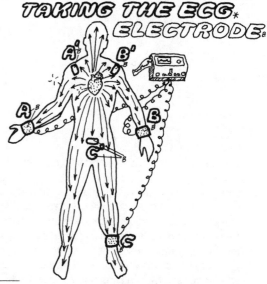

In the ECG, electrodes are placed on the arms and left leg. Body fluids conduct electrical signals from the surface of the heart to the electrodes. Measurements are taken as the difference between two of the three electrodes, the legs and arms serving as simple extensions of the electrodes. Measurements from the ankle (C) approximate electrical variations that would be measured with an electrode placed in the groin (C^1). Similarly for A and A^1, and B and B^1.

THE ECG.*

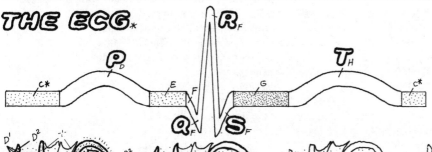

RESTING c*
P WAVE D
SA NODE EXCITATION D¹
ATRIAL DEPOLARIZATION D² E
AV NODE, PURKINJE EXCITATION E¹
QRS WAVE F
PURKINJE EXCITATION F¹
VENTRICULAR DEPOLARIZATION F²
TOTAL DEPOLARIZATION G
T WAVE H
VENTRICLE REPOLARIZATION H¹

In a typical ECG the P wave corresponds to atrial depolarization, the QRS complex occurs as the impulse invades the ventricle, the flat interval between S and T signifies complete ventricular depolarization, and the T wave corresponds to repolarization of the ventricle.

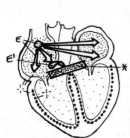

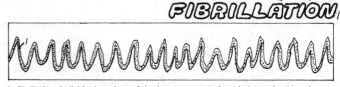

HEART BLOCK*
ATRIAL BEAT E
VENTRICULAR BEAT E¹

In 2nd degree heart block, the A-V node fails to pass every impulse so that only 1 out of 2, or 1 out of 3 impulses pass. In these cases the ECG contains 2 or 3 P waves for every QRS complex.

FIBRILLATION

In fibrillation, individual portions of the heart appear to beat independently and without any coordination. The heart is reduced to a quivering mass of tissue. This fatal condition is reflected in a chaotic ECG.

EXCITATION-CONTRACTION COUPLING IN CARDIAC MUSCLE

Just as in excitation, the contractile properties of the heart are very similar to skeletal muscle (see plate 20). Like skeletal muscle contraction, heart muscle contraction is based on *actin* and *myosin* filaments that interdigitate and slide closer together in the presence of *free Ca++ in the cytoplasm*. In both cases, the sliding is mediated by myosin cross bridges, which reach out to contact special sites on the actin filaments. Both skeletal and cardiac muscles contain *T tubules*, which conduct impulses perpendicular to the cell surface toward its interior; both contain a well-developed tubular network, the *sarcoplasmic reticulum* (abbreviated as SR), which releases Ca++ to trigger contraction and sequesters it for relaxation; and both contain the regulatory proteins, *tropin* and *tropomyosin*, which keep the actin and myosin cross bridges apart in the absence of free Ca++.

There are also important differences. When skeletal muscle is excited, enough Ca++ is released to react with all of the troponin so that all reactive sites on actin become available and all *cross bridges* become activated. Normally in contracting cardiac muscle, this is not the case; troponin is not fully covered by Ca++. This is important because it implies that anything that increases Ca++ availability inside the cell will increase the number of cross bridges that can form, and just as increasing the number of persons pulling on a rope in a "tug of war" will increase the tension or pull on the rope, this increase in the number of active cross bridges will increase the strength of cardiac contraction. Whatever controls internal Ca++ will control cardiac performance.

How is free Ca++ controlled? Free Ca++ levels in the cytoplasm (i.e., in the non-enclosed watery space inside the cell) are about 20,000 times lower than external free Ca++. Most Ca++ inside the cell is either bound to proteins or is sequestered inside mitochondria or the SR. Ca++ is poised at higher concentrations both outside the cell and in the SR waiting to enter the cytoplasm, where it will have easy access to the troponin and the contractile filaments. During activity, action potentials travel over the surface membrane, invade the T tubules, where the excitatory waves come in close proximity to the SR (more precisely, the cisternae), and, by some unknown mechanism, cause release of Ca++. In addition, some Ca++ enters the cell with each action potential. Although this is not nearly enough to activate the filaments, it does stimulate the release of more Ca++ from internal stores in the SR. Somehow Ca++ triggers its own release. The Ca++ level rises rapidly, cross bridges are activated, and these levels persist throughout the prolonged action potential.

During relaxation, the level of internal free Ca++ is reduced primarily because it is pumped back into the SR by an *ATP driven Ca++ pump*. If there is more Ca++ in the cell, more Ca++ will be loaded back into the SR, and more Ca++ will be released at the next beat to cause a more forceful contraction. The Ca++ that continually leaks into the cell (e.g., via the action potential) is removed by two routes:

1. It is pumped out of the cell by an ATP driven Ca++ pump.
2. It is pumped out of the cell by a *Na+-Ca++ exchanger.*

Although a complete description of free Ca++ balance within the heart cell is essential for understanding cardiac performance, the details still elude us. The story of the common cardiac drug *digitalis* provides an example of how known details have been used to interpret clinical experience. This drug has been successfully used for many years in cardiac patients to strengthen the force of cardiac contraction, yet experiments failed to reveal any effect of the drug on the contractile machinery. All that could be shown was that digitalis is a potent inhibitor of the *Na+-K+ pump*. What does the Na+-K+ pump have to do with cardiac contraction? Our current interpretation involves the Na+-Ca++ exchanger. This exchanger works because Na+ is more concentrated outside the cell than inside and because Na+ movements into the cell along this route are tightly coupled to Ca++ movements out. The energy for moving Ca++ from low internal to high external concentrations (i.e., for pumping the Ca++ out) is provided from the energy loss accompanying the movement of Na+ from high external to low internal concentrations. When digitalis is administered, it inhibits the Na+-K+ pump, less Na+ is pumped out of the cell, its internal concentration rises, and this inhibits the Na+-Ca++ exchanger (which requires low internal Na+). Internal Ca++ rises so that more is available to activate the cross bridges, resulting in more forceful contractions.

CN: Use red for M, and dark colors for A and K.
1. Begin with the second illustration from the top: an anatomical description of cardiac muscle cells.
2. Color the schematic diagram of the same subject just below it. Color the cartoon on the left.
3. Go to bottom of the page and work your way up through the relaxation sequence. The two rectangels at the top of this panel illustrate the operation of the Na+-K+ pump and the Na+-Ca++ exchanger. Do the left one first, then the other which suggests how digitalis enhances cardiac muscle contraction by preventing Ca++ removal.

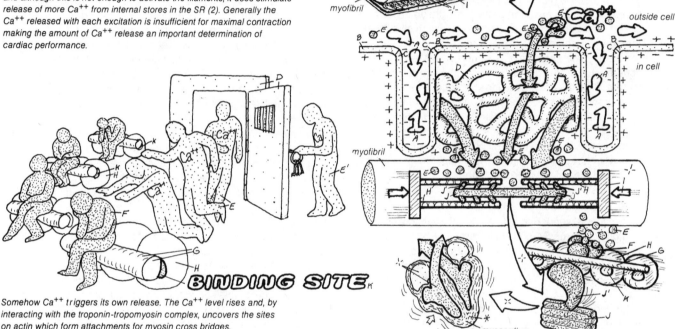

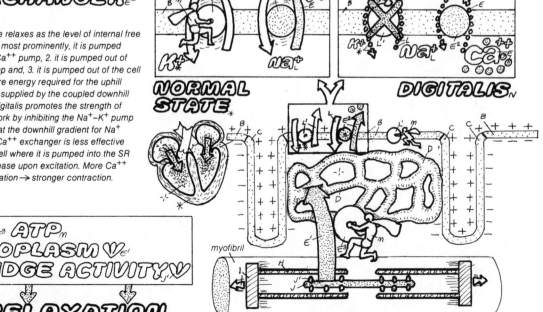

ACTION POTENTIAL _A_
PLASMA MEMBRANE _B_
T TUBULE _C_
SARCOPLASMIC RETIC. _D_
Ca^{++} IN CYTOPLASM $\wedge$ _E'_
TROPONIN _F_ TROPOMYOSIN _G_
ACTIN _H_ Z LINE _I_
MYOSIN _J_ CROSS BRIDGE $\wedge$ _M_

MUSCLE CONTRACTION *

During activity, action potentials travel over the cell membrane, invade T tubules where the excitatory waves comes close to the SR and by some unknown mechanism cause release of Ca^{++} (1). In addition some Ca^{++} enters via the cell membrane with each action potential, and although this is not enough to activate the filaments, it does stimulate release of more Ca^{++} from internal stores in the SR (2). Generally the Ca^{++} released with each excitation is insufficient for maximal contraction making the amount of Ca^{++} release an important determination of cardiac performance.

CARDIAC MUSCLE *

intercalated disk

mitochondrion

intercalated disk

myofibril

outside cell

in cell

myofibril

myocardium

BINDING SITE _K_

Somehow Ca^{++} triggers its own release. The Ca^{++} level rises and, by interacting with the troponin-tropomyosin complex, uncovers the sites on actin which form attachments for myosin cross bridges.

NA^+/K^+ PUMP _L_ ATP _M_
NA^+/CA^{++} EXCHANGER _E²_

After the action potential, the muscle relaxes as the level of internal free Ca^{++} is reduced via three routes: 1. most prominently, it is pumped back into the SR by an ATP driven Ca^{++} pump, 2. it is pumped out of the cell by an ATP driven Ca^{++} pump and, 3. it is pumped out of the cell by the a Na^+–Ca^{++} exchanger where energy required for the uphill movement of Ca^{++} out of the cell is supplied by the coupled downhill flow of Na^+ into the cell. The drug digitalis promotes the strength of cardiac contraction. It appears to work by inhibiting the Na^+–K^+ pump and lowering intercellular Na^+ so that the downhill gradient for Na^+ entry is depressed. Thus, the Na^+–Ca^{++} exchanger is less effective and Ca^{++} accumulates inside the cell where it is pumped into the SR making more Ca^{++} available for release upon excitation. More Ca^{++} release $\rightarrow$ more cross bridges activation $\rightarrow$ stronger contraction.

NORMAL STATE

DIGITALIS _N_

myofibril

Ca^{++} PUMP _E³_ ATP _M_
Ca^{++} IN CYTOPLASM $\vee$ _E'_
CROSS BRIDGE ACTIVITY _J,V_

MUSCLE RELAXATION *

NEURAL CONTROL OF THE HEART

During activity, the heart beats faster and stronger; during rest, it slows down. These alterations occur largely through the action of *sympathetic* and *parasympathetic nerves* on the heart, and to a lesser extent by *catecholamine* (adrenalin) secretions from the *adrenal medulla* gland. The physiological effects of these two nerves are simple: the sympathetic nerves liberate norepinephrine (noradrenalin), which stimulates the heart, increasing its rate and force of contraction; the parasympathetic nerves (carried by vagus nerves) liberate acetylcholine, which inhibits the heart, slowing its rate. Both sets of nerves innervate the *SA* and *AV* *nodes*, and both affect the heart rate by their influence on the primitive pacemaker activity of the SA node. However, unlike the sympathetic nerves, parasympathetic nerves have no effect on the strength of ventricular contraction; in fact, the ventricular musculature is virtually free of any parasympathetic innervation, but is profusely innervated by sympathetic fibers.

How do these nerves (or more precisely these neurotransmitters) work? Recall (plate 27) that heart excitation occurs in the SA node as a result of Ca^{++} *ions* moving down their concentration gradient into the cell and depolarizing the cell (i.e., making the inside less negative) until the *membrane potential* reaches *threshold*. Any K^+ leakage out of the cell during this time does just the opposite; it tends to repolarize the cell, driving the membrane potential away from threshold. The key to our problem lies in the balance between the opposing actions of K^+ moving out of and Ca^{++} moving

into the cell.

Acetylcholine liberated by the vagus nerve acts on the SA node primarily by increasing its *permeability to K^+*. Consequently, the resting potential of the SA node becomes more negative, pushing it further away from the threshold potential. The outward moving K^+ slows the normal rate of depolarization (the pacemaker potential) caused by inward moving Ca^{++}. This lengthens the time required to reach threshold and slows the heart. In addition to its effect on the rate of firing of the pacemaker (SA node), the outward moving K^+ impedes excitability in other cells, and this tends to slow conduction of the impulse through the atrium and AV node.

Norepinephrine acts in several ways. It increases heart rate by increasing Ca^{++} *permeability*. In the SA node this increases the rate of rise of the pacemaker potential, shortening the time to reach threshold and increasing the heart rate. In ventricular muscle, the increased amounts of Ca^{++} that enter with each beat not only trigger Ca^{++} release, but also build up the internal Ca^{++} stores so that more is available for release with each contraction, and, after a few beats, the contractions are stronger. Finally, norepinephrine also increases the rate of *re-uptake of Ca^{++}* by the *sarcoplasmic reticulum*; this speeds up the relaxation process and consequently shortens the duration of contraction. With a fast heart rate, it is important to curtail the contraction period to allow sufficient time for the heart to fill between beats.

CN: Use red for J, and dark colors for A, B, H and I.
1. Begin with the upper right diagram.
2. Color the three stages of autonomic nerve activity.
3. Finish with the two phases of Ca^{++} involvement in the contraction process.

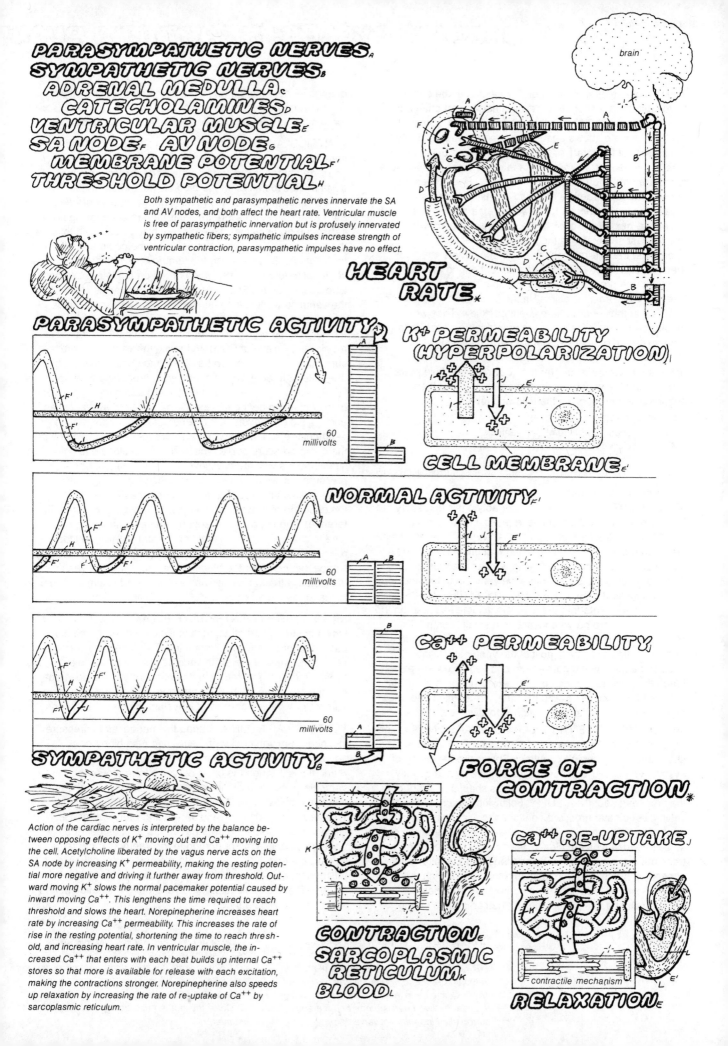

PARASYMPATHETIC NERVES A
SYMPATHETIC NERVES B
ADRENAL MEDULLA C
CATECHOLAMINES D
VENTRICULAR MUSCLE E
SA NODE F AV NODE G
MEMBRANE POTENTIAL F'
THRESHOLD POTENTIAL H

Both sympathetic and parasympathetic nerves innervate the SA and AV nodes, and both affect the heart rate. Ventricular muscle is free of parasympathetic innervation but is profusely innervated by sympathetic fibers; sympathetic impulses increase strength of ventricular contraction, parasympathetic impulses have no effect.

HEART RATE *

PARASYMPATHETIC ACTIVITY A

60 millivolts

K⁺ PERMEABILITY (HYPERPOLARIZATION)

CELL MEMBRANE E'

NORMAL ACTIVITY F'

60 millivolts

Ca⁺⁺ PERMEABILITY

60 millivolts

SYMPATHETIC ACTIVITY B

FORCE OF CONTRACTION *

Action of the cardiac nerves is interpreted by the balance between opposing effects of K⁺ moving out and Ca⁺⁺ moving into the cell. Acetylcholine liberated by the vagus nerve acts on the SA node by increasing K⁺ permeability, making the resting potential more negative and driving it further away from threshold. Outward moving K⁺ slows the normal pacemaker potential caused by inward moving Ca⁺⁺. This lengthens the time required to reach threshold and slows the heart. Norepinepherine increases heart rate by increasing Ca⁺⁺ permeability. This increases the rate of rise in the resting potential, shortening the time to reach threshold, and increasing heart rate. In ventricular muscle, the increased Ca⁺⁺ that enters with each beat builds up internal Ca⁺⁺ stores so that more is available for release with each excitation, making the contractions stronger. Norepinepherine also speeds up relaxation by increasing the rate of re-uptake of Ca⁺⁺ by sarcoplasmic reticulum.

Ca⁺⁺ RE-UPTAKE J

CONTRACTION E
SARCOPLASMIC RETICULUM K
BLOOD L

contractile mechanism

RELAXATION E

CARDIAC CYCLES: HEART AS A PUMP

The pumping action of the heart is reflected in the changes in volume and pressure that occur in each heart chamber and in the great arteries as the heart completes a single cycle. This plate shows changes that occur on the left (systemic) side of the heart. (Changes on the right [pulmonary] side of the heart are similar with the one exception that pressures are only about one-eighth as large.) Five curves are shown. The top three are obtained by inserting pressure measuring instruments into the *aorta, left atrium*, and *left ventricle*. The next curve depicts the volume of the left ventricle, and the last curve shows the *ECG*. Our object is to appreciate the interrelationships of these curves and how they relate to blood flow at each moment during the cardiac cycle. To interpret these curves, we note that each of the heart valves normally "points" in the direction of the flow. This means that they operate to prevent backflow. They work because whenever the downstream pressure builds up larger than upstream pressure (a condition for backflow), this pressure difference forces the valve closed. Similarly, when upstream pressure is greater than downstream, fluid flows forward and the valves are forced open. Applying these notions to the heart gives the following table:

VALVES	STATE	CONDITION
AV	open	$P(atrium) > P(ventricle)$
AV	closed	$P(atrium) < P(ventricle)$
aortic	open	$P(ventricle) > P(aorta)$
aortic	closed	$P(ventricle) < P(aorta)$ — see exception below

The cycle begins with:

ATRIAL CONTRACTION. Atrial contraction is signaled by the P wave of the ECG. Atrial pressure rises, and blood is thrust into the ventricles through the open *AV valves*. These valves are open (as they have been throughout the diastole) because pressure in the atrium is higher than pressure in the quiescent ventricle. Blood enters the ventricle but cannot leave because the *aortic valves* are closed (P[aortic] > P[ventricular]). Note that the resulting volume increase on the ventricular volume curve appears as a small "bump." The atrium serves as a "booster" pump, but its contribution to ventricular filling is small; most of the ventricular filling occurred earlier, when both atrium and ventricle were at rest. When the heart rate goes up, as in exercise, there is less time between beats for filling, and the atrial contribution becomes more significant. Atrial contraction is followed by:

ISOVOLUMETRIC VENTRICULAR CONTRACTION. Now the impulse invades the ventricles (QRS in the ECG), and, after a short delay, they begin to contract. This is the beginning of systole. Ventricular pressure builds up steeply and quickly exceeds atrial pressure. The AV valves snap shut, producing the first heart sound — "LUPP." Following closure of the AV valves, ventricular pressure continues to rise steeply until it exceeds aortic pressure. Pressure rises rapidly because both sets of heart valves are closed; the heart continues to contract, but there is no place for the blood to go to relieve the ascending pressure. (Contraction of the heart during this period is similar to an isometric contraction in skeletal muscle.) During this period, the ventricular volume cannot change — note the flat horizontal trace on the ventricular volume curve. The constant ventricular volume is the reason for naming this period "isovolumetric ventricular contraction."

VENTRICULAR EJECTION. As soon as the ventricular pressure exceeds aortic pressure, the aortic valves are thrust open, and blood is ejected into the aorta. Pressure in the aorta begins to rise because blood is entering from the ventricles faster than it can leave through the smaller arteries. Prior to this time, pressure in the aorta had been falling because the aortic valves were closed; blood continued to leave the aorta though smaller arteries, but none could enter from the ventricle. Blood leaving the ventricles is reflected in the ventricular volume curve, which drops precipitously as soon as ejection begins. Soon afterward, the contractile force of the ventricle wanes; the ventricular pressure ascent slows and begins to reverse while the initial rapid change in ventricular volume begins to level off. As the ventricles begin to repolarize (T wave of ECG) and relax, the ventricular pressure curve crosses the aortic curve and goes below it. Shortly thereafter, the aortic valve snaps shut, producing a sharp "DUP" sound (the second heart sound) and bringing the ventricular ejection period as well as the period of systole, to an end. (Systole = isovolumetric ventricular contraction period + ventricular ejection period.) It also produces a bump notch on the aortic pressure curve. The aortic valve closure is not simultaneous with the crossover of the ventricular and aortic pressure curves because the blood flowing through the valves has an appreciable momentum (mass × velocity) in the direction of forward flow. Applying a force (pressure difference) in the opposite direction requires a small amount of time to stop or reverse the motion. (Imagine trying to stop a rolling automobile with a hand push in the opposite direction.) Notice that not all of the blood contained within the ventricle is ejected with each beat. The residual blood is almost equal to the amount ejected.

ISOVOLUMETRIC VENTRICULAR RELAXATION. Now, as in isovolumetric contraction, both valves are closed, and blood cannot enter or leave the ventricles. This time, however, the ventricular muscles relax; it is the beginning of diastole. Pressure falls precipitously, but ventricular volume does not change. Soon the ventricular pressure falls below atrial pressure, the AV valves open, and isovolumetric relaxation ends.

VENTRICULAR FILLING. During this period, atrial pressure is higher than ventricular pressure because blood continues to flow into the atrium from the pulmonary veins. Blood flows through the open AV valve from atrium to ventricle. This filling of the ventricle continues throughout diastole, not just when the atrium contracts. The ventricular volume curve during diastole shows that early ventricular filling is most prominent and that contraction of the atrium contributes only a minor portion to the ventricular contents. Toward the end of this period, atrial contraction ensues, and this period, as well as diastole, ends with closure of the AV valves. (Diastole = isovolumetric ventricular relaxation period + ventricular filling period.)

CN: Use red for F and dark colors for B and D.
1. Begin in the upper left corner, by coloring titles A-E. Color all structures in each left heart illustrated across the top. Note that the pulmonary arteries have been left out of the last three figures. Include the two heart sounds and the sound bars surrounding the relevant valve.

2. Color the pressure section and refer to the relevant phase above.
3. Do the same for the blood volume of the left ventricle.
4. Color the ECG results gray.
5. Color the bottom figures representing the time intervals.

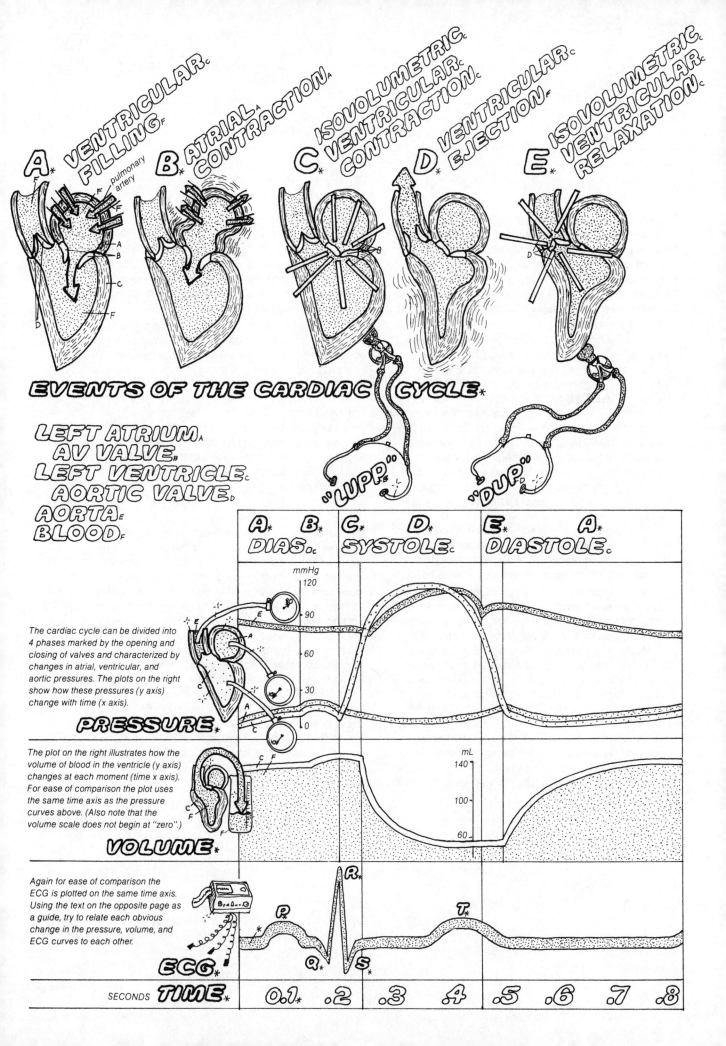

EVENTS OF THE CARDIAC CYCLE

VENTRICULAR FILLING
ATRIAL CONTRACTION
ISOVOLUMETRIC VENTRICULAR CONTRACTION
VENTRICULAR EJECTION
ISOVOLUMETRIC VENTRICULAR RELAXATION

A B C D E

pulmonary artery

"LUPP"   "DUP"

LEFT ATRIUM
AV VALVE
LEFT VENTRICLE
AORTIC VALVE
AORTA
BLOOD

| A B | C D | E A |
| DIAS | SYSTOLE | DIASTOLE |

The cardiac cycle can be divided into 4 phases marked by the opening and closing of valves and characterized by changes in atrial, ventricular, and aortic pressures. The plots on the right show how these pressures (y axis) change with time (x axis).

mmHg
120
90
60
30
0

PRESSURE

The plot on the right illustrates how the volume of blood in the ventricle (y axis) changes at each moment (time x axis). For ease of comparison the plot uses the same time axis as the pressure curves above. (Also note that the volume scale does not begin at "zero".)

mL
140
100
60

VOLUME

Again for ease of comparison the ECG is plotted on the same time axis. Using the text on the opposite page as a guide, try to relate each obvious change in the pressure, volume, and ECG curves to each other.

R
P   T
Q   S

ECG

SECONDS TIME   0.1   .2   .3   .4   .5   .6   .7   .8

THE PHYSICS OF BLOOD FLOW

Blood flows from the arteries through the capillaries to the veins because the pressure is higher in the arteries than in the veins. Pressure is the force exerted on each square centimeter; it is a measure of "push." Blood flows from arteries to veins because the blood in the arteries "pushes" harder than the blood in the veins. It is the *difference in pressure* that constitutes the driving force for movement.

How do we measure this force or push? Think about a fluid particle at the bottom of a tube where there is no motion (see diagram). It is subjected to the force exerted by the weight of the column of fluid on top of it. If, for some reason, this force was less than the pressure on the particle, it would move upward, and fluid would flow into the pipe until the weight exactly balanced the pressure. The fact that there is no motion means that the weight of the column is just equal to the pressure on the particle. We use the height of a fluid column as a convenient way to measure pressure. (The weight of the column depends on what the fluid is. *Mercury* is denser than water, so a given weight will require a much smaller column of mercury as compared to water. For this reason, it is simply more convenient to use mercury as a reference. To convert a millimeter of mercury to a millimeter of water, multiply by 13.6.)

In the illustration with the stopcock closed, the fluid rises to the same level in all standpipes; the pressure is the same at each point in the horizontal tube — no pressure difference, no driving force, no motion. When the stopcock is open, the fluid flows out of the tube, and the different levels in each standpipe indicate different pressures at each point along the horizontal. The pressure falls uniformly from left to right along the horizontal. In the figure below, a partially open stopcock is placed closer to the left, where it obstructs but does not stop the flow. Now fluid piles up behind the stopcock until the pressure difference across the stopcock becomes large enough to maintain the flow despite the obstruction. The pressure still falls from left to right, but the fall is no longer uniform. The greatest decrease in pressure occurs across the obstructing stopcock. When we examine the pressures in the circulation, we find that the greatest fall in pressure occurs across those terminal arteries, the *arterioles*, that enter the capillaries. In the blood circulation, the arterioles offer the greatest resistance to flow.

The idea of frictional *resistance* can be made quantitative. For any given vessel or system of vessels, we simply divide the pressure difference between any two points by the flow; the quotient is defined as resistance between those points. (Think of pressure difference as "cost" and flow as "payoff.") It follows that:

flow = pressure difference ÷ resistance.

Resistance is very sensitive to the radius of a tube — much more so than to its length. For a given difference in pressure, doubling the radius of a tube will increase the flow (decrease the resistance) sixteen times! The flow is proportional to the radius raised to the fourth power.

Examination of the pressure in different parts of the circulation shows that the greatest drop occurs across the arterioles. Because the flow is the same through all sections of the circulatory tree, that section with the greatest pressure drop, the arterioles, has the most resistance to flow. (This follows from the above equation.) In other words, the arterioles are the bottleneck, or rate-limiting step, in circulation. Bottlenecks are strategic places for regulation, and the arterioles appear to be a chief site for regulation of both blood pressure and flow to specific tissues. This is accomplished by smooth muscles that are wrapped circularly around the walls of the arterioles. These muscles are controlled by nerves and hormones. When the muscles relax, the radius increases; when they contract, the radius decreases. Remember that the resistance of a tube is very sensitive to its radius. By controlling the radii of arterioles, the body exercises tight control over the flows and pressures in its own circulation.

CN: Use red for C.
1. Begin at the top working from left to right.
2. Color the next series of diagrams noting the long bars (B¹) that reflect the overall level of pressure.
3. In the large chart below, the pressure line (B¹) represents the mean blood pressure. Note the change (for emphasis) in color of blood at the arteriole level.

MERCURY, FORCE, PRESSURE, FLUID, BLOOD, RESISTANCE.

Pressure is a measure of "push" or force per unit area. Pressures can be measured with a "U" tube half filled with mercury (Hg). The man pushes on the mercury. The harder he pushes, the higher the mercury rises. Its height (mm Hg) measures his push (pressure). Similarly blood pressure can be measured by connecting an artery to a "U" tube; the "push" arises in the heart.

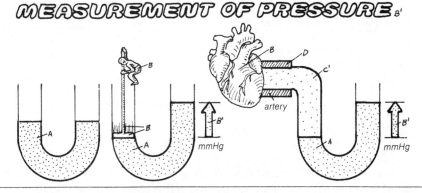

MEASUREMENT OF PRESSURE.

When fluid in a horizontal pipe is not moving, the pressure is the same at each point along the horizontal. We verify this by placing standpipes along the path; the level of fluid rises to the same height in each standpipe (now we use the fluid itself instead of Hg to measure pressure). When we open the stopcock, fluid flows and the pressure falls uniformly along the horizonal. The difference in pressure from point to point pushes the fluid along. When a half open stopcock is placed in the middle of the pipe it resists the flow. Fluid piles up behind the stopcock until the pressure difference across the stopcock becomes large enough to maintain the flow despite the resistance. Pressure still falls from left to right, but the fall is no longer uniform. The greatest fall in pressure occurs across the obstructing stopcock. To maintain flow, fluid must be continually forced through a pipe to overcome frictional resistances (depicted in bottom panel) which oppose the motion.

PRESSURE AND FLOW.

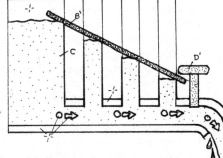

FLOW = PRESSURE DIFFERENCE / RESISTANCE.

Resistance to flow is increased by increasing viscosity, increasing pipe length, and decreasing pipe radius. Resistance is by far most sensitive to changes in radius. Halving the radius decreases flow 16 times!

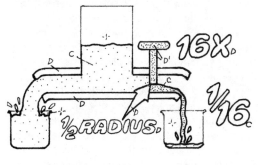

In the circulation, flow is constant in each total cross-section, and the largest fall in pressure occurs in the arterioles. This means that arterioles offer the greatest resistance to flow; they are "the bottleneck". Arterioles act like stopcocks. Contraction of smooth muscles in the arteriole walls change the vessel radius and alter resistance.

ARTERIOLES

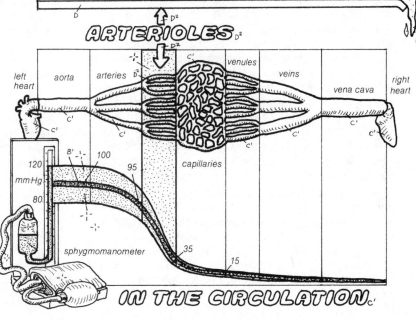

IN THE CIRCULATION.

Resistance to flow occurs because of frictional forces which oppose the motion of two layers of fluid sliding past each other. In a blood vessel, the blood layer immediately adjacent to the vessel wall is held back as it tends to adhere to the stationary wall. This layer retards the next layer which retards the next and so on. This results in telescoping layers of fluid. Fluid in the center of the pipe moves the fastest, fluid at the wall does not move at all. The overall resistance to flow arises from the frictional interactions of these layers as they slide past each other. The drop in pressure that occurs during flow through a vessel reflects the energy lost to these frictional interactions.

LAYERS OF FLUID DURING FLOW.

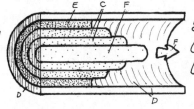

STATIONARY
MOVING
FASTEST

ARTERIAL PRESSURE AND ITS MEASUREMENT

The heart pumps blood intermittently; during *systole*, some 70 mL of blood is thrust into the aorta, but during *diastole*, no blood leaves the heart. Despite this choppy, discontinuous flow of blood through the root of the aorta, blood flows out of the arteries into the capillaries in a smooth and continuous motion. This is possible because the aorta and other arteries are not rigid pipes; instead, they have *elastic walls*, which can passively *expand or recoil*, much like a simple rubber band. During systole, blood enters the arteries faster than it leaves through the capillary beds. Packing more fluid into the arteries tends to increase arterial pressure, which forces the elastic arterial walls to expand the same as a balloon does when more air is forced into it. The excess fluid is taken up by the expanding arteries, and this relieves some of the pressure increase that would have occurred if the walls were more rigid and could not expand as much. In contrast, during diastole, blood still leaves the arteries for the capillaries, but none enters from the heart. At this time, blood stored in the expanded arteries leaves, propelled in part by the recoil of the arterial walls; it is this stored blood that prevents the pressure from falling as low as it would if the arteries were rigid. The elastic arterial walls minimize fluctuations in pressure that would otherwise occur: i.e., they buffer changes in pressure. In a system with rigid walls, the pressure would rise to very high values during systole and fall to near zero during diastole. Similarly, the blood would spurt into the capillary bed with each systole and virtually come to a standstill during diastole. Imagine what would happen to the flow out of a faucet if you intermittently turned the tap on and off. In a healthy arterial system, the arterial pressure fluctuates with each beat, but not nearly as much as in the rigid system. The maintenance of a reasonable pressure level throughout the entire cycle has two advantages: first, it sustains a smooth and continuous flow into the capillaries, and second it relieves the heart of work that would be required to eject blood against the enormous systolic pressure that would develop.

Arterial pressure does pulsate. With each heartbeat, the arterial pressure in a normal young adult varies between 80 and 120 mm Hg. The minimum pressure occurs just at the end of diastole; it is called *diastolic pressure* (80 mm Hg in our example). The maximum occurs midway into systole; it is called *systolic pressure* (120 mm Hg in our example). The difference between systolic and diastolic is called the *pulse pressure* (120 – 80 = 40 mm Hg). From the discussion above,

it follows that a person with more rigid arteries (e.g., an older person) will have a higher pulse pressure.

Rather than dealing with fluctuating pressures, it is sometimes useful to have a single measure that represents the average pressure or driving force within the arterial tree. (Plate 32 shows that this number is approximately the same for any artery.) Simply taking the average between systolic and diastolic pressures ([120 + 80]/2 = 100 mm Hg) is not strictly correct, because inspection of the pressure contour shows that arterial pressure spends more time around diastolic pressure than around systolic pressure. This is taken into account by the *mean pressure* which is represented by the horizontal line in the upper lefthand figure. The position of this line can be determined by the property that it splits the area under the pressure contour into two equal parts, one area lying above the horizontal line drawn at the diastolic pressure level and below the mean pressure line, the other area lying above the mean pressure and contained by the upper parts of the pressure curve. The mean pressure can be approximated by the formula: *mean pressure = diastolic pressure + (1/3) pulse pressure.*

To measure human arterial blood pressure (lower figure), an inflatable rubber bag constrained by a cloth cuff is wrapped around an arm. The bag is inflated and compresses the blood vessels in the arm. It is assumed that the pressure in the bag is transmitted to the arm so that the bag pressure equals the actual pressure within the tissue of the arm. It follows that when the pressure in the bag just exceeds the pressure in the artery, the compression will be sufficient to collapse the artery. The procedure is to inflate the bag above the arterial pressure so that blood flow stops. Air is then released from the bag so that bag pressure falls very slowly. At a certain bag pressure, flow resumes, but only for the short time that arterial pressure is at its maximum. During this time, sounds are produced that can be easily heard with the aid of a stethoscope placed near the artery. The pressure at which the sounds first occur is a measure of systolic pressure. As more and more pressure is released, a point is reached where the sounds become very muffled. This pressure is the diastolic pressure. The sounds arise from the turbulent blood flow through the narrowed (partially collapsed) artery under the cuff, just as sounds arise from the turbulent flow in a stream when it passes through a narrow bed.

CN: Use red for A, dark colors for B, C, and D.
1. Begin with the comparison of rigid versus elastic tubes in the upper right corner.
2. Color the arterial pressure contour on the left.
3. Color the measurement of pressure, working from left to right. Complete each section before going on to the next.

ARTERIAL PRESSURE *

BLOOD A
TUBE B ARTERY B2 RECOIL B2
SYSTOLE C
DIASTOLE D
PULSE PRESSURE E
MEAN PRESSURE F

The heart pumps blood intermittently; during systole blood is thurst into the aorta, but during diastole no blood leaves the heart. In vessels with rigid walls the pressure would rise to very high values during systole and fall to near zero during diastole. Blood would spurt into the capillaries with each systole and almost stop during diastole. But arteries are not rigid pipes, they have elastic walls. During systole, part of the stroke volume is stored by the expanding arteries; during diastole blood stored in the expanded arteries leaves for the capillaries, propelled in part by the recoil of arterial walls. The elastic walls buffer changes in pressure and flow caused by the intermittent heart beat.

RIGID TUBE B

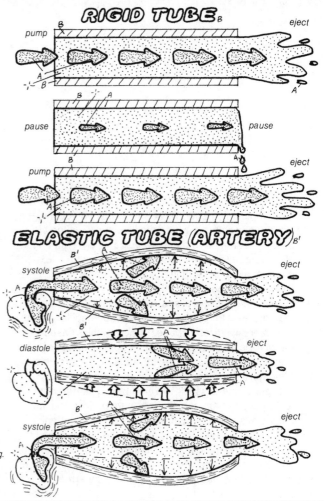

ELASTIC TUBE (ARTERY) B'

ARTERIAL PRESSURE *

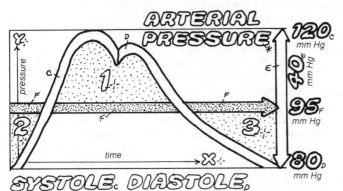

120 C mm Hg
40 E mm Hg
95 F mm Hg
80 D mm Hg

SYSTOLE C. DIASTOLE D

The arterial pressure (y axis) at each moment (time on x axis). With each heartbeat, the arterial pressure in a young adult varies between 80 (diastolic) and 120 (systolic) mm Hg. Pulse pressure = systolic − diastolic pressure = 120 − 80 = 40 mm Hg. The mean pressure, represented by the horizontal line is determined by splitting the area under the pressure curve into two equal parts (areas 1 = 2 + 3) in the illustration.

MEASUREMENT OF PRESSURE *

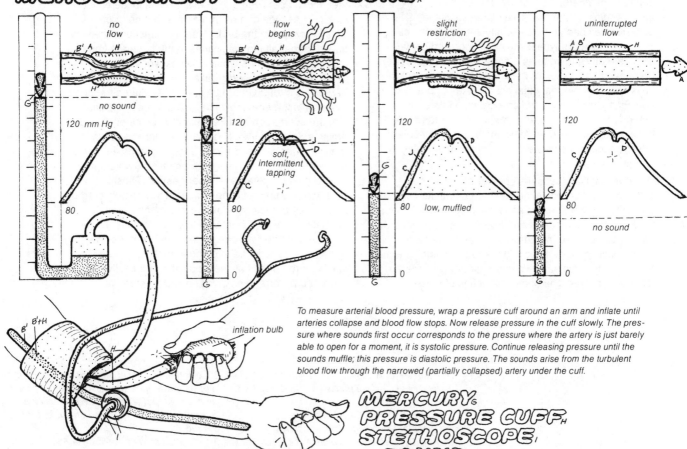

To measure arterial blood pressure, wrap a pressure cuff around an arm and inflate until arteries collapse and blood flow stops. Now release pressure in the cuff slowly. The pressure where sounds first occur corresponds to the pressure where the artery is just barely able to open for a moment, it is systolic pressure. Continue releasing pressure until the sounds muffle; this pressure is diastolic pressure. The sounds arise from the turbulent blood flow through the narrowed (partially collapsed) artery under the cuff.

MERCURY G
PRESSURE CUFF H
STETHOSCOPE I
SOUND J

CAPILLARY STRUCTURE AND SOLUTE DIFFUSION

The amount of blood that fills the *capillary bed* at any moment is only about 5% of the total blood volume. Nevertheless, this is where the "business" of the circulation is transacted. It is the place where exchange of O_2 and nutrients for CO_2 and wastes occurs. Exchange takes place in the capillaries because their walls are composed of only one layer of very thin porous *endothelial cells*, which permit solutes smaller than proteins to rapidly diffuse between capillary blood and interstitial tissue spaces. (The granular *basement membrane* that surrounds each capillary does not present any special barrier to diffusion.) Before entering the capillaries, blood must pass through *arterioles*, the resistance vessels. They range in diameter from 5 to 100 µm and are surrounded by thick, smooth muscular walls that can contract to constrict the arteriole and regulate blood flow delivery to the capillary bed. Blood leaving the capillaries enters the *venules*, which serve as collecting vessels. Their walls are thinner than arterioles, but thicker and much more impermeable than capillaries.

In some tissues, blood goes directly from arteriole to capillary; in others, the blood is delivered to *metarterioles*, which then give rise to capillaries. Metarterioles can serve as supply vessels to the capillaries, or they can bypass the capillaries and convey blood directly into venules. Capillaries that arise as side branches of arterioles or metarterioles have muscle cells around their origin that act as gates or *precapillary sphincters*, the last control on local blood flow before entering the capillary bed. Sometimes a second type of bypass vessel called an *AV shunt* is found. These are direct connections between arterioles and venules that do not give rise to capillaries.

Not all capillary beds are open at any one time. Smooth muscles regulating the microcirculation are controlled by both nerves and local metabolites (chemicals involved in metabolism). Arteriolar smooth muscle has a rich supply of nerves and is less sensitive to metabolites; metarterioles and precapillary sphincters have a poor nerve supply and are largely governed by local metabolites. The combined action of these muscular controls produces an intermittent flow through any particular capillary bed. First one bed opens, then it closes while another opens.

Most solutes diffuse freely through the capillary walls. The concentration of O_2 and nutrients physically dissolved in the blood plasma is higher than in the tissues because they are *consumed* in the tissue; the concentration gradient promotes nutrient diffusion from blood plasma to tissue. In contrast, CO_2 and waste products are constantly *produced* in the tissues; their concentration gradient promotes diffusion from tissue to blood plasma. If they can pass through the capillary walls, there is no need for special transport systems to exchange materials between blood and tissue.

How do solutes get through the capillary walls? The respiratory gases O_2 and CO_2 are lipid soluble so that permeation is no problem; they permeate all cell membranes (including the endothelial cells that make up the capillary walls) with ease. In addition, capillary walls behave as though they contain large pores to permit anything smaller than a protein to pass through. In many tissues like skeletal, cardiac, and smooth muscle, the *junction between endothelial cells* is loose enough to allow passage of most molecules, but not proteins. This is not true in the brain. Here the junctions are very tight and restraining; capillaries in the brain are impermeable to many small molecules as well as protein. This barrier to exchange, called the "blood-brain barrier," is circumvented by special facilitated transport systems in the endothelial cells of the brain capillaries that transport such required nutrients as glucose and amino acids. In contrast, capillary endothelial cells in the intestines, kidneys and endocrine glands are riddled with large "windows" called *fenestrations*, which provide large surface areas for permeation. These fenestrations are not simple holes. They are covered by a thin, very porous, very permeable membrane that permitts passage of relatively large molecules. Finally, capillaries in the liver can be extremely porous. The endothelial cells do not provide a continuous covering, leaving large gaps between cells that are easily traversed by *large molecules, including proteins.*

CN: *Use red for A and a closely related color for red blood cell (G). Use purple for D and blue for F. Use a dark color for C.*
1. *Begin with the large illustration of a capillary bed. Note that in the right half, the capillary network (D) is left uncolored because precapillary sphincters (C) are tightened. Next color the inset.*

2. *Color the diagram of the capillary. Note that the titles for the various arrows shown entering and leaving are listed at the bottom and only the heavily outlined portion of the arrows should be colored.*
3. *Color the capillaries below, starting with continuous and completing it before going on.*

CAPILLARY BED*
ARTERIOLE A
METARTERIOLE B
PRECAPILLARY SPHINCTER C
CAPILLARY D
AV SHUNT E
VENULE F
RED BLOOD CELL G

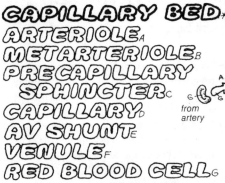

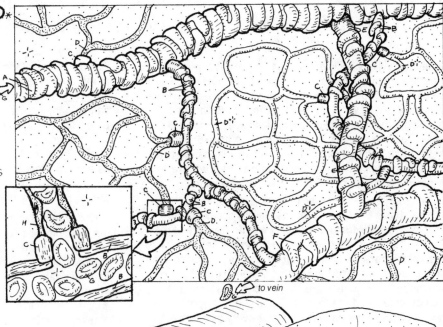

from artery

to vein

In many tissues blood is delivered to metarterioles which serve as supply vessels to capillaries. Capillaries which arise as side branches of either arterioles or metarterioles have muscle cells around their origin which act as gates (precapillary sphincters). If the gates are shut, blood in the metarteriole is conveyed directly to the venule. Sometimes a second type of bypass vessel called an AV shunt is found. These are direct connections between arterioles and venules which do not give rise to capillaries. Note the occurrence of smooth muscle in the different vessels.

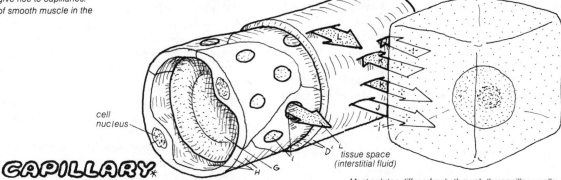

cell nucleus

tissue space (interstitial fluid)

CAPILLARY*
ENDOTHELIAL CELL H
BASEMENT MEMBRANE D'
FENESTRATION I

Most solutes diffuse freely through the capillary walls. The concentration of O_2 and nutrients dissolved in the blood plasma is higher than in the tissues because they are **consumed** in the tissue; the concentration gradient promotes nutrient diffusion from blood plasma to tissue. In contrast, CO_2 and waste products are constantly **produced** in the tissues; their concentration gradient promotes diffusion from the tissue to blood plasma.

TYPES OF CAPILLARIES*

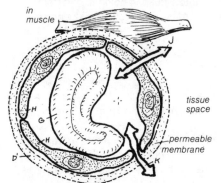

in muscle

tissue space

permeable membrane

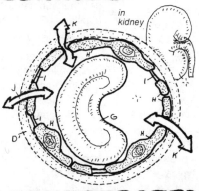

in kidney

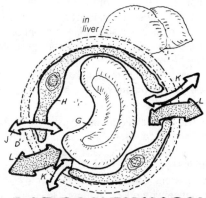

in liver

CONTINUOUS H
In many tissue like skeletal, cardiac, and smooth muscle, the junction between endothelial cells is loose enough to allow passage of most molecules, but not proteins.

FENESTRATED I
Capillary endothelial cells in intestines, kidney, and endocrine glands are riddled with large "windows" called **fenestrations** which provide large surface areas for permeation. These fenestrations are not simple holes. They are covered by a thin, porous, very permeable membrane.

DISCONTINUOUS H
Capillaries in the bone marrow, liver, and spleen can be extremely porous. The endothelial cells do not provide a continuous covering leaving large gaps between cells which are easily traversed by large molecules including proteins.

SOLUTES*
LIPID SOLUBLE MOLECULE (O_2, CO_2) J
WATER SOLUBLE MOLECULE (GLUCOSE, IONS, H_2O) K
MACROMOLECULE (PLASMA PROTEIN) L

FILTRATION & REABSORPTION IN THE CAPILLARIES

Because capillary walls are porous, you might expect fluid to filter through the walls from the lumen of the capillary, where pressure is higher, to tissue spaces, where pressure is lower. Tissues would fill with fluid, swelling and stretching surrounding structures until pressures in tissue spaces build up to a level where they balance the blood pressure and prevent further flow. Normally this does not happen because an additional important force, the *osmotic pressure exerted by plasma proteins*, counteracts the *blood pressure*.

The figure at the top of the plate reviews osmotic pressure, with emphasis on plasma proteins. Strictly speaking, all dissolved solutes contribute to the osmotic pressure of the solution. However, the composition of the solutions on the two sides of the capillary wall (i.e., blood plasma in capillaries and intercellular fluids in tissue spaces) are practically identical, with the major exception that plasma contains large quantities of protein and intercellular spaces have very little. This follows because the capillary pores are very permeable to small molecules and hinder only giant molecules like proteins from passing. The small molecules do not contribute much to water movements because (1) they are almost evenly distributed on both sides of the capillary and (2) in porous membranes, extremely permeable molecules hardly influence osmotic water flow even when they are not uniformly distributed. This simplifies our problem. For osmotic flow across capillaries, we have to consider only the proteins. The osmotic pressure exerted by the plasma proteins is sometimes called *oncotic pressure* and sometimes called *colloid osmotic pressure*. The other important force, *capillary blood pressure*, is often called *hydrostatic pressure*.

Fluid flow through the capillary wall depends on a balance between the hydrostatic (blood pressure) pressure gradient, $\triangle P_{hydro}$, forcing fluid out of the capillary, and the osmotic (oncotic) pressure gradient, $\triangle P_{osm}$, pulling fluid in. (The symbol $\triangle$ represents "difference" or "gradient.") As shown by the top plate, the oncotic pressure of the plasma is approximately 25 mm Hg. Most of this pressure is due to albumin, the smallest, most abundant protein in plasma. Because of its small size, a tiny amount of albumin leaks out into tissue spaces, so that the oncotic pressure of tissue spaces may be approximately 2 mm Hg. The net osmotic gradient drawing fluid into the capillary is 25 - 2 = 23 mm Hg. How does this compare with the hydrostatic pressure gradient pushing fluid out?

The hydrostatic pressure gradient, $\triangle P_{hydro}$, is equal to the difference between the blood pressure within the capillary, P_{cap}, and the hydrostatic pressure in tissue spaces, P_{tiss}. Both of these vary from tissue to tissue, and P_{cap} even varies at different locations within the same capillary. A typical situation where P_{tiss} = 2 mm Hg is illustrated in the plate. At the arterial end of the capillary, $\triangle P_{hydro}$ = P_{cap} - P_{tiss} = 35 - 2 = 33 mm Hg. The oncotic pressure pulling fluid in is only 23 mm Hg. Therefore, a net force of 33 - 23 = 10 mm Hg pushes fluid out into the tissues at the arterial end, of the capillary. Now examine the venous end where blood pressure is lower, about 15 mm Hg. (It has fallen as a result of friction encountered in pushing blood through the narrow capillaries.)

The properties of the tissue space as well as the composition of the plasma remain practically the same. Now we have only $\triangle P_{hydro}$ = 15 - 2 = 13 mm Hg pushing fluid out while the same oncotic pressure gradient of 23 mm Hg pulls fluid in. In other words, at the venous end of the capillary, we have 23 - 13 = 10 mm Hg pulling fluid in! In this capillary, fluid flows out at the arterial end but flows right back in at the venous end; as a result, blood leaves the capillary containing the same amount of fluid as when it entered. The tissue has neither gained nor lost fluid from this capillary.

Not every capillary is as precisely balanced as the one in our example. The figures used for the pressures were rough averages, and there can be considerable variation from vessel to vessel. In one capillary, filtration may dominate while in another nearby capillary, reabsorption prevails. On the whole, in most organs, they nearly balance, and whatever small imbalance exists (usually filtration is slightly larger than reabsorption) is taken care of by the lymphatic vessels (plate 37). These vessels drain tissue spaces of excess fluid together with plasma proteins that leak out of capillaries and return them to the circulation via the large veins. The final result is no net fluid exchange.

Capillary beds in the lungs are a notable exception. Blood pressure in the pulmonary circulation, including lung capillaries, is low. This favors reabsorption of fluid, which is advantageous because it keeps the lungs drained of congesting fluids, which could hamper respiration. In the kidneys, one set of capillaries (glomerular capillaries) has an unusually high blood pressure; they filter large quantities of fluid along their entire length. Another set (peritubular capillaries) has an unusually low blood pressure together with an elevated oncotic pressure; they reabsorb fluid.

Fluid balance can be upset by a number of factors, resulting in the accumulation of large amounts of fluid in tissue spaces, a condition called *edema*. Edema poses a threat because it compromises the circulation by increasing distances that substances have to diffuse in going to and from capillaries. This is particularly true in the lungs; pulmonary edema is serious.

Causes of edema are revealed by a review of factors involved in fluid transfer:

1. *Increase in capillary blood pressure.* This could result from dilation of arterioles or from venous congestion extending back to the capillaries, as may occur in heart failure.

2. *Decreased plasma oncotic pressure.* This can result from starvation (dietary protein deficiency), disturbed protein synthesis, as in liver disease, and loss of plasma protein caused by kidney disease.

3. *Increased permeability of capillary walls to proteins.* This reduces the oncotic pressure gradient across the capillary and decreases reabsorption. Increased capillary permeability occurs in response to allergies, inflammation, and burns.

4. *Disturbance of lymph drainage.* This occurs following obstruction of lymph vessels, and sometimes follows operations.

CN: Use very light colors for A and C, and dark colors for B and D.
1. Begin with the upper diagram, color the background areas of blood plasma (A) and interstitial fluid (C) first, and then color the proteins (D¹) and pressure numbers (B & D) afterwards, with darker colors. Follow this procedure for the other diagrams. Don't color the capillary membrane.
2. Color the large diagram beginning with the three examples above the main drawing.
3. Color the two examples of edema caused by excessive filtration at the bottom.

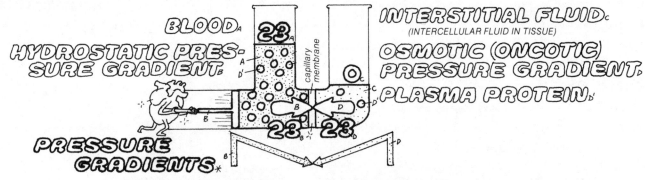

BLOOD

HYDROSTATIC PRESSURE GRADIENT.

PRESSURE GRADIENTS

INTERSTITIAL FLUID
(INTERCELLULAR FLUID IN TISSUE)

OSMOTIC (ONCOTIC) PRESSURE GRADIENT. PLASMA PROTEIN.

The composition of blood plasma and interstitial fluid is almost identical except for the presence of considerable amounts of protein in the plasma and very little in the interstitial fluid. With plasma on one side and interstitial fluid on the other side of a membrane made up of a capillary wall, we find fluid flowing from interstitial fluid to plasma. The flow is due to inequality of plasma proteins in the two fluids. To prevent flow, we have to apply a pressure of 23 mm Hg on the plasma. This is the osmotic gradient due to the proteins (the oncotic gradient). In the body, this pressure is applied by the heart, it is the capillary blood pressure.

FILTRATION AND REABSORPTION IN THE CAPILLARY

Capillary blood pressure is not the same in all parts of a capillary; it is higher at the arterial end and falls continuously through the capillary until reaching the lowest value at the venous end. Keeping books for fluid balance in an average capillary shows at the arterial end: Hydrostatic pressure gradient = blood pressure – tissue pressure = 35 – 2 = 33 mm Hg pushing fluid out (filtration). Oncotic pressure gradient = plasma oncotic pressure – tissue oncotic pressure = 25 – 2 = 23 mm Hg pulling fluid in (reabsorption). The net result at the arterial end is 33 – 23 = 10 mm Hg pushing fluid out. At the venous end, only the capillary pressure has changed. Now it is 15 mm Hg. Repeating the analysis shows a net force of 10 mm Hg pulling fluid in at the venous end.

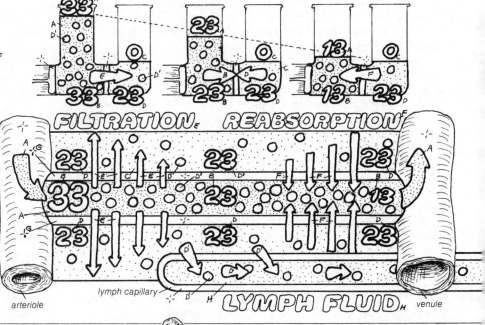

FILTRATION · **REABSORPTION**

arteriole
lymph capillary
LYMPH FLUID
venule

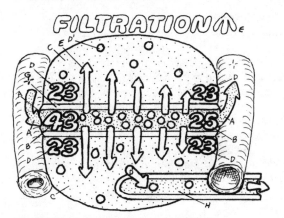

FILTRATION ↑

EDEMA DUE TO CAPILLARY PRESSURE

Edema (swollen tissue) often results from an increased capillary blood pressure. Forces favoring filtration are increased while oncotic pressures may not change very much so that filtration prevails along the entire length of the capillary. This could result from dilation of the arterioles, or from venous congestion extending back to the capillaries as may occur in heart failure.

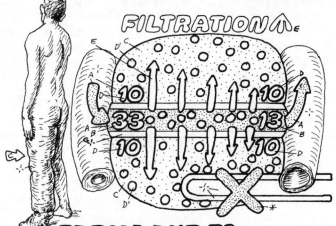

FILTRATION ↑

EDEMA DUE TO HIGH TISSUE PROTEINS

Edema can also result from disturbances of lymph drainage. Lymphatics serve as an overflow system, returning excess tissue fluid and proteins to the circulation. When lymph drainage is poor, proteins accumulate in tissue spaces and compromise the oncotic pressure gradient while the hydrostatic pressure gradient remains unchanged. These disturbances occur following obstruction of lymph vessels that can result from operations. It can also result from parasitic invasion (and obstruction) of the lymph vessels of the limbs, a condition known as elephantiasis.

THE CIRCULATION OF LYMPH

Most tissues contain enormous numbers of tiny lymph vessels called *lymph capillaries*. One end of the lymph capillary is blind (closed off); the open ends coalesce into larger vessels called collecting ducts, which in turn merge into still larger vessels (called *lymph ducts*) and so on until the largest ducts drain into the circulation via connections with large veins (e.g., the subclavians at the junction with the jugular veins). The lymph ducts resemble veins: both have smooth muscle embedded in their walls, and both contain *one-way valves* directed toward the heart. Although both lymphatic and ordinary circulatory capillaries are constructed of similar endothelial cells, there are important differences between them. Lymph capillaries have no basement membranes, and the junctions between their endothelial cells are often open, with no tight intercellular connections. This makes them very permeable to proteins as well as smaller molecules and water. When the tissue spaces fill with fluids, the increased pressure does not compress and close the lymph capillaries because they are held open by *anchoring filaments* attached at one end to the endothelial cells and to surrounding connective tissue at the other end. The edges of the endothelial cells overlap slightly so that they form "flap valves," which allow fluid to enter the lymph capillary but not to leave it.

Lymph flow is propelled through the periodic contractions of the smooth muscle embedded in the walls of the ducts. These contractions, which "milk" the lymph along, occur on the average some two to ten times per minute. One-way flow (out of the tissues and toward the veins) is ensured by the numerous valves that occur every few millimeters. In addition to these contractions, lymph flow is aided by the same factors that promote venous return (plate 37). These include contractions of nearby skeletal muscle compressing lymph vessels (the skeletal muscle pump), periodic changes in intrathoracic pressure associated with breathing (respiratory pump), and possibly the pulsations of nearby arteries. Finally, the high velocity of flow in the veins where the lymph ducts terminate produces a suction that draws lymph toward it.

During each twenty-four hours the heart pumps 8400 L of blood. Of this, 20 L filter out of the capillaries into the tissues; and of this 20 L, some 16 to 18 L are returned to the circulation via capillary reabsorption, leaving 2 to 4 L to be returned by the lymphatic system. Compared to the blood, lymph flows very slowly. The lymphatic system also returns plasma protein that has leaked into the tissue from the capillary. Although this leakage is slow on a time scale of minutes, when viewed over an entire day, the amount of protein returned by the lymphatics is equal to 25 to 50% of all the plasma proteins of the body. Seen from this perspective, the lymphatics seem to be a simple overflow system for correcting fluid imbalance and for recovering protein that capillaries were unable to restrain. However, this perspective is limited; the lymphatic circulation fulfills a real function and is not simply a means of compensating for the apparent capillary inefficiency. Passage of plasma proteins into tissue spaces is an important means of transport for antibodies and for many hormones that are bound to plasma albumin. In addition, the lymphatic circulation provides a path for transport of long chain fatty acids and cholesterol that have been absorbed from the intestine and for the entrance of lymphocytes (a type of white blood cell) into the circulation.

As lymph travels from the tissues toward the veins, it flows through one or more enlarged structures called *lymph nodes*. The lymph nodes contain phagocytic cells, which engulf and destroy foreign matter that is brought to them in the circulating lymph. The nodes also sequester lymphocytes which can transform into antibody-producing plasma cells. Lymph nodes are powerful defense stations, guarding against foreign materials and invading bacteria. When there is a local infection, the regional lymph nodes frequently become inflamed as a result of the accumulation of toxins or bacteria carried to the node by the lymph. The efficiency of this system has been demonstrated by animal experiments in which bacteria have been injected into a lymph duct leading to a node; lymph collected from the duct leading away from the node was virtually free of bacteria.

CN: Use red for A, purple for B, blue for C, and dark colors for E, G, and I.
1. Begin with upper simplified diagram for blood and lymph circulation, including the two enlargements of the fluid exchange process.
2. Color the larger diagram of blood circulation, below, starting with the left heart and ending with the right heart, then color the lymphatic portion beginning with the lymph capillary (D) at the bottom.
3. With the person on the right, color the three areas of superficial lymph nodes (H) and the cut away diagram of the deeper lymph structures. It isn't necessary to color the superficial vessels, drawn in very light lines.

LYMPHATIC SYSTEM*
LYMPH CAPILLARY D
ANCHORING FILAMENT E
LYMPH FLUID F
LYMPH VESSEL F'
VALVE G
LYMPH NODE H
FLUID EXCHANGE I

BLOOD CIRCULATION*
LEFT HEART A
ARTERIES A'
ARTERIOLES A²
CAPILLARIES B
VENULES C
VEINS C'
RIGHT HEART C²

daily flow in Liters

8400*

2-4 F

16-18L

20L

2-4L

During each 24 hours the heart pumps 8400 liters of blood. Of the 20 liters that filter out of the capillaries into the tissues, some 16 to 18 liters are returned to the circulation via capillary reabsorption leaving 2 to 4 liters to be returned by the lymphatic system.

When the tissue spaces fill with fluids, the lymph capillaries are held open by anchoring filaments attached at one end to the endothelial cells and to surrounding connective tissue at the other end. Tissue fluid containing small amounts of plasma protein that has escaped from the blood capillaries freely permeates the gaps between these endothelial cells of the lymph capillary.

The network of lymph vessels returns fluid to the circulation via connections with large veins. Lymph flow is propelled largely through the periodic contractions of the smooth muscle embedded in the walls of the ducts. One-way flow (out of the tissues and toward the veins) is insured by the numerous valves which occur every few mm. As lymph travels through the lymphatics, it flows through lymph nodes which serve as powerful defense stations guarding against foreign materials and invading bacteria.

developing plasma cells

lymphocytes and phagocytes

sinus

cervical nodes

axillary nodes

thoracic duct

thoracic duct

right lymph duct

right heart

left heart

inguinal nodes

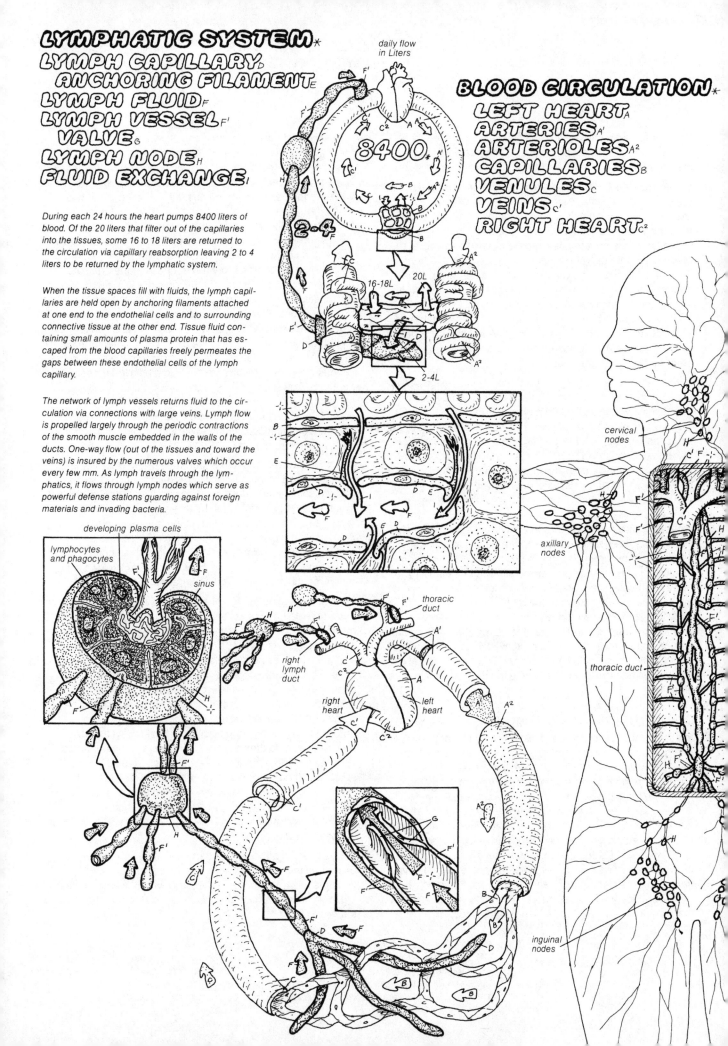

VENOUS STORAGE & RETURN OF BLOOD TO THE HEART

Veins have two main functions: (1) they provide a *low-pressure storage system* for blood and (2) they serve as *low-resistance conduits* to return the blood to the heart.

LOW-PRESSURE STORAGE. The walls of the veins are thin; they contain very little elastic tissue and are difficult to stretch. At first sight, it appears difficult to expand or contract the veins to accommodate changes in blood storage. However, this is not the case. Normally, veins are easily distended because they are partially collapsed. On the other hand, the volume of the veins can be reduced through the contraction of smooth muscle cells embedded in their walls. At rest, veins contain about two-thirds of the blood in the body, although this can vary. In response to hemorrhage or exercise, for example, sympathetic nerves stimulate venous smooth muscles, constricting the veins and shunting blood to other parts of the circulation.

To study the capacity of veins to store blood, take isolated segments of veins, tie off all possible exits, and inflate them with fluid as you would inflate a bicycle tire with air. (See figure at top of plate.) The question is, "How much fluid (air) can you push into the vein (tire) before the pressure in the vein (which will oppose your effort) rises 1 mm Hg?" This amount is called the *compliance*. The larger the compliance, the larger the capacity for storage. You can see from the plate that veins have a much larger compliance than arteries; at normal venous pressures, they are only partially filled and can easily distend. When they are filled far beyond normal, the slack is taken up and the pressure begins to rise very fast; the compliance falls. The easy distensibility (high compliance) of normal veins serves the storage function very well; however, it can create problems. For example, when we rise from the supine to the upright position, blood tends to pool in the veins (especially in the legs and feet); it is essentially withdrawn from the arterial side of the circulation. Without any compensatory response, such as activation of sympathetic nerves, the result would be disastrous. The nature of these challenges and the compensatory responses are covered in plate 41.

LOW-RESISTANCE CONDUITS. Blood flows from regions where its mechanical energy is high to regions where it is low. When we are in a recumbent position, most of this energy is in the form of pressure. As blood passes through the narrow arterioles and capillaries, the pressure falls substantially. In many venules, blood pressure is around 15 mm Hg. In the atria, the average pressure is close to 0 mm Hg. It follows that there is a small but definite *pressure gradient* available to force blood back to the heart. The fact that this small gradient (approximately 15 mm Hg) is sufficient to drive large volumes of blood demonstrates the low resistance of the venous pathway. Even veins that appear to be collapsed have a low resistance because the apparent "creases" in the vessel are never really flat; they always leave some space that can be easily traversed by circulating blood.

In addition to pressure gradients, there are other mechanisms that aid venous return of blood to the heart. These include "pumping actions" of noncardiac muscles as well as movements of the heart itself, and they depend on the *valves* in the veins, which point in the direction of the heart. This orientation ensures a forward flow toward the heart: blood flowing forward forces the valves open; backflow snaps them shut. The third figure on the plate shows this action in a vein lodged between two *skeletal muscles*. When the muscles are relaxed, blood flows forward because of the pressure gradient described above, and the vein fills with blood. When the muscles contract, they squeeze on the vein, forcing blood in all directions. Blood flowing backward closes the bottom valve, but forward-flowing blood keeps the upper valve open so that blood spurts in the forward direction. When the muscle relaxes, there is no longer any external force pushing on the venous walls; the presure gradient from below (farthest from the heart) forces blood flow in the forward direction, opening the lower valve and reestablishing our initial condition. Thus, each time the muscle contracts and relaxes, a spurt of venous blood is sent toward the heart. This action is called the *muscle pump.*

A good illustration of the importance of the muscle pump in exercise is provided if a runner remains motionless just after finishing a strenuous race. His cardiac output is still high and his capillaries and small blood vessels are still dilated in response to the exercise. Without the muscle pump the veins are quickly drained, venous return to the heart decreases, and the cardiac output may falter sufficiently to compromise the blood supply to the brain. Fainting can be avoided if the runner continues mild exercise for a few minutes.

An additional pumping action, *the respiratory pump*, occurs during breathing. Each breath is preceded by an enlargement of the *chest cavity (the thorax)*. The enlarging thorax acts as a bellows and "sucks" air into the lungs (see plate 44). The same expansion also "sucks" blood into the thoracic veins. The thoracic cage not only expands its lateral dimensions (a result of skeletal muscles pulling the rib cage upward), it also expands its vertical dimensions as a result of the dome-shaped *diaphragm* contracting and pushing downward on the *abdomen*. Pushing on the abdominal contents then squeezes the veins in the abdomen. Thus, each time a breath is drawn, the expanding thorax sucks and the compressed abdomen squeezes blood toward the heart. During expiration, the reverse occurs, and, although there are no valves in the great veins of the thorax or abdomen, backflow is checked by valves in the large veins of the extremities.

Finally, the motion of the heart itself aids venous return. Each time the ventricles beat, the upper portions of the ventricles near the valves move downward toward the apex. This results in expansion of the atria, which draws blood into the heart.

CN: Use blue for A, red for C, and purple for D. Use a dark color for H.
1. Begin with the top panel, coloring the diagrams of the circulatory system with a person at rest and then during exercise. Color the graph to the right, comparing the relationship between pressure and volume.
2. Color the remaining panels.

VENOUS STORAGE *

VEINS A
HEART B
ARTERIES C
CAPILLARIES D
SYMPATHETIC NERVE E

Veins provide a low-pressure storage system for blood. At rest the veins contain about 63% of the blood in the body, but this can vary. In response to hemorrhage, or exercise, for example, sympathetic nerves become active and stimulate the smooth muscles which constrict the veins. As a result blood is shunted to arteries and other parts of the circulation.

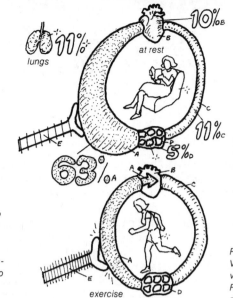

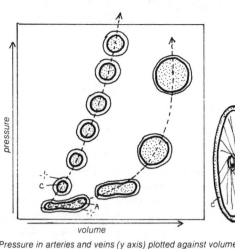

Pressure in arteries and veins (y axis) plotted against volume (x axis). Veins have a much larger storage capactiy than arteries, at normal venous pressures they are only partially filled and can easily distend. Filling them far beyond normal takes up the slack and the pressure rises fast.

VENOUS RETURN *

PRESSURE GRADIENT A'-

There is a small but definite pressure gradient availble to force blood back to the heart. The fact that this small gradient (e.g. 15 mm Hg) is sufficient to drive large volumes of blood to the heart demonstrates the low resistance to the venous pathway.

ARTERIAL PRESSURE C'
VENOUS PRESSURE A'

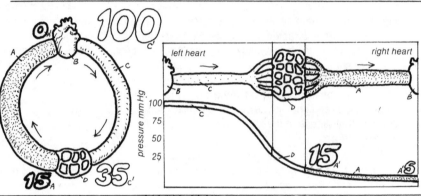

MUSCULAR PUMP G-

When the muscles are relaxed, the venous pressure gradient drives the blood forward toward the heart, filling the vein. When the muscles contract, they squeeze the thin-walled vein, squishing blood in all directions. Blood flowing backwards closes the bottom valve, but the forward-flowing blood keeps the upper valve open. Blood spurts forward toward the heart. A pulsating artery next to the vein may exert the same "milking" action as the muscle.

VALVE F
SKELETAL MUSCLE G

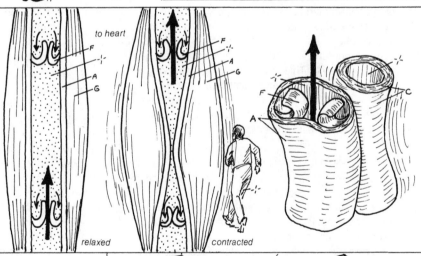

RESPIRATORY PUMP H-

With each breath, the thorax acts as a bellows. During inspiration it "sucks" air into the lungs and blood into the thoracic veins. In addition, the diaphragm pushes on the abdominal contents, squeezing the abdominal veins and forcing blood into the thoracic veins. During expiration the reverse occurs. Backflow is checked by valves in veins of the extremities.

DIAPHRAGM H
CHEST CAVITY & AIR PASSAGES I
ABDOMINAL CAVITY J

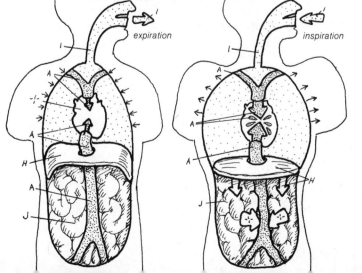

LOCAL & SYSTEMIC CONTROL OF SMALL BLOOD VESSELS

LOCAL REGULATION OF TISSUE PERFUSION. When a tissue becomes active, its blood vessels dilate. Local blood flow increases, bringing the active tissue more nourishment and washing away waste products. This response is clearly beneficial; more blood is supplied to active tissue, less to quiescent tissue. How does it occur? How does the tissue signal its blood vessels that it has become active? A simple experiment provides a clue: a tourniquet is applied to an arm, blocking blood flow for a few minutes, and then released. Following release, local blood vessels dilate and blood flow to the impoverished arm is temporarily much higher than the norm, as if to compensate for the period when flow was obstructed. If extracellular fluids collected from the starved tissues during the blockage are injected into the opposite normal arm, its vessels dilate and its blood flow increases, just as though it too were recovering from the tourniquet. Apparently, the impoverished tissue releases into its surrouding fluids substances that diffuse to the local blood vessels and induce dilation.

A search for these dilating substances has shown that any of the following characteristics of the fluid can cause *dilation* of *arterioles* and *precapillary sphincters*: high concentration of *acids*, CO_2, *potassium*, and *adenosine* and *low concentration of* O_2. A common feature of all these conditions is that they are produced by cells when their blood supply is inadequate to support their activity. This provides a very simple *negative feedback* scheme to match local blood flow to cellular activity: *increased tissue activity* (*inadequate blood supply*)—→*accumulation of metabolites* (*acids, CO_2, low O_2, etc.*)—→*dilation of blood vessels* (*increased blood flow*).

NEURAL REGULATION OF BLOOD FLOW. In addition to local chemical control, smooth muscle in blood vessel walls is also controlled by *sympathetic nerves*. These nerves are generally active, constantly barraging the blood vessels with impulses that liberate *norepinephrine*, causing the vascular smooth muscle to contract and *constrict* the vessels. When the *frequency of sympathetic impulses increases*, blood vessel constriction is more intense; when the frequency decreases,

the vascular smooth muscle is more relaxed, and blood vessels dilate. This description has covered the primary mechanism for neural control of blood vessels. In addition, there appear to be a number of minor pathways that operate on a different basis. Parasympathetic fibers supply blood vessels of the head and viscera, but not skeletal muscle or skin. These fibers release acetylcholine, causing vasodilation. Acetylcholine is also liberated by a small number of vasodilator fibers that are carried in the sympathetic nerve trunks going to skeletal muscle. Their significance is not apparent; they are activated by excitement and apprehension, and it has been suggested that they are involved in the vasodilation that occurs with the anticipation of exercise. They have been found in cats and dogs and are probably present in humans.

The density of sympathetic innervation varies widely from tissue to tissue. Arterioles and veins of the viscera and skin have a rich supply of nerves and show intense vasoconstriction upon sympathetic stimulation. In contrast, blood vessels in the brain and coronary circulation are nonresponsive to sympathetic stimulation. Fortunately, circulation to these two organs is rarely compromised by vasoconstriction; neither the brain nor the heart can sustain O_2 deprivation for any significant amount of time.

While chemical control matches blood flow to metabolic activity, sympathetic vasoconstrictors play a major role in the control of vascular resistance (and therefore blood pressure). Situations in which these two mechanisms oppose each other (e.g., blood pressure falling while an organ has inadequate blood supply) can easily arise. Through reflexes discussed in plate 40, the sympathetic nerves are activated in response to the low blood pressure, causing vasoconstriction. At the same time, the deprived organ starts producing vasodilator substances. Although the net result depends on the particular organ, the vasodilator response most often predominates. In fact, there is evidence that vasodilator substances act not only on blood vessels but also directly on sympathetic nerve endings to inhibit the amount of norepinephrine released by sympathetic impulses.

CN: *Use red for A, purple for C, and dark colors for B and G.*
1. *Begin with the upper panel. Note that the upper capillary (C) whose sphincter (B) is completely contracted, is left uncolored to emphasize the lack of blood flow in that vessel. Color the panel to the right in which both arteriole (A) and precapillary sphincter (B) are dilated by metabolites (E¹). Color the metabolites within the arrows, but not the arrows.*
2. *Color the three states of arteriole constriction below. Then color the two illustrations at the left.*

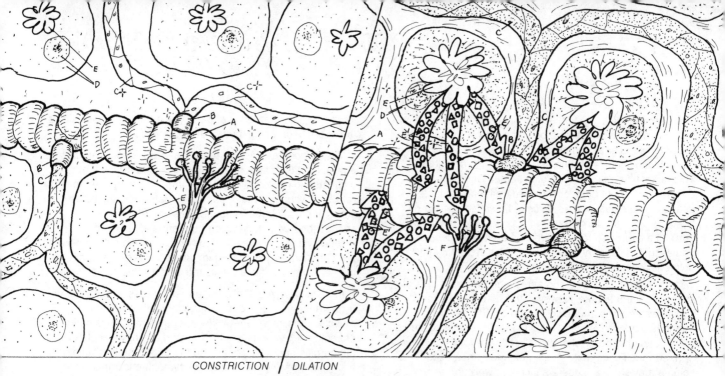

CONSTRICTION | DILATION

ARTERIOLE $_A$
PRECAPILLARY SPHINCTER $_B$
CAPILLARY $_C$
TISSUE CELL $_D$
METABOLISM $_E$
VASODILATING METABOLITES $_{E'}$
SYMPATHETIC NERVE $_F$
NOREPINEPHRINE $_G$
NEUROTRANSMITTER RECEPTOR $_{G'}$

LOCAL (CHEMICAL) REGULATION $_{E'}$

When a tissue becomes active, it releases a number of metabolic products which dilate local arterioles and precapillary sphincters. Local blood flow increases, bringing the active tissue more nourishment and washing away waste products. High concentration of acids, CO_2, K^+ and adenosine, and low concentration of O_2, are all produced by cells when their blood supply is inadequate to support their activity. These substances all dilate blood vessels. This provides a negative feedback to match local blood flow to cellular activity.

Smooth muscle, in blood vessel walls are also controlled by sympathetic nerves. These nerves constantly barrage blood vessels with impulses that liberate norepinepherine, causing the vascular smooth muscle to contract and constrict the vessels. When the frequency of sympathetic impulses increases, blood vessel constriction is more intense; when the frequency is decreased, the muscle is more relaxed and blood vessels dilate.

SYSTEMIC (NEURAL) REGULATION

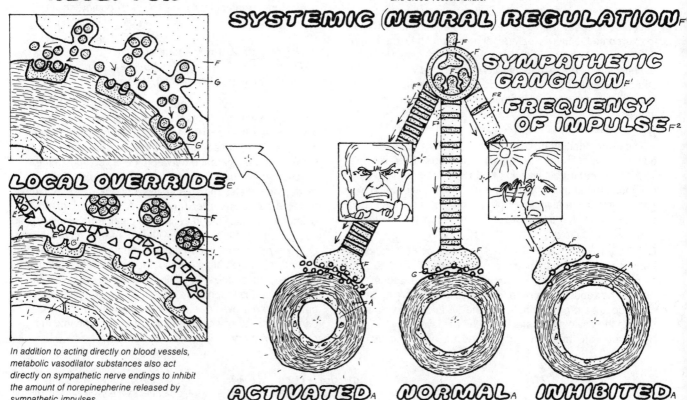

SYMPATHETIC GANGLION $_{F'}$
FREQUENCY OF IMPULSE $_{F^2}$

LOCAL OVERRIDE $_{E'}$

In addition to acting directly on blood vessels, metabolic vasodilator substances also act directly on sympathetic nerve endings to inhibit the amount of norepinepherine released by sympathetic impulses.

ACTIVATED $_A$
(constricted)

NORMAL $_A$

INHIBITED $_A$
(dilated)

CONTROL & MEASUREMENT OF CARDIAC OUTPUT

The cardiovascular system is intricate and complex, yet its function is simple: it moves blood. The most important index of cardiovascular performance, the bottom line, is: "How much blood is moved to the tissues during each minute?" This quantity is called the *cardiac output*. Cardiac output equals the amount of blood expelled from one ventricle during a single beat (*stroke volume*) times the number of beats per minute (*cardiac output = stroke volume × heart rate*). In a steady state, the cardiac output of the left heart equals that of the right heart (flows in the systemic and pulmonary circulations are equal).

In an average sized person at rest, the cardiac output is about 5 L per minute. However, this figure fluctuates; it rises with activity, reaching as high as 25 L per minute during heavy exercise, and even higher in athletes. Cardiac output can change by alterations in either stroke volume or heart rate. During exercise, for example, the stroke volume may show a moderate increase while the heart rate rises about three times. These changes in stroke volume and heart rate are brought about by intracardiac mechanisms (*response of the contractile machinery to stretch*) and to extracardiac mechanisms (action of *sympathetic* and *parasympathetic* nerves).

INTRACARDIAC MECHANISM. As in skeletal muscle contraction, the strength of contraction of heart muscle depends on its length. Under normal resting conditions, the length of an average heart muscle fiber may be only about 20% of its optimum length for maximal force. Stretching the fiber beyond its norm reveals a reserve of additional power for forceful contractions. This response to stretch, called the *Frank-Starling mechanism*, has important implications. If more blood is returned to the heart, the walls of the ventricles are stretched, and the Frank-Starling mechanism ensures that the heart can develop the extra strength required to empty itself. If arterial pressure suddenly rises, the stroke volume will decrease because the ventricle will not have sufficient force to overcome the increased arterial pressure. The extra blood that remains in the heart (the *residual volume*) just following the beat will increase, and this increased blood will help stretch the walls prior to the next beat. Consequently, the force of the next beat will increase, helping the heart meet the increased load imposed by increased arterial pressure. This will increase the stroke volume back toward normal.

The Frank-Starling mechanism is particularly important in adjusting the output of the right and left hearts. If, for example, your right heart output was just 1 mL/min. greater than your left, then after about 15 min. some 1000 mL of fluid would accumulate in the pulmonary circulation. The increased pressure would force fluid out of the capillaries into the lungs, and you would drown!

EXTRACARDIAC MECHANISM. The action of the autonomic cardiac nerves has been considered in plate 30. The *parasympathetic* nerves to the heart are carried in the vagus nerve. The vagus nerve is generally active, discharging a continuous barrage of impulses at the SA and AV nodes and slowing the basic heart rate. When the parasympathetic nerve supply to the heart is interrupted, the heart speeds up.

Increasing the frequency of parasympathetic impulses slows the heart; decreasing the frequency speeds it.

The *sympathetic* nerves to the heart are also continually active, but their effect on rate is opposite to that of the parasympathetic nerves. Sympathetic impulses increase the heart rate, and when this nerve supply is interrupted, the heart slows. Increasing the frequency of sympathetic impulses speeds the heart; decreasing the frequency slows it. Generally, the activities of these two opposing sets of nerves are coordinated; when the sympathetic nerves are excited, the parasympathetic are inhibited, and vice versa.

In addition, sympathetic impulses increase stroke volume by increasing the force of contraction of the ventricular muscle. Thus, there are two independent mechanisms for changing stroke volume: (1) changing the initial length of the cardiac fibers (i.e., changing the end diastolic volume) and (2) increasing the barrage of sympathetic impulses to the ventricular musculature (or, similarly, by releasing catecholamine hormones from the *adrenal medulla*).

THE FICK PRINCIPAL — MEASUREMENT OF CARDIAC OUTPUT. Blood flow through any organ can be measured by a simple application of the conservation of matter known as the *Fick principle*. Applying this to O_2 consumption, this principle relies on the following facts, which hold whenever the organ is in the steady state. During each minute:

1. (O_2/min.) = amount of oxygen consumed
   = oxygen delivered by blood
2. oxygen delivered by blood = amount carried in by artery
   − amount carried away by veins
3. amount carried in by artery = L of blood flowing in (F)
   × the amount in each L $[O_2]$ art.
   = F × $[O_2]$ ven.
4. amount carried out by veins = L of blood flowing out (F)
   × amount in each L $[O_2]$
   = F × $[O_2]$ ven.

Putting steps 1, 2, 3, and 4 together:
   (O_2/min) = F × $[O_2]$ art. − F × $[O_2]$ ven. Solving for F:
   F = (O_2/min)/($[O_2]$ art. − $[O_2]$ ven.)

To measure cardiac output, simply measure the blood flow through the lungs. (This is the flow out of the right heart, which equals the flow into and out of the left heart.) In this case, the O_2 removed from the blood is obtained by measuring the difference between the O_2 content in the inspired air and the O_2 content in the expired air. The measurement also requires analysis of $[O_2]$ in blood samples from the pulmonary artery and vein. A sample representing pulmonary venous blood can be obtained from any systemic *artery*. Pulmonary veins and systemic arteries have the same O_2 content because blood has no opportunity to exchange O_2 until it reaches the systemic capillaries via the left heart and systemic arteries. Obtaining a sample of pulmonary arterial blood is much more difficult. It requires passing a catheter (a narrow, flexible, hollow tube) into a vein and carefully threading it through the right heart and into the pulmonary artery, a nontrivial routine!

CN: Use red for B and blue for G.
1. Begin with the upper panel.
2. Color the heart rate panel.
3. Color the stroke volume panel.
4. Color the elements of cardiac content measurement below. Note carefully the letter labels.

CARDIAC OUTPUT: $HR_B \times SV_C = CO$

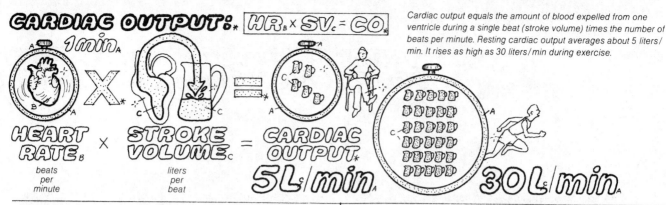

Cardiac output equals the amount of blood expelled from one ventricle during a single beat (stroke volume) times the number of beats per minute. Resting cardiac output averages about 5 liters/min. It rises as high as 30 liters/min during exercise.

HEART RATE $_B$ × STROKE VOLUME $_C$ = CARDIAC OUTPUT

beats per minute   liters per beat

5 L/min   30 L/min

HEART RATE $_B$

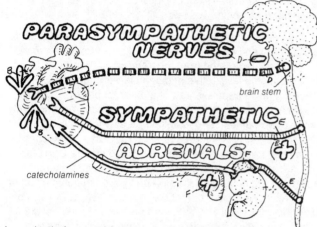

PARASYMPATHETIC NERVES

brain stem

SYMPATHETIC $_E$

ADRENALS $_F$

catecholamines

spinal cord

Increasing the frequency of vagus (parasympathetic) impulses slows the heart, decreasing the frequency speeds it. When the parasympathetic nerve supply is interrupted, the heart speeds up. Effects of sympathetic nerves on heart rate are the opposite. Increasing the frequency of sympathetic impulses speeds the heart, decreasing the frequency (or cutting the nerves) slows it.

STROKE VOLUME $_C$
VENTRICULAR FILLING PRESSURE $_G$
SYMPATHETIC INFLUENCE $_E$

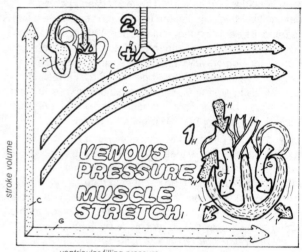

VENOUS PRESSURE MUSCLE STRETCH

stroke volume

ventricular filling pressure

There are two independent mechanisms for increasing stroke volume: 1. Stretch the initial length of the cardiac fibers (e.g., increasing the venous pressure that forces blood into the relaxed heart) and 2. Increase the barrage of sympathetic impulses to the ventricular musculature. The plot shows how the stroke volume increases with increased ventricular filling pressure. When the sympathetic nerves are more active, strength of contraction increases and a new plot (upper curve) is obtained.

MEASURING CARDIAC OUTPUT
(THE FICK PRINCIPLE)*

LITER $_C$ OXYGEN $_J$

COMING IN

PULMONARY ARTERY $_K$ PULMONARY VEIN $_L$

The problem is to calculate the number of liters flowing past the broken line per min. Each liter coming in contains 3 circles (oxygen), but each one leaving contains 4 circles so that there is a discrepancy of 2 for every liter that flows. How many liters must flow to account for the amount (10 circles) entering via the lungs during each minute?

$$\frac{10 \text{ (entering lungs per min)}}{2 \text{ (discrepancy per liter)}} = 5 \text{ liters per min}$$

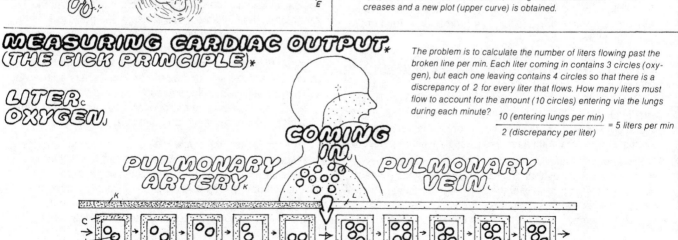

ONE MIN.   ONE MIN.

COMING IN →   GOING OUT →

$$\text{CARDIAC OUTPUT} = \frac{O_2/min}{[O_2] - [O]} = \frac{10}{[4] - [2]} = 5 L/min$$

BARORECEPTOR REFLEXES & CONTROL OF BLOOD PRESSURE

A tissue can get its nutrients only from its blood supply. During activity, it consumes more nutrients, and it will be able to sustain the increased activity only if it receives more blood. Plate 38 illustrated how each tissue was able to regulate its own perfusion (i.e., the amount of blood flowing through it) to satisfy its metabolic requirements. When metabolism increases, its products, potent dilators in the microcirculation, accumulate. The result is an opening of local capillary beds so that the tissue receives more blood. The opposite occurs during quiescence. This scheme will work only if there is a reasonably high pressure in the arteries; opening or widening blood vessels hardly helps if there is no pressure head to propel the blood. Further, the blood supply to a particular tissue can increase only by compromising the blood supply to other tissues or by increasing the cardiac output, or both. There are no alternatives.

How does the heart "know" to speed up or to increase its stroke volume? How does smooth muscle in the arterioles of quiescent tissue "know" to contract and constrict vessels so that blood can be shunted to more active areas where it is needed? Somehow the nervous system must be involved, but how does it "know"? The missing piece to this puzzle is provided by the arterial pressure. Recall that:

Pressure difference = flow × resistance.

Applying this to the entire circulation, (top illustration in plate) the flow is simply the cardiac output. If we neglect the small pressure in the blood just prior to its entering the ventricles, we can equate the pressure difference (arteries – right atrium) with pressure in the arteries. Finally, recall that the primary bottleneck or resistance in the vascular tree is in the arterioles, so that the approximate formula becomes:

Arterial pressure = cardiac output × arteriolar resistance.

Although only an approximation, this expression is fundamental to our interpretations.

When a tissue becomes active, metabolic products accumulate and dilate the local microcirculation, which reduces the resistance. Looking at our formula, we should expect the reduced resistance to produce a decrease in arterial pressure. This does not happen (or at least the drop in pressure is minimized) because the body has a number of mechanisms to maintain a relatively constant blood pressure.

The primary control over sudden changes in blood pressure involves reflexes that originate in special areas (called baroreceptors) in the walls of the aortic arch and the internal carotid arteries. Receptors in these areas are sensitive to stretch. At normal pressure, the walls are stretched, and the receptors are active, sending impulses via sensory nerves to centers in the brain that are responsible for coordinating information and regulating the cardiovascular system. These cardiovascular centers control the autonomic nerve supply to the heart and blood vessels.

When arterial pressure drops, arterial walls are subjected to less stretch, and the sensory nerves coming from the carotid sinus (sinus nerve) and from the aortic arch (depressor nerve) become less active and send fewer impulses. Upon receiving fewer impulses from the baroreceptors (signaling the fall in pressure), the cardiovascular centers respond by exciting sympathetic and inhibiting parasympathetic nerves. This results in (1) an increased heart rate, (2) an increased strength of contraction (stroke volume) so that cardiac output increases, (3) a general increased constriction of arterioles (but not in brain or heart), and (4) an increased constriction of veins. All these factors contribute to a compensatory raising of the blood pressure back toward normal. Factors 1 and 2 both act to raise the cardiac output (flow), factor 3 raises the resistance, and factor 4 raises venous return to the heart as it redistributes blood, shifting it from the venous reservoir to the arterial side of the circulation. When the pressure rises, just the reverse occurs (see plate).

The cardiovascular centers receive detailed information about the high-pressure (arterial) side of the circulation. Careful study of the nerve impulse patterns on the sinus and depressor nerves show that the baroreceptors respond not only to the actual pressure in the carotid sinus and aortic arch, but also to the rate of change of that pressure. It appears that the signals (patterns of nerve impulses) sent to the cardiovascular centers contain information about the mean pressure, the steepness of rise of the pulse curve, the pulse pressure, and the heart rate. In addition, cardiovascular centers receive information regarding the low-pressure side of the circulation. Baroreceptors similar to those of the arteries are also found in the atria and pulmonary arteries, but their significance in rapid regulation of blood pressure is not clear. (They appear to be more involved in slower, long-term regulation of blood volume and pressure — see plate 42).

The cardiovascular centers are also influenced by higher brain centers. The hypothalamus sends impulses that are connected with vascular responses to temperature regulation, defense, and rage. Examples of influence from the cerebral cortex appear in the vascular responses of fainting at the sight of blood and blushing from embarrassment.

CN: Use red for A and dark colors for H and I.
1. Begin with the upper panel, coloring each example completely before going on to the next line of illustrations.
2. Color the large illustration beginning with (1) in the aortic arch above the heart. Note that the sympathetic nerves (I) are represented by dotted lines to indicate that they are not active. Although the parasympathetic nerves (H) are active, they are represented by a broken line to emphasize that they exert an inhibiting effect on the heart.
3. Color the summary diagram below.

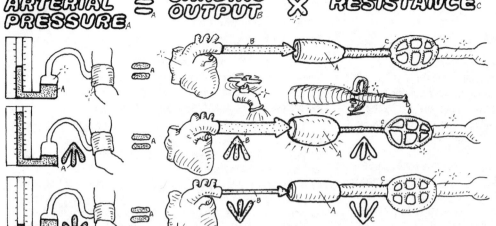

In the circulation, total flow is equated to cardiac output, driving force for this flow is equated to arterial pressure and resistance is attributed to the arterioles. The arterial pressure is given by the product of cardiac output × arteriolar resistance. It is easy to see how an increase of either cardiac output or arteriolar resistance will cause an increased arterial pressure. Imagine pinching a garden hose. Turning more water on causes more water to pile up behind the pinch and raises the force or pressure. Pinching harder causes a similar pileup. Decreasing either cardiac output or resistance decreases pressure.

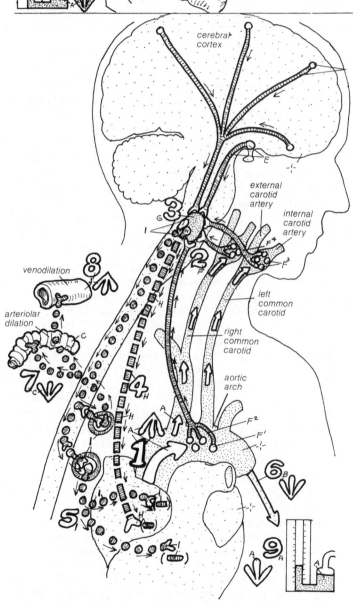

HIGHER BRAIN CENTERS_D HYPOTHALAMIC THERMOREGULATION CENTER_E

When aterial pressure rises (1), arterial walls are stretched and the sensory nerves coming from the carotid sinus (sinus nerve) and from the aortic arch (depressor nerve) become more active, and send more impulses (2). Upon receiving more impulses from the baroreceptors (signaling the rise in pressure), the cardiovascular centers (3) respond by exciting parasympathetic (4), and inhibiting sympathetic (5) nerves. This results in (6) a decreased cardiac output (via decreased heart rate and stroke volume), (7) a general decreased constriction of arterioles, (8) a decreased constriction of veins. All of these factors contribute to a compensatory lowering of the blood pressure back toward normal.

1 BLOOD PRESSURE RISE_A
2 BARORECEPTORS_F–
AORTIC ARCH_{F¹}
DEPRESSOR NERVE_{F²}
CAROTID SINUS_{F³}
SINUS NERVE_{F⁴}
3 MEDULLA CARDIO-VASCULAR CENTER_G
4 PARASYMPATHETIC NERVE (VAGUS)_H
5 SYMPATHETIC NERVE_I
6 CARDIAC OUTPUT_B
7 PERIPHERAL RESISTANCE_C
8 VENOUS RESERVOIR_J
9 BLOOD PRESSURE DROP

A summary of the baroreceptor reflex (described above) emphasizing its negative feedback character. Begin on the left and follow the two loops that feedback to compensate for the disturbed (in this case increased) pressure.

HEMORRHAGE AND POSTURE

The response to a sudden loss of blood provides a good example of regulation in the cardiovascular system. The "leak" could occur in the arteries or in the veins. When blood is withdrawn from the arteries faster than it is replaced by the heart, the mean arterial pressure falls. When blood is suddenly withdrawn from the veins faster than it is replaced by capillary flow, the mean venous pressure falls. A drop in venous pressure results in a decreased venous return, and this decreases cardiac output, which in turn decreases mean arterial pressure. In both types of leak, the mean arterial pressure falls unless some regulatory compensation occurs. This fall in arterial pressure is rapidly offset by the regulatory mechanisms described in plate 40. The following familiar sequence is set into motion: *decreased arterial pressure → reduced stretch of baroreceptors in aortic arch and carotid sinus → reduced frequency of nerve impulses traveling on sensory nerves (vagus and glossopharyngeal) to the cardiovascular centers in the medulla → inhibition of parasympathetic and activation of sympathetic nerves.*

Parasympathetic nerves (vagus nerve in this case) normally slow the heart; inhibiting them will speed the heart. Activating the sympathetic nerves also speeds the heart and makes it beat more forcefully. In addition, the sympathetic nerves cause intense constriction of the arterioles (raising resistance) and veins (reducing the volume of the vascular tree). All these responses occur within moments after blood is lost, and all tend to raise the arterial pressure back toward normal.

Constriction of blood vessels by sympathetic nerves is particularly interesting. Constriction of the veins decreases the proportion of blood held in the veins, shifting it to the arteries; the total vascular volume is decreased so that less blood is required to fill the system. Constriction of the arterioles increases peripheral resistance to raise blood pressure. It also diminishes flow into the capillary beds so that blood pressure in the capillaries falls. Fluid balance across the capillary walls (plate 35) is upset, and fluid filters from tissues into capillaries. After several minutes, the total amount of fluid transferred becomes significant; it helps replace blood lost during the hemorrhage. Of course, this tissue fluid is not blood; it doesn't contain plasma proteins or blood cells, and as a result, it dilutes the plasma proteins and the blood cells.

Constriction of the blood vessels is most intense in organs like the skin, kidneys, and liver, but hardly occurs at all in the brain, heart, or lungs under these conditions. Nourishment is maintained in organs whose moment-to-moment performance is essential. Should the intense vasoconstriction persist, or more blood be lost, dire consequences called circulatory shock may result. When the oxygen supply to any organ is inadequate, metabolic acids, which accumulate and impair organ function, are produced. Tissue damage can occur, vasodilator substances are released, and capillary walls may become leaky, allowing protein to leak into tissue spaces. Vasodilator substances expand the vascular tree, pooling blood in tissues and veins and thus reducing venous return, cardiac output, and arterial pressure. Loss of plasma protein into tissue spaces again upsets fluid balance across the capillary walls, but now in the direction from capillary to tissues. Thus, fluid is *lost* from the vascular tree, and the blood becomes more viscous, sludges, and eventually may even stop as a result of intravascular coagulation.

The simple act of changing from a recumbent to an upright position presents some of the same challenges as a hemorrhage! A new force, gravity, must be accounted for. When we are in a recumbent position the effect of gravity is insignificant. When we are in a standing position, the weight of our blood becomes important; blood in a capillary in a foot, for example, may have to support the weight of the column of fluid contained within the veins reaching all the way, several feet, from the foot to the heart. The pressure on a fluid particle within that capillary will rise to the same level it would experience at the bottom of a water tank filled to the same height (several feet). It is important to realize that this does not directly influence flow within the closed circulatory system. This follows because the increase in pressure on the particle tending to push it upward, where the pressure is less, is just counterbalanced by the weight of the column of fluid. Thus, the forces operative in propelling the blood in a recumbent subject are not disturbed. However, the increased pressures due to gravity are significant because they redistribute fluids in two ways: (1) Veins are more extensible than arteries. As shown in the figure, the increased pressure expands the venous system, and blood pools in the systemic veins. (As much as 600 mL may pool in the lower extremities upon quiet standing.) (2) The high hydrostatic pressures in the capillaries force fluid out of the capillaries into the tissue spaces.

Because of venous pooling, the sudden change in position from recumbent to upright resembles hemorrhage — the subject bleeds into his own vascular system. The same compensatory responses (activation of sympathetics, inhibition of the parasympathetics) occurs. However, in contrast to hemorrhage, filtration of fluid from capillaries to tissues occurs. Venous pooling and edema can be counteracted by moving about (plate 37). Contracting muscles compress veins and lymph vessels to help empty them and temporarily relieve local venous pressures. Valves close, preventing back flow and supporting the weight of blood above them until the vein refills with blood from the capillaries. This provides temporary relief from the high hydrostatic capillary pressure and begins to alleviate the edema.

About 10% of the human population find it difficult to stand up after lying down for a long time. They experience dizziness, impaired vision, and buzzing in the ears, all signs of the inadequate cerebral circulation that arises from the drop in blood pressure following the sudden change to the upright position. In more severe cases, fainting may occur (a fortunate response in this case, because it restores the recumbent position and relieves the stress). Similar reactions may occur even in healthy persons, especially when blood vessels in the skin or muscles are dilated due to heat or exercise. In these cases, the regulatory responses may fail because the intense demands of heat regulation and metabolism have priority.

CN: *Use red for A and dark colors for G and H.*
1. Begin with the upper right diagram and the systemic response to hemorrhage, directly beneath it.

Then color the diagrams to the left.
2. Color the four examples below, completing each left to right.

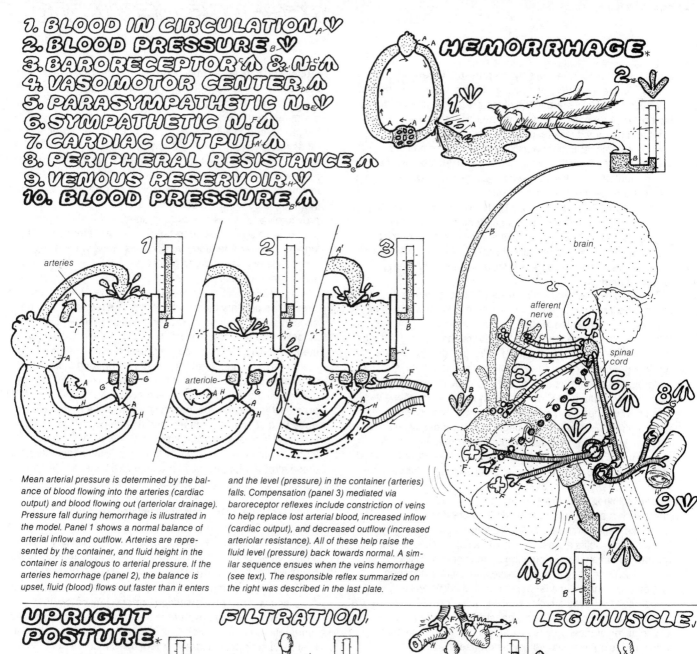

1. BLOOD IN CIRCULATION.ᴀ ⱱ
2. BLOOD PRESSURE.ʙ ⱱ
3. BARORECEPTOR ᴄ ⋀ & N.ᴄ'⋀
4. VASOMOTOR CENTER.ᴅ ⋀
5. PARASYMPATHETIC N.ᴇ ⱱ
6. SYMPATHETIC N.ꜰ ⋀
7. CARDIAC OUTPUT.ᴀ' ⋀
8. PERIPHERAL RESISTANCE.ɢ ⋀
9. VENOUS RESERVOIR.ʜ ⱱ
10. BLOOD PRESSURE.ʙ ⋀

HEMORRHAGE.*

arteries

brain

afferent nerve

spinal cord

Mean arterial pressure is determined by the balance of blood flowing into the arteries (cardiac output) and blood flowing out (arteriolar drainage). Pressure fall during hemorrhage is illustrated in the model. Panel 1 shows a normal balance of arterial inflow and outflow. Arteries are represented by the container, and fluid height in the container is analogous to arterial pressure. If the arteries hemorrhage (panel 2), the balance is upset, fluid (blood) flows out faster than it enters

and the level (pressure) in the container (arteries) falls. Compensation (panel 3) mediated via baroreceptor reflexes include constriction of veins to help replace lost arterial blood, increased inflow (cardiac output), and decreased outflow (increased arteriolar resistance). All of these help raise the fluid level (pressure) back towards normal. A similar sequence ensues when the veins hemorrhage (see text). The responsible reflex summarized on the right was described in the last plate.

arteriole

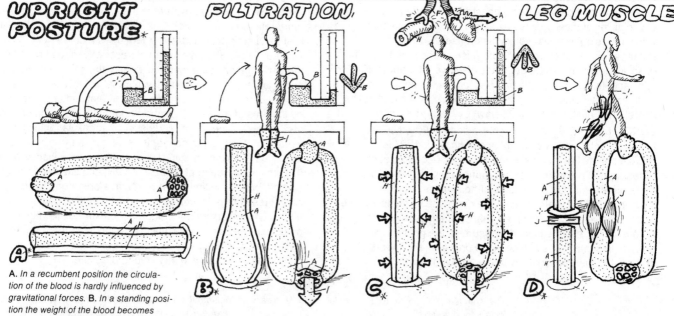

UPRIGHT POSTURE.*

FILTRATION.

LEG MUSCLE.

A. In a recumbent position the circulation of the blood is hardly influenced by gravitational forces. **B.** In a standing position the weight of the blood becomes important. It increases pressures below the level of the heart, particularly in the lower parts of the body. The increased pressures redistribute fluid in two ways: 1. They expand the venous system and blood pools in systemic veins.

2. They force fluids out of capillaries into tissue spaces. Because of venous pooling, the sudden change in position from recumbent to upright resembles hemorrhage — the subject bleeds into his own vascular system. **C.** The same compensatory response (activation of sympathetics, inhibition of the parasympathetics) occurs. However,

in contrast to hemorrhage, we now have filtration of fluid from capillaries. The extremities become edematous. **D.** Venous pooling and edema are counteracted by moving about. Contracting muscles compress veins and lymph vessels to help empty them and temporarily relieve local venous pressures.

BLOOD PRESSURE REGULATORS

Regulation of blood pressure by the pressure receptors described in plate 40 is very effective and rapid. It does not persist, however, and if the alteration in blood pressure persists for hours the pressure receptor mechanism "adapts" to the new conditions and becomes less responsive. Fortunately, there are a number of mechanisms, classified as rapid, intermediate, and long-term regulators, which help stabilize blood pressure. Rapid regulators work within seconds; *arterial pressure receptor reflexes* are the most important example of these.

INTERMEDIATE-TERM REGULATION. These regulators begin in a matter of minutes following a sudden change in pressure, but they are not fully effective until hours later. We include three mechanisms under this classification: (1) transcapillary volume shifts, (2) vascular stress relaxation, and (3) the renin-angiotensin mechanism.

Transcapillary volume shifts (plate 35) occur when capillary blood pressure rises. If capillary pressure is high, fluid leaves the vascular tree which tends to lower the blood pressure. When capillary pressure is low, the reverse happens. Through this mechanism, extracellular fluid in tissue spaces forms a reserve pool of fluid that is available to the vascular system.

Vascular stress relaxation refers to a peculiar property of blood vessels that is well developed in veins. When these vessels are stretched by increased pressure, they very slowly expand so that the pressure becomes correspondingly less. Conversely, when the intravascular volume decreases, the opposite occurs. The net effect is to return pressures toward normal after some 10-60 min. following a change in vascular volume.

The renin-angiotensin system (plate 66) is activated whenever blood flow through the kidneys is reduced, as would occur with a sharp drop in arterial blood pressure. The response begins with secretion of the hormone *renin* by the kidney. Renin splits a plasma protein called *angiotensinogen* (produced in the liver), producing a small peptide called *angiotensin I*. A converting enzyme present in plasma changes angiotensin I into a smaller peptide called *angiotensin II*, which gives rise to an even smaller peptide, *angiotensin III*. Angiotensins II and III cause intense constriction of arterioles, raising vascular resistance. To a lesser extent, they also constrict veins, reducing vascular volume. Increased vascular resistance and reduced volume both raise arterial pressure. The renin-angiotensin system becomes effective after about 20 min., and its effects can persist for a long time. More importantly, angiotensins II and III also stimulate *thirst* as well as the secretion of *aldosterone* (see below) by the *adrenal cortex*.

LONG-TERM REGULATION. This is accomplished by the *kidney*, which regulates the *volume of body fluids*. This volume represents the balance between fluid intake and excretion. When arterial pressure rises, the kidney responds by excreting more urine, reducing the volume of body fluids (including both plasma and interstitial volume). The diminished plasma volume decreases venous return to the heart, reducing the cardiac output so that elevated blood pressure is brought back toward normal. In some forms of high blood pressure, this mechanism is exploited therapeutically by the administration of diuretic

drugs, which increase urine excretion. A decrease in blood pressure elicits the opposite response; urine excretion is decreased. These long-term responses to disturbances are mediated by two hormones, aldosterone and ADH (vasopressin).

Aldosterone (plates 65, 66) is secreted by the adrenal cortex in response to angiotensins II and III. It acts on kidney tubules to retain sodium that would have been excreted in the urine. In these sections of the kidney, water follows the sodium, maintaining osmotic equilibrium. The result is water retention: i.e., an increase in body (and blood) fluid volume. A drop in arterial pressure→renin secretion→angiotensins II and III production→aldosterone secretion→sodium retention by the kidney→water retention→increased blood volume → compensatory rise in blood pressure.

ADH (anti-diuretic hormone) is produced in the *hypothalamus* (plates 62, 65). It travels through nerve fibers to storage sites in the pituitary gland, from which it is released into the circulation. This hormone acts on the kidney (independently of aldosterone) to promote water retention. When blood volume is markedly higher, the resulting increased venous return stretches the atria. *Stretch receptors* embedded in the *atrial walls* are stimulated, sending to the hypothalamus impulses that inhibit the formation and secretion of ADH. With less ADH present, there is more urine excreted (less water retention). Body fluid volume decreases, and this helps compensate for the initial increase in blood volume. Conversely, a decrease in stretch of the atrial walls will withdraw any inhibitory effects of these stretch receptors and promote ADH release. Because blood volume is closely related to blood pressure, it is not surprising to find that regulating blood volume often regulates blood pressure. ADH is also called vasopressin because, in high concentrations, it causes strong vasoconstriction. Recent evidence suggests that these concentrations occur in cases where blood pressure falls markedly, so that in addition to its effect on the kidney ADH may act directly on blood vessels.

Over the years, many physiologists have speculated about an unknown *sodium-regulating hormone*. Animals with excessive body fluids excrete in the urine a substance called *natriuretic hormone* which inhibits Na^+ transport in epithelial tissues similar to the kidney. Where does it come from? What is it doing? Many attempts, using extracts of various tissues like the pineal gland, were made to identify the substance and its origin, but they failed. Finally, within the last few years, the mysterious hormone, a peptide, has been identified. It is secreted by the atria of the heart! The heart not only pumps blood; it is also an endocrine gland. Whenever extracellular fluid volume is expanded, the plasma concentrtion of this hormone increases and causes an increase in Na^+ excretion by the kidney. It also inhibits the secretion of renin and ADH, and it desensitizes the adrenal cortex to stimuli that increase aldosterone secretion. All of these promote water excretion, helping to compensate for the original disturbance (increased extracellular fluid). Study of this elusive hormone is finally possible, and there are high hopes that it will be a key to puzzles associated with regulation of normal, and causes of high blood pressure.

CN: Use red for A, purple for E, and blue for G. Use dark colors for K and N.
*1. Begin with the short term regulation column and work your way down. For consistency, all the examples on this page (with the exception of the final one) show the response to a **decrease** in blood pressure. The bottom right illustration (natriuretic hormone) shows the response to an **increase** in pressure.*
2. Color the intermediate regulators. At the bottom of this

column the adrenal cortex is shown secreting aldosterone (K). Color that arrow as it leads to an action in the third column. Color the title, aldosterone (K) but then go to the top of the column and work downwards.
3. In long-term regulators (response 1), atrial receptors are shown responding to a drop in blood pressure by shutting off afferent nerves that normally have an inhibiting effect on the neurosecretory cells of the hypothalamus (i.e., the cells are released from inhibition).

SHORT-TERM REGULATORS*

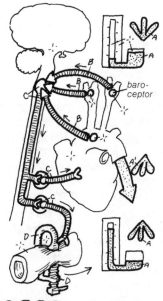

BLOOD PRESSURE.
SENSORY N.
AUTONOMIC N.
BLOOD FLOW.
VASOCON-STRICTION.

baroceptor

Short-term changes in blood pressure are buffered by arterial pressoreceptor (baroreceptor) reflexes. These work within a few seconds but they do not persist and are ineffective in regulating chronic changes.

Intermediate regulators begin in minutes but are not fully effective until hours later. These include 1. transcapillary volume shifts in responses to changes in capillary blood pressure 2. vascular stress relaxation, the slow constriction (expansion) of blood vessels in response to decreased (increased) stretch, and 3. secretion of renin in response to decreased blood flow to the kidney. Renin produces angiotensins from plasma proteins which cause intense vasoconstriction, raising vascular resistance and blood pressure.

Long-term regulation is accomplished by the kidney. It regulates body fluid volume by tipping the balance between fluid intake and excretion. When arterial pressure rises, the kidney excretes more urine, reducing body fluid volume, decreasing venous return and cardiac output so that the elevated blood pressure is reduced. **A decrease in blood pressure** elicits the opposite response; these responses are mediated by: (1) ADH (secreted by the posterior pituitary in response to a **decrease** in nerve impulses arising from stretch receptors in atrial walls) which reduces water excretion by the kidney, and (2) aldosterone (secreted by the adrenal cortex in response to angiotensins) which promotes Na^+ and water retention by the kidney. (3) In addition, natriuretic hormone, a newly discovered factor secreted by the atrium in response to increased volume also plays a role. This hormone promotes Na^+ (and water) excretion and inhibits renin, ADH, and aldosterone secretion. If pressure and volume decrease, the hormone secretion slows and its water excreting actions diminish.

INTERMEDIATE REGULATORS*

1 TRANS-CAPILLARY SHIFT.

CAPILLARY. BODY FLUID.

2 VASCULAR STRESS RELAXATION.

VENOUS SYSTEM.

3 RENIN: ANGIO-TENSIN SYSTEM.

ANGIOTEN-SINOGEN.

adrenal

liver

kidney

blood circulation

ANGIOTENSIN I.

ANGIOTENSIN II, III.

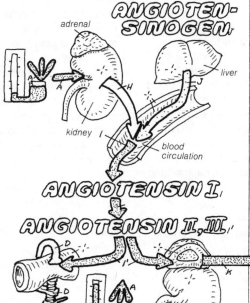

excretion

LONG-TERM REGULATORS*

EFFECT OF BODY FLUID VOLUME.

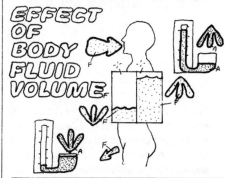

THIRST.

HYPOTHALAMUS.

neurosecretory cell

afferent nerve

posterior pituitary

ATRIAL RECEPTOR.

ADH. 1

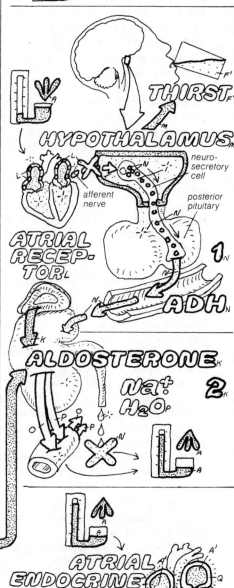

ALDOSTERONE.

Na^+ H_2O 2

ATRIAL ENDOCRINE CELLS.

NATRIURETIC HORMONE. 3

RENIN ADH ALDOSTERONE.

STRUCTURE OF THE RESPIRATORY TRACT

We live at the bottom of a vast sea of air comprised primarily of oxygen and nitrogen. By living in air rather than water, we enjoy surroundings 50 times richer in oxygen. By breathing, we give our body fluids access to this reservoir as they continually exchange both oxygen (O_2) and carbon dioxide (CO_2) with air. There are no long-term storage sites for oxygen within the body; they are not necessary as long as this exchange between body fluids and air remains unimpeded.

Efficient contact of body fluids with air is mediated by the *respiratory tract*, which begins in the *nasal* and *oral cavities* and ends in a huge number of microscopic blind end sacs called *alveoli* in the deepest recesses of the lungs. During inspiration, air travels from the atmosphere through the nasal (or oral) passages, through the *pharynx*, and into the *trachea*. During this time, it is warmed and takes up water vapor. After passing down the trachea, it flows through the *bronchi*, *bronchioles*, and *alveolar ducts* and finally reaches the microscopic alveoli, where exchange of oxygen and carbon dioxide takes place. Following a single O_2 molecule along this tortuous route, we find about 23 forks in the path as the airways birfurcate into finer and finer branches. During expiration, the same path is traversed, but in the opposite direction.

The widest tubes (trachea and bronchi) contain stiff cartilage together with some smooth muscle. They are lined by a layer of epithelial cells that often have minute hairlike structures, called cilia, projecting from their surface. These cells also secrete over their surface a *mucus* sheet that is continuously transported like an escalator in an upward direction (away from the lungs) by the coordinated movement of the cilia. This process serves as an efficient filter for dust particles that strike the walls as the turbulent air flows in and out of the air passage. Once the upward traveling mucus reaches our pharynx we unconsciously swallow it. The smaller branches (bronchioles) also contain smooth muscle, but no cartilage, cilia, or mucus glands. Particles deposited in the bronchioles and alveoli are removed by wandering alveolar macrophages.

The extensive branching pattern of the air passages results in an enormous number of alveoli, approximately 300 million. The diameter of each alveolar sphere is only about 0.3 mm, but adding all their surface area together gives a total alveolar surface area available for gas exchange with blood of about 85 sq m (close to the size of half a tennis court!). Yet this enormous surface is contained within a maximum total volume of only 5 to 6 L which fits very nicely into the thorax. However, this device is not without problems. The tiny alveoli are at the dead end of narrow brochial tubes in a complex branching tubular network. Left to itself, air would stagnate within them. This does not occur because the alveoli are intermittently flushed with fresh air as we breathe.

The enlarged view of a single alveolus in the plate shows the actual interface between body fluids and air where *gas exchange* takes place. Alveolar walls, like blood capillaries, are made of extremely thin cells. Despite the fact that O_2 and CO_2 have to traverse two cell layers in passing between alveolus and capillary, the total distance is very short, and diffusion is correspondingly rapid. Efficient gas exchange is also enhanced by the dense supply of capillaries in the lungs, one of the most profuse networks of blood vessels in the entire body.

The pulmonary circulation that transports blood from the right heart to this alveolar exchange interface also has peculiarities that appear well adapted to its function. Most notably, the pressures in the pulmonary circulation are small; the mean pressure in the pulmonary artery is about 15 mm Hg, only about one-seventh the 100 mm Hg mean pressure in the aorta. Thus, the forces driving blood through the pulmonary circulation are relatively small, and because the blood flows through the pulmonary and systemic circulations are equal, it follows that the resistance of the pulmonary circulation must also be small. Keeping the pulmonary pressures and resistance low so that flow can be maintained reduces the work required of the right heart. In addition, the low pressure in the pulmonary capillary pushing fluids out into the alveolar spaces is overbalanced by the oncotic pressure of the plasma proteins (see plate 35) drawing fluids in. The net force favors reabsorption of fluid from the alveolus so that the normal lung has no tendency to fill with fluid. Further, lung blood vessels have an atypical response to low concentrations of O_2 dissolved in blood plasma. Unlike arterioles of the systemic circulation, which dilate, lung arterioles constrict in response to low local plasma O_2 concentrations. This has the advantage of shunting blood away from areas of the lung that are poorly ventilated and cannot serve as adequate sources of O_2.

CN: Use red for A, blue for B, purple for N, and a very light color for L.
1. Begin with the large illustration. The edge of the right lung is colored gray, and the left lung is colored completely gray. Only the bronchi (I) are colored in the right lung.
2. Color the enlargement on the lower left showing circulation to and from the alveoli.
3. Color the enlargement (lower right) of the gas exchange between an alveolus and a lung capillary.
4. Color the schematic diagram (to the left of the large figure) of external and internal respiration.

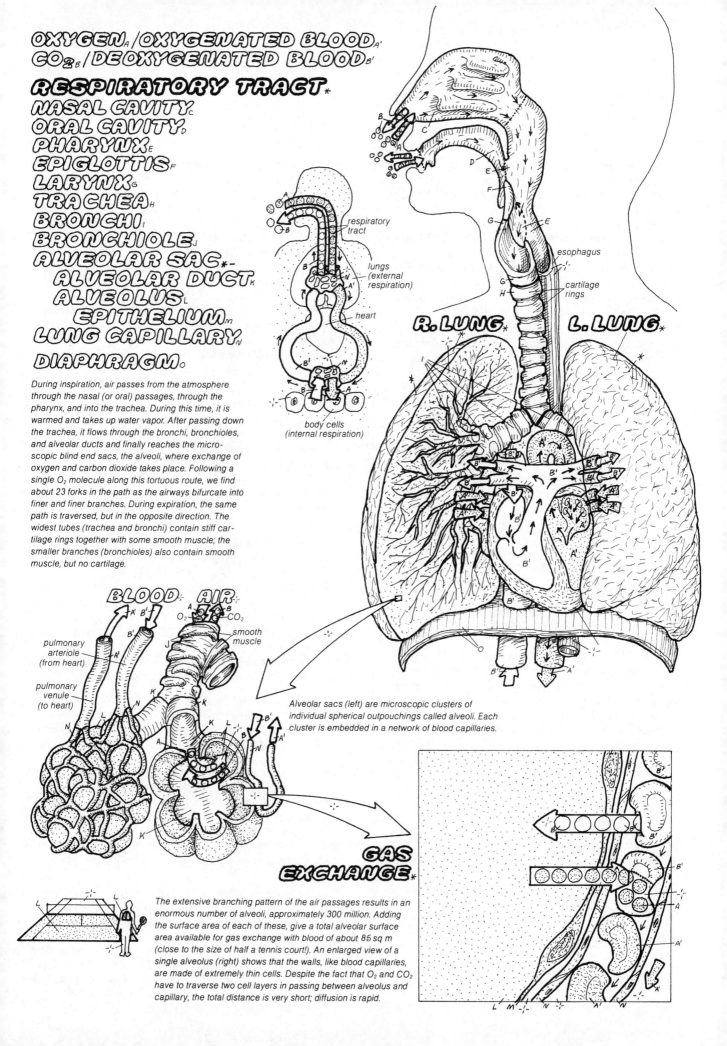

OXYGEN /OXYGENATED BLOOD
CO₂ /DEOXYGENATED BLOOD
RESPIRATORY TRACT
NASAL CAVITY
ORAL CAVITY
PHARYNX
EPIGLOTTIS
LARYNX
TRACHEA
BRONCHI
BRONCHIOLE
ALVEOLAR SAC
ALVEOLAR DUCT
ALVEOLUS
EPITHELIUM
LUNG CAPILLARY
DIAPHRAGM

During inspiration, air passes from the atmosphere through the nasal (or oral) passages, through the pharynx, and into the trachea. During this time, it is warmed and takes up water vapor. After passing down the trachea, it flows through the bronchi, bronchioles, and alveolar ducts and finally reaches the microscopic blind end sacs, the alveoli, where exchange of oxygen and carbon dioxide takes place. Following a single O_2 molecule along this tortuous route, we find about 23 forks in the path as the airways bifurcate into finer and finer branches. During expiration, the same path is traversed, but in the opposite direction. The widest tubes (trachea and bronchi) contain stiff cartilage rings together with some smooth muscle; the smaller branches (bronchioles) also contain smooth muscle, but no cartilage.

respiratory tract

lungs (external respiration)

heart

body cells (internal respiration)

esophagus

cartilage rings

R. LUNG   L. LUNG

BLOOD   AIR

O_2 CO_2

smooth muscle

pulmonary arteriole (from heart)

pulmonary venule (to heart)

Alveolar sacs (left) are microscopic clusters of individual spherical outpouchings called alveoli. Each cluster is embedded in a network of blood capillaries.

GAS EXCHANGE

The extensive branching pattern of the air passages results in an enormous number of alveoli, approximately 300 million. Adding the surface area of each of these, give a total alveolar surface area available for gas exchange with blood of about 85 sq m (close to the size of half a tennis court!). An enlarged view of a single alveolus (right) shows that the walls, like blood capillaries, are made of extremely thin cells. Despite the fact that O_2 and CO_2 have to traverse two cell layers in passing between alveolus and capillary, the total distance is very short; diffusion is rapid.

MECHANICS OF BREATHING

Efficient gas exchange will occur only if the alveoli are regularly flushed with fresh air; this happens with each breathing cycle as air is pumped into and out of the lungs. The lungs themselves are passive structures. They are in contact with the chest wall through a thin layer of fluid, the *pleural fluid*, which allows them to glide easily on the chest wall. The lungs resist being pulled off the wall in the same way that two pieces of wet plate glass slide on each other but are not easily separated. The pressure within this pleural fluid, the *intrapleural pressure*, is subatmospheric (between 3 and 6 mm Hg less than atmospheric) at the end of respiration, when the system is at rest.

During inspiration, the thoracic cage enlarges, and the lungs expand so that air is drawn in via the air passages. This enlargement is produced primarily by contraction of the dome-shaped *diaphragm*, which flattens and increases the vertical length of the cage. In normal breathing, the diaphragm moves about 1 cm, but during forced breathing, this excursion may reach 10 cm. Contraction of *external intercostal muscles* also contributes to the expansion of the chest by pulling the sagging rib cage upward into a more horizontal position, increasing its width.

During quiet breathing, expiration is passive. The inspiratory muscles relax; the diaphragm assumes its resting curved shape and pushes upward on the thoracic cage while the relaxing external intercostals allow the rib cage to sag downward under its own weight. The lungs and chest wall are elastic, so they return to their former position, driving air out of the lungs. During forced expiration, new sets of muscle become active. Muscles of the abdominal wall contract, pushing the diaphragm upward while the contraction of the *internal intercostal muscles*, whose orientation is opposite to that of the external intercostals, pulls the rib cage downward. These actions accelerate the expulsion of air.

If we follow the pressure changes in the intrapleural and alveolar spaces during a single quiet breathing cycle, we arrive at the results shown at the bottom of the plate. At rest, the pressure of the thin layer of pleural fluid is about -3 mm Hg relative to atmospheric pressure (i.e., it is 3 mm Hg below atmospheric); pressure in the lungs is atmospheric (0 mm Hg). This negative intrapleural pressure (-3 mm Hg) reflects the elastic recoil properties of the lungs. As the lungs attempt to pull away from the chest wall, there is no air to fill the potential gap, and the slightest move away from the wall creates a negative pressure ("pulls a vacuum"). The "suction" reflected by the -3 mm Hg pulls the lungs toward the chest wall, and it just balances the elastic recoil pulling them away. (If air is introduced into this space, raising the intra-pleural pressure to atmospheric — e.g., by opening the chest wall — the lungs pull inward and collapse.) Alveolar pressure is atmospheric at this time, reflecting the fact that there is no pressure gradient between the atmosphere and the lungs so that there is no air flowing in or out. Now inspiration begins. The thoracic cage expands, and the falling intrapleural pressure pulls the lungs along with it. As a result, intrapleural pressure falls to -5 mm Hg, and intrapulmonary pressure falls to -1 mm Hg. Air flows down the pressure gradient from the atmosphere (0 mm Hg) to the lungs (-1 mm Hg) until this gradient is finally dissipated at the height of inspiration, when 0.5 L of air has been added. Now expiration begins; the lungs become compressed, raising intrapulmonary pressure and forcing air out until the added 0.5 L is expelled. The system returns to its initial state.

These figures for normal quiet breathing are subject to great variation. For example, at the end of a deep inspiration, intrapleural pressure may be as low as -14 mm Hg, and during a particularly forceful expiration, it may reach as high as +50 mm Hg. Nevertheless, it is remarkable that a pressure gradient of only 1 mm Hg is sufficient to move the required 0.5 L of air in and out of the lungs during normal quiet breathing. It illustrates the lungs' easy distensibility (high compliance). In contrast, a toy balloon may require up to 200 mm Hg for the same increase in volume.

The alveolar spaces are in constant contact with the atmosphere via the airways (nose and mouth, trachea, bronchi, and bronchioles). The fact that the pressure in the alveoli is not equal to that in the atmosphere at various times in the respiratory cycle (i.e., at the beginning of inspiration and at the beginning of expiration) reflects the *resistance* to air flow offered by the airways. The major site of resistance lies in the medium-sized bronchi. (Although the smaller bronchioles have narrower tubes, they are much more numerous, and this factor more than compensates for their small size.)

Airway resistance changes during the normal respiratory cycle. During inspiration, both lungs *and* airways expand in response to the decreased intrapleural pressure; the widened airways offer less resistance. During expiration, the reverse occurs, and resistance increases. This explains why persons with constricted airways (e.g., *asthma*) have much more difficulty exhaling than inhaling. The resistance can also be altered by contractions of bronchial smooth muscle, which narrow the passages and increase resistance. These muscles are under the control of autonomic nerves: sympathetic stimulation (norepinephrine) dilates them, and parasympathetic stimulation (acetylcholine) constricts them.

CN: Use dark colors for C and D.
1. Begin with the diagram in the upper left corner.
2. Color the figure labeled "inspiration" on the right, and include the diagram showing intercostal muscles below it. Do the same for "expiration."
3. Color the lower diagrams, beginning on the left and completing each in sequence before moving on to the next.

RESPIRATORY STRUCTURES *

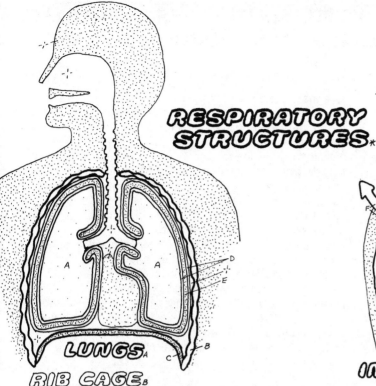

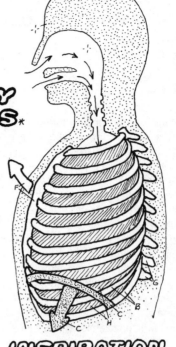

LUNGS A
RIB CAGE B
DIAPHRAGM C
PLEURAL MEMBRANE D
INTRAPLEURAL FLUID E

During inspiration, the thoracic cage enlarges; air is drawn into the lungs via the air passages. This enlargement is produced by contraction of the dome-shaped diaphragm, which flattens and increases the vertical length of the cage, and by contraction of external intercostal muscles, which pull the sagging rib cage upward into a more horizontal position, increasing its width. During expiration, these muscles relax: the diaphragm assumes its resting curved shape and pushes upward on the thoracic cage while the relaxing external intercostals allow the rib cage to sag downward under its own weight. Air is driven out of the lungs. During forced expiration, a new set of muscles, the internal intercostals, becomes active. Their orientation is opposite to that of the external intercostals. When they contract, they pull the rib cage downward and accelerate the expulsion of air.

INSPIRATION H
EXPIRATION I

movable / fixed

STERNUM F SPINAL COLUMN G
EXTERNAL INTERCOSTAL MUSCLES H
INTERNAL INTERCOSTAL MUSCLES I

INTRAPLEURAL PRESSURE E'
INTRAPULMONARY PRESSURE A'
NORMAL ATMOSPHERIC PRESSURE J
RESTING LUNG VOLUME K (FUNCTIONAL RESIDUAL CAPACITY)

RESPIRATORY PRESSURES *

INSPIRATION J

atmosphere
lungs

+.25L K

+.5L K

EXPIRATION A'

+.25L K

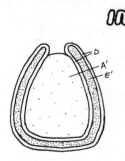

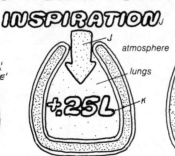

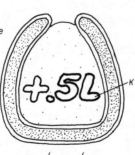

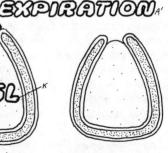

-3 | 0 | 0
-5 | -1 | 0
-6 | 0 | 0
-5 | +1 | 0
-3 | 0 | 0

The lungs and chest wall are lined by pleural membranes. At rest, the pressure of the thin layer of pleural fluid that lies in between these membranes is about -3 mm Hg relative to atmospheric pressure (i.e. it is 3 mm Hg below atmospheric). Pressure in the lungs is atmospheric (0 mm Hg). This pressure difference makes the lungs adhere to the chest wall, keeping them inflated. During inspiration, the thoracic cage expands, intrapleural pressure falls (to -5 mm Hg), and so does intrapulmonary pressure (to -1 mm Hg). Air flows down the pressure gradient from the atmosphere (0 mm Hg) to the lungs (-1 mm Hg) until this gradient is finally dissipated at the height of expiration when 0.5 L of air has been added. Now expiration begins: the lungs become compressed, raising intra-pulmonary pressure and forcing air out until the added 0.5 L is expelled, and the system returns to its initial state.

SURFACTANT, SURFACE TENSION, AND LUNG COMPLIANCE

Although it is important that the lungs can be distended by small forces, it is equally important that they show elastic behavior and return to their original volume when distending forces are relaxed. Two components are responsible for this elastic behavior. First, *elastic tissue*, consisting of elastic and collagen fibers embedded in alveolar walls and around bronchi, resists stretching. Second, *surface tension*, which arises at any air-water interface, resists expansion of the surface. The importance of these two components is illustrated in the top panel, which shows that it requires much less pressure (force) to inflate the lungs with water (more precisely, with physiological saline) than with air. When inflating with water, there is no air-water interface, therefore no surface tension; the only resisting force comes from the elastic tissue. When inflating with air, both forces are operative. By taking the difference of the two measurements, we can estimate that forces arising from surface tension account for two-thirds of the lung's elastic behavior; the remaining one-third arises from elastic tissue.

How does surface tension arise? As shown in the second panel, water molecules attract each other. If they did not, the molecules would fly apart, and water would not be a liquid; it would be a gas. Those water molecules in the bulk of the fluid have neighbors in every direction, and they are pulled in every direction. Molecules on the surface have neighbors only in the interior of the fluid. Accordingly, they are continually pulled off the surface toward the interior. In other words, the water molecules tend to avoid the surface, and as a result, the surface behaves like a thin sheet of rubber that resists expansion. This property is called surface tension; it is a force that acts tangentially to the surface and resists expansion of it.

Surface tension can be reduced by introducing solute molecules called surface active agents or *surfactants*. In contrast to water, surfactants are attracted to the surface; they displace water molecules there and allow the surface to expand. Phospholipids are common surfactants; they have a polar, hydrophilic head that is attracted to the water and a hydrophobic tail that is squeezed out of the water phase (plate 7). Unless they form micelles or bilayers, the only place that can accommodate both the hydrophobic and hydrophilic properties of the molecule is the air-water interface, with the heads immersed in the water and the tails in the air. This allows for easy expansion of the surface. The surface tension is determined by the relative proportions of water and surfactants that occupy the surface. Surfactants, particularly phospholipids, are secreted by some of the cells lining the alveoli. These secretions are important because they reduce surface tension in the alveolar air-water interface, decreasing both resistance to stretch and the work of breathing.

Further complications arise from the relation between the surface tension and the internal pressure required to keep an alveolus inflated. In a spherical structure like an alveolus or a soap bubble, surface tension acts to collapse the bubble, and the pressure required to keep it inflated depends on both the surface tension and the *size* of the bubble. The smaller the bubble, the larger the pressure — remember how difficult it is to begin blowing up a balloon, but once it attains a reasonable size, the task is much easier. This follows because the curvature of the bubble modified the surface force so that part of it pulls inward toward the center of the sphere. The smaller the sphere (the greater the curvature), the larger the force pulling inward. This inward component operates to compress the bubble and requires an oppositely directed pressure. If you imagine a small patch on the surface (see plate), you will notice that the larger the bubble, the less curved the patch will be and the less inward pull there will be from surface forces. As the bubble gets very large, the patch becomes practically flat, and there is no inward-directed component. The mathematical relation between sphere size (radius R), tension T, and pressure P is $P = 2T/R$.

The lungs can be regarded as a collection of 300 million minute bubbles connected to each other. If there were no surfactant, the surface tension in each bubble would be the same, and the system would be unstable because, as shown in the bottom panel (top figure), the smaller bubbles would have a larger pressure and would blow up the larger ones and collapse in the process. When surfactant is present (lower figure), this does not occur because the smaller bubbles have a higher proportion of surfactant on their surfaces and, therefore, smaller surface tensions than larger ones. This follows because, as bubbles become smaller, their surface areas decrease, largely by losing surface *water* molecules (not surfactant) to the interior. Thus, the proportion of surfactant to water in the surface increases so that the decrease in alveolar size is accompanied by a decrease in surface tension. By this mechanism, the surface tension of the smaller alveoli is lowered so that the pressure need not rise to keep it inflated. In our example with no surfactant, the surface tension T is 20 (arbitrary units) in both bubbles. The pressure of each bubble is given by $P = 2T/R$; so the large bubble (R = 2) has P = 20, the smaller bubble (R = 1) has P = 40. Air will move from the small bubble to the large one. Further, the more air that moves, the smaller the bubble gets and the greater the imbalance. With surfactant, both bubbles have a larger surface tension, but the smaller one has less (T = 5) than the larger (T = 10). Now the two pressures balance at 10 each, and the system is stable.

The importance of lung surfactant is apparent in infants born with deficient secretion of it, giving rise to "respiratory distress syndrome." In these cases, the lungs are "stiff," areas are collapsed, and breathing requires extraordinary effort.

CN: Use light blue for E and dark colors for A and H.
1. Begin with the upper panel and the chart on the left. Then color the two diagrams of the lungs being filled with water and air. Note the enlargement of a lung alveolus (G) in which a band of surface tension (A) separates the water-lined alveolus and the air (F).
2. Color the next panel. Note that the band of surface tension in the upper left beaker represents the alignment of water molecules (E) along the surface of the enlarged area. Also note that only the upper band of molecules and a single molecule in the center are colored. In the example to its right, the band of tension is weakened by the presence of surfactant molecules (H) that displace water along the surface of the enlarged portion.
3. Color the chart in the next panel, noting the great amount of pressure required to fill a small balloon. Color the two examples of how surface tension is affected by bubble size.
4. Color the effects of surfactant below.

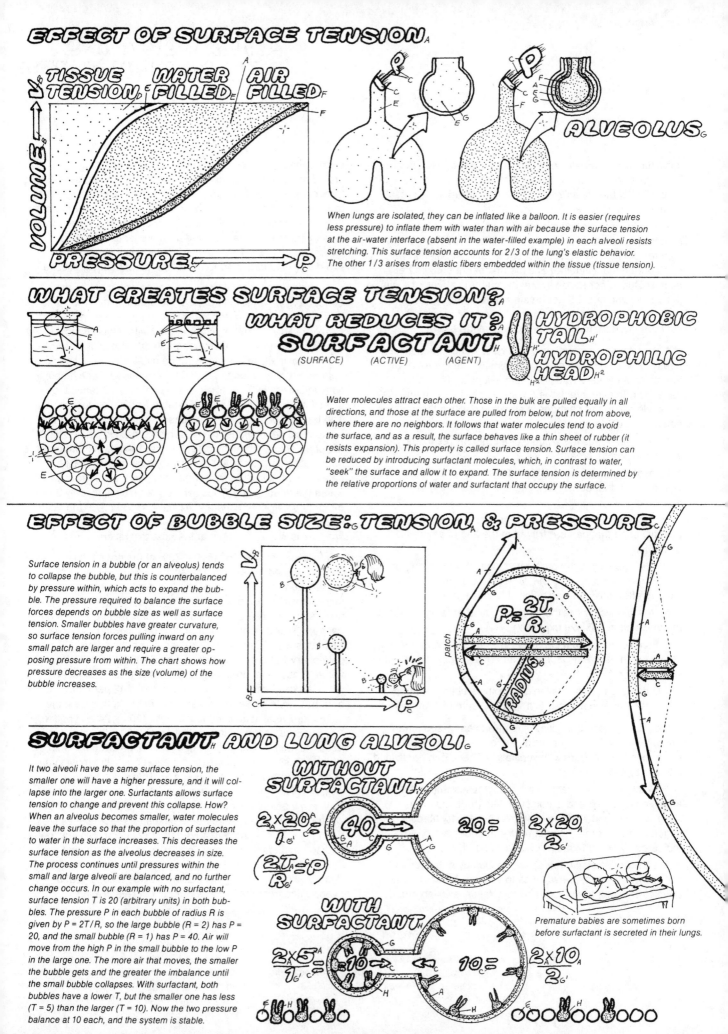

EFFECT OF SURFACE TENSION.

TISSUE TENSION · **WATER FILLED** · **AIR FILLED**

VOLUME → V

PRESSURE → P

When lungs are isolated, they can be inflated like a balloon. It is easier (requires less pressure) to inflate them with water than with air because the surface tension at the air-water interface (absent in the water-filled example) in each alveoli resists stretching. This surface tension accounts for 2/3 of the lung's elastic behavior. The other 1/3 arises from elastic fibers embedded within the tissue (tissue tension).

ALVEOLUS.

WHAT CREATES SURFACE TENSION? WHAT REDUCES IT? SURFACTANT
(SURFACE) (ACTIVE) (AGENT)

HYDROPHOBIC TAIL
HYDROPHILIC HEAD

Water molecules attract each other. Those in the bulk are pulled equally in all directions, and those at the surface are pulled from below, but not from above, where there are no neighbors. It follows that water molecules tend to avoid the surface, and as a result, the surface behaves like a thin sheet of rubber (it resists expansion). This property is called surface tension. Surface tension can be reduced by introducing surfactant molecules, which, in contrast to water, "seek" the surface and allow it to expand. The surface tension is determined by the relative proportions of water and surfactant that occupy the surface.

EFFECT OF BUBBLE SIZE: TENSION & PRESSURE.

Surface tension in a bubble (or an alveolus) tends to collapse the bubble, but this is counterbalanced by pressure within, which acts to expand the bubble. The pressure required to balance the surface forces depends on bubble size as well as surface tension. Smaller bubbles have greater curvature, so surface tension forces pulling inward on any small patch are larger and require a greater opposing pressure from within. The chart shows how pressure decreases as the size (volume) of the bubble increases.

$$P = \frac{2T}{R}$$

SURFACTANT AND LUNG ALVEOLI.

It two alveoli have the same surface tension, the smaller one will have a higher pressure, and it will collapse into the larger one. Surfactants allows surface tension to change and prevent this collapse. How? When an alveolus becomes smaller, water molecules leave the surface so that the proportion of surfactant to water in the surface increases. This decreases the surface tension as the alveolus decreases in size. The process continues until pressures within the small and large alveoli are balanced, and no further change occurs. In our example with no surfactant, surface tension T is 20 (arbitrary units) in both bubbles. The pressure P in each bubble of radius R is given by P = 2T/R, so the large bubble (R = 2) has P = 20, and the small bubble (R = 1) has P = 40. Air will move from the high P in the small bubble to the low P in the large one. The more air that moves, the smaller the bubble gets and the greater the imbalance until the small bubble collapses. With surfactant, both bubbles have a lower T, but the smaller one has less (T = 5) than the larger (T = 10). Now the two pressure balance at 10 each, and the system is stable.

WITHOUT SURFACTANT

$$\frac{2 \times 20}{1} = 40 \rightarrow 20 = \frac{2 \times 20}{2}$$

$$\left(\frac{2T}{R} = P\right)$$

WITH SURFACTANT

$$\frac{2 \times 5}{1} = 10 \rightarrow 10 = \frac{2 \times 10}{2}$$

Premature babies are sometimes born before surfactant is secreted in their lungs.

LUNG VOLUMES AND VENTILATION

If the function of breathing is to flush the alveoli with fresh air, it is natural to ask how much air is moved. How efficient is the ventilation of the alveoli? What common disturbances result from this scheme?

The volume of air that moves in (or out) of the lungs per minute is called the *pulmonary ventilation* or sometimes the *minute volume*. It is the product of the amount taken in with each breath (tidal volume) and the number of breaths per minute. During normal quiet breathing, this is about 6 L/min. (a tidal volume of 0.5 L per breath × 12 breaths per min.), but both the depth of each breath and the rate of breathing can vary greatly, depending on the body's needs.

At rest, the tidal volume is a small fraction of the total lung capacity, and even the deepest expiration cannot expel all the air; some always remains in the alveoli and in the air passages. To evaluate these relations in both health and disease, we divide the changes in air volume within the lungs at different stages of breathing into the following categories:

1. *Tidal volume* is the amount of air that moves in and out with each normal breath.

2. *Inspiratory reserve volume* is the maximal additional volume of air that can be inspired at the end of a normal inspiration.

3. *Expiratory reserve volume* is the maximal additional quantity of air that can be expired at the end of a normal expiration.

4. *Vital capacity* is the greatest volume of air that can be moved in a single breath. The largest portion that can be expired after maximal inspiration, it is the sum of 1, 2, and 3.

5. *Residual volume* is the amount of air that remains within the lungs after maximal expiration.

6. *Functional residual capacity* is the "resting volume." The volume of the system just before a normal inspiration, it is the sum of 2 and 5.

7. *Total lung capacity* is the lung volume at its maximum (i.e., after a maximal inspiration). It is the sum of 4 and 5.

Measuring these quantities is relatively easy (see plate) and often provides diagnostic clues for respiratory tract disturbances that interfere with ventilation. These can be divided into two types:

1. *Restrictive disturbances* are those cases where the lungs' ability to expand is compromised (reduced *compliance*). This occurs, for example, in pulmonary fibrosis or in fusion of the pleurae. Restrictive disturbances are often indicated by an abnormally low *vital capacity*.

2. *Obstructive disturbances* are caused by constriction of the airway (increased *resistance* to airflow). These contractions often result from mucus accumulation, swollen mucus membranes, and bronchial muscle spasms as occurs in bronchial asthma or in spastic bronchitis. Because these disturbances are due to changes in resistance, identifying them requires measuring flow rather than volume (i.e., a rate rather than an equilibrium property). This can be accomplished by measuring the volume expelled from the lungs by forced expiration *in 1 sec.* This quantity, called the FEV_1 *(forced expiratory volume)*, is abnormally low in obstructive disease.

In addition to lung volumes, the space occupied by the conducting airways, the trachea, the bronchi, and the bronchioles — the *anatomical dead space* — also requires attention. The 150 mL of air contained within this "dead" space moves in and out with each breath. But unlike alveolar air, it is not in close contact with the capillaries; so it has no opportunity to exchange O_2 or CO_2 with blood. Each time a tidal volume of 500 mL of air is exhaled, 500 mL leave the alveoli, but only 500 - 150 = 350 mL reach the atmosphere. The trailing 150 mL is still contained within the airways, the anatomical dead space. When a fresh breath is inhaled, 500 mL of air enters the alveoli, but the first 150 mL that enters is not atmospheric. It is the "old" alveolar air from the last exhalation that never reached the atmosphere and was trapped within the dead space. Thus, with each inspiration, only 350 mL of fresh air enters the alveoli, the last 150 mL of the fresh inspired air never makes it because it is held up in the dead space and will be expelled at the next expiration.

It follows that only 350/500 = 70% of the normal tidal volume is used to ventilate the alveoli. Instead of using *pulmonary ventilation = tidal volume × breaths per min.* as a physiological index of effective lung ventilation, we more accurately use *alveolar ventilation = (tidal volume - anatomical dead space) × breaths per min.* The following example illustrates why. Consider two subjects with the same pulmonary ventilation: subject A has a small tidal volume (say 250 mL) but a fast breathing rate of 24 per min.; subject B, with a tidal volume of 500 mL and a rate of 12 per min., breathes twice as deep but half as often. In both cases, the pulmonary ventilation is 6000 mL/min. (250 × 24 and 500 × 12). But B has an alveolar ventilation of (500 - 350) × 12 = 4200 mL/min. A has only (250 - 150) × 24 = 2400 mL/min. Clearly, B is better off; most of A's effort goes into moving air back and forth in the dead air space. This result holds in general: given the same pulmonary ventilation, alveolar ventilation will be enhanced by deeper breaths (even though they will be less frequent). In extreme cases (e.g., sometimes during circulatory shock), the breathing becomes so shallow and so rapid that hardly any ventilation takes place, and the subject is in acute danger. Dogs, however, can use this rapid shallow breathing in a controlled way to lose heat by evaporation from the airways without *over-ventilating*.

CN: Use a dark color for I.
1. Begin with the upper drawing, coloring all the cubes; each one represents 500 mL of air.
2. Color the chart, including the two vertical titles.
3. Color the spirometer on the right.
4. Color the anatomical dead space. Note that the drawings on the right are schematics of the more accurate anatomical drawing on the left.

AIR VOLUMES DURING RESPIRATION*

NORMAL, QUIET BREATHING:*
TIDAL VOLUME (500 mL)A
DEEPEST INSPIRATION:*
INSPIRATORY RESERVE VOLUME (2500-3000 mL)B
DEEPEST EXPIRATION:*
EXPIRATORY RESERVE VOLUME (1000 mL)C
REMAINING AIR:*
RESIDUAL VOLUME (1000 mL)D

The volume of air (500 mL) that moves in (or out) of the lungs with each inspiration (or expiration) during quiet breathing is called the tidal volume. During strenuous breathing, the amount of air moving with each breath increases. The maximum amount of additional air that can be inspired above the tidal volume is called the inspiratory reserve volume; the maximal volume of additional air that can be expired is called the expiratory reserve volume. The maximal amount of air that can be moved with each breath, the vital capacity, equals the sum of the inspiratory reserve, tidal, and expiratory reserve volumes. However, the lungs never empty completely; the volume of remaining air following a maximal expiration is called the residual volume. Finally the total lung capacity equals the sum of all these volumes.

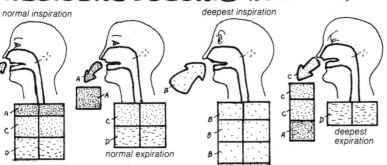

normal inspiration   deepest inspiration   deepest expiration

normal expiration

Respiratory volumes are measured with a spirometer, consisting of an inverted container (bell) floating on water. Using a connecting hose, the subject expires (inspires) into (from) the bell as if it were a partially inflated balloon. The bell moves up (or down) with each breath, and its movements, which are proportional to changes in volume, are recorded on a rotating drum.

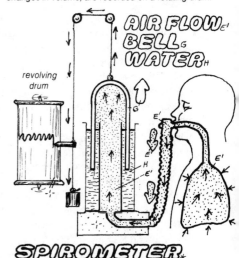

TOTAL LUNG CAPACITY*

VITAL CAPACITY*

liters

time axis

revolving drum

AIR FLOWE'
BELLG
WATERH

SPIROMETER*

FUNCTIONAL RESIDUAL CAPACITY (RESTING LUNG VOLUME)F

ALVEOLAR AIR AFTER RESPIRATIONA'

ANATOMIC DEAD SPACEI	150 mLI
FRESH AIRJ	350 mLJ
TIDAL VOLUMEA	500 mLA

During inspiration, some stale air reaches the alveoli. Close to 1/3 of the tidal volume is nonfunctional and is required simply to fill the air passages of the head, neck, bronchi, etc. The total volume of these passageways (about 150 mL) is called the anatomical dead space. Each time 500 mL of air is drawn into the lungs, the first 150 mL comes from the dead space, with the following 350 ml arising from fresh atmospheric air. If your tidal volume were only 150 mL, you would never get any fresh air! You would simply exchange the 150 mL back and forth between dead space and alveoli. Similarly, if you use a snorkel tube with a 350 mL volume, then you will increase your dead space to 500 mL! In this case, normal tidal volume of 500 mL will be useless. Dogs lose heat by fluid evaporating from their dead space during panting. By restricting the amount of air moved, they bring fresh dry air to the dead space without allowing it to reach the alveoli. Thus, their rapid breathing movements do not interfere with normal respiration; they do not over-ventilate.

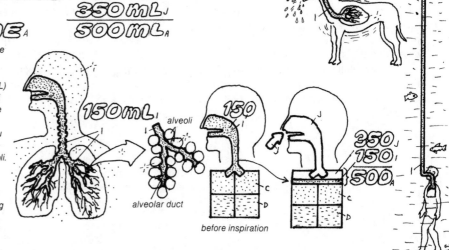

150 mL

alveoli

alveolar duct

150

350
150
500

before inspiration

DIFFUSION OF O_2 AND CO_2 IN THE LUNG

Diffusion of O_2 and CO_2 in the lung alveoli is complicated because these molecules move across an air-water interface. To describe these movements, we need a vocabulary equally applicable to both the liquid and air (gaseous) phases. We begin with a review of the properties of a gas.

In a gas, *pressure* (force/unit area) results from gas molecules colliding with the walls of the container. It is determined by the frequency and force of the collisions. Each gas molecule is oblivious to the presence of any other; it strikes the container walls just as frequently as if it were all alone. Increasing the temperature of a gas raises pressure, because the higher the temperature the greater the velocity of the molecules, causing more frequent collisions and greater force to be imparted. Decreasing the volume occupied by the gas also increases pressure because the gas molecules are confined to a smaller space and collide with the walls more frequently.

The pressure of air (or any gas) is measured by bringing it in contact with a pool of mercury (Hg) connected to a closed-ended tube containing no air (or gas). Force exerted by air pressure is not opposed by the vacant tube; therefore, the Hg rises until its weight just counterbalances the air pressure. The height of this column (mm Hg) is a measure of the pressure of the air (gas). Atmospheric air has a pressure of 760 mm Hg at sea level.

In a mixture of gases, each component acts independently of the others, and each molecule makes the same contribution to the pressure. Air (a mixture of gases) consists of approximately 20% O_2 and 80% N_2. If we remove the N_2, we measure a pressure of 20% of 760 = 152 mm Hg. Similarly, retaining the N_2 but removing the O_2 yields a pressure of 80% of 760 = 608 mm Hg. In the mixture of the two, O_2 contributes 152 mm, and N_2 contributes 608 mm Hg pressure. These are the *partial pressures* of O_2 and N_2, respectively. They are abbreviated as PO_2 and PN_2. Knowing the partial pressure of a gas is useful because at constant temperature (which is always the case in the alveoli) the partial pressure is a measure of the concentration of the gas and indicates the driving force available to dissolve the gas in a liquid.

Now suppose we bring the air in contact with a gas-free liquid, say water. The higher the partial pressure of O_2 (PO_2) in the gas, the more often O_2 will strike the surface of the water, and the more often some of the O_2 molecules will enter and dissolve in the liquid. But the dissolved O_2 molecules will also strike the surface from below, and some of these will tend to escape into the gas phase. As the concentration of O_2

builds up in the liquid, more and more will tend to escape until we reach an *equilibrium*, where the number leaving exactly balances the number entering the liquid. The O_2 concentration in the liquid is *directly proportional* to the partial pressure of the O_2 that it is equilibrated with, and we often use partial pressure as a measure of the concentration of O_2 in the liquid. If the PO_2 in the air were 152 mm Hg, then the PO_2 in solution would also be 152 mm Hg.

What has been described for O_2 applies equally to all gases, particularly CO_2. When the partial pressures between any two points are not equal, the two points are not in equilibrium; given the opportunity, gas will diffuse from one to the other. If the partial pressure in a gaseous phase (e.g., alveolus) is greater than in the water (e.g., plasma), gas will move into the water; if it is less, gas will move out of the water. Gas molecules move down partial pressure gradients.

With each inspiration, air moves by bulk flow into the alveoli, as described in plate 44. From the alveoli, O_2 diffuses down its partial gradient into the blood while CO_2 diffuses in the opposite direction. Alveolar air loses O_2 and gains CO_2, together with some water vapor that has evaporated from the walls of the moist respiratory passages. As a result, the partial pressures of these gases in the alveoli differ from those in the atmosphere, as shown in the illustration. The circulation (bulk flow) carried O_2 contained in the blood to systemic capillaries, where once again it diffuses down its partial pressure gradient, this time into the tissues. Again CO_2 diffuses in the opposite direction, this time into the blood, which will carry it by bulk flow via the venous system and the pulmonary artery to the lungs.

Three important variables determine the speed of gas diffusion in the body: (1) the gradient in partial pressure, (2) the surface area available for diffusion, and (3) the magnitude of the diffusion distance. Although the bottom diagram shows that O_2 always moves down its partial pressure gradient from the atmosphere to mitochondria, movement over the long distances (between atmosphere and alveoli and between lungs and systemic tissue) is driven by the pumping action of respiratory and cardiac muscles. In these cases, transport occurs by bulk flow. Diffusion is the effective transport mechanism only over the short distances between alveolus and blood and between blood and tissue. Similar remarks apply to CO_2. Gas transport can be compromised if diffusion distances are lengthened, as in pulmonary edema, and if the surface area available for diffusion is reduced, as in emphysema.

CN: Use red for B and a dark color for I.
1. Begin with the upper panel, top line first. The titles for B and C are O_2 and N_2 in the equation on the right.
2. Do the middle panel next.

3. Color the lower panel, following the numbered sequence.
4. Color the numbers in the diagram of PO_2 at various stages in its journey to its site of utilization, the mitochondria. Do the same for PCO_2.

WHAT IS PRESSURE?
GAS MOLECULE

In a gas, pressure (force/unit area) results from gas molecules colliding with the walls of the container. It is determined by the frequency of collisions and the force imparted by each collision.

WHAT CHANGES PRESSURE?

Increasing temperature raises pressure because it increases the velocity of molecules which increases frequency of collision and the force imparted. Compression also raises pressure because gas molecules collide with the walls more frequently.

PARTIAL PRESSURE (P) OF GAS

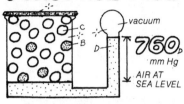

760 mm Hg
AIR AT SEA LEVEL

vacuum

152 mm Hg
AIR LESS NITROGEN

MERCURY

$$PO_2 = 1/5 \times 760 = 152$$
$$PN_2 = 4/5 \times 760 = 608$$

TOTAL PRESSURE = 760

Force (pressure) exerted by air is not opposed by the vacant (air-free) tube. Therefore the Hg rises until its weight just balances the air pressure. The height of this column (mm Hg) is the pressure of the air. Atmospheric air has a pressure of 760 mm Hg at sea level.

The total gas pressure reflects the sum of the collisions exerted by all gas molecules. If air has 20% O_2, then O_2 exerts $.20 \times 760 = 1/5 \times 760 = 152$ mm Hg. O_2 has a partial pressure (PO_2) of 152 mm Hg. Partial pressure is proportional to the concentration of the gas.

SOLUBILITY OF GAS: P IN AIR VS. P IN GAS

When O_2-free water is first brought into contact with air, O_2 enters (dissolves in) the water until equilibrium is reached, when the rate of O_2 leaving the water just equals the rate of O_2 entering. We measure concentrations of gas in solution in terms of partial pressure. If the final PO_2 in the air were 152 mm Hg, then PO_2 in solution would also be 152 mm Hg. By definition, the partial pressure of a gas in solution equals the partial pressure of the same gas in the gaseous phase that would be required if the water and gaseous phases were in equilibrium.

ALVEOLUS. BLOOD CAPILLARY

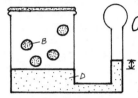

O_2

CO_2

If the partial pressure in the gaseous phase (e.g., alveolus) is greater than in the water (e.g., plasma), gas will move into the water. If it is less, gas will move out of the water. Gas molecules move down partial pressure gradients.

FACTORS AFFECTING TRANSPORT OF GAS IN LUNG AND TISSUES

1. P GRADIENTS

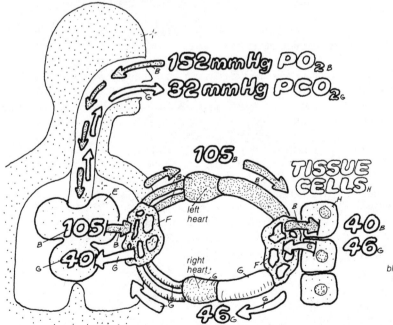

152 mm Hg PO_2
32 mm Hg PCO_2

105

left heart

right heart

105
40

40
46

TISSUE CELLS

40
46

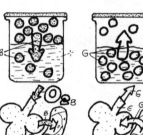

2. SURFACE AREA
EMPHYSEMA

Partial pressure gradients provide the driving force for gas transport, but the pathway is also important. Gas transport can be compromised if the surface area available for diffusion is reduced, as happens in emphysema, and also if diffusion distances (interstitial fluid) are lengthened, as happens in pulmonary edema.

3. THICKNESS OF DIFFUSION DISTANCE

red blood cell

capillary wall   wall alveolus

INTERSTITIAL FLUID

Follow the partial pressure of O_2 as it moves from external air to its point of utilization inside the cells, the mitochondria. Movement over the long distances between atmosphere and alveoli and between lungs and tissue occurs by bulk flow. Diffusion is the effective transport mechanism over the short distances between alveolus and blood and between blood and tissue. Corresponding remarks apply to CO_2, but recall that PCO_2 is highest in the cells, where it is produced, and lowest in the atmosphere; so it moves in the opposite direction.

OXYGEN GRADIENTS

105 → **100** → **40** → **30** → **15** → **5-2**

alveoli   arteries   capillaries   interstitial fluid   cytosol   mitochondria

THE FUNCTION OF HEMOGLOBIN

Like any solute, O_2 can simply dissolve in the watery fluids of the blood, but the amount that can dissolve is very small. At the PO_2 (partial pressure of O_2) = 100 mm Hg that exists in arterial blood and with a normal cardiac output, the amount dissolved could supply only about 6% of the body's requirements at rest. During activity, it would fall even shorter. Clearly, there has to be, and is, another way. Most O_2 carried by the blood is combined with *hemoglobin* (Hb), an iron-containing protein within the red blood cell. Hb can carry nearly 70 times the O_2 held in simple solution.

Although CO_2 is more soluble than O_2, it too is carried primarily in different combined forms in the plasma and red cells. Most CO_2 reacts with water to form carbonic acid (H_2CO_3), which dissociates into H^+ and bicarbonate (HCO_3^-) according to the reaction

$$H_2O + CO_2 \rightarrow H_2CO_3 \rightarrow H^+ + HCO_3^-$$

Another fraction of CO_2 combines with some of the amino groups on polypeptide portions of Hb to form *carbaminohemoglobin*.

Hb's ability to bind O_2 depends on the presence of a *heme* group within the molecule. Heme, a nonpolypeptide, consists of an organic part and an *iron* atom; it gives Hb (and red cells) its characteristic red color. The iron can be in one of two states; the ferrous state (charge = +2) or the ferric state (charge = +3). Only the Hb with iron in the ferrous state binds O_2. Hb in the ferric state is a darker color, called *methemoglobin*, and cannot bind O_2.

The heme group is embedded in a large polypeptide chain, and together (heme + polypeptide chain) they are called a subunit. The entire Hb molecule, which has a molecular weight of 64,450, consists of four of these subunits. The size of an O_2 molecule makes up only about 0.01% of the size of one of these subunits. It thus seems natural to wonder whether the large structure has any significance and whether the combination of subunits into groups of four has advantage. Would an iron molecule or a heme by itself suffice? How about a subunit by itself?

When isolated heme is dissolved in water, it binds O_2 but only momentarily because it is rapidly converted from the ferrous (+2) to the ferric (+3) state. But this does not happen to the heme in Hb or even in a subunit because the heme is embedded in a crevice that has a distinctive nonpolar character so that water is excluded. Apparently, the polypeptide

protects the heme from water and helps keep it in the reduced ferrous (+2) state. Even here some conversion takes place at a slow rate, but the red cell contains an enzyme that can keep pace and convert the methemoglobin back to Hb.

Given that iron in a subunit will be reasonably stable in the ferrous state and bind O_2, why bother to string four of them together? The answer appears to be "too much of a good thing"; a subunit binds O_2 too well. This can be demonstrated by studying *myoglobin*, a very close relative of Hb containing heme and a similar polypeptide chain, but consisting of only one subunit. Myoglobin takes up O_2 well at a very low PO_2, much lower than the PO_2 of venous blood. But this also means that the myoglobin won't give it up until the PO_2 is correspondingly low. Myoglobin functions well as an O_2 storage compound in muscle, where it releases its O_2 only when the PO_2 drops very low during strenuous exercise, but it would not suffice as an O_2 carrier in the blood. We could imagine other single subunits that have lower affinities for O_2, but they would present a new problem: they would not pick up enough O_2 in the lungs. Thus, one type of molecule binds too tightly; it works well in the lungs but not in the tissues. The other binds too loosely; it gives up O_2 readily in the tissues but can't pick up enough in the lungs. Ideally, we would like a molecule that switches between the two types as it goes from lungs to tissue. By stringing four subunits together so that the heme sites can interact, Hb approximates that ideal.

Hb exists in more than one state. When none of the iron-binding sites is occupied, Hb is in a T ("tense") state and not receptive to O_2. However, once an O_2 does bind to one site, the iron moves slightly and so do parts of the polypeptide chain attached to it. This loosens the stucture, making it easier for the next O_2 to attach to one of the remaining empty sites. The sequence repeats, making it still easier for the next O_2, etc., until (in the lungs) all four sites are occupied by O_2, and the Hb is in an R ("relaxed") state. Conversely (in the tissues), as one O_2 frees itself from the Hb, the Hb changes slightly, making it easier for the next to unload. This behavior is called *cooperative*. A simple analogy in the plate shows a boat (Hb) with room for four people (O_2). They swim in the water; but as one gets on the boat, he helps the next, etc. The physiological significance of this cooperative behavior is discussed in plate 49.

CN: Use red for B and blue for C (both for venous blood and CO_2 transport).
1. Begin with the upper panel. Note that the symbol for oxyhemoglobin is a further simplification of the symbol used at the bottom of the page, which is a simplification of the hemoglobin model (the large illustration below).

2. Color the lower panel, beginning with the large illustration. Note that the lower left alpha chain (E) shows the polypeptide chain of which it and the other chains are composed. Color the three examples below. Note that the boat on the far right is given the oxyhemoglobin red because it is holding the four O_2 molecules.

O₂ SATURATION IN BLOOD CIRCULATION

Like any solute, O_2 can simply dissolve in blood plasma, but the amount that can dissolve is very small and cannot supply the body's needs. Most O_2 carried by the blood is not in simple solution; rather it is combined with hemoglobin (Hb), an iron-containing protein within the red blood cell. Hb is represented by the squares in the beaker.

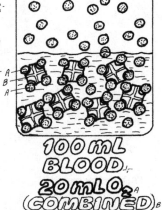

TRANSPORT OF O₂

97% AS OXYHEMOGLOBIN
3% DISSOLVED IN PLASMA

DEOXYHEMOGLOBIN (HHb) +

$$HHb + O_2 \rightleftharpoons HbO_2 + H^+$$

OXYHEMOGLOBIN (HbO₂)

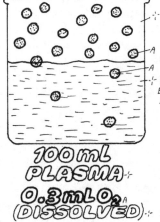

100 mL PLASMA
0.3 mL O₂ (DISSOLVED)

100 mL BLOOD
20 mL O₂ (COMBINED)
+0.3 mL O₂ (DISSOLVED)

TRANSPORT OF CO₂

67% AS BICARBONATE
24% AS CARBAMINOHEMOGLOBIN
9% DISSOLVED IN PLASMA

CO_2 is also carried in different combined forms in the plasma and red cells. Most CO_2 reacts with water to form carbonic acid (H_2CO_3), which dissociates into H^+ and bicarbonate HCO_3^-). Some of the remaining CO_2 combines with amino groups on polypeptide portions of Hb to form carbaminohemoglobin.

MITOCHONDRION

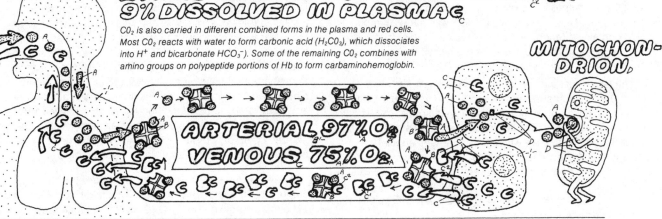

ARTERIAL 97% O₂
VENOUS 75% O₂

HEMOGLOBIN MOLECULE (Hb)

2 ALPHA PEPTIDE CHAINS
2 BETA PEPTIDE CHAINS
4 HEMES:
4 PORPHYRINS
4 IRON ATOMS

Hemoglobin (Hb) consists of 4 polypeptide chains called subunits. One heme, a nonpeptide, is embedded in a crevice of each chain. Each heme contains an iron atom, which is the binding site for O_2. Keeping the iron "hidden" from water helps prevent deterioration of Hb into methemoglobin which can not bind O_2.

HOW O₂ BINDS

In deoxy-Hb the subunits are held together very tightly by electrical forces (salt bridges). This is called the T (tense) state. In the T state, it is very difficult for O_2 to gain access to iron-binding sites. However, oxygen binding is cooperative. Once an O_2 does bind to one site, the iron moves slightly and so do parts of the peptide chain attached to it. This breaks some of the salt bridges, loosening the structure and making it easier for the next O_2 etc. until all 4 sites are occupied by O_2 and the Hb is in an R (relaxed) state.

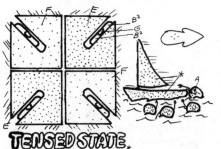

TENSED STATE

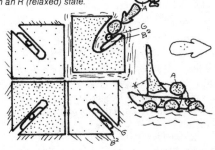

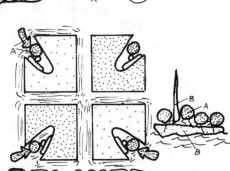

RELAXED

OXYGEN TRANSPORT BY THE BLOOD

When *hemoglobin* (Hb) is exposed to O_2, the O_2 molecules continually collide with it. If there is an empty binding site on the Hb, a colliding O_2 may bind to it. But bound O_2s are continually shaking loose from their sites. Equilibrium is reached when the number being bound just equals the number shaking loose. In Hb, this equilibrium is reached very fast, and its position is determined largely by the PO_2. The higher the PO_2 (the more concentrated the O_2), the more frequent the collision with Hb and the more frequently an O_2 will bind. As the O_2 concentration increases, more and more binding sites are filled, until finally every site is filled, with each Hb molecule containing four bound O_2 molecules. At this point, we say the Hb is 100% *saturated*; when only half are occupied, the Hb is 50% saturated.

The large illustration in the plate shows how Hb takes up O_2 at the partial pressures that exist in the lungs and in the tissues. In the lungs, PO_2 = 105 mm Hg; the curve shows that Hb is 97% saturated. The illustration also shows that Hb will unload O_2 in the tissues where PO_2 averages about 40 mm Hg and may fall even lower to 20 mm Hg in active muscles. The vertical arrows show the difference between the percentage of Hb saturation of blood just after leaving the lungs and the percentage of Hb saturation in the tissues. This difference is the O_2 delivered to tissues.

Hb "works" because its saturation curve is S shaped; it unloads most of its O_2 in a very narrow range of PO_2 between 20 and 40 mm Hg. This behavior is due to the fact that Hb is made of *four interacting subunits that "cooperate" in binding O_2*. The first portion of the curve at very low PO_2 is flat because Hb is in the tense state and not receptive to O_2. As more O_2 molecules are introduced, the likelihood of one of them binding goes up. Once it binds, it influences the other vacant binding sites on the same Hb molecule, increasing the probability of binding a second O_2, which will increase the chances for a third, etc. Thus, the binding (saturation) curve rises very steeply and fortunately in just the right region!

Contrast this behavior with that of *myoglobin*, the O_2 storage protein in muscle cells. It is similar to Hb, but it contains only one subunit; one molecule binds only one O_2, and there is no possibility of a T state or of *cooperative binding*. Its binding curve is not S shaped, and rather than giving up its O_2 at the PO_2 found in the venous blood, it takes it up. But this fits its function; myoglobin stores O_2 and will give it up in the tissues only when the PO_2 falls very low.

The PO_2 is not the only variable that influences the binding of O_2 to Hb. The last diagram in the plate shows several percentage of saturation curves for Hb under different conditions. In one of them, the concentration of CO_2 has increased, and the O_2 saturation curve for Hb has shifted to the right (i.e., it lies below the "normal" curve). In this case, a higher PO_2 is required to achieve the same percentage of saturation, and this means the Hb has a lower affinity for O_2. If the Hb were just sitting there, exposed to a constant PO_2, and CO_2 suddenly increased, shifting the curve to the right, then the Hb would release some of its O_2. This actually happens as blood passes through a capillary, and CO_2 diffuses into the blood from the tissues. In addition to CO_2, two other important substances shift the curve to the right. These are H^+ and a phosphorous-containing metabolite, *2,3 DPG*. These each bind at separate locations on the Hb molecule, but they all act in similar ways by strengthening linkages between Hb subunits, which promotes the tense state with low O_2 affinity. Tissues commonly produce CO_2 and H^+. This helps drive O_2 off the Hb, making it more available to tissue cells.

When the curve is shifted to the left, above the "normal" curve, the Hb has more affinity for O_2; it takes some up. This will occur whenever the 2,3 DPG level falls. In fact, when all the 2,3 DPG is removed, Hb's affinity for O_2 increases to such an extent that it begins to resemble myoglobin. The *Hb in fetal red cells* is different from adult Hb; in particular, fetal Hb does not bind 2,3 DPG as readily as adult Hb. In other words, it is less sensitive to 2,3 DPG. As a result, the O_2 saturation curve for fetal Hb lies above the curve for maternal Hb, showing that fetal Hb has a greater affinity for O_2. This is an advantage for the fetus because when fetal Hb comes in proximity to maternal Hb (in the placenta), it will draw O_2 from the maternal blood.

The role of 2,3 DPG has attracted a good deal of attention because it is not simply an essential "ingredient" whose presence is required for normal Hb function. Rather, its level can vary considerably, and it is involved in regulating O_2 transport in both health and disease. Its level rises when O_2 uptake in the lungs is compromised, and this helps the Hb unload a larger portion of the O_2 that it does carry when it gets to the tissues. This rise in 2,3 DPG occurs, for example, during the first day's adaptation to high altitude (plate 53) and during obstructive lung diseases.

CN: Use the same color for O_2 (B) as used on previous plates. Note that the Hb is shown by two different symbols, a dump truck and a four-unit structure.
1. Begin with the graph in the upper right. First color the percentage and PO_2 coordinates.

Then color the curve and the corresponding O_2 concentrations below on the horizontal axis.
2. Color the myoglobin example (F).
3. Color the factors influencing the curve. Note that the dump truck receives a different color in two of the examples.

HEMOGLOBIN/OXYGEN DISSOCIATION CURVE

The more concentrated the O_2 (i.e., the higher the PO_2), the more it will fill up empty sites on Hb. When all possible sites are occupied by O_2, we say the Hb is 100% saturated; when only half are occupied, Hb is 50% saturated. The illustration shows that Hb will take up O_2 in the lungs; here PO_2 = 105 mm Hg, and the curve shows that Hb is 97% saturated. The figure also shows that Hb will unload (dump) O_2 in the tissues where PO_2 averages about 40 mm Hg but may fall as low as 20 mm Hg in active muscles. Hb "works" because its saturation curve is S shaped; it unloads most of its O_2 in a very narrow range of PO_2 (between 20 and 40 mm Hg). This behavior is due to the co-operative nature of O_2 binding to Hb. The first portion of the curve at very low PO_2 is flat because Hb is in the tense state and not receptive to O_2. As more O_2 are introduced, the likelihood of one of them binding goes up. Once it binds, it increases the probability of a second one, which increases the chances for a third, etc. Thus, the binding (saturation) curve rises very steeply and fortunately in just the right region!

MYOGLOBIN

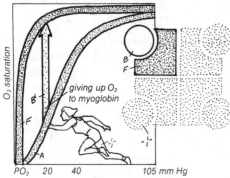

Contrast this with myoglobin, the O_2 storage protein in muscle cells. It is similar to Hb but contains only one subunit. One molecule binds only one O_2, and there is possibility of a T state or of cooperative binding. Its binding curve is not S shaped, and rather than giving up its O_2, it takes it up. But this fits its function; it stores O_2 and will give it up in the tissues only when the PO_2 falls very low.

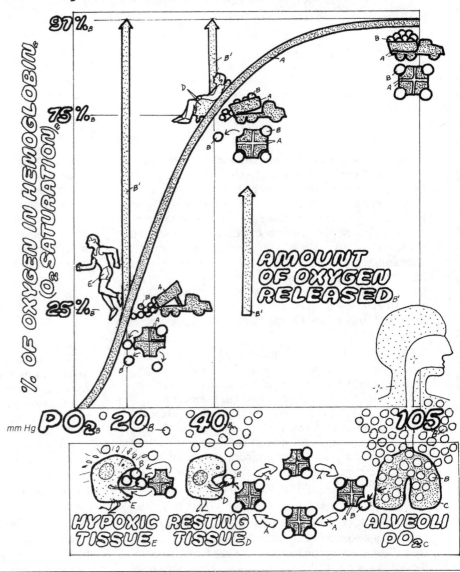

% OF OXYGEN IN HEMOGLOBIN (O_2 SATURATION)

97% 75% 25%

AMOUNT OF OXYGEN RELEASED

mm Hg PO₂ 20 40 105

HYPOXIC TISSUE   RESTING TISSUE   ALVEOLI PO₂

FACTORS AFFECTING THE CURVE

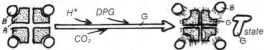

NORMAL CURVE: MATERNAL Hb

SHIFT TO THE RIGHT: pH⁺ CO₂ & DPG

When an O_2 saturation curve for Hb is shifted to the right (i.e., when it lies below the "normal" curve), it has a lower affinity for O_2; it gives some up. Three important substances shift the curve to the right. These are H^+, CO_2, and a phosphorous-containing metabolite, 2,3 DPG. They all act in similar ways by creating bridges between the Hb subunits that promote the tense state with low O_2 affinity. Tissues commonly produce CO_2 and H^+. This helps drive O_2 off the Hb, making it available to tissue cells.

H^+ DPG CO_2 T state

SHIFT TO THE LEFT: FETAL Hb ABSENCE OF DPG

When the curve is shifted to the left, the Hb has more affinity for O_2; it takes some up. The O_2 saturation curve for fetal Hb lies above the curve for normal maternal Hb; the fetal Hb has a greater affinity for O_2. This is an advantage for the fetus. It means that when fetal Hb comes in proximity to maternal Hb (in the placenta), it will draw O_2 from the maternal blood. Fetal blood has this property because it is less reactive to affinity-lowering 2,3 DPG than normal adult Hb.

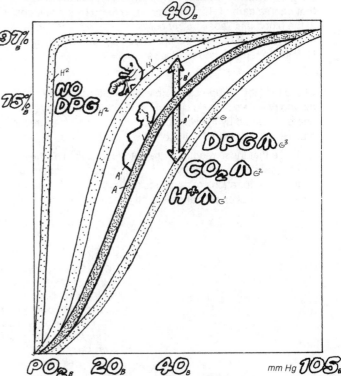

97% 75% 40

NO DPG

DPG↓ CO₂↓ H⁺↓

PO₂ 20 40 mm Hg 105

TRANSPORT OF CO_2, H^+ AND O_2

In plate 49, we saw how the subunit structure of Hb introduces into the molecule new properties that are not shared by the simpler single unit analog, myoglobin. In particular, increasing the concentrations of CO_2 and H^+ drives O_2 off the Hb molecule. The converse also holds: increasing the concentration of O_2 drives off both CO_2 and H^+. At first, this unusual sensitivity of Hb to its environment may seem undesirable in a molecule whose function is to stabilize the PO_2 in body fluids. However, the function of Hb goes beyond this; it not only transports O_2, it also transports both CO_2 and H^+. Further, Hb reacts with these three substances in a remarkable way so that just the "right" thing happens at the "right" time.

Like O_2, CO_2 transport is passive. PCO_2 is high in the tissues because it is produced there. It is low in the lung alveoli because it is swept out with each breath, and therefore it is also low in the arterial blood that enters tissue capillaries. CO_2 moves down its partial pressure gradient from tissue to capillary blood to lung alveoli (plate 48). Although blood holds a small amount of CO_2 (about 9%) in simple solution and another fraction (about 27%) in combination with Hb, the major portion (64%) reacts with water, forming *bicarbonate* (HCO_3^-) and *hydrogen ions* (H^+).

$$CO_2 + H_2O \rightleftharpoons H_2CO_3 \rightleftharpoons HCO_3^- + H^+$$

Because PCO_2 is high in the tissues, this reaction proceeds to the right, and CO_2 is carried as bicarbonate. However, there is a major problem with this reaction; it leads to the accumulation of H^+ ions. Not only are H^+ ions acid, but their accumulation will slow down and block the reaction of CO_2 with water, which severely limits the amounts of CO_2 that can be carried. The dilemma is resolved by substances in the blood that "soak up" or *buffer* excess H^+ ions. Hb is one of the most important of these buffers; its reaction with H^+ can be represented as follows:

$$H^+ + HbO_2^- \rightleftharpoons HHb + O_2$$

where the HbO_2^- represents Hb with O_2 attached (*oxyhemoglobin*), and the (-) sign signifies one of the many (-) charges carried by the Hb molecule. Similarly, HHb represents Hb with an extra H^+ attached.

Notice that these reactions are both reversible (i.e., they can proceed from left to right or from right to left depending on the concentrations of reactants and products). At *equilibrium*, the reaction proceeds in both directions, but at equal

rates so that no noticeable change takes place. However, when concentrations of substances on the right are decreased, the reaction gets "pulled" from left to right. Increasing concentrations on the left will "push" the reaction from left to right. Conversely, decreasing the concentrations of substances on the left, or increasing them on the right, moves the reaction from right to left.

In the tissues, the reactions involving Hb and bicarbonate are coupled because H^+ ions are a common participant in both. In the tissues:

$$CO_2 + H_2O \longrightarrow H_2CO_3 \longrightarrow HCO_3^- + H^+$$
$$H^+ + HbO_2^- \longrightarrow HHb + O_2$$

The first reaction proceeds in the indicated direction because (1) CO_2 is produced in tissues so its concentration is high, and (2) as soon as excess H^+ begins to accumulate, it is consumed by the second reaction. The second reaction proceeds in the indicated direction because (1) a steady supply of H^+ is liberated by the first reaction, (2) a steady supply of HbO_2^- at high concentration is coming from the lungs, (3) HHb is continually swept away in the venous blood, and (4) O_2 is consumed by the tissues, so its concentration is low. Note that as soon as H^+ is produced, it is picked up by the Hb, so free H^+ does not accumulate to dangerous levels. In the process, the tissues receive an extra dividend: more O_2 is driven off the Hb than would be without the H^+ binding.

In the lungs, these same reactions occur, but now in reverse:

$$O_2 + HHb \longrightarrow HbO_2^- + H^+$$
$$H^+ + HCO_3^- \longrightarrow H_2CO_3 \longrightarrow H_2O + CO_2$$

The first reaction proceeds in the direction of the arrow because (1) PO_2 is high in the lungs, (2) there is a steady supply of HHb at high concentration coming from the tissues (via systemic venous blood), and (3) as soon as excess H^+ accumulates, it is consumed by the second reaction. The second reaction proceeds as shown because (1) there is a steady supply of H^+ liberated by the first reaction, (2) there is a steady supply of HCO_3^- at high concentration coming from the tissues, and (3) breathing keeps CO_2 at a low level.

Thus, H^+ ions, which at first appeared to be a problem, actually play a very useful role: in the tissues they drive O_2 off of Hb, and in the lungs they help drive CO_2 off of HCO_3^-. They never accumulate in the free state because they are passed back and forth like a "hot potato" between Hb and HCO_3^-.

CN: Use the same colors as on previous page for O_2 (I). Use red for C, blue for D, and light blue for F. Use a dark color for H.
1. Begin by coloring the tissue cell and the titles at the top of the page and the lung alveolus and titles at the bottom. Then color the red blood cell section. Color the two horizontal bands (where gas exchanges occur) gray, and color the vertical bands, of arterial (C) and venous (D) circulation.
2. Start with number 1 at the top (under CO_2 produced), and follow the numbered sequence. Continue down the right side, coloring all symbols. Then color all the processes of gas exchange in the lungs, beginning with number 5 in the lower right corner.
3. Color the overview diagram within the rectangle.

INTERNAL RESPIRATION

CO₂ PRODUCED   O₂ CONSUMED

TISSUE CELL

RED BLOOD CELL

In tissues, CO₂ production promotes the reaction CO₂ + H₂O → H₂CO₃ → HCO₃⁻ + H⁺.
The consumption of O₂ promotes the reaction H⁺ + HbO₂ → HHb + O₂.

This is shown above as (1) CO₂ diffuses from tissue cells where it is produced into
the plasma and then into red blood cells. (2) In the red cells, combination of the CO₂
with water to form H₂CO₃ is accelerated by the enzyme carbonic anhydrase. (3) The
H₂CO₃ rapidly dissociates into HCO₃⁻ (bicarbonate) and H⁺ ions (acid). (4) The H⁺ are
not left free, a large portion of them combines with oxyhemoglobin. This provides two advantages: blood does not become intolerably acid, and the combination of H⁺
with oxyhemoglobin helps unload the O₂ in the tissues.

CARBON DIOXIDE
WATER
CARBONIC ANHYDRASE
CARBONIC ACID
BICARBONATE
HYDROGEN ION
OXYHEMOGLOBIN
OXYGEN
DEOXYHEMOGLOBIN

In the alveoli, high O₂ and low CO₂ are continually
maintained through the act of breathing, so the
reactions described above are reversed. Here
O₂ + HHb → HbO₂ + H⁺
and H⁺ + HCO₃⁻ → H₂CO₃ → CO₂ + H₂O.

This is shown as (5) O₂ diffuses from the alveoli
into the plasma and then into red cells. (6) O₂
combines with HHb to form HbO₂, releasing H⁺.
(7) H⁺ combines with HCO₃⁻, forming H₂CO₃ and
then (8) H₂O and CO₂. Again the liberated H⁺
does not accumulate; it reacts with HCO₃⁻ and
helps drive off CO₂, which is (9) expelled from the
alveoli with each breath.

ARTERIAL CIRCULATION

VENOUS CIRCULATION

RED BLOOD CELL

LUNG ALVEOLUS

CO₂ EXPIRED   O₂ INSPIRED

EXTERNAL RESPIRATION

NEURAL CONTROL OF RESPIRATION

Skeletal muscles provide the motive force for respiration. Unlike cardiac or smooth muscle, they have no rhythmic "beat" of their own; they depend entirely on the nervous system for a stimulus to contract. Two separate neural systems control respiration: (1) *Voluntary control* originates in *cerebral cortex* neurons, which send impulses down the corticospinal nerve tracts to motor neurons located in the spinal cord, which relay excitatory impulses to the muscles of respiration, the intercostal muscles and the diaphragm. This voluntary system can interrupt or modulate the normal automatic breathing pattern; it is most apparent during speech and while playing wind instruments, where the lungs serve as air reservoirs to be emptied at controlled rates. (2) *Automatic control* originates in lower brain centers, in the *pons* and the *medulla*. Impulses arising in this system also descend in the spinal cord to the motor neurons controlling respiratory muscles, but they travel along nerve tracts lying in the lateral and ventral parts of the cord, separate from the corticospinal tracts. In general, motor neurons to expiratory muscles are inhibited during inspiration and vice versa.

The medulla contains a diffuse network of neurons involved in respiration. Although they are collectively referred to as the *respiratory "center"* (or "centers"), they are not located in nice discrete packages. There are two types of these neurons: the *I neurons*, which fire during inspiration, and the *E neurons*, which fire during expiration. During inspiration, E neurons are actively inhibited; during expiration, I neurons are inhibited.

The primitive rhythm for involuntary breathing is apparently generated by the I neurons. They show bursts of spontaneous activity interspersed with quiet periods about 12 to 15 times/min. In contrast, the E neurons are not self-excitatory; they are excited only by other neurons (including the I neurons) that send impulses to them. When the activity of the inspiratory neurons increases, the rate and depth of breathing increase. The primitive activity of the I neurons, like that of all pacemakers, is modulated by a number of outside influences, including nerve impulses from centers in the pons and from receptors in the lungs. These influences are dramatically revealed after injuries and are outlined in the plate.

If the brainstem is transected below the medulla (at D in the plate), all breathing stops, showing that the brain drives respiration and that communication between brain and respiratory muscles takes place via the spinal cord. But if the transection is made lower in the cord, at E, breathing is not interrupted because the connections between brain and respiratory neurons remain intact, as do motor nerves (i.e., the *phrenic* nerve) that carry the impulses to the muscles of respiration. Regular breathing also continues when all the cranial nerves, including the vagi, are severed, and the brain is transected above the pons at A. These results locate the centers for automatic breathing somewhere between the top of the pons and the lower medulla — clearly, higher brain centers like those in the cortex are not necessary.

Given this localization, we can dissect the respiratory centers even further. If the vagus nerves are cut and two transections are made, one at the top of the pons as before at A and the other in the middle at B, the I cells discharge continuously, arresting respiration in inspiration. This stopping of respiration in sustained inspiration is called *apneusis*, and the neurons in the lower pons, which apparently shower I neurons with excitatory impulses and keep them firing, are collectively referred to as the *apneustic center*. Apneusis occurs only when influences from the upper pons are removed (transection at B). This suggests that neurons in the upper pons continually inhibit the apneustic center, holding its inspiratory drive in check. These neurons are members of another collection called the *pneumotaxic center*. When all pons influence is removed by a transection at C, respiration continues. Although it may be irregular and punctuated with gasps, it is rhythmic, and it demonstrates that the neurons of the respiratory centers themselves have a spontaneous rhythmicity. The role of the pontine centers appears to be to make these rhythmic discharges smooth and regular.

All these responses depend to some extent on whether the vagus nerves are intact. Apneusis, for example, cannot be demonstrated by transection of the mid pons (B) unless the vagi are also severed because vagus nerves carry impulses that originates in stretch receptors located in the lung airways. When the lungs expand during inspiration, these receptors initiate impulses that reflexively inhibit the inspiratory drive, reinforcing the actions of the pneumotaxic center and protecting the lungs from overexpansion. This response is called a Hering-Breuer reflex. In humans, it does not appear to be activated until the tidal volume reaches 1 L, so it plays no part in regulating ventilation during normal quiet breathing.

Several additional factors influence the respiratory centers so that their activity is commensurate with the body's metabolic needs. These include reflexes originating in receptors (proprioceptors) located in muscles, tendons, and joints that are sensitive to movement. They send to the respiratory centers stimulating impulses that presumably help increase ventilation during exercise. Other important reflexes are initiated by low PO_2, low pH, and high PCO_2 in the plasma; these are taken up in detail in plate 52.

CN: Use dark colors for E, G, and S.
1. Begin at the top with voluntary controls from the cerebral cortex (A).
2. Color the large involuntary control diagram of the pons, medulla, and spinal cord. Do not include material on the far right (transections) at this time. The diagram deals with breathing during inspiration; this is emphasized by not coloring the titles and structures from the inhibited expir. neur. (I)

down to the right half of the diaphragm (O¹).
3. To gain anatomical perspective, color the diagram in the lower right.
4. Color the effects (outlined letters A-E) of transections (indicated by a broken line) on breathing patterns along the right margin.
5. Color the title and arrows pointing from chemical controls (T) in the panel at the far left. This inset summarizes material on the next plate.

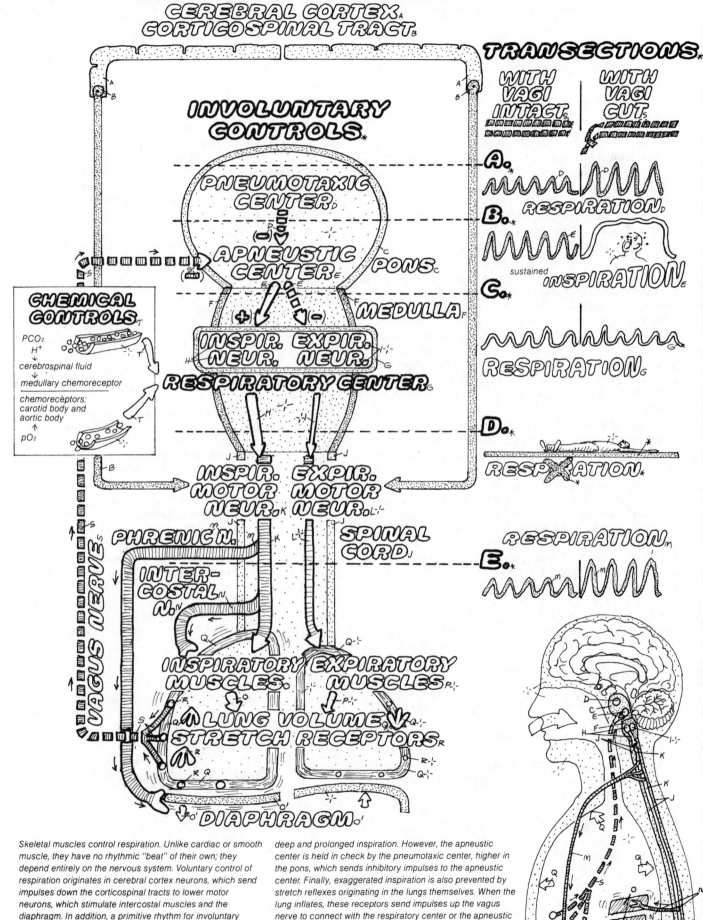

Skeletal muscles control respiration. Unlike cardiac or smooth muscle, they have no rhythmic "beat" of their own; they depend entirely on the nervous system. Voluntary control of respiration originates in cerebral cortex neurons, which send impulses down the corticospinal tracts to lower motor neurons, which stimulate intercostal muscles and the diaphragm. In addition, a primitive rhythm for involuntary breathing is generated by inspiratory neurons located in the respiratory center of the medulla; these also send impulses down the spinal cord to lower motor neurons. The respiratory center is influenced from above by the apneustic center, which excites inspiration and, if left unchecked, would drive deep and prolonged inspiration. However, the apneustic center is held in check by the pneumotaxic center, higher in the pons, which sends inhibitory impulses to the apneustic center. Finally, exaggerated inspiration is also prevented by stretch reflexes originating in the lungs themselves. When the lung inflates, these receptors send impulses up the vagus nerve to connect with the respiratory center or the apneustic center and inhibit inspiration. This reflex links neural control with the actual mechanical response of the lungs. This scheme for neural control was arrived at by the breathing patterns observed when the nervous system is transected at various levels (see text).

CHEMICAL CONTROL OF RESPIRATION

Despite the fact that O_2 consumption and CO_2 production by body tissues vary enormously during daily activities, the PO_2 and PCO_2 are held remarkably constant. Pulmonary ventilation rises and falls to match metabolic needs. How do chemical activities regulate breathing? How do muscles involved in breathing become more active when other parts of the body consume more O_2?

Breathing is regulated by reflexes that respond to the CO_2, O_2, and H^+ levels (PCO_2, PO_2, and pH) of the blood. Of these, PCO_2 is the most important. Whenever plasma PCO_2 rises (as it does during increased metabolism), it is met by a compensatory increase in ventilation, which returns the PCO_2 toward normal. Conversely, when PCO_2 falls, ventilation slows, allowing CO_2 to accumulate until PCO_2 approximates the normal level. This regulation is very sensitive and precise; an increase of arterial PCO_2 by only 1 mm Hg will stimulate an increase in ventilation of about 3 L/min. In common daily activities of rest and exercise, arterial PCO_2 does not appear to vary by more than 3 mm Hg.

The response to PCO_2 is mediated by special areas called *central chemoreceptors* located on the ventral surface of the medulla. These are anatomically distinct from the respiratory centers and are bathed in cerebrospinal fluid, which is separated from blood by the *blood-brain barrier* (i.e., blood capillary membranes that are highly permeable to CO_2, O_2, and water, but only slowly permeable to most other substances). Local application of H^+ ions to these areas rapidly stimulates ventilation. The connection with CO_2 arises because CO_2 easily diffuses through the barrier into the cerebrospinal fluid, where it is converted into HCO_3^- and H^+. Consequently, a rise (fall) in CO_2 is followed by a rise (fall) in H^+ ion concentration in the cerebrospinal fluid. The CO_2 level in the blood regulates respiration by its effect on the H^+ ion concentration in cerebrospinal fluid. (The effect of arterial CO_2 is much stronger than that of arterial H^+ concentration, presumably because CO_2 diffuses through the blood-brain barrier much more easily than H^+.)

When arterial PO_2 drops to very low levels, compensatory increases in ventilation act to return PO_2 toward normal. This response is mediated by a reflex that begins in O_2-sensitive receptors called *peripheral chemoreceptors* located close to the aortic arch and the bifurcation of the carotid arteries. Known as the *aortic* and *carotid bodies*, these receptors are small nodules of tissue containing epithelial-like cells in contact with nerve terminals, together with a profuse blood supply. A drop in PO_2 in the arterial blood supplying these receptors stimulates them. This increases the frequency of impulses sent to the respiratory center, which responds by increasing its discharge along those motor nerves, which increase ventilation. Normally, the PO_2 in alveolar blood can be reduced considerably before this reflex becomes activated so that it does not appear to play a significant role in the day-to-day management of ventilation. However, in cases where arterial PO_2 is markedly reduced ($PO_2 < 60$ mm Hg), for example at high altitudes, in lung disease, or in hypoventilation, this reflex becomes significant.

Increasing the H^+ ion concentration in the plasma also stimulates ventilation. In practice, it is difficult to separate the effects of H^+ ions from PCO_2 because the reaction of H^+ with HCO_3^- produces CO_2. However, experiments where the PCO_2 is artifically maintained at a constant level while the H^+ ion concentration is changed leave no doubt that H^+ ions by themselves stimulate ventilation. This response to H^+ ions is believed to be mediated by the peripheral chemoreceptors.

Under normal circumstances, we rarely encounter a situation where only one of the three chemicals (CO_2, O_2, and H^+) that drive respiration changes. Each time ventilation changes, we can anticipate changes in all three. Because the response to CO_2 is so strong, its regulation most often dominates and sometimes obscures other responses. For example, if the PO_2 of inspired air is suddenly depressed, there will be an increased ventilation due to the peripheral chemoreceptor reflex, but this increased ventilation will also "blow off" CO_2, depressing the PCO_2 in the blood. The decreased PO_2 stimulates respiration, but the secondary decreased PCO_2 inhibits respiration; the two stimuli conflict. As a result, the increased respiration is not nearly as large as it would have been if PCO_2 were held constant. In some instances, the respiratory gases interact in synergistic ways. Depressing PO_2 and elevating PCO_2 both stimulate respiration, but somehow the response due to the two stimuli is greater than the sum of the responses to each alone.

We might anticipate that the large increase in ventilation during exercise is brought about by a lower arterial PO_2 and elevated PCO_2, but this does not seem to be the case. Careful measurements show that PO_2 and PCO_2 remain nearly constant during exercise and can hardly provoke the immense increases in ventilation. Somehow, during exercise, ventilation keeps pace with metabolism so that CO_2 is eliminated as fast as it is produced, and arterial O_2 is supplied as fast as it is consumed. The detailed mechanism for this response is not known.

CN: Use red for D (blood plasma found to the left of the second panel).
1. Begin with the upper panel.
2. Color the CO_2 control of ventilation, beginning with the blood vessel in the upper left corner and continuing clockwise. Note that the curve superimposed near the bottom of the ribs is the diaphragm, a breathing muscle along with the intercostals.
3. Color the bottom panel, starting at the upper left and continuing clockwise but excluding the illustration on the far right. Note that the numbers 1 and 2 (but not their titles) are colored gray.
4. Color the summary diagram on the far right.

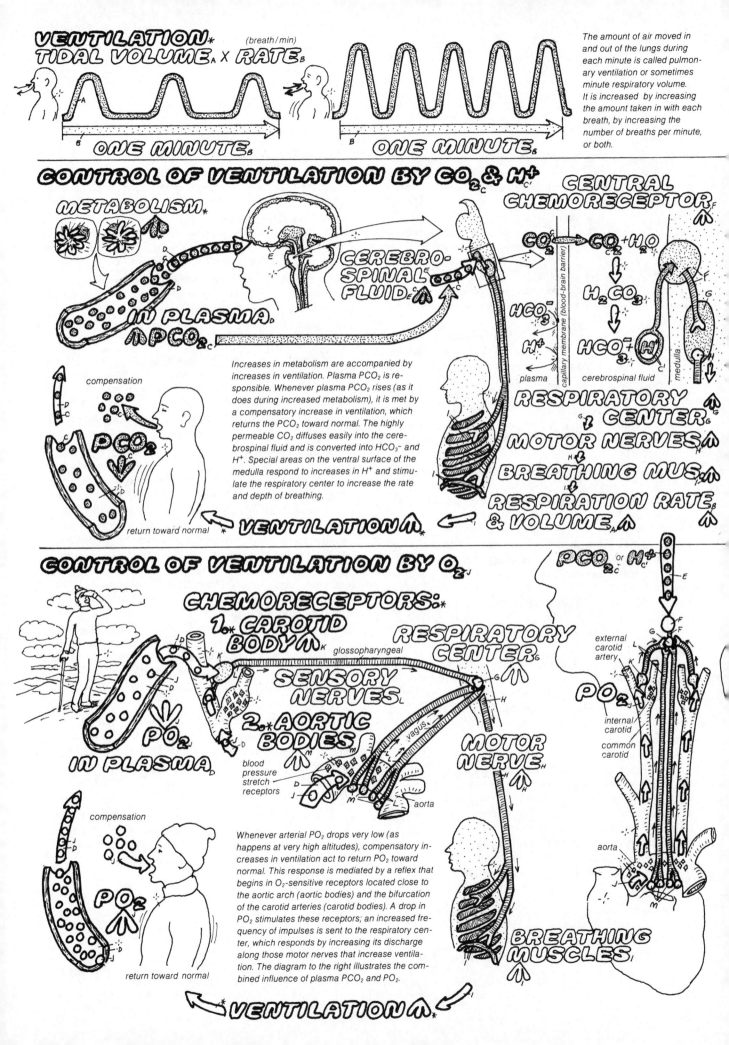

VENTILATION* = TIDAL VOLUME_A × RATE_B

(breath/min)

ONE MINUTE_B

ONE MINUTE_B

The amount of air moved in and out of the lungs during each minute is called pulmonary ventilation or sometimes minute respiratory volume. It is increased by increasing the amount taken in with each breath, by increasing the number of breaths per minute, or both.

CONTROL OF VENTILATION BY CO_2 & H^+

CENTRAL CHEMORECEPTOR

METABOLISM*

IN PLASMA_D

↑ PCO_2_C

compensation

return toward normal

PCO_2

Increases in metabolism are accompanied by increases in ventilation. Plasma PCO_2 is responsible. Whenever plasma PCO_2 rises (as it does during increased metabolism), it is met by a compensatory increase in ventilation, which returns the PCO_2 toward normal. The highly permeable CO_2 diffuses easily into the cerebrospinal fluid and is converted into HCO_3^- and H^+. Special areas on the ventral surface of the medulla respond to increases in H^+ and stimulate the respiratory center to increase the rate and depth of breathing.

CEREBRO-SPINAL FLUID

CO_2 → $CO_2 + H_2O$

HCO_3^-

H^+

H_2CO_3

$HCO_3^- + H^+$

capillary membrane (blood-brain barrier)

plasma — cerebrospinal fluid — medulla

RESPIRATORY CENTER

MOTOR NERVES

BREATHING MUS.

RESPIRATION RATE & VOLUME_A

VENTILATION ↑*

CONTROL OF VENTILATION BY O_2

CHEMORECEPTORS:*

1.* CAROTID BODY_K

glossopharyngeal

RESPIRATORY CENTER_G

SENSORY NERVES

2.* AORTIC BODIES

blood pressure stretch receptors

vagus

aorta

MOTOR NERVE_H

IN PLASMA_D

PO_2

compensation

return toward normal

PO_2

Whenever arterial PO_2 drops very low (as happens at very high altitudes), compensatory increases in ventilation act to return PO_2 toward normal. This response is mediated by a reflex that begins in O_2-sensitive receptors located close to the aortic arch (aortic bodies) and the bifurcation of the carotid arteries (carotid bodies). A drop in PO_2 stimulates these receptors; an increased frequency of impulses is sent to the respiratory center, which responds by increasing its discharge along those motor nerves that increase ventilation. The diagram to the right illustrates the combined influence of plasma PCO_2 and PO_2.

PCO_2_C or H^+_C

PO_2

external carotid artery

internal carotid

common carotid

aorta

BREATHING MUSCLES

VENTILATION ↑*

HYPOXIA

Hypoxia means there is an O_2 deficiency in the tissues. In most cases of severe hypoxia, the brain is the first organ to be affected. If, for example, cabin pressure is suddenly lost in an aircraft flying above 50,000 ft., the inspired PO_2 will fall to less than 20 mm Hg, consciousness will be lost in about 20 sec., and death will follow 4-5 min. later. Less severe hypoxia also affects the brain, producing an inebriated type of behavior, including impaired judgment, drowsiness, disorientation, and headache. Other, non-mental symptoms of hypoxia may include anorexia, nausea, vomiting, and rapid heart rate. Hypoxia has been classified into 4 different types, depending on the cause.

1. HYPOXIC HYPOXIA. This refers to a reduced PO_2 in arterial blood. It occurs in normal people at high altitudes, where the O_2 content of the air is low, and it is also found in lung diseases like penumonia. Symptoms of "mountain sickness" are seen in many people 8-24 hr. after they arrive at high altitudes. These symptoms, which include headache, irritability, insomnia, breathlessness, nausea, and fatigue, gradually disappear in the course of 4-8 days through a process called *acclimatization*.

Acclimatization begins with an increase in ventilation stimulated by the low arterial PO_2. At first, this increase in ventilation is small because it drives off CO_2, so the normal stimulating action of PCO_2 on ventilation has been diminished. However, ventilation steadily increases over the next 4 days as the central chemoreceptor response to low PCO_2 gradually subsides. To understand this gradual reduction of sensitivity to the lowered PCO_2, recall that CO_2 regulates respiration by its effect on the H^+ ion concentration in cerebrospinal fluid (plate 52). Apparently, the body adjusts to a chronically low PCO_2 by raising the H^+ ion concentration in the cerebrospinal fluids back toward normal despite the low PCO_2.

However, the low CO_2 creates another problem: it shifts the bicarbonate reaction in a direction that uses up H^+ ions causing other body fluids to become alkaline (see plate 59). Fortunately, this problem is also handled within the next few days, this time by the *kidneys* as they *excrete more HCO_3^-* (plate 60).

Acclimatization also involves the enhanced production of *2,3 DPG* in red cells. Recall (plate 49) that 2,3 DPG lowers the O_2 affinity of hemoglobin (shifting the saturation curve to the right) so that it releases more O_2 to the tissues. This shift occurs within a day. However, in severe hypoxia, its usefulness is limited because the lowered affinity also makes it harder for Hb to pick up the O_2 in the lungs.

An increase in red blood cell concentration of the circulating blood also begins during the first few days of acclimatization. This raises the concentration of Hb in the blood, thus increasing the blood's capacity to carry Hb even though the PO_2 is low. The stimulus for the enhanced production and release of red cells by the bone marrow is provided by a hormone, *erythropoietin*, which the kidneys secrete in response to hypoxia (plate 136). Although the increased red cell production begins in 2-3 days, it may take several weeks before this response is complete.

In addition, long-term acclimatization also involves a *growth of new capillaries*, thus reducing the distance O_2 must diffuse to move from blood to tissue cell. The myoglobin content of muscle, the number of mitochondria, and the tissue content of oxidative enzymes also increase.

In summary, acclimatization promotes the O_2 supply to tissues in 3 ways: (1) greater delivery of O_2 to the blood via increases in ventilation, (2) enhanced O_2-carrying capacity of the blood due to increases in red cell production, and (3) easier delivery of O_2 to the tissues by means of the 2,3 DPG response and the raised vascularization.

2. ANEMIC HYPOXIA. This occurs when arterial PO_2 is normal, but there is a deficiency in the amount of Hb available to carry O_2. Because *arterial PO_2* is normal, there is little if any stimulation of peripheral chemoreceptors. However, compensatory increases in 2,3 DPG are often sufficient to remove hypoxia distress during rest. Difficulties arise during exercise because the ability to enlarge O_2 delivery to active tissues has been reduced. Anemias arise from a variety of causes; some are nutritional (e.g., *iron deficiency*), others are genetic (e.g., *sickle cell anemia*). The symptoms of anemic hypoxia also appear in *carbon monoxide poisoning* because carbon monoxide competes with O_2 for binding sites on the Hb molecule, reducing the amount of Hb available to carry O_2. (Hb's affinity for carbon monoxide is about 200 times larger than its affinity for O_2!) An additional handicap arises because, in the presence of carbon monoxide, any "surviving" HbO_2 binds its O_2 more tenaciously, making it less available to the tissues.

3. STAGNANT (OR ISCHEMIC) HYPOXIA. In this condition PO_2 and Hb are normal, but O_2 delivery to the tissue is impaired because of poor circulation. This is particularly a problem in the kidneys and heart during shock and may become a problem for the liver and possibly the brain in congestive heart failure.

4. HISTOTOXIC HYPOXIA. This arises when the tissue cells are poisoned and cannot utilize the O_2, even though the O_2 delivery rate is adequate. Cyanide poisoning, which inhibits oxidative enzymes, is the most common source of this syndrome.

CN: Use the same color as on previous page for O_2 (D).
1. Begin in the upper panel, starting on the right with O_2 loading onto Hb.
2. Color the vertical panel on hypoxic hypoxia on the left. Then color the material within the rectangle dealing with acclimation.
3. Color the remaining cartoon panels below.

NORMAL OXYGEN TRANSPORT_D

Oxygen is loaded from the lungs (dock) onto hemoglobin molecules (boats), which are moved by the circulation (stream) to the tissues (dock), where it is unloaded.

TISSUE CELLS_A   HEMOGLOBIN_B
CARDIAC OUTPUT_C   OXYGEN_D

(TISSUE)   (LUNGS)

HYPOXIC HYPOXIA ⌄

When there is a deficiency of O_2, or O_2 utilization at the tissues, the condition is called hypoxia. There are many different types of hypoxia, a common one, hypoxic hypoxia, occurs whenever the arterial PO_2 is low (e.g., at high altitudes).

SYMPTOMS OF ALTITUDE ILLNESS *

headache
dizziness
confusion
disorientation
insomnia
fatigue
vomiting
diarrhea

OTHER CAUSES: *

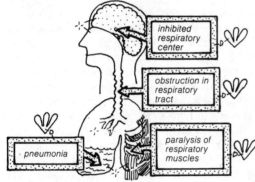

inhibited respiratory center ⌄_D

obstruction in respiratory tract ⌄_D

pneumonia

paralysis of respiratory muscles ⌄

HIGH ALTITUDE ACCLIMATIZATION *

TEMPORARY -:- (minutes or days)

VENTILATION ⋀_D-
CARDIAC OUTPUT ⋀_C
BLOOD REDISTRI-BUTION_C-
HEMOGLOBIN RELEASE OF_B O_2 ⋀_D

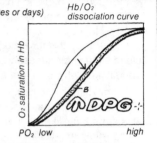

Hb/O_2 dissociation curve

O_2 saturation in Hb

⋀ DPG -:-

PO_2 low   high

Compensatory responses to hypoxia involve an increased ventilation, an increased delivery rate of Hb to the tissues, a redistribution of blood flow favoring the heart and brain, an increased hemoglobin and red cell content of the blood, a shift in the O_2 saturation curve of Hb (to the right) so that it releases more O_2 to the tissues, and tissue changes that facilitate O_2 diffusion into the tissues.

LONG TERM -:- (days or months)

RED BLOOD CELLS ⋀_B'
HEMOGLOBIN ⋀_B
CAPILLARY GROWTH ⋀_E

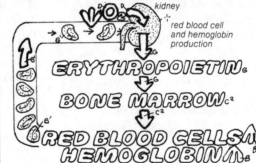

kidney
red blood cell and hemoglobin production

ERYTHROPOIETIN_G
BONE MARROW_C2
RED BLOOD CELLS ⋀_B'
HEMOGLOBIN ⋀_B

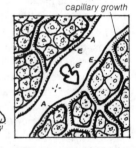

capillary growth

HYPOXIC_D ⌄_D HYPOXIA_D

(TISSUE)   (LUNGS)

In hypoxic hypoxia, the cells, the circulation, and the Hb content are normal, but the primary O_2 supply is deficient — each boat carries 3 O_2 instead of 4.

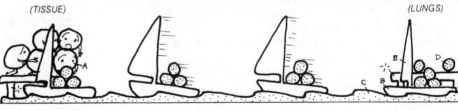

ANEMIC_B ⌄_B HYPOXIA_D

In anemic hypoxia, there is a deficiency of Hb — not enough boats.

iron deficiency -:-

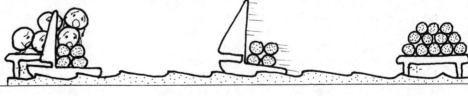

STAGNANT_C ⌄ HYPOXIA_D

In stagnant (ischemic) hypoxia, the circulation is failing — the stream is stagnant.

shock

HISTOTOXIC_A ⌄ HYPOXIA_D

In histotoxic hypoxia, O_2 delivery is normal, but the cells can't utilize it (e.g., as when they are poisoned with cyanide).

poisoning -:-

INTRODUCTION TO KIDNEY STRUCTURE

Kidneys produce urine. Under normal resting conditions, the kidneys, which comprise less than 0.5% of the body weight, receive 25% of the cardiac output! Each minute some 1300 mL of blood enter the kidneys through the *renal arteries*, and approximately 1298-1299 mL leave via *renal veins*, with the difference, 1-2 mL, leaving as urine via the *ureter*. Why all this fuss (hogging one quarter of the body's blood supply) for a measly 2 mL of urine? What does urine contain and why is its formation so important?

At first glance, the composition of the urine is not impressive: water, salt, small amounts of acid, and a variety of waste products like urea. What is impressive is how urine composition and volume *change* to compensate for any fluctuation in volume or composition of body fluids. The composition of the body fluids is apparently determined not by what the mouth takes in but by what the kidneys keep. The design of the gastrointestinal tract appears to maximize absorption indiscriminately without regard for quantities. The kidneys are the guardians of the internal environment; they rework the body fluids fifteen times a day. When the body is dehydrated, the volume of water excreted decreases; when body fluids become more acid, kidneys excrete more acid; if the K^+ content of body fluids rises, the kidneys excrete more K^+. "We have the kind of internal environment we have because we have the kidneys we have" — Homer Smith.

The kidneys are about the size of a clenched fist. They lie against the back abdominal wall, just above the waistline. The outer covering of the kidney, called the *capsule*, is thin but tough and fibrous. When it is cut open, two regions appear: an outer zone (the *cortex*) and an inner region (the *medulla*). These gross sructures do not provide much of a clue to how the kidney works. However, a microscopic view reveals the unit of kidney function, the *nephron*. Each kidney has about 1 million nephrons, which are tubular structures about 45 to 65 mm long and about .05mm wide. Their walls are made of a single layer of epithelial cells.

A funnel-like structure about 0.2 mm in diameter called *Bowman's capsule* comprises the top end of the nephron. These capsules are always found in the cortex. Fluid flows through the lumen of the tubule from the Bowman's capsule into the next section, the *proximal tubule*, which has a "curly" or convoluted section and then a straight portion that dips into the medulla. This section, about 15 mm long, is called the proximal tubule because it is near the origin of the nephron (Bowman's capsule). Fluid then flows into a long, thin tube that plummets straight toward the depths of the medulla. This is the descending limb of the *loop of Henle*. At its lowest point, the loop makes a hairpin turn and begins to ascend out of the medulla back toward the

cortex, becoming considerably thicker toward the later portions of its ascent. In the cortex, the ascending limb of the loop becomes continuous with the distal tubule. Finally, the distal tubule empties into the *collecting duct*, a tube that gathers fluid from several nephrons.

There are two major classes of nephrons. The majority, called cortical nephrons, originate in the outer portions of the cortex and are characterized by short loops of Henle that reach only the outer regions of the medulla. The remaining nephrons, which comprise only about 15% of the total, originate closer to the medulla and are known as juxtamedullary nephrons. These have very long loops of Henle that reach deep into the medulla; they are important for water conservation in the body.

Individual collecting ducts coalesce into larger tubular structures, and this pattern repeats until several of the larger tubes empty into a still larger funnel structure, the *renal pelvis*. Fluid in the renal pelvis is identical to urine. The renal pelvis is continuous with the ureter, which leaves each kidney to convey urine to the bladder, where it is stored until elimination via the urethra.

The blood supply to the nephrons is special because it consists of *two* capillary beds in series. Each Bowman's capsule has its own capillary bed called a *glomerulus*. (Sometimes the combined structure, Bowman's capsule + glomerulus, is referred to as the glomerulus.) The vessel bringing blood to the glomerulus is called the *afferent arteriole*. Blood leaving the glomerulus does not enter a venule; rather, it enters another arteriole, the *efferent arteriole*, which serves as a conduit to the second capillary bed, called *peritubular capillaries*. The peritubular capillaries are so interconnected that it is difficult to tell which capillary came from which efferent arteriole; the tubules of any one nephron probably receive blood from several efferent arterioles. Efferent arterioles from juxtamedullary nephrons also form peritubular capillaries in much the same way, but, in addition, they send off branches, which are straight tubes that follow the descending limbs of Henle's loops deep into the medulla, turn at the bend of the loop, and ascend back toward the cortex. These hairpin loops of blood vessels are called *vasa recta*; their design is important for water conservation.

By the time the fluid in the nephron has passed through the collecting ducts to reach the pelvis, it has become urine. Plate 55 shows how fluid simply filters out of the glomerular capillaries into Bowman's capsule. From here, it flows along the lumen of the nephron and is modified by the epithelial cells of the tubules and the collecting ducts until it finally becomes urine.

CN: Use red for A structures, blue for B, purple for R, and yellow for H. Use dark colors for J and T.
1. Begin with the cut-away drawing of the kidney in the upper right. Color the section of a kidney showing the location of two types of nephrons. Note that these have been greatly enlarged for diagrammatic purposes.
2. Color the enlarged view of a kidney section in the lower right corner. Begin with the entry of blood (A¹) at the bottom

and color the arteries and arterioles. Note that the afferent and efferent arterioles have been given different colors (P & Q) to distinguish them from the other vessels. Color the blood circulation before coloring the structures of the nephron: K-O.
3. Color the lower left diagram of the glomerulus and the flow of filtrate through the nephron. Color in the Bowman's capsule (K) first, and then the related structures.

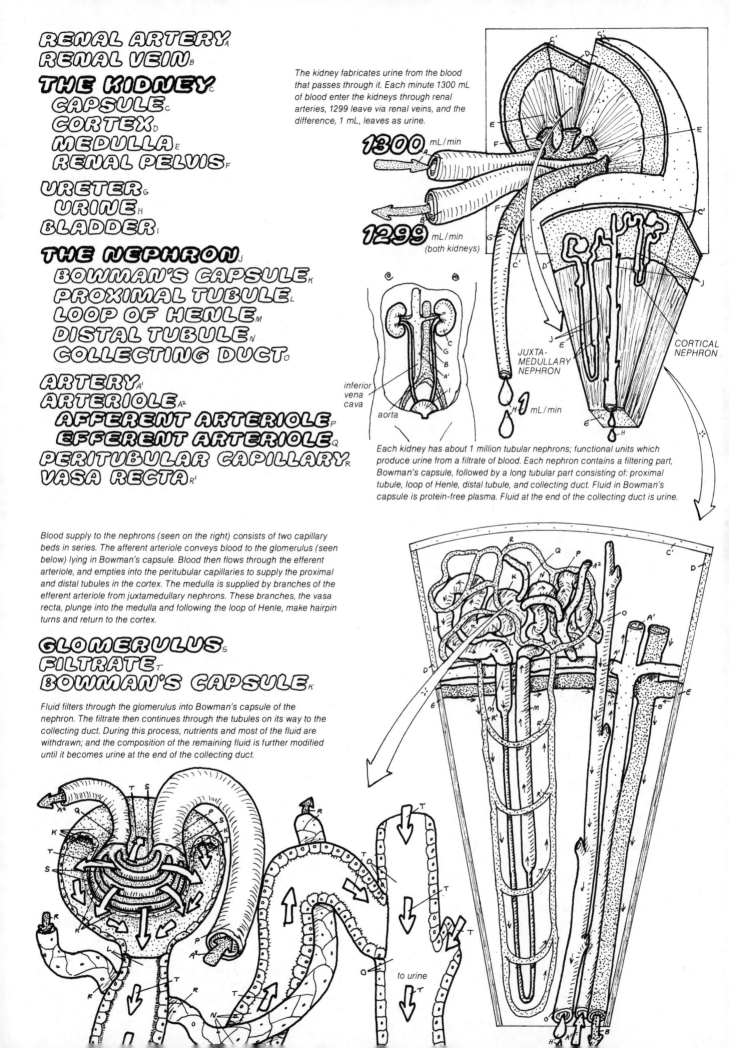

RENAL ARTERY. A
RENAL VEIN. B

THE KIDNEY C
CAPSULE. C
CORTEX. D
MEDULLA. E
RENAL PELVIS. F

URETER. G
URINE. H
BLADDER. I

THE NEPHRON. J
BOWMAN'S CAPSULE. K
PROXIMAL TUBULE. L
LOOP OF HENLE. M
DISTAL TUBULE. N
COLLECTING DUCT. O

ARTERY. A¹
ARTERIOLE. A²
AFFERENT ARTERIOLE. P
EFFERENT ARTERIOLE. Q
PERITUBULAR CAPILLARY. R
VASA RECTA. R¹

The kidney fabricates urine from the blood that passes through it. Each minute 1300 mL of blood enter the kidneys through renal arteries, 1299 leave via renal veins, and the difference, 1 mL, leaves as urine.

1300 mL/min

1299 mL/min (both kidneys)

inferior vena cava

aorta

JUXTA-MEDULLARY NEPHRON

CORTICAL NEPHRON

1 mL/min

Each kidney has about 1 million tubular nephrons; functional units which produce urine from a filtrate of blood. Each nephron contains a filtering part, Bowman's capsule, followed by a long tubular part consisting of: proximal tubule, loop of Henle, distal tubule, and collecting duct. Fluid in Bowman's capsule is protein-free plasma. Fluid at the end of the collecting duct is urine.

Blood supply to the nephrons (seen on the right) consists of two capillary beds in series. The afferent arteriole conveys blood to the glomerulus (seen below) lying in Bowman's capsule. Blood then flows through the efferent arteriole, and empties into the peritubular capillaries to supply the proximal and distal tubules in the cortex. The medulla is supplied by branches of the efferent arteriole from juxtamedullary nephrons. These branches, the vasa recta, plunge into the medulla and following the loop of Henle, make hairpin turns and return to the cortex.

GLOMERULUS. S
FILTRATE. T
BOWMAN'S CAPSULE. K

Fluid filters through the glomerulus into Bowman's capsule of the nephron. The filtrate then continues through the tubules on its way to the collecting duct. During this process, nutrients and most of the fluid are withdrawn; and the composition of the remaining fluid is further modified until it becomes urine at the end of the collecting duct.

to urine

FILTRATION, REABSORPTION, AND SECRETION

By the time fluid in the nephron passes through the collecting ducts to reach the pelvis, it has become urine. What is the fluid in the nephron and how did it get there in the first place? The unusual pattern of blood circulation to the kidney provides a clue. Nowhere else in the body do we find one capillary bed (the *glomerulus*) leading into an arteriole (*efferent arteriole*), which in turn leads into another capillary bed (*peritubular capillaries*). The pressure in a typical capillary located elsewhere in the body begins around 35 mm Hg and falls some 20 mm until it reaches 15 mm Hg at the venous end of the vessel. If glomerular capillaries were typical, pressure in blood delivered to the efferent arteriole would be only 15 mm Hg, hardly enough to drive blood through the next vessel, the narrow efferent arteriole. Thus, the pressures in these capillary beds must be atypical.

Pertinent data for normal kidney blood pressures in humans is unavailable, and it is difficult to obtain accurate values even in animals, where measurements are compromised by anesthesia, surgical trauma, and blood loss. The best figures for monkeys, dogs, and rats indicate glomerular blood pressure is high, around 50 mm Hg. The pressure drop in passing through these capillaries is small, only a few mm Hg; apparently, glomerular capillaries have a low resistance. However, like that of any arteriole, the resistance of the efferent arteriole is considerable; by the time blood reaches the peritubular capillaries, the pressure has fallen to about 15 mm Hg.

What physiological significance can we attach to these aberrant pressures? Glomerular pressures are abnormally high; peritubular pressures are abnormally low. Fluid transfer across capillary walls is determined by the balance of capillary *blood pressure* and *oncotic pressure*, so the high glomerular pressure suggests that a net fluid *filtration* occurs at these capillaries and that fluid within Bowman's capsule is simply a *filtrate* of blood plasma (i.e., fluid that would be obtained from blood if it were strained through a porous filter, in this case the porous walls of the glomerular capillary). This has been verified experimentally. Fluid at the beginning of the nephron does not arise out of any active transport process; proteins and cells are simply separated from the plasma by a passive filtration process.

This filtration is the first step in modifying a portion of blood plasma that will eventually be excreted as urine. As the fluid flows along the nephron past the cells making up the tubular walls, substances may be withdrawn from the fluid and returned to the blood via the peritubular capillaries; this process is called *reabsorption*. Alternatively, some substances may be removed from the blood and added to the tubular fluids in a process called *secretion*. Glomerular filtration followed by tubular reabsorption and secretion are the fundamental processes by which the kidney regulates the internal environment.

The funnel like structure of *Bowman's capsule* allows it to collect the filtrate and convey it to the lumen of the proximal tubule. Fluid that filters from the blood into the lumen of the nephron must pass through three potential barriers: (1) the thin cell layer making up the capillary wall (the capillary endothelium), (2) the *basement membrane* associated with the capillary, and (3) the epithelial cell layer making up Bowman's capsule. The capillary endothelium is riddled with *fenestrations*, which are easily penetrated by most molecules but not by cells. The basement membrane and the outer surfaces of both cell layers are embedded with glycoproteins that contain a strong negative charge. The last barrier, those epithelial cells of Bowman's capsule that are in direct contact wth the capillaries, have a peculiar structure; they are called *podocytes*. Podocytes send out foot processes that interdigitate with foot processes of other cells. Vacant spaces or *slits* remain between these foot processes. The filtration pathway through the capillary fenestrations, across the basement membrane, and through the slit passages does not hinder the passage of small molecules like salts, glucose, and amino acids. As the size of the molecule increases, the pathway begins to offer some resistance, depending on the molecule's size, shape, and electrical charge. Electrically neutral molecules, the size of the plasma protein albumin, can permeate this barrier to a limited extent. However, albumin, which is negatively charged, is restrained by the negative charge on the basement membrane and cell surfaces.

The glomerular capillaries not only have a higher pressure, they are also more permeable than many capillaries. Both factors promote filtration. Gomerular capillaries filter twenty times more fluid than ordinary capillaries. Fully one-fifth of the fluid entering the capillary is delivered to the nephron via the filtration path. This loss of fluid from the blood concentrates the remaining proteins. By the time blood enters the efferent arteriole and peritubular capillaries, the oncotic pressure (osmotic pressure exerted by the plasma proteins) has risen from a normal value of 25 to 30 mm Hg.

Just as high pressure prepares the glomerulus for filtration, low presssure in the peritubular capillary promotes fluid reabsorption over its entire length. The major force for reabsorption from interstitial spaces arises from the osmotic pressure of the plasma proteins, and we have seen that this is unusually high in peritubular capillaries. Because the opposing filtration pressure (blood pressure in the capillary) is low, the peritubular capillary is well adapted to reabsorb fluid from its environment. However, these arguments apply only to reabsorption between capillary and interstitial fluids. Fluid reabsorption between nephron and interstitial space is governed by a more complex set of osmotic forces that are determined by the concentrations of small solutes, especially NaCl. This follows because, compared to capillaries, nephron walls are much less permeable to small solutes, so the contribution of these small solutes to effective osmotic pressure gradients becomes large.

CN: Use red for A, light blue for G, and dark colors for H, J, and L.
1. Begin with the diagram of filtration, reabsorption and secretion by coloring the horizontal blood vessels, blood and plasma proteins before coloring the upper diagrams.

2. Color the elements of the renal corpuscle, beginning with the small exterior diagram, then the functional diagram on the far left, and then the interior view. Notice that the slit pore (I'), represented by arrows, refer to a narrow space between the cells of the podocytes (P).

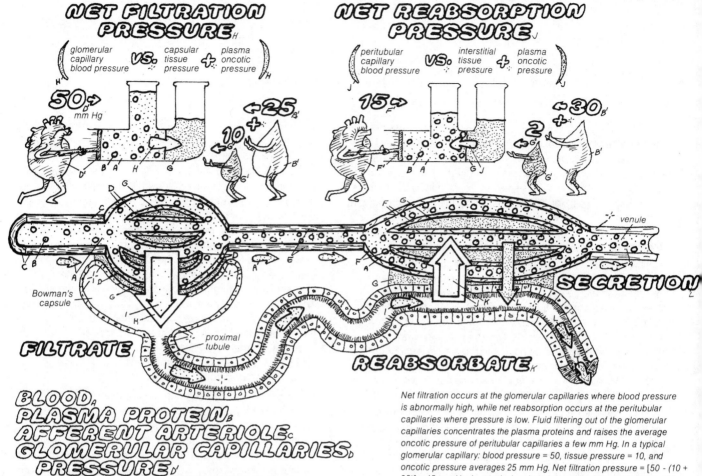

NET FILTRATION PRESSURE

glomerular capillary blood pressure **VS.** capsular tissue pressure **+** plasma oncotic pressure

50 mm Hg → ← 25, 10 +

NET REABSORPTION PRESSURE

peritubular capillary blood pressure **VS.** interstitial tissue pressure **+** plasma oncotic pressure

15 → 2 + ← 30

venule

SECRETION

Bowman's capsule

FILTRATE

proximal tubule

REABSORBATE

BLOOD, A
PLASMA PROTEIN, B
AFFERENT ARTERIOLE, C
GLOMERULAR CAPILLARIES, D
PRESSURE, D'
EFFERENT ARTERIOLE, E
PERITUBULAR CAPILLARIES, F
PRESSURE, F'
INTERSTITIAL FLUID, G PRESS, G'
ONCOTIC (OSMOTIC) GRADIENT, B'

Net filtration occurs at the glomerular capillaries where blood pressure is abnormally high, while net reabsorption occurs at the peritubular capillaries where pressure is low. Fluid filtering out of the glomerular capillaries concentrates the plasma proteins and raises the average oncotic pressure of peritubular capillaries a few mm Hg. In a typical glomerular capillary: blood pressure = 50, tissue pressure = 10, and oncotic pressure averages 25 mm Hg. Net filtration pressure = [50 - (10 + 25)] = 15 mm Hg. In a typical peritubular capillary: blood pressure = 15, tissue pressure is small, possibly 2, and oncotic pressure averages 30 mm Hg. Net reabsorption pressure = [(2 + 30) - 15] = 17 mm Hg. Fluid obtained from Bowman's capsule is identical to blood with no cells and no proteins. As fluid flows in the nephron, substances are withdrawn and returned to blood via peritubular capillaries, a process called **reabsorption**. Other substances may be removed from blood and added to tubular fluids, a process called **secretion**.

THE RENAL CORPUSCLE. *

GLOMERULUS, L
CAPILLARY FENESTRATIONS, M
BASEMENT MEMBRANE, N
BOWMAN'S CAPSULE, O
PODOCYTE, P
SLIT PORE, I'

To filter from blood to nephron, a substance must pass through two cell layers plus the basement membrane that separates them. Blood cells and proteins are too large to pass, but smaller substances can. Capillary walls have many fenestrations and are very permeable. The basement membrane is also very permeable; it contains fixed negative charges which help repel plasma proteins (also negatively charged). The last cell layer, called podocytes, has slit like pores between cells. Podocytes are part of Bowman's capsule.

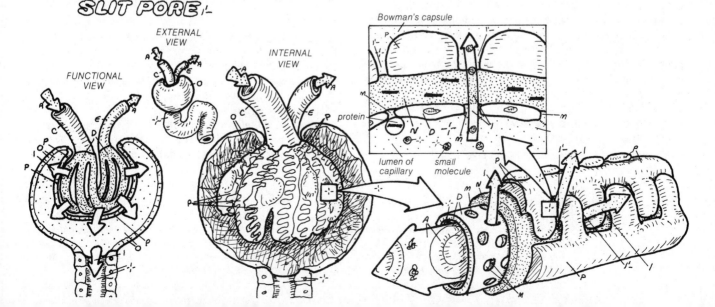

FUNCTIONAL VIEW

EXTERNAL VIEW

INTERNAL VIEW

Bowman's capsule

protein

lumen of capillary

small molecule

FUNCTIONS OF THE PROXIMAL TUBULE

Approximately 120 mL of protein-free plasma filter into the nephrons each minute. If this fluid were excreted as urine, it would take only 25 min. (3000 mL plasma ÷ 120) to exhaust the entire plasma volume. This fluid would carry with it everything dissolved in plasma (glucose, amino acids, minerals, vitamins, etc.). The fact that you are reading this page is living proof that this does not happen. The tubules recapture (reabsorb) most of the fluid, practically all the nutrients, and some minerals before the filtrate reaches the ends of the collecting ducts.

The nephron is primarily a regulatory organ. Faced with a torrent of filtered fluid at its origin, it must almost immediately reduce the volume of the filtrate to manageable levels and reclaim essential nutrients. The responsibility for this task falls primarily on the *proximal tubule*. By the time the filtrate reaches the end of the proximal tubule, two-thirds of the water and virtually all of the nutrients have invariably been reabsorbed. Of the original 120 mL of fluid that entered through the filter, only 40 mL pass on to the loop and distal tubule, where more subtle regulatory processes take place.

This massive transport requires asymmetric tubular cells. Note in the bottom diagram that the cell membrane facing the lumen is convered with fingerlike projections called *microvilli*. They resemble bristles in a brush; hence, the membrane is called the *brush border*. The membrane surrounding the remaining three quarters of the cell has no microvilli; it is called the *baso-lateral* membrane (plate 2). These two membranes have different properties; they contain different proteins, enzymes, and transport systems. The two membranes are separated by tight junctions that prevent migration of any proteins from one membrane to the next. The baso-lateral membrane resembles membranes of most cells; it contains an abundance of Na^+-K^+ *pumps* and *facilitated diffusion systems* for glucose and amino acids (plates 9, 10). The brush border does not contain these transport systems, but it contains others.

The prime mover for most proximal tubular transport is the *active transport of Na^+*, which keeps intracellular Na^+ concentration more than ten times lower than extracellular. Because Na^+ concentration is higher in the lumen than in tubular cells, it moves down its concentration gradient into the cell. But it cannot be pumped back out into the lumen, because the brush border has no Na^+ pump; it can be pumped out of the cell into the interstitial spaces only by the baso-lateral membrane. The result is a stream of Na^+ diffusing from lumen to cell, only to be pumped out into the interstitial space, where it can diffuse into the peritubular capillary. In other words, Na^+ is reabsorbed. But Na^+ carries a positive charge; it thus attracts negatively charged ions, which move along with it. Because Cl^- is the most abundant permeable negative ion, we finally reabsorb large quantities of Na^+ and Cl^-.

Although the tubular walls are permeable to Na^+ and Cl^-, the transport pathways are fairly restrictive; so that both Na^+ and Cl^- are important determinants of the effective *osmotic gradients* across the tubular cell. Each time a Na^+ and Cl^- are transported from lumen to interstitial space, the lumen loses two osmotically active particles while the interstitial space gains two. This creates an osmotic gradient favoring reabsorption of water. For each Na^+ and Cl^- moved, about 370 water molecules follow to maintain osmotic equilibrium. Once the water arrives in the interstitial space, the high oncotic pressure (and low blood pressure) in the peritubular capillaries are sufficient to absorb the water back into the blood. The loss of water from the tubular lumen concentrates the remaining solutes, and those that are freely permeable to tubular membranes will diffuse down the resulting concentration gradient from lumen to interstitial space. So in addition to reabsorption of Na^+, the asymmetric active transport of Na^+ is also responsible for reabsorption of Cl^-, copious amounts of water, and some fraction of other diffusible solutes.

But our Na^+ story does not end there; Na^+ transport is also coupled to the reabsorption of glucose and amino acids. The brush border contains one system that *co-transports* Na^+ and glucose and another that co-transports Na^+ and amino acids (plate 9). The operation of the two systems is similar; we shall describe only Na^+-glucose. This system transports Na^+ and glucose together but will not operate with either alone. The system is symmetric; it does not require ATP, and it is capable of transporting the pair into the cell or out. In practice, the co-transport system always transports the pair into the cell because of the Na^+-K^+ pump, which keeps intracellular Na^+ scarce and makes it difficult for glucose to find a Na^+ partner to ride the co-transport system in the reverse direction. By keeping intracellular Na^+ low, the cell creates a one-way system for glucose transport. As a result, glucose accumulates inside the cell even above its concentration in the lumen or plasma; it is as though glucose has been actively transported into the cell. And in a way, it has; only now the energy for transport has come from the Na^+ gradient and only indirectly from the splitting of ATP. Once glucose is inside the cell in higher concentrations, it moves out of the baso-lateral membrane toward the blood via a facilitated transport system that does not require Na^+. The transport of amino acids is analogous.

Proximal tubules also play a role in acid-base balance and in regulating calcium, magnesium, and phosphorus. In addition, they have active transport systems for *secretion* of organic acids and bases from blood to lumen. This system is important clinically because many drugs fall into this category. The secretory pump is located on the baso-lateral membrane, so the secreted material is accumulated in the cell. The brush border is permeable to these substances, and they move from the cell, where they are highly concentrated, to the lumen.

CN: Use the same colors as on the previous plate for proximal tubule (A), filtration (B), secretion (E), and capillary (C).
1. Begin with the upper diagram, coloring the filtrate (B¹) entering the proximal tubule (A) on the far left.
2. In the lower diagram, first color the filtrate arrow, the tubular cells (A²) with their brush borders (F), and the capillary wall on the right. Then follow the order of the titles and color the various transport mechanisms. Note that the substances being transported receive the color of that particular mode of transport, so that it is possible for Na^+ to receive four different colors. Start with the ATP-driven sodium pump at the asterisk in the lower right corner of the center cell.

THE PROXIMAL TUBULE.

The proximal tubule invariably reabsorbs 2/3 of the water and virtually all of the nutrients in the filtrate. It also secretes organic acids and bases into the lumen. Of the original 120 mL of fluid that enters through the filter, only 40 mL is passed on to the loop and distal tubule where more subtle regulatory processes take place.

FILTRATION.

Bowman's capsule

PROTEIN-FREE PLASMA:
water
salts and minerals
glucose, amino acids
vitamins, hormones
urea & other small solutes

FILTRATE

VOLUME IN
120 mL / min

PERITUBULAR CAPILLARY.

BLOOD.

SECRETION.

fatty acids, prostaglandins,
uric acid, bile salts
cyclic AMP, acetyl choline,
epinepherine, histamine, dopamine
saccharin, aspirin, penicillin,
morphine, cimetidine
others

loop of Henle

VOLUME OUT
40 mL / min

REABSORPTION.

water (2/3)
glucose, amino acids, vitamins
some salts and minerals
some urea and other permeable
solutes

80 mL / min

THE TUBULAR CELL

BASOLATERAL MEMBRANE
BRUSH BORDER (MICROVILLI)
ACTIVE TRANSPORT
CO-TRANSPORT
DIFFUSION
FACILITATED DIFFUSION
COUNTER TRANSPORT
OSMOSIS

Most reabsorptive processes are coupled to movements of Na^+. Na^+ is pumped out of the tubular cell by a Na^+-K^+ pump located in the basolateral membrane, but not in the brush border. This provides a one way movement from lumen to blood. Cl^- follows Na^+ because of electrical attraction. Movement of Na^+ and Cl^- creates an osmotic gradient which drags water with it. By linking with downhill Na^+ movement into the cell, other solutes are pumped uphill. Glucose and amino acids are co-transported with Na^+ into the cell. H^+ is counter-transported out of the cell (in exchange for Na^+).

tight junction

channel

tubular lumen

FILTRATE

CAP.

red blood cell

Na^+
GL
(also amino acids)

GL

GL

GL

Na^+

Na^+

Na^+
H^+

Na^+

H^+

Na^+

K^+

K^+

K^+

ATP

large protein

interstitial fluid

capillary lumen

$NaCl$

H_2O

Na^+

Cl^-

H_2O

H_2O

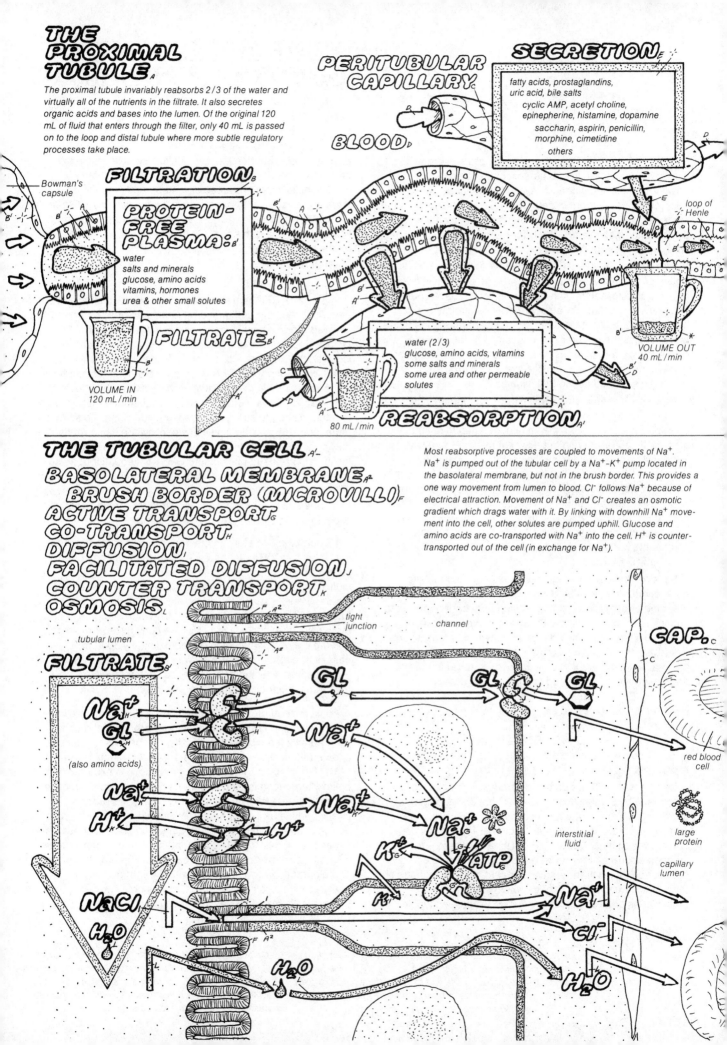

MEASURING FILTRATION, REABSORPTION, & SECRETION

In modern times it has been possible to micro-dissect the kidney in anesthetized animals, collect fluid at different positions in individual nephrons, and tease out portions of nephrons to study them in detail. However, techniques for the study of quantitative aspects of the whole kidney have been available for many years. The latter techniques are particularly valuable because they are non-invasive and can be readily applied to unanesthetized humans.

The principle involved is simple: what goes in must come out. It is an application of the conservation of matter. Suppose you knew that 100 mg of a sugar were filtering into the nephrons each minute, but only 60 mg were appearing in the urine. Unless the nephrons were destroying the sugar, 40 mg (100 - 60) were reabsorbed. Alternatively, if 120 mg appeared in the urine, you would conclude that 20 mg (100 - 120 = –20) had been secreted during that minute.

How do we estimate how much goes through the filter and how much comes out in the urine during each minute? The latter is easy. Collect the urine over a period of time, say an hour. Analyze it to find out how much sugar there is in each milliliter (i.e., determine the concentration of the sugar in the urine), then multiply this figure by the total number of milliliters of urine collected. This gives the amount excreted per hour. To find the amount excreted per minute, divide by 60. Letting E = *the amount* of a solute excreted per minute, U_S = the concentration of the solute in the urine, and V = the volume of urine that is excreted per minute, we have:

$$E = U_S \times V. \qquad (1)$$

Estimating the amount of *solute* going through the filter each minute (called *filtered load*) is trickier. A related quantity, the number of milliliters of fluid flowing through the filter each minute, is called the *glomerular filtration rate*, abbreviated as GFR. If we knew the GFR, the problem would be easier. Let P_S = the concentration of any solute, say sugar, in the blood plasma. Then the amount of sugar coming through the filter each minute (i.e., the filtered load), F, will be given by

$$F = P_S \times GFR. \qquad (2)$$

For our final bookkeeping on tubular activities (*reabsorption* or *secretion*), which we denote by RS_S,

$$RS_S = F - E = [P \times GFR] - [U_S \times V]. \qquad (3)$$

If RS_S is positive, it represents reabsorption. If it is negative, it represents secretion.

Using equation 3, we can calculate RS_S, provided we can measure all the quantities on the right-hand side. Three of these, P_S, U_S, and V, are routine. The fourth, GFR, is not. Turning the problem inside out, if we knew RS: for any substance, we could solve for GFR. Fortunately, these substances exist; *inulin* is one of them. Inulin is a nontoxic polysaccharide that is small enough to pass freely through the filter but too large to pass through solute channels in cell membranes or through the tight junctions between tubular

epithelial cells. Further, inulin is not lipid soluble so it won't permeate the lipid bilayer portion of the cell membrane. Finally, inulin is not produced or metabolized in the body; there are no special transport systems that will carry it. In particular, the tubules neither secrete nor reabsorb inulin; $RS_{inulin} = 0$. Using this fact, we rewrite the above expression for the special case of inulin: $0 = [P_{in} \times GFR] - [U_{in} \times V]$. Solving for GFR:

$$GFR = [U_{in} \times V] / P_{in}.$$

In practice, GFR is measured by injecting a small amount of inulin, collecting and analyzing blood and urine samples at intervals, and using this last expression for calculation. For historical reasons, the ratio $[U_{in} \times V] / P_S$ for any substance s is called the *clearance* of s. The GFR is equal to the *inulin clearance*. Notice that GFR is simply the amount of fluid flowing through the filter per minute. It really has nothing to do with inulin. Inulin is merely an artificial substance that we use to trace the filtrate so we can measure its volume. To study a more interesting solute, call it S, we follow the same routine; only now we analyze the blood and urine for both inulin and S. Inulin data are used to calculate GFR from equation 3 as before, and this result, together with the blood and urine data for S, is used in equation 3 to calculate RS_S. These procedures have been used both clinically to test renal function and experimentally to study renal mechanisms.

Through the use of inulin clearance, an estimate of a normal value for GFR = 120 mL/min. (both kidneys) has been obtained. This means that each day $120 \times 60 \times 24 = 172,800$ mL of fluid pass through the glomerular filter into the lumens of the nephrons, a space that is essentially outside the body. That is an enormous amount of fluid, a volume approximately three times the total volume of all the body fluids. It means that the entire plasma volume (approximately 3000 mL) passes through the nephrons every (3000 ÷ 120) = 25 min.! Put in another way, by selective reabsorption and secretion, the renal tubular cells renew the plasma every 25 min.

Application of clearance techniques to glucose excretion is illustrated in the lower diagram. Various amounts of glucose were administered to systematically change plasma glucose concentrations from normal to very high. At normal levels (70-110 mg/100 mL) and below, no glucose is excreted; the entire filtered load is reabsorbed. As plasma concentration is increased, so is the filtered load. Eventually, we reach a threshold plasma concentration where almost all reabsorption sites are working to maximal capacity, and some glucose escapes reabsorption, spilling over into the urine. The maximal capacity to reabsorb glucose is called the T_M (tubular max). The diagram shows how reabsorption RS for glucose changes with plasma concentration. It is obtained by subtracting E from F at each concentration.

CN: Use the same colors as on the previous page for filtration (C), reabsorption (E), and excretion (D). Use light blue for A and a dark color for B. Note that the two shapes at the top are Bowman's capsule (filtration) and a drop of urine (excretion).
1. Begin with the formula at the top.

2. Color the three central panels, starting on the left. Note that in the first two panels the number of boxes filtering is kept artifically small for purposes of simplicity.
3. Color the procedure for the measurement of glucose reabsorption into the blood stream.

WATER (mL)ₐ SOLUTE CONCENTRATION (mg/mL)ᵦ

Kidney performance can be assessed by simple bookkeeping; i.e., measuring the net balance between inflow through the filter (filtration), and outflow into the urine (excretion). For each substance (S): Inflow + number of mL of plasma filtering in per min (GFR) × amount of S contained in each mL. Outflow = number of mL of urine excreted per min × amount of S contained in each mL of urine.

FILTRATION꜀ EXCRETION (URINE)꜀

$$GFR \times P - V \times U = \text{the amount of } REABSORPTION \text{ or } SECRETION$$

filtrate

If the difference, RS, between filtration and excretion is positive, RS represents net reabsorption, if negative, then net secretion.

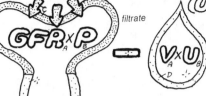

$$\left(\text{jug}_A \times \text{box}_B \right)_C - \left(\text{jug}_D \times \text{box}_B \right)_D = \text{tray} \quad mg/min$$

FILTRATION & REABSORPTION꜀

$$(5 \times 2)_C - (2 \times 3)_A = 4_E$$
REABSORPTION_E

Count the number of solute particles (circles) filtering in during each minute. Each box of water contains 2 and there are 5 boxes per min. (GFR) so that filtered load = 5 × 2 = 10 per min. Similarly, there are 2 × 3 = 6 particles leaving (excreted) per min. The difference, RS = 10 − 6 = 4 particles per min., is reabsorbed.

FILTRATION & SECRETION꜀

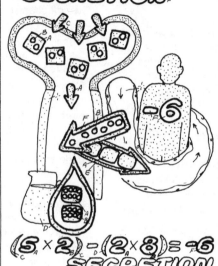

$$(5 \times 2)_C - (2 \times 8)_A = -6$$
SECRETION_F

Count the number of particles filtering in per min: 5 × 2 = 10 per min. Similarly there are 2 × 8 = 16 particles leaving per min. The balance (difference), RS = −6 is negative. 6 particles are secreted per min.

FILTRATION ALONE꜀

INULINᵦ'

$$(120 \times 2)_C - (1 \times 240) =$$
REABSORPTION = 0
SECRETION = 0_F

Inulin is a special substance; it is not reabsorbed and not secreted $RS_{in} = 0$ and filtered load = excretion. We use this fact to measure GFR. Inulin is injected into plasma, and samples of plasma and urine are taken for analysis. If 240 inulin particles are excreted per min. and each mL of plasma only contains 2 inulin, then we require 120 mL of plasma per min. to deliver the 240 particles being excreted GFR = 120 mL per min. The algebra is shown below.

$$(GFR \times P)_C - (V \times U) = 0_E$$

$$GFR = \frac{V \times U}{P_{in}}$$

$$GFR = C_{IN}$$
(CLEARANCE OF INULIN)ᵦ'

MEASURING GLUCOSE REABSORPTION (RS_GL)_E

To study how glucose reabsorption depends on plasma glucose.

1. INJECT: GLUCOSEᵦ & INULINᵦ'

3. GFR = $\frac{V \cdot U_{IN}}{P_{IN}}$

2. TAKE SAMPLES & MEASURE

plasma   urine

P_{GL} U_{GL}
P_{IN} U_{IN}
V

Calculate the GFR from measured values of P_{in}, U_{in}, and V. Use this value of GFR together with measured values of P_{gl}, U_{gl} and V to calculate RS_{gl}.

4. $(GFR \times P_{GL}) - (V \times U_{GL}) = RS_{GL}$_E

A plot of results of these measurements at different plasma glucose concentrations (P_{gl}) is shown here. RS_{gl} is obtained by subtracting the amount excreted from the amount filtered at each concentration. At normal levels (A) and below, no glucose is excreted; the entire F is reabsorbed. Increasing P_{gl} increases F and RS_{gl}. Eventually a P_{gl} is reached where almost all reabsorption sites are working to maximal capacity (B) and some glucose spills over into the urine (C). Increasing F beyond this level cannot increase RS_{gl}.

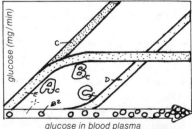

glucose (mg/min)

glucose in blood plasma

A   B   C

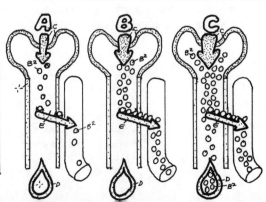

REGULATION OF THE GFR

Control of the *glomerular filtration rate (GFR)* is crucial to kidney performance. An abnormally fast filtration will swamp the tubules, allowing filtrate to speed by the cells before they have time to modify the fluid. Abnormally slow rates will compromise the kidneys' ability to process adequate amounts of fluid during each minute. Nevertheless, blood flow to the kidney does change, often in response to stresses not directly related to kidney function (e.g., a sudden drop in arterial pressure—plate 40). How can blood flow to the kidney undergo significant changes without upsetting GFR and renal function?

Panel A shows that renal blood flow is reduced by *sympathetic nerve* impulses, which constrict arterioles, but the effect of these impulses on GFR depends on which arterioles are most constricted. Constricting the *afferent arteriole* reduces renal blood flow, causing downstream (glomerular) pressure to decrease, thereby decreasing GFR. Constricting the *efferent arteriole* also reduces renal blood flow, but now the glomerulus is upstream. Its pressure rises, and GFR increases. Because the afferent arterioles contain more smooth muscle, we may expect their constriction to be the more forceful. But even in these cases, the simultaneous constriction of the efferent arteriole can be expected to diminish changes in GFR that might otherwise occur.

Panel B illustrates an important property of renal blood vessels: both renal blood flow and GFR are very insensitive to changes in systemic arterial blood pressure in the range of 80 to 180 mm Hg. (Compare the flat part of the curves with the dotted diagonal line that would be expected if the blood vessels were simple passive structures.) Shared by most vascular beds, this behavior is most pronounced in the kidneys. Due to properties of the smooth musculature of the vessel walls, this behavior persists when all nerve supplies are cut but disappears when the smooth muscle is paralyzed with drugs. Apparently, the blood vessel smooth muscles are sensitive to pressure. When pressure rises, flow would ordinarily increase, but the smooth muscle in the arteriolar walls contracts, reducing the radius of the vessel and increasing its resistance. As a result, flow does not increase as much, and energy (pressure) is lost flowing through the high resistance. Thus, capillary pressure and the ensuing GFR do not increase as much.

The kidneys' capacity to regulate body fluids is especially sensitive to the rate at which fluid is delivered to the *distal tubule*. This is where regulation of salt, water, and acidity occurs. If flow is too fast, the distal tubule cells will be overwhelmed; if it is too slow, there is danger of overcompensation. The lower diagram on the left shows a *feedback mechanism* that adjusts the GFR *in each single nephron* to maintain a constant load delivery to the distal tubule. The

beginning of the distal tubule of each nephron is located next to its corresponding glomerulus and makes contact with the afferent arteriole in a specialized structure called the *juxtaglomerular (JG) apparatus*. As flow increases, solute delivery (probably Cl^-) to the JG apparatus increases and in some unknown way stimulates constriction of the afferent arteriole so that GFR in the same nephron decreases. Conversely, as flow decreases, GFR increases. In this way, the GFR is matched to the reabsorption capacity of the proximal tubule. The mechanism is particularly interesting because, unlike the two mechanisms described above, it is a discrete, local regulation; each nephron has its own independent control system. If, for example, the glomerulus of a particular nephron becomes damaged and leaky so that the filtration rate in that nephron increases, the feedback will constrict the afferent arteriole of that nepron and no others.

Finally, we describe two simple *physical* mechanisms that operate to match proximal fluid reabsorption to GFR. If, for some reason, GFR increases, so does *proximal tubular reabsorption*. If GFR goes down, reabsorption decreases. To understand the first mechanism, illustrated on the bottom of the plate, recall that fluid reabsorption is determined by net Na^+ reabsorption. But net Na^+ reabsorption is given by the difference between active pumping of Na^+ (lumen to interstitial space) and the back leak of Na^+ through tight junctions (TJ) in the reverse direction. If GFR decreases, compensatory reductions in proximal fluid reabsorption occur because of the following. With a small GFR, less fluid is removed from glomerular capillaries, so the plasma proteins become less concentrated as they flow through the glomerulus. This means that the oncotic pressure delivered to the peritubular capillaries is lowered, reducing the forces favoring fluid reabsorption by these capillaries from the interstitial fluid. The buildup of fluid in the interstitial space will increase the tissue pressure, which may force the seal between cells (the tight junction) to leak so that both water and Na^+ leak back into the tubular lumen. The steps are reversed when GFR increases, resulting in a compensatory increase in reabsorption.

The second mechanism that helps match changes in tubular reabsorption to changes in GFR depends on the coupling of fluid reabsorption to solute reabsorption, particularly to Na^+, which is co-transported with glucose and amino acids. With normal GFR, these co-transported nutrients are completely reabsorbed before they reach the end of the proximal tubule. With higher GFR, more solute is filtered, and the more distant reaches of the tubule begin to be utilized. More solute is reabsorbed so more fluid is also reabsorbed. Those distant portions of the proximal tubule not used to transport glucose or amino acids during normal GFR supply a reserve for reabsorption under increased loads.

CN: Use red for the D letters including RBF (renal blood flow —D³). Use dark colors for A and B.
1. After noting the question posed at the top of the page, color the titles: 1. above and 2. below, which are the two major categories covered. Begin coloring in the upper left corner, which defines the area of the nephron that deals with the subject matter of this page. Note the area enclosed in the rectangle in the upper part of the illustration. These are the

cells of the juxtaglomerular apparatus (I) which is composed of part of the distal tubule (G) and the afferent arteriole (C).
2. Color the systemic sympathetic (J) control of the GFR.
3. Color example B., noting the sharp rise in both RBF and GFR when the arterial blood pressure exceeds 180 mm Hg.
4. Color the response of distal tubule feedback to a rise in the GFR. Begin with the GFR.
5. Color the bottom panel, completing the left example first.

1. REGULATE GLOMERULAR FILTRATION RATE (GFR)

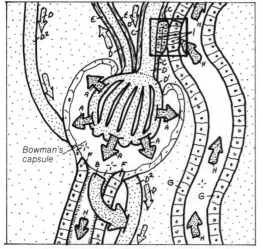

Bowman's capsule

AFFERENT ARTERIOLEc
BLOOD,D **GLOMERULUS**,D'
EFFERENT ARTERIOLEE
PERITUBULAR CAPIL.D²
PROXIMAL TUBULEF
DISTAL TUBULE,G
SOLUTESH
JUXTAGLOMERULAR APPARATUS (JG),

A. SYSTEMIC SYMPATHETIC CONTROLS.

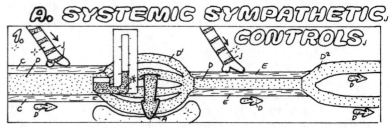

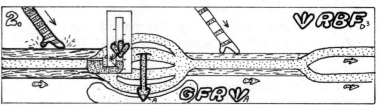

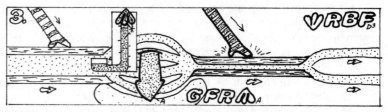

Renal blood flow (RBF) is controlled by sympathetic nerve impulses which constrict arterioles, but their effect on GFR depends on which arterioles are constricted the most. Constricting the afferent arteriole reduces RBF, causing downstream (glomerular) pressure to decrease, thereby decreasing GFR (2). Constricting the efferent arteriole also reduces RBF, but now the glomerulus is upstream. Its pressure rises and GFR increases (3).

C. DISTAL TUBULAR FEEDBACK.G

GFR in each single nephron is adjusted to maintain a constant delivery to the distal tubule by negative feedback. As flow increases, solute delivery (probably Cl⁻) to the JG apparatus increases and stimulates constriction of the afferent arteriole, so that GFR in the same nephron decreases.

FEEDBACK LOOP*

cells of the juxtaglomerular apparatus

B. LOCAL RESPONSE TO ARTERIAL PRESSURE CHANGE*

RBF and GFR are insensitive to changes in systemic arterial pressure (compare with dotted line that would be expected if blood vessels were simple passive structures). Blood vessels regulate flow by constricting when pressure within goes up. This provides more resistance and prevents the elevated pressure from causing a proportional increase in flow.

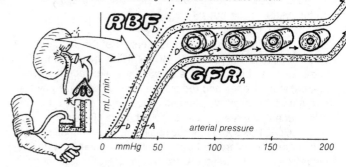

RBF

GFR

mL/min.

0   mmHg 50   100   150   200
arterial pressure

2. REGULATE PROXIMAL TUBULE REABSORPTION.B

Fluid reabsorption is determined by net Na⁺ reabsorption, which is the balance between active pumping of Na⁺ (lumen to interstitial fluid, ISF) and back leak of Na⁺ through tight junction (TJ) in the reverse direction. If GFR decreases, compensatory decreases in proximal fluid reabsorption occur because: less filtration → lower concentration ISF plasma proteins (oncotic pressure) delivered to peritubular capillaries.→ less capillary reabsorption of fluid from ISF → increased pressure in ISF→ → increased back leak of Na⁺ through TJ → → less net Na⁺ reabsorption →less fluid reabsorption. Conversely an increase in GFR induces a compensatory increase in reabsorption.

NORMAL GFRA

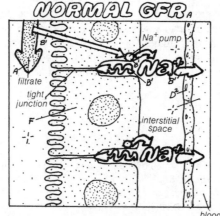

Na⁺ pump
filtrate
tight junction
interstitial space

▼GFRA

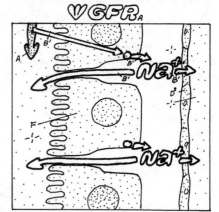

blood plasma

INTRODUCTION TO ACID-BASE BALANCE

Acid-base balance refers to the complex array of mechanisms employed to regulate the concentration of *hydrogen ions* in the body fluids, even though H^+ ions are present only in trace amounts. There is almost four thousand times more Na^+ in blood plasma than H^+; yet H^+ is important because it is so reactive. It is simply a positive charge (a proton) that can easily attach to a variety of molecules, especially proteins, changing their charge and how they interact. Pure water contains 0.0001 mM H^+ (pH = 7.0). Any aqueous solution that contains more H^+ is *acidic*; if it contains less, it is called *alkaline*. Blood contains 0.00004 mM H^+ (pH = 7.4); it is slightly alkaline.

Although *free* H^+ ions are scarce, there are huge numbers of potential H^+ lurking in the background bound to other substances. To understand acid-base balance, we must take into account substances that are sources of H^+ and those that may absorb H^+, as well as following the concentration of free H^+ ions. Sources are called acids; an acid is a substance that gives up H^+. A strong acid gives up most of its H^+; a weak one gives up only part. A base is a substance that takes up H^+.

A *buffer* is a pair of substances that resist changes in acidity of a solution. It works by storing (binding) the H^+. When H^+ is added to a solution containing buffer, it is "soaked up" by empty storage sites on some of the buffer molecules. When H^+ is removed, it is replaced by H^+ that had been stored on other buffer molecules. In order to work, a buffer must have some molecules with storage sites that are occupied by H^+ while other molecules have empty (storage) sites. Those buffer molecules with occupied sites are acids (they can give up H^+); those that are unoccupied are bases (they can take up H^+).

The pair *bicarbonate / carbonic acid* forms an important buffer system in the body:
$$H^+ + HCO_3^- \rightleftharpoons H_2CO_3 \rightleftharpoons H_2O + CO_2.$$
H_2CO_3 (carbonic acid) is the acid member of the pair because it can release H^+. Bicarbonate, HCO_3^-, is the base member because it can bind H^+. In water, this step takes about a minute, but in the kidney and red blood cells, it is catalyzed by the enzyme carbonic anhydrase and is completed within a fraction of a second. The reaction is so rapid that we often identify CO_2 with H_2CO_3. This system is especially important because two of its components are rigorously controlled by the body: the *lungs control CO_2*, and the *kidneys control HCO_3^-*. Although there are other buffers in the body, this simple chemical reaction links the lungs and kidneys and allows them to maintain a viable H^+ concentration in the body fluids.

Each day an average person on a mixed diet produces 60 mM of H^+ in the form of sulfuric, phosphoric, and organic acids. These are called *metabolic acids* because they do not arise from CO_2, and the disturbances in H^+ they create must eventually be corrected by the kidney. When metabolic H^+ is produced in any organ, most of it is picked up by HCO_3^- in the blood and forms CO_2. The increased CO_2 plus the increased H^+ stimulate respiration, which helps eliminate the increased CO_2. In this case, the bicarbonate reaction shown above moves to the right, downhill, because one of the reactants, H^+, is continually produced while one of the products, CO_2, is continually removed.

The respiratory regulation of H^+ described above will work only if the bicarbonate that is used can be replenished. This task is accomplished by the kidneys, where the bicarbonate reaction takes place in the reverse direction. This reversal in direction occurs because the kidneys remove the H^+ as fast as it forms and excrete it in urine. In the process, the newly formed HCO_3^- is reabsorbed. Thus, the kidney manufactures HCO_3^- without retaining the attendant H^+ (see plate 60). The result is that for every H^+ produced by metabolism, one H^+ is excreted in the urine, and one bicarbonate is reabsorbed.

A H^+ concentration below 0.00002mM (pH = 7.7) or above 0001 mM (pH = 7.0) is incompatible with life. If plasma becomes more acid than normal, the condition is called *acidosis*; when it is less acid, the condition is *alkalosis*. In either case, it is useful to recognize whether the disturbance arises from respiratory or other (metabolic) causes. The best clues come from studies of the buffer pair HCO_3^-/H_2CO_3. An increase of H_2CO_3 will tend to increase H^+; an increase in HCO_3^- will "soak up" free H^+ and reduce its concentration.

Respiratory acid-base disturbances are reflected by changes in plasma CO_2 or the equivalent H_2CO_3. If these are depressed, as in rapid breathing, there is a diminution of suppliers of H^+; the H^+ concentration goes down; the condition is respiratory alkalosis. Compensation by the kidney requires an excretion of HCO_3^- to rid the plasma of a disproportionate amount of substances that soak up the scarce H^+. Conversely, in pneumonia or polio, there is a failure to eliminate CO_2 (and H_2CO_3); plasma acidity rises; the condition is respiratory acidosis. Renal compensation consists of elevating the plasma HCO_3^- to a level commensurate with the elevated H_2CO_3.

Nonrespiratory acid-base disturbances are called metabolic disturbances. When plasma HCO_3^- decreases and plasma H^+ increases, the condition is called metabolic acidosis. The increased H^+ signifies acidosis, and the decreased HCO_3^- implicates its nonrespiratory origin. These occur, for example, in renal failure and in diabetes. Respiratory compensation occurs because the increased H^+ stimulates breathing, which reduces CO_2 and H_2CO_3. Finally in metabolic alkalosis, which sometimes occurs during vomiting of HCl from the stomach, there is an increased HCO_3^- with decreased H^+. Respiratory compensation consists of CO_2 and H_2CO_3 retention.

CN: Use a dark color for A. Notice that three different carbon substances receive the B color (B^1, B^2, B^3).
1. Color the two examples of variation of blood pH level.
2. Color the bicarbonate reaction at the top of the lower panel.
3. Starting at the asterisk in the liver, follow hydrogen and color the downhill reaction to the release of CO_2 into the kidneys and color the downhill process to the release of acid in the urine and the passage of bicarbonate to the plasma.

STRONG ACID
HYDROGEN ION, H^+

BUFFER PAIR
BICARBONATE, HCO_3^-
CARBONIC ACID, H_2CO_3

WHAT IS A BUFFER?

A buffer is a pair of substances which resists changes in acidity of a solution.

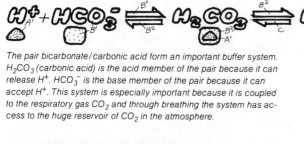

holding one's breath

ACID
pH 7.4-7.1

When acid (H^+) is added to the blood the pH decreases. Then increased acidity (decreased pH) is minimized by buffers which bind some of the added H^+.

hyperventilation

BASE
pH 7.4-7.7

When acid is taken away, blood becomes more alkaline (pH increases). This change is minimized by buffers which release H^+ and replace some of the acid that had been lost.

NORMAL PLASMA pH LEVEL 7.4

PLASMA pH LEVEL

BUFFER ACTION

$$H^+ + HCO_3^- \rightleftharpoons H_2CO_3 \rightleftharpoons H_2O + CO_2$$

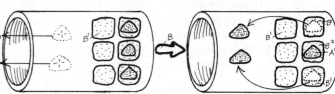

The pair bicarbonate/carbonic acid form an important buffer system. H_2CO_3 (carbonic acid) is the acid member of the pair because it can release H^+. HCO_3^- is the base member of the pair because it can accept H^+. This system is especially important because it is coupled to the respiratory gas CO_2 and through breathing the system has access to the huge reservoir of CO_2 in the atmosphere.

RESPIRATORY REGULATION (MINUTES)
LUNGS
RESPIRATORY CENTER
NERVE

Each day an average person on a mixed diet produces 60 mM of H^+ in the form of sulfuric, phosphoric, and organic acids. These are called metabolic acids because they do not arise from CO_2. The figures shows metabolic H^+ produced in the liver. Most of it is picked up by HCO_3^- in the blood and forms CO_2. The increased CO_2 plus the increased H^+ stimulate respiration which helps eliminate the increased CO_2. The bicarbonate reaction moves to the right (downhill) because H^+ is continually produced while CO_2 is continually removed.

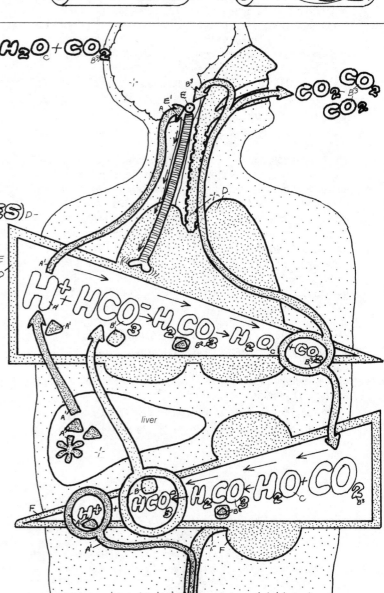

RENAL REGULATION (HOURS)
KIDNEYS

Respiratory regulation shown above will work only if the bicarbonate that is used can be replenished. This task is accomplished by the kidneys where the bicarbonate reaction takes place in the reverse direction. This reversal in direction can occur because the kidneys continually remove the newly formed H^+ and excrete it in the urine. In the process, the newly formed HCO_3^- is reabsorbed. Thus, the kidney is able to manufacture HCO_3^- without retaining the attendant H^+.

RENAL REGULATION OF ACID-BASE BALANCE

Regulation of plasma acidity begins in the *proximal tubule* (top diagram), where some 80-90% of the filtered HCO_3^- is reabsorbed. The process involves:

1. CO_2 and *water* combine to form *carbonic acid* inside the cell. The reaction is catalyzed by the enzyme *carbonic anhydrase*. Carbonic acid dissociates into H^+ and HCO_3^-, two products with separate fates.

2. HCO_3^- diffuses down its concentration gradient through the baso-lateral membrane into the blood; it is reabsorbed. The mechanism used by HCO_3^- to cross the baso-lateral membrane is not clear. Apparently, the luminal membrane (brush border) is much less permeable to HCO_3^-, accounting for its one-way transport into the blood.

3. H^+ moves in the opposite direction and crosses the luminal membrane via a special Na^+-H^+ exchanger located in the luminal, but not the baso-lateral, membrane. Intracellular H^+ is exchanged for luminal Na^+ and is *counter-transported* into the lumen of the nephron. This counter-transporter works in the indicated direction because of the high concentration gradient for Na^+ tending to drive it into the cell.

4. The secreted H^+ then combines with the HCO_3^- that has filtered into the nephron, forming H_2CO_3. The external membrane surfaces of the proximal tubule contain carbonic anhydrase, so the H_2CO_3 is quickly converted to water and CO_2. The CO_2 simply diffuses down its gradient into the cell, where it can enter the cycle at step 1. Note that the H^+ that gets secreted in the above cycle never finds its way into the urine. The principal role of the proximal tubule in acid-base balance is to reclaim the HCO_3^- that comes through the filter.

To focus on HCO_3^- reabsorption, it is instructive to use the plate (after coloring) to trace the fate of the carbon dioxide in its various disguises, beginning with its entrance in the filtrate in the form of HCO_3^-. The HCO_3^- is converted to H_2CO_3 and then to CO_2 in the lumen. Moving into the cell, this or other CO_2 is then transformed back into H_2CO_3 and finally into HCO_3^-, which is reabsorbed. The diagram shows that the process is driven by the continuous flow (secretion) of H^+, which in turn is driven by the continuous flow (reabsorption) of Na^+. Energy for the entire process comes from the ATP energy expended by the Na^+-K^+ pump, which is responsible for the Na^+ gradient.

Similar processes are at work inside *distal tubule* cells, but the filtrate no longer contains much HCO_3^- (most of it has been reabsorbed in the proximal tubule). Further, H^+ is still secreted, but now instead of being driven by Na^+ exchange, newer evidence suggests that it is directly coupled to ATP splitting. In any case, the ability of kidney cells to secrete H^+ is limited; if the free H^+ concentration in the lumen gets too high (pH$<$4.5), H^+ secretion stops. The luminal H^+ is prevented from rising too high by buffers, especially *phosphate buffers*, which are present in the filtrate and which bind the free H^+.

$$H^+ + HPO_4^{2-} \rightarrow H_2PO_4^-$$

The kidneys also manufacture *ammonia*, NH_3. Ammonia, like CO_2, is soluble in lipids and passes through cell membranes with ease; it diffuses into the lumen, where it binds H^+ to become NH_4^+ (*ammonium*). Ammonium is positively charged and cannot get through cell membranes very easily; it is "trapped" in the lumen and will be excreted along with the attached H^+.

$$H^+ + NH_3 \rightarrow NH_4^+$$

The H^+ excreted into the urine bound to ammonia and to phosphate compensate for metabolic acids. Note that for each H^+ excreted, a HCO_3^- that wasn't present in the fitrate is reabsorbed. Some call it "new" HCO_3^- to distinguish it from reabsorbed HCO_3^- that simply replicates the HCO_3^- that came through the filter.

In *acidosis*, plasma H^+ rises, and the kidneys compensate by excreting acid in the urine. (In respiratory acidosis, there is an increased CO_2, which promotes H^+ formation and secretion by the kidney cells. In metabolic acidosis, the filtered load of HCO_3^- is reduced so that it is less effective in trapping secreted H^+ in the lumen. Once secreted, H^+ has more of a chance to escape into the urine.) All the HCO_3^- is reabsorbed, and most of the HPO_4^{2-} is converted to $H_2PO_4^-$. Chronic acidosis stimulates the kidneys to synthesize more NH_3. This provides more buffering capacity in the filtrate so that more secreted H^+ can be carried (in the form of NH_4^+) without substantially decreasing the pH of the filtrate. Thus, the H^+ gradient from cell to lumen does not increase to prohibitive levels despite the increased H^+ secretion.

In *alkalosis*, both plasma and intracellular H^+ fall. This includes kidney cells, making less intracellular H^+ available for secretion. As a result, HCO_3^- reabsorption does not go to completion and some HCO_3^- escapes into the urine. The urine becomes alkaline so that blood leaving the kidney is more acidic than blood entering.

CN: Use red for B and the same colors that were used on the previous page for water (D), hydrogen ion (F), carbon dioxide (C), carbonic acid (C¹) and bicarbonate (C²).
1. Begin with the proximal tubule by first coloring the cell membrane (A²), (A³) and capillary (B). Next color the role of Na^+ (G) in this process by following it down the tubule lumen as part of the filtrate, through the membrane via a counter transport mechanism (G¹) and out of the cell via the sodium-potassium pump (G³). Then follow the numbered sequence starting at the top of the cell.
2. Do the same with the distal tubule by coloring the Na^+ material first. Note that because of space limitations, the step preceding the formation of carbonic acid (C¹), the combination of carbon dioxide and water plus enzyme action, has been omitted in the bottom cell.
3. Color the center panel summarizing these processes.

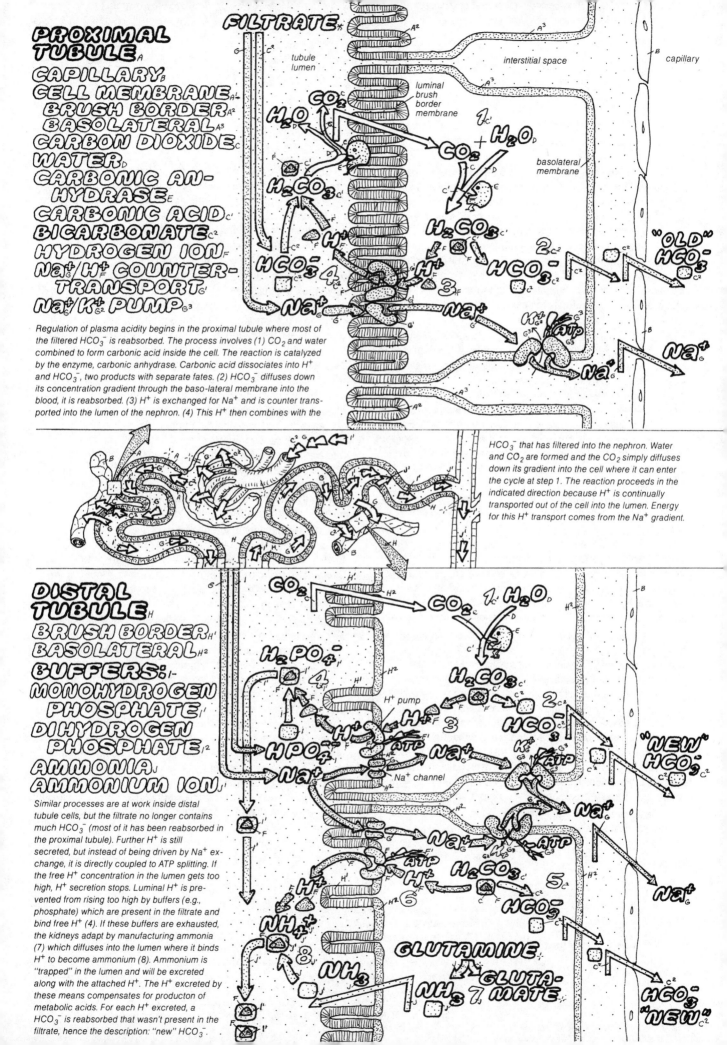

PROXIMAL TUBULE $_A$

CAPILLARY $_B$
CELL MEMBRANE $_{A^1}$
  BRUSH BORDER $_{A^2}$
  BASOLATERAL $_{A^3}$
CARBON DIOXIDE $_C$
WATER $_D$
CARBONIC AN-HYDRASE $_E$
CARBONIC ACID $_{C^1}$
BICARBONATE $_{C^2}$
HYDROGEN ION $_F$
Na$^+$/H$^+$ COUNTER-TRANSPORT $_{G^1}$
Na$^+$/K$^+$ PUMP $_{G^3}$

Regulation of plasma acidity begins in the proximal tubule where most of the filtered HCO_3^- is reabsorbed. The process involves (1) CO_2 and water combined to form carbonic acid inside the cell. The reaction is catalyzed by the enzyme, carbonic anhydrase. Carbonic acid dissociates into H^+ and HCO_3^-, two products with separate fates. (2) HCO_3^- diffuses down its concentration gradient through the baso-lateral membrane into the blood, it is reabsorbed. (3) H^+ is exchanged for Na^+ and is counter transported into the lumen of the nephron. (4) This H^+ then combines with the

HCO_3^- that has filtered into the nephron. Water and CO_2 are formed and the CO_2 simply diffuses down its gradient into the cell where it can enter the cycle at step 1. The reaction proceeds in the indicated direction because H^+ is continually transported out of the cell into the lumen. Energy for this H^+ transport comes from the Na^+ gradient.

DISTAL TUBULE $_H$

BRUSH BORDER $_{H^1}$
BASOLATERAL $_{H^2}$
BUFFERS: $_I$
MONOHYDROGEN PHOSPHATE $_{I^1}$
DIHYDROGEN PHOSPHATE $_{I^2}$
AMMONIA $_J$
AMMONIUM ION $_{J^1}$

Similar processes are at work inside distal tubule cells, but the filtrate no longer contains much HCO_3^- (most of it has been reabsorbed in the proximal tubule). Further H^+ is still secreted, but instead of being driven by Na^+ exchange, it is directly coupled to ATP splitting. If the free H^+ concentration in the lumen gets too high, H^+ secretion stops. Luminal H^+ is prevented from rising too high by buffers (e.g., phosphate) which are present in the filtrate and bind free H^+ (4). If these buffers are exhausted, the kidneys adapt by manufacturing ammonia (7) which diffuses into the lumen where it binds H^+ to become ammonium (8). Ammonium is "trapped" in the lumen and will be excreted along with the attached H^+. The H^+ excreted by these means compensates for producton of metabolic acids. For each H^+ excreted, a HCO_3^- is reabsorbed that wasn't present in the filtrate, hence the description: "new" HCO_3^-.

REGULATION OF POTASSIUM IN THE DISTAL TUBULE

Potassium is the most abundant solute inside cells. Its high concentration is required for optimal growth and DNA and protein synthesis; it is an important factor in the performance of many enzyme systems; and it plays a role in the maintenance of cell volume, pH, and membrane potentials. Because most of the body's K^+ lies within cells, with only about 2.5% in the extracellular fluid, a small K^+ shift between intra- and extracellular fluids could cause a huge change in extracellular K^+. If, for example, only 5% of the body K^+ moved into the extracellular fluids, the extracellular K^+ would triple (going from 2.5 to 2.5 + 5 = 7.5%).

Alterations of extracellular fluid or plasma K^+ are important because cell *excitability* (*membrane potential*) is sensitive to extracellular K^+. Increasing extracellular K^+ depolarizes membranes and raises excitability; in the heart, fibrillation may occur. Decreasing K^+ hyperpolarizes and lowers excitability. Skeletal and smooth muscle disturbances may include flaccid paralysis, abdominal distension, and diarrhea. Shifts of K^+ in and out of cells can easily occur in acid-base disorders, disturbances of hormone balance, and in response to drugs. Further, on a normal diet, the amounts of K^+ absorbed from the intestine into the plasma each day exceeds the total K^+ content of the entire extracellular fluid! Disaster is prevented by the kidneys, which continuously regulate the level of K^+ in the body fluids.

Most K^+ is *reabsorbed in the proximal tubule* and in the loop of Henle; by the time it reaches the distal tubule, only 5-10% of the filtered load remains. From here on, depending on conditions, it may be reabsorbed further, but most often it is *secreted*. Most regulatory changes in excretion are due to variations in secretion in these latter portions of the nephron.

Distal tubule and collecting duct cells accumulate high concentrations of intracellular K^+ via the *Na^+–K^+ pump*, which is located primarily in the baso-lateral membrane. The *electrical gradient* (membrane potential) across the baso-lateral membrane is sufficiently high (70 mv) to oppose the *K^+ concentration gradient* and prevent leakage from cell to interstitial space, but the membrane potential across the apical membrane (facing the lumen) is smaller and cannot prevent leakage. The result is a simple pathway for K^+ secretion; K^+ is pumped into the cell from the blood side and leaks out the lumen side. This idea can be used to interpret renal control of body K^+ in a number of different contexts.

1. Regulation of cellular K^+ occurs. According to the previous discussion, K^+ excretion will increase whenever distal tubular (or collecting duct) cell K^+ increases because the concentration gradient driving K^+ leakage into the lumen will increase. But the K^+ content of these cells often reflects the K^+ content of body cells in general. This provides a mechanism for regulating intracellular K^+; changes that increase internal K^+ will increase leakage and secretion.

2. Intracellular K^+ (and consequently K^+ secretion) has a tendency to rise and fall with plasma K^+, providing some regulation of plasma K^+. However, plasma K^+ is guarded by another potent feedback mechanism. A rise in plasma K^+ stimulates the *adrenal cortex* to secrete *aldosterone*, which promotes secretion and excretion of K^+ (and reabsorption of Na^+). Unlike other feedback paths that involve aldosterone secretion, K^+ stimulates the adrenal cortex directly and does not utilize the renin-angiotensin system as an intermediary.

3. Tubular cell leakage of K^+ also helps explain the frequent positive correlation between excretion of Na^+ and K^+. As more Na^+ is delivered to the distal tubule, the excess Na^+ causes an increase in both Na^+ reabsorption and Na^+ excretion. K^+ leaks faster because more positive charge (in the form of Na^+) is available to exchange with K^+ across the luminal membrane; this allows more K^+ to escape down its concentration gradient without building up a membrane potential.

4. When fluid flow in the distal tubule increases, there is generally an increase in K^+ excretion. This can be explained by the more efficient "washing away" of the secreted K^+ by the faster moving tubular stream. This reduces the K^+ concentration in the luminal fluid adjacent to the tubular cells and promotes leakage from cells to lumen.

5. K^+ excretion commonly increases during acute alkalosis and decreases during acute acidosis. This is consistent with the fact that alkalosis is often associated with K^+ entry into cells and acidosis with its departure. It is almost as though K^+ and H^+ exchange across the cell membranes. For example, in acidosis, H^+ enters the cell and reacts with negatively charged proteins, reducing the charge on the protein. K^+, the most abundant intracellular cation (positively charged ion), suddenly finds itself in excess; it is in an environment with too few negative charges to support all the K^+. Being the most permeable cation, some of the K^+ moves out. Renal cells are no exception. In acidosis, distal tubular and collecting duct cells lose K^+ to the plasma; the intracellular K^+ decreases, as does the leakage and secretion into the tubular lumen. During alkalosis, the reverse occurs.

CN: *Use red for blood vessels (E), which include both arterioles and capillaries.*
1. Begin with the upper panel. Note the large area within the cell on the left marked by the letter K^+. This symbolically represents the larger amounts of K^+ present within the cell.
2. Color the illustration (top of bottom panel) of filtrate flow through the nephron, beginning in the upper left corner of the rectangle. Note the separation of K^+ (A) from the filtrate (F) as it is first reabsorbed into the proximal tubule (uncolored).
3. Color the enlargement of the secretion process. Note the two symbols representing K^+ concentration gradients in the lower tubule cell and the way in which they compare to the two electrical potentials (C).
4. Color the summary below, beginning with the circle representing elevated K^+ in the blood.

ROLE OF K⁺ IN NERVE & MUSCLE EXCITABILITY.

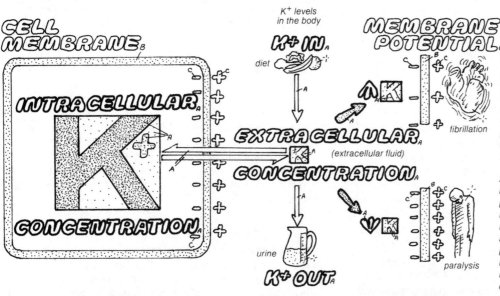

CELL MEMBRANE.

K⁺ levels in the body

K⁺ IN. diet

INTRACELLULAR **K⁺** CONCENTRATION.

EXTRACELLULAR (extracellular fluid) CONCENTRATION.

urine

K⁺ OUT.

MEMBRANE POTENTIAL.

fibrillation

paralysis

Most body K⁺ (about 90%) lies within cells, with only about 2.5% in the extra-cellular fluid (the rest is in bone). A small decrease say -5% in cellular K⁺ could cause a huge change in extra-cellular K⁺ (in our example it could triple, going from 2.5 to 7.5%). These large changes do not occur because the kidney regulates plasma K⁺.

Alterations of plasma K⁺ are important because cell excitability (membrane potential) is sensitive to extracellular K⁺. Increasing extracellular K⁺ depolarizes membranes and raises excitability in the heart and fibrillation may occur. Decreasing K⁺ hyperpolarizes membranes and lowers excitability. Skeletal and smooth muscle disturbances may include flaccid paralysis, abdominal distension and diarrhea.

K⁺ LEVELS REGULATED BY SECRETION.

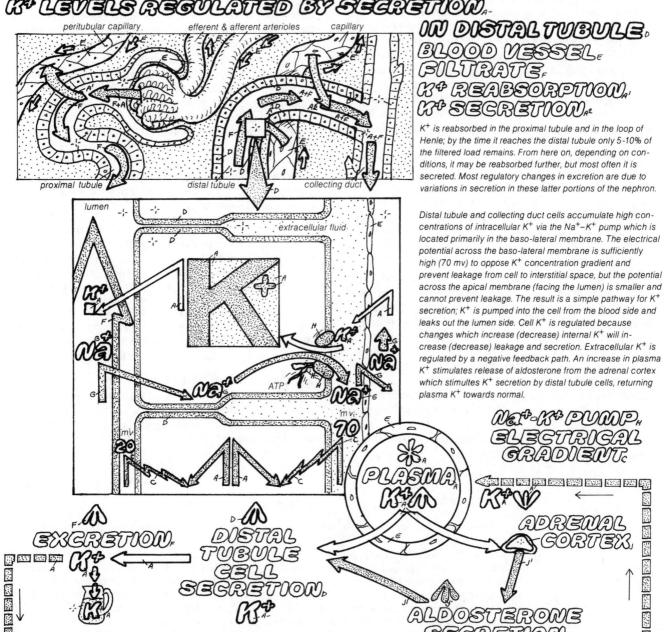

peritubular capillary. efferent & afferent arterioles capillary

proximal tubule distal tubule collecting duct

IN DISTAL TUBULE. BLOOD VESSEL. FILTRATE. K⁺ REABSORPTION. K⁺ SECRETION.

K⁺ is reabsorbed in the proximal tubule and in the loop of Henle; by the time it reaches the distal tubule only 5-10% of the filtered load remains. From here on, depending on conditions, it may be reabsorbed further, but most often it is secreted. Most regulatory changes in excretion are due to variations in secretion in these latter portions of the nephron.

Distal tubule and collecting duct cells accumulate high concentrations of intracellular K⁺ via the Na⁺–K⁺ pump which is located primarily in the baso-lateral membrane. The electrical potential across the baso-lateral membrane is sufficiently high (70 mv) to oppose K⁺ concentration gradient and prevent leakage from cell to interstitial space, but the potential across the apical membrane (facing the lumen) is smaller and cannot prevent leakage. The result is a simple pathway for K⁺ secretion; K⁺ is pumped into the cell from the blood side and leaks out the lumen side. Cell K⁺ is regulated because changes which increase (decrease) internal K⁺ will increase (decrease) leakage and secretion. Extracellular K⁺ is regulated by a negative feedback path. An increase in plasma K⁺ stimulates release of aldosterone from the adrenal cortex which stimultes K⁺ secretion by distal tubule cells, returning plasma K⁺ towards normal.

lumen

extracellular fluid

K⁺ **Na⁺** ATP

Na⁺–K⁺ PUMP. ELECTRICAL GRADIENT.

PLASMA K⁺

ADRENAL CORTEX.

EXCRETION. **K⁺**

DISTAL TUBULE CELL SECRETION. **K⁺**

ALDOSTERONE SECRETION.

WATER CONSERVATION & ANTIDIURETIC HORMONE

Animals living in fresh water are continuously challenged with water balance problems. Their plasma has a high solute concentration (*osmolarity*) and tends to draw water by osmosis from its surroundings. They cope with a continuous inundation of water by excreting large volumes of it. Animals, including humans, living on land face the opposite problem. Their environment is arid, and they face the threat of drying up. To conserve water, birds and mammals excrete very small volumes of concentrated urine, but how?

Only birds and mammals excrete urine that is *hypertonic* (more concentrated than their plasma). Only birds and mammals have long loops of Henle. Further, those species with the more highly developed loops are capable of excreting more concentrated urine. These observations led earlier investigators to suggest that the formation of hypertonic urine takes place in the loops of Henle. This idea was shattered by the first micropuncture studies of the distal tubule, which contains the fluid just after it has passed through the loop of Henle. This fluid is always *hypotonic* or at most isotonic, but *never* hypertonic, as required by the hypothesis. Apparently, hypertonic urine must be formed in the collecting duct. The loops of Henle are involved in a more subtle way. By actively pumping NaCl without allowing water to follow, the loops of Henle create a unique *hypertonic interstitial fluid* in the deep portions of the *medulla*. Collecting ducts pass through this fluid on the way to the ureter and take advantage of their hypertonic surroundings by allowing water to be withdrawn by osmosis from the lumen of the duct to the interstitial space. Although this general scheme is universally accepted, the exact details have baffled physiologists for years and are the source of continuing controversy.

The loops of Henle of juxtamedullary nephrons plunge into the depths of the medulla. These descending limbs are fairly permeable to Na^+ and water and do not appear to have any special properties. Once around the bend in the loop, however, the tubules become *water impermeable*, a property that extends well into the distal tubule. Further, the ascending limb actively transports NaCl from lumen into the interstitial fluid. The exact nature of this transport, both its location and its specificity, is controversial. At first, it was assumed that the entire ascending limb took part, but now it appears that at least the major portion of the active transport takes place in the thick (upper) portions of the ascending limb, which has cells richly endowed with mitochondria (ATP producers). It

was also assumed that Na^+ was actively transported, with Cl^- following to maintain electrical neutrality. Later experiments indicated the reverse; Cl^- is actively pumped, and Na^+ follows. Finally, more recent work once again favors primary active transport of Na^+ with Cl^- following. In any case, the net result is transport of NaCl.

The ascending limbs actively transport NaCl but prevent the usual concomitant transport of water; so fluid delivered to the distal tubule is hypotonic regardless of the final composition of the urine. This transport of NaCl (out of the water-impermeable ascending limb) without water creates a unique hypertonic interstitial fluid in the medulla. Collecting ducts pass through these fluids on their way to the ureter. If the hormone *ADH* (*antidiuretic hormone, vasopressin*) is present, the latter portions of the distal tubule and the entire collecting duct become water permeable. As fluid flows through these sections of the nephron (distal tubule and collecting ducts), water equilibrates with the surrounding interstitial fluids. Therefore, as fluid descends via the collecting ducts into the medulla, it becomes more and more concentrated (hypertonic) until urine leaving the collecting duct has the same hypertonic osmolarity as the interstitial fluid in the lower medulla. In fact, the osmolarity of the medulla sets the limit to which urine can be concentrated. ADH also makes the last portions of the collecting duct permeable to urea, which becomes trapped in the interstitial fluid and makes a substantial contribution to osmolarity.

When ADH is absent, the distal tubule and collecting duct are practially impermeable to water. The hypotonic fluid delivered to the distal tubule becomes even more hypotonic as salts are reabsorbed (without water being able to follow). Urine reaching the end of the collecting duct is hypotonic.

In times of water deprivation, the kidneys conserve water; they excrete a low-volume, concentrated (hypertonic) urine. With water intoxication, they release the excess water by excreting a high-volume, dilute (hypotonic) urine.

A rise in osmolarity of the extracellular fluids (and thus blood plasma) stimulates cells in the hypothalamus to increase ADH production and to cause release of ADH from the posterior pituitary. ADH travels via the bloodstream to the kidney, where it promotes water retention to relieve the rise in osmolarity.

CN: Use light blue for D, a dark color for G, yellow for H, red for I, and purple for J.
1. Begin by coloring the rectangular borders (A) and (B) of the two large diagrams representing a kidney nephron. This will demonstrate which part of the nephron lies within the cortex (A) and which part is in the medulla (B).
2. Color the state of low osmolarity on the left by beginning with the cartoon figure above and then coloring the borders of the nephron itself (D^1 and E). Do not color the isotonic title, but do color the other titles within the diagram. Color

all the circles and arrows present. Include the large drop representing excessive hypotonic urine.
3. Do the same for the diagram on the right, noting the ADH influence (G^1) on the collecting duct and a portion of the distal tubule (making the membranes water permeable).
4. Starting with step one, color the lower right illustration which shows how a rise in osmolarity results in ADH secretion and water reabsorption. Do the summary diagram to the left. Note that the numbers below the symbols of increase refer to the steps of the illustration on the right.

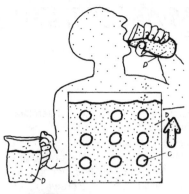

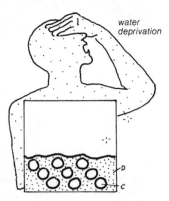

water deprivation

In times of water deprivation (right diagram), the kidneys conserve water; they excrete a low-volume, concentrated (hypertonic) urine. With water intoxication (left diagram) they release the excess water by excreting a high-volume, dilute (hypotonic) urine. To accomplish this task the loops of Henle of juxtamedullary nephrons plunge into the depths of the medulla. The ascending limbs actively reabsorb NaCl, but prevent the usual concomitant reabsorption of water (due to impermeable membranes) so that fluid delivered to the distal tubule is hypotonic regardless of the final composition of the urine. Further, this transport of NaCl (without water following) creates a unique hypertonic interstitial fluid in the medulla. Collecting ducts pass through these fluids on their way to the ureter. If the hormone ADH is present (right diagram) the latter portions of the distal tubule and the entire collecting duct become water permeable. Water equilibrates with interstitial fluid in these sections of the nephron, fluid leaving the distal tubule is isotonic, while fluid leaving the collecting duct has the same hypertonic osmolarity as the interstitial fluid in the lower medulla. When ADH is absent (left diagram) the distal tubule and collecting duct are practically impermeable to water. The hypotonic fluid delivered to the distal tubule becomes even more hypotonic as salts are reabsorbed (without water being able to follow). Urine reaching the end of the collecting duct is hypotonic. ADH also makes the last portions of the collecting duct permeable to urea which becomes trapped in the interstital fluid and makes a substantial contribution to its osmolarity.

LOW OSMOLARITY ∨ADH

HIGH OSMOLARITY ∧ADH

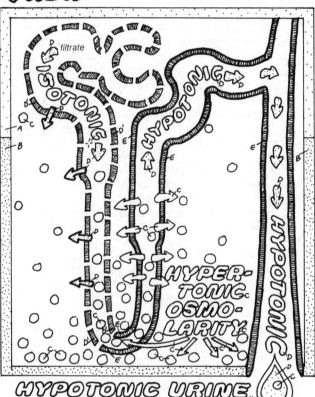

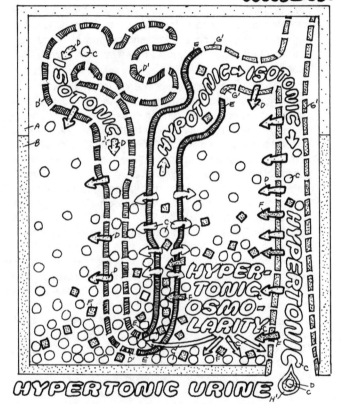

KIDNEY CORTEX
KIDNEY MEDULLA
SODIUM CHLORIDE (NaCl)
WATER
WATER PERMEABLE MEMBRANE
IMPERMEABLE MEM.
UREA
ANTIDIURETIC HORMONE (ADH)
INFLUENCE ON MEMBRANE

HIGH OS.

ARTERY
CAPILLARY
RECEPTOR
ADH

OSMOTIC PRESSURE (1)	OSMOTIC RECEPTORS (2)	ANTIDIURETIC HORMONE (ADH) (3)	WATER REABSORPTION (5)

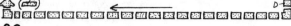

(6)

A rise in osmolarity of the extracellular fluids (and thus blood plasma) (1) stimulates cells in the hypothalamus to increase ADH production (2) and to release ADH from the posterior pituitary (3). ADH travels via the bloodstream (4) to the kidney (5) where it promotes water retention to relieve the rise in osmolarity (6).

NORMAL OS.

THE COUNTER-CURRENT MULTIPLIER IN THE LOOP OF HENLE

The kidney regulates the internal environment by judicious excretion of water-soluble plasma constituents and water. It also excretes waste products, the most notable being *urea*. Urea is produced in the liver and contains the nitrogen derived from amino acids or proteins. When these compounds are broken down by metabolism, they yield ammonia. Free ammonia is very soluble in water and very toxic. Fortunately, the liver quickly converts it to the relatively harmless urea. Metabolism of protein produces about 30 g of urea per day, which is excreted in the urine. Because ions and urea are water soluble, their excretion necessarily draws water with them. Excretion of water in the urine is obligatory, and it behooves the kidney to conserve water whenever it is in short supply by excreting a concentrated urine. What do we mean by "concentrated" urine?

Ordinarily, we express the concentration of a solutelike urea by the number of moles (1 mole = 6×10^{23} molecules) of urea contained in each liter of solution. This is the *molar concentration* of urea. When this number is small, we reduce the unit by 1000 and call it a millimole (mM, 1000 millimoles = 1 mole). In every solution, each specific solute has its own molar (or millimolar) concentration. When we are dealing with osmotic water movements, all molecules and ions make an almost equal contribution to osmotic pressures. A 100 mM solution of urea exerts the same osmotic force as a 100 mM glucose solution because they both contain the same number of molecules per liter. A solution containing both (100 mM urea + 100 mM glucose) contains twice as many molecules per liter and exerts twice as much osmotic force. The sum of the molar concentration of all the molecules and ions in a given solution is called the *osmolar* concentration (Osm). Sometimes we use *milliosmolar* (mOsm) instead (1000 mOsm = 1 Osm). The "total solute concentration" of a solution containing 100 mM urea + 100 mM glucose is 200 sOsm. (Note that 100 mM NaCl would be 200 mOsm because it contains 100 mM Na^+ + 100 mM Cl^-.) The total concentration of blood plasma is consistently about 300 mOsm; urine is commonly around 950 but can range from 50 to 1400 mOsm.

Excretion of a concentrated urine requires an *interstitial fluid space* in the *medulla* four to four and one-half times more concentrated (1200 to 1400 mOsm) than blood plasma. To create this space, the kidneys rely on Na^+ *pumps* in the *ascending loop of Henle* that can create 200 mOsm gradients across the tubular cells. Because proximal tubule fluid delivered to the loop is isotonic (300 mOsm), the most concentrated interstitial space possible should be 500 mOsm. How does the kidney manage to get 1400 mOsm?

The ability of the Na^+ pump to create a 200 mOsm gradient is called the "*single effect*." The single effect is multiplied severalfold by imbedding the pumps in the ascending limb of the two streams moving in opposite directions (*counter-current*) through the loop of Henle. The ascending limb is impermeable to water; NaCl is pumped out into the interstitial fluid (ISF), but water cannot follow. The NaCl that has been pumped creates a small gradient of 200 mOsm, so the ISF becomes slightly hypertonic. The descending limb is permeable to both NaCl and water; NaCl diffuses down its concentration gradient into the descending limb while water is drawn out of the descending limb into the hypertonic environment. This loss of water and gain of solute makes the contents of the descending limb hypertonic, like the ISF. But the slightly concentrated fluid in the descending limb moves! It flows toward the ascending limb where the pumps are located, giving the pumps an opportunity to create the same 200 mOsm gradient — only this time they begin with a higher concentration and can create a correspondingly higher concentration in the ISF. The cycle repeats, with elevated concentrations delivered to the descending limb, which in turn delivers these elevated concentrations to pumps in the ascending limb; *the single effect is multiplied*. The concentration of solutes in the ISF builds up until a steady state is reached where the amounts delivered to the ISF are just balanced by the amounts taken away by the blood supply.

The diagram on the right illustrates the scheme in a steady state. Note that the proximal tubule continues to deliver isotonic fluid (300 mOsm) to the loop, but, as it descends, the fluid becomes more concentrated as NaCl enters and water leaves. The greatest concentration is at the tip. Upon ascending, the fluid becomes less concentrated as NaCl is pumped out without any water. Finally, fluid leaves the loop less concentrated (100 mOsm) than when it came in. Because the ascending limb is impermeable to water, relatively more NaCl than water is left behind in the medullary ISF.

In the presence of ADH, urea also makes a substantial contribution to the ISF solute concentration in the medulla. Urea becomes trapped in the lower medullary ISF as it flows in a circle along the following path (lower left illustration): lower collecting duct → lower medullary ISF → thin ascending limb → thick ascending limb → distal tubule → collecting duct → lower collecting duct → ... This circulation and trapping occur because the upper portions of the collecting duct are impermeable to urea, and as water is reabsorbed, the remaining urea becomes concentrated. When it reaches the lower portions, the collecting duct becomes permeable, and urea diffuses to the ISF. From here, some of it diffuses into the lower thin ascending limb, which is urea permeable. The thick ascending limb and distal tubule are urea impermeable. As water is withdrawn from these portions, the urea becomes even more concentrated, only to be delivered to the collecting duct, where the cycle begins anew. In this way, the urea circulates and becomes more and more concentrated in all sections of its route (including the ISF) until it reaches a steady state where the delivery of "new" urea is just balanced by the amounts of urea the blood circulation carries away.

CN: Use light blue for E and darker colors for B, C, and D.
1. Begin by coloring the entire title in the upper left corner, noting that the NaCl pump (B) receives a different color. Then color the anatomical drawing of the loop of Henle.
2. Color the counter-current multiplier by starting in the upper left corner with the NaCl solutes (A) entering the descending limb. Work your way down the limb, coloring the numbers and solutes in the interstitial fluid and the diffusion/gradient symbol (A) moving into the descending limb. The broken line suggest that the membranes of the cells of descending limbs are permeable to water (E) (arrows entering ISF). Then work your way up the ascending limb.
3. Color the way in which urea becomes trapped in the ISF. The drawing of the ascending limb of Henle, distal tubule, and collecting duct is highly simplified. Note that here the water permeable membranes (broken lines) are not colored.

PROBLEM:

HOW TO CREATE A 300-1200 mOsm CONCENTRATION GRADIENT IN THE MEDULLA WITH ONLY A 200 mOsm NaCl PUMP?

By excreting a concentrated urine, the kidney conserves body water. This requires an interstitial fluid (ISF) space in the medulla some 4 to 4.5 × more concentrated (1200 to 1400 mOsm) than blood plasma (300 mOsm). To create this space, the kidneys rely on Na^+ pumps that can only create 200 mOsm gradients across the cells. Since proximal tubule fluid delivered to the loop is isotonic (300 mOsm), it would appear that the most concentrated ISF is 500 mOsm. How does the kidney manage to get 1200 mOsm?

SOLUTION:

THE COUNTER-CURRENT MULTIPLIER

The ability of the Na^+ pump to create a 200 mOsm gradient is multiplied several fold by embedding the pumps in the ascending limb of the two streams moving in opposite directions (counter-current) through the loop of Henle. The ascending limb is impermeable to water. NaCl is pumped out into the ISF creating a small gradient (200 mOsm) and making the ISF hypertonic. The descending limb is permeable to both NaCl and water. They equilibrate passively so that the contents of the descending limb match the ISF. But the slightly concentrated fluid in the descending limb moves! It flows toward the ascending limb where the pumps are located, giving the pumps an opportunity to create the same 200 mOsm gradient — only this time they begin with a higher concentration and can create a correspondingly higher concentration in the ISF. The cycle repeats with elevated concentrations delivered to the descending limb which in turn delivers these elevated concentrations to pumps in the ascending limb: the single effect is multiplied. The diagram on the right illustrates the scheme in a steady state. Note that the loop continues to receive isotonic fluid (300 mOsm), but as fluid descends it becomes more concentrated as NaCl enters and water leaves. The greatest concentration is at the bend. Upon ascending, the fluid becomes less concentrated because NaCl is pumped out without water. Finally, fluid leaves the loop less concentrated (100 mOsm) than when it came in. Relatively more NaCl than water is left behind in the medullary ISF because the ascending limb is impermeable to water. (Note: Pumps are probably located exclusively in the thick portion of the ascending limb.

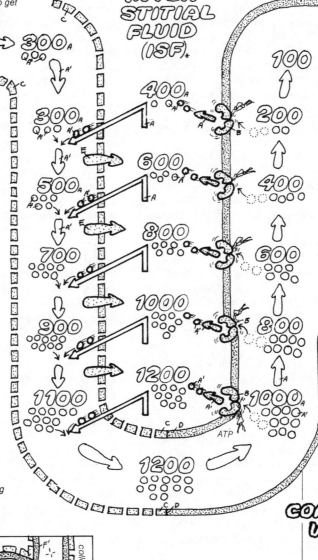

LOOP OF HENLE:

DESCENDING LIMB, ASCENDING LIMB, NaCl SOLUTE, WATER

MEDULLA CORTEX

INTERSTITIAL FLUID (ISF)

The desert rat does not need to drink water because his counter-current multiplier can establish a hypertonic medullary ISF that is 20 times more concentrated than blood plasma. His urine is so concentrated that he can maintain his body fluids with water obtained from carbohydrate breakdown.

ISF CONCENTRATION VARIATIONS

DAILY FLUID INTAKE

Man can only establish a medullary ISF concentration that is 4 times as concentrated as blood plasma. Therefore, the highest urine concentration is also 4 times more concentrated than blood plasma.

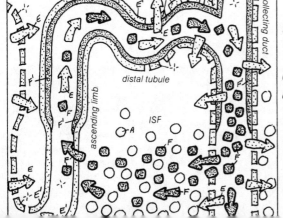

distal tubule

ascending limb

ISF

UREA TRAPPING

H₂O BARRIER
UREA BARRIER

collecting duct

In the presence of ADH, urea is trapped in the ISF and contributes to its solute concentration. Follow the route taken by urea as it circulates from collecting duct, through the ISF to ascending limb distal tubule, and back again to collecting duct. Note places which are impermeable to urea and where water is reabsorbed. They are responsible for concentrating the urea.

THE COUNTER-CURRENT EXCHANGER IN THE MEDULLARY BLOOD SUPPLY

Like any tissue, the renal medulla must be supplied with blood, and if solutes in the *medullary ISF* are highly concentrated, we might expect them to be washed away as blood within the capillary beds equilibrates with the ISF. A capillary exchange vessel entering an impermeable venule at the tip of the medulla would carry away fluid as concentrated as 1200-1400 mOsm! This does not happen because of the peculiar shape of the exchange vessels, the *vasa rectae*. These are long, highly permeable vessels that exchange materials with their surroundings along their entire length just as though they were capillaries. They enter from the cortex, descend into the medulla, form loops, and return to the cortex. The important point is that they leave the medulla at the level of the cortex. Few, if any, exchange vessels enter an impermeable collecting vein in the depths of the medulla, so few if any collecting veins contain highly concentrated (1200 mOsm) fluid.

Follow the exchange of solute and water in the diagram on the far right as the vasa rectae travels from the cortex, makes a hairpin turn in the depths of the medulla, and returns to the cortex before entering an impermeable collecting vein. Solute (*NaCl* and/or *urea*) flows passively down the concentration gradient, from regions of high to regions of low concentration (i.e., from higher to lower numbers in the diagram). Water flows passively in the opposite direction, from regions where the solute is less concentrated to regions of higher concentration (i.e., from lower to higher numbers in the diagram). Note that water always moves from descending to ascending limb (left to right in the diagram), and solute always moves from ascending to descending limb (right to left in the diagram).

Fluid entering the descending vasa rectae is isotonic; it comes from the general circulation. Fluid leaving the ascending vasa rectae is slightly hypertonic; it has been in contact with the hypertonic fluids in the medullary ISF. Hence, water flows across from ascending to descending limb, and solutes flow in the reverse direction. Similar arguments apply to the two limbs at each level of the medulla: water takes a shortcut, flowing from descending to ascending limb so that not much of it reaches the depth where it could dilute the hypertonic ISF. Solutes take a similar shortcut in the reverse direction, flowing from ascending to descending limb so that not much is allowed to escape with the fluid entering the veins. Although both water and solute flow in the indicated

directions at every level, some of the solute flow arrows have been omitted from the top portions of the vasa rectae to emphasize the entering water that never reaches the bottom. Similarly, some of the water flow arrows have been omitted from the bottom to emphasize solutes that are trapped in the depths and do not escape.

Note that fluid leaving the medulla at the top of the ascending vasa rectae is slightly more hypertonic than fluid entering at the top of the descending vasa rectae (350 mOsm compared to 300 mOsm). The counter-current exchange system is not 100% efficient. The vasa recta carries away more solute from the medulla than it brings in. It also carries away water that has been reabsorbed from the collecting duct, but because the nephron continually transfers more solute than water into the medullary ISF, the system will reach a steady state only when the blood supply carries this excess solute away as fast as it forms. Thus, fluid leaving the medulla in the ascending vasa recta must be hypertonic.

At first, the conclusion of the above paragraph (blood leaving the medulla is hypertonic) seems to challenge the assertion that the counter-current multiplier and exchanger act to conserve water. The apparent contradiction is resolved by the fact that the medulla receives only a tiny fraction of the total blood supply to the kidney and that considerable water reabsorption occurs in the distal tubule (see plate 63), which more than compensates for the small hypertonic blood flow that leaves the medulla.

The bottom diagram on the plate shows how activities of the nephron (on the right) and its blood supply (on the left) are integrated to provide the hypertonic ISF required for water conservation. The nephron (more specifically, the loop of Henle) acts as a *counter-current multiplier*; it creates the hypertonic environment. This is an active process requiring metabolic energy that becomes apparent through the active transport of NaCl. The blood supply to the medulla (the vasa recta) acts as a *counter-current exchanger*; it maintains the stability of the hypertonic environment by minimizing the likelihood of excess solutes being washed away by the circulation. This is a purely passive process where much of the entering water is shunted across the top of the exchanger and never reaches the depths, while solutes are shunted across the bottom and are trapped as they simply recirculate from ascending vasa recta to ISF to descending vasa recta and around the loop again to ascending vasa recta.

CN: Use the same colors as were used on the previous page for water (B) and the NaCl concentration solutes (C). Use purple for A, red for D, and blue for F.
1. Begin with the problem in the upper left corner; note that a different color is given to each line of this title. Color the long, straight blood vessel (capillary) which illustrates the problem.
2. Color the title, "solution," and the counter-current exchanger illustration on the far right. Note that the numbers reflecting the osmolarity within the vasa recta and the

numbers within the ISF are not to be colored. Begin wth the entry of arterial blood (D) in the upper left corner and first color the diffusion of water across to the ascending limb. Then color the diffusion in the opposite direction of NaCl (C) and the build-up of the concentration gradient.
3. Color the anatomical illustration on the left side, noting that only the vasa recta (A) and the loop of Henle (F) are colored.
4. Color the lower diagram which summarizes the mechanisms discussed on this and the previous page.

PROBLEM:
HOW TO KEEP THE BLOOD CIRCULATION FROM WASHING AWAY THE CONCENTRATION GRADIENT?

All parts of the renal medulla require blood circulation for nutrition. But, a capillary leaving the deepest parts of the medulla would equilibrate with the ISF, leaving water behind and carrying concentrated solute (NaCl and urea) with it. The work of the counter-current multiplier would be dissipated.

SOLUTION:
FOLD THE BLOOD VESSEL (VASA RECTA) BACK ON ITSELF.

The kidney solves this problem by not allowing blood vessels to leave from the inner (deep) portions of the medulla; they all exit near the cortex where solutes are nearly isotonic. The blood supply is carried by straight tubes that follow the loop of Henle deep into the medulla, turn and ascend back toward the cortex. These hairpin loops of blood vessels, are called vasa rectae; they behave like capillaries, exchanging solutes and water freely with surrounding ISF.

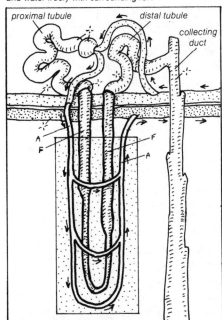

VASA RECTA
LOOP OF HENLE

proximal tubule · distal tubule · collecting duct

TWO COUNTER CURRENT MECHANISMS OF THE MEDULLA

This panel summarizes the interrelated counter-current movements of solutes and water which create and stabilize the unique hypertonic ISF required for water conservation. The multiplier (2) shows NaCl trapping within the loop of Henle which allows pumps in the ascending limbs to effectively power the whole process. The exchanger (1) shows how the medulla can be supplied with blood without undermining solute concentrations in the ISF. By shunting some water across the top, water is kept out of the medulla, by shunting some solute across the bottom, solute is not allowed to escape.

ARTERIAL BLOOD.

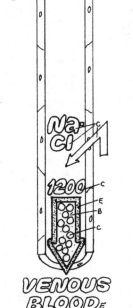

300
H₂O
NaCl
1200

VENOUS BLOOD

THE COUNTER-CURRENT EXCHANGER

INTERSTITIAL FLUID (ISF)

(isotonic)   (slightly hypertonic)

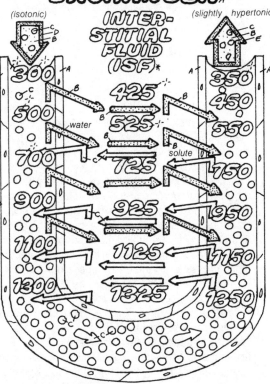

300 — 425 — 350
500 — 525 — 450
water — 550
700 — 725 — 750
solute
900 — 925 — 950
1100 — 1125 — 1150
1300 — 1325 — 1350

The vasa recta exchanges solute and water along its entire course. Fluid entering the descending vasa recta is isotonic; it comes from the general circulation. Fluid leaving the ascending vasa recta is slightly hypertonic; it comes from equilibrated fluid at the hypertonic bend of the loop deep in the medulla. Hence water flows across the descending to ascending limb while solutes flow in the reverse direction. Similar arguments apply to these two limbs at each level of the medulla; water takes a shortcut, flowing from descending to ascending limb so that not much of it reaches the depths where it could dilute the hypertonic ISF. Similarly, solutes flow from ascending to descending limb so that not much is allowed to escape with the fluid entering the veins.

EXCHANGER: PASSIVELY MAINTAINS GRADIENT 1.

MULTIPLIER: ACTIVELY CREATES GRADIENT 2.

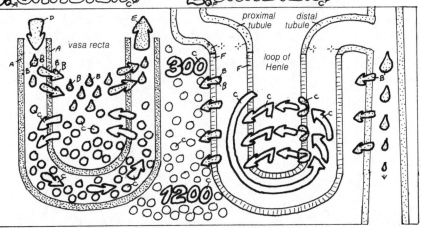

vasa recta · 300 · proximal tubule · distal tubule · loop of Henle · 1200

diffusion

PASSIVE

200 mOsm · ATP · NaCl pump

ACTIVE

REGULATION OF EXTRACELLULAR VOLUME: ADH & ALDOSTERONE

One of the major functions of the kidney is to regulate the *volume of extracellular fluid*. This is important because plasma volume is largely determined by extracellular volume; plasma and other extracellular spaces continually exchange fluid across capillary walls. When plasma volume and extracellular volume fall, the amount of fluid filling the vascular tree can become inadequate, and despite short-term compensations (increase in heart rate and increase in vascular resistance), the long-term effect is likely to be a decrease in blood pressure. On the other hand, a rise in extracellular volume may fill the vascular tree with too much fluid; it becomes tense, and in the long run pressure will increase. Normally, these events do not occur because, despite the huge variations in daily water and salt intake, the extracellular fluid and plasma volumes remain fairly constant; they are regulated by the kidney so that responsibility for long-term regulation of blood pressure also resides with the kidney (see plate 42).

The most important factor that determines extracellular volume is the total *amount* (not concentration) of NaCl in the extracellular spaces. This follows because the NaCl concentration is closely regulated by mechanisms illustrated in the plate and explained below. Increasing NaCl causes water retention by the kidney, which dilutes the NaCl but raises extracellular fluid volume. Conversely, decreasing NaCl is accompanied by extra water excretion and a decreased extracellular volume. These responses take place because (1) *NaCl is the most abundant solute in the extracellular fluids*, so it largely determines extracellular *osmotic pressure* (concentration of solutes), and (2) the *hormone ADH* closely regulates osmotic pressure. The *"quick osmotic response"* of the ADH system to an increase in salt is illustrated in the plate, where the response has been artificially broken into two steps for purpose of illustration. In stage B, NaCl is suddenly introduced so that there is an exaggerated increase in total amount of NaCl without change of fluid volume. The result is increased NaCl concentration and increased osmotic pressure. In stage C, the ADH

mechanism responds (plate 62), releasing ADH, which promotes water reabsorption until the NaCl concentration is practically back to normal. The excess NaCl has not been removed, but the extracellular volume has been increased. In practice, these events take place continuously. Compensation by the ADH system is relatively rapid and precise, so the mass of NaCl and fluid volumes generally appears to rise and fall together, with only small changes in NaCl concentrations.

The action of ADH explains the linkage between NaCl and extracellular volume, but it does not account for volume regulation. These are accounted for by the *"slow volume response"* illustrated in the plate. The increased fluid volume initiates a series of steps (described in plate 66) that results in the inhibition of *aldosterone* release by the *adrenal cortex*. Without aldosterone, reabsorption of NaCl by the *distal tubule* is reduced; more NaCl spills over into the urine, carrying water along with it. The increased ADH secretion that caused the original water retention is no longer operative because the solute concentration has been corrected; the original stimulus for ADH secretion has been removed.

How do ADH and aldosterone exert their characteristic effects on the cells of the kidney? ADH acts by opening *channels* in the collecting ducts and in the distal tubule. The hormone reacts with a *receptor* on the basal membrane activating *adenyl cyclase*, the enzyme that converts *ATP to cyclic AMP*. Cyclic AMP acts as a second messenger, initiating a sequence of steps that culminates in the opening of *water channels*.

Aldosterone promotes Na^+ reabsorption in the distal tubule and collecting ducts. The hormone is lipid soluble; it passes through the plasma membrane and reacts with a *receptor protein* in the cytoplasm, which acts on the *nucleus* and leads to the synthesis of new protein. The new protein may be involved in the supply of (1) new Na^+/K^+ pumps on the basal membrane, (2) *more ATP* to power the pumps, and (3) *new Na^+ channels* on the luminal membrane.

CN: *Use the same colors as were used on the previous page for water (A) and NaCl (D).*
1. Color the upper panel first.
2. Color the quick osmotic response to an increase to the solute concentration of the plasma.
3. Color the slow volume response which deals with the resulting increase in body fluid volume shown in figures C and C[1].
4. In the lower panel, color the actions of ADH (E) and aldosterone (F).

BODY WATER CONTENT

The total body water comprises 60% of the body weight. 2/3 of this water lies within cells (intracellular), 1/3 lies outside (extracellular). Most cell membranes permit free exchange of water between intra- and extracellular spaces.

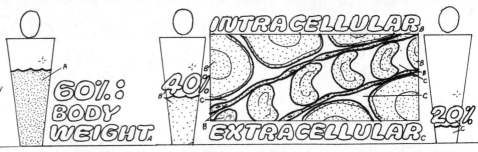

60%: BODY WEIGHT

INTRACELLULAR 40%

EXTRACELLULAR 20%

EXPANSION & CONTROL OF EXTRACELLULAR VOLUME.

SOLUTE CONCENTRATION (NaCl)

URINE
ADH
ALDOSTERONE

By responding quickly to alterations in the solute concentrations (osmolarity) of the plasma, ADH keeps the body fluids practically isotonic at all times. If the solute concentration goes up (more solute dissolved in the same water volume as shown in the middle figure), ADH is secreted, less water is lost in urine so that the isotonic condition is restored. But now the volume of body fluids has increased. This ADH mechanism ensures that a proportionate amount of water will be retained (or lost) whenever there is a gain (or loss) of solute (principally NaCl). Body fluid volume faithfully follows changes in total solute.

1. QUICK OSMOTIC RESPONSE: ADH

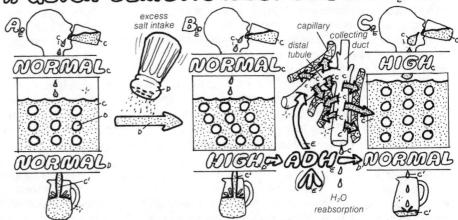

NORMAL — NORMAL

excess salt intake

capillary — collecting duct — distal tubule

HIGH

HIGH → ADH → NORMAL

H_2O reabsorption

2. SLOW VOLUME RESPONSE: ALDOSTERONE

The increased fluid volume initiates a series of steps which results in the inhibition of aldosterone release by the adrenal cortex. Without aldosterone, reabsorption of NaCl by the distal tubule is reduced; more NaCl spills over into the urine carrying water along with it. The increased ADH secretion that caused the original water retention is no longer operative because the solute concentration has been corrected; the original stimulus for ADH secretion has been removed (Increased fluid volume may also inhibit ADH secretion).

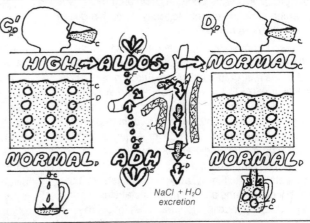

HIGH → ALDOS. → NORMAL

NORMAL — NORMAL

ADH

NaCl + H_2O excretion

ANTIDIURETIC HORMONE "WATER REABSORPTION"

ADH RECEPTOR
ATP → CYCLIC AMP

ADH acts by opening channels in the collecting ducts and in the distal tubule. The hormone reacts with a receptor on the basal membrane, activating adenyl cyclase, the enzyme which converts ATP to cyclic AMP. Cyclic AMP acts as a second messenger initiating a sequence of largely unknown steps which culminates in the opening of water channels.

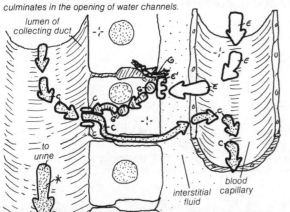

lumen of collecting duct

to urine

interstitial fluid

blood capillary

ALDOSTERONE "NaCl REABSORPTION"

RECEPTOR PROTEIN
NUCLEUS
SYNTHESIZED PROTEIN

Aldosterone promotes Na^+ reabsorption; it passes through the plasma membrane and reacts with a receptor protein in the cytoplasm which acts on the nucleus and leads to the synthesis of new protein. The new protein may be involved in the supply of:
1. new Na^+/K^+ pumps on the basal membrane, 2. more ATP, 3. new Na^+ channels on the luminal membrane.

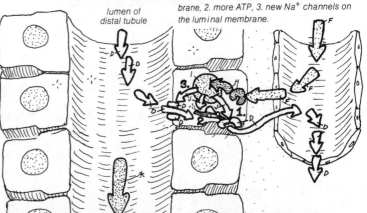

lumen of distal tubule

REGULATION OF EXTRACELLULAR VOLUME: ANGIOTENSIN-RENIN SYSTEM

Plate 65 illustrated how the total amount of NaCl determines the extracellular volume. Attention is focused primarily on Na⁺ because regulatory mechanisms act primarily on it and because changes in Cl⁻ are, to a large extent, secondary to Na⁺ movements. Our example showed how the body fluids expand whenever the amount of Na⁺ (or NaCl) increases, and how compensatory changes help return the volume toward normal. In this plate, the theme is continued as we examine how *extracellular volume is regulated by the kidney through the hormonal control of Na⁺ excretion.* This time our example concerns the reverse situation: the response to body fluid depletion.

Depletion of the extracellular volume is a common clinical event. It occurs in severe vomiting, in diarrhea, and in the sweating response to intense heat (heat prostration). In each of these cases, considerable Na⁺ is lost from the body, and compensatory processes are set in motion to restore the Na⁺ and water loss. The plate emphasizes the *renin-angiotensin-aldosterone* response, one of the most important of these compensatory processes. This system is activated by several stimuli, all of which arise directly or indirectly from changes in extracellular volume (see below).

Renin is released from specialized secretory cells in the wall of the afferent arteriole where it butts up against the distal tubule and forms a structure called the *juxtaglomerular apparatus* (see plate 58). The released renin is an enzyme that acts on the plasma protein *angiotensinogen* (produced by the *liver*) and splits off a small, ten-amino-acid fragment called *angiotensin I*. Angiotensin I is converted into a smaller peptide (eight amino acids), *angiotensin II*, by action of a "converting enzyme" that is especially prominent in the lungs but also occurs elsewhere. Finally, angiotensin II is split into an even smaller peptide, *angiotensin III*. Angiotensins II and III are active products. In addition to vasoconstriction, they both stimulate secretion of aldosterone, and they both stimulate thirst.

Aldosterone reaches the kidney via the circulation and promotes reabsorption of Na⁺ by the distal tubule and the upper collecting ducts. Cl⁻ follows the Na⁺, preserving electrical neutrality, and water follows, preserving osmotic equilibrium. The net result is the reabsorption of NaCl and water. In addition, angiotensins II and III stimulate thirst. The volume of body water and the NaCl content rise toward normal. The relative proportions of NaCl and water gained is "finely tuned"

by the *ADH* feedback mechanism, which operates on water reabsorption to maintain a constant solute concentration in the body fluids.

We have yet to account for the linkage between changes in extracellular volume and renin secretion. Stimuli giving rise to renin secretion have been identified, but details of the steps leading from stimulus to final response have remained elusive and speculative. In our example, the depleted volume depresses venous and arterial pressures. These lowered pressures may reduce cardiac filling and cardiac output so that arterial pressure also falls. Pressoreceptors imbedded in the walls of these structures normally send nerve impulses to the brain stem, where they inhibit sympathetic nerves. When pressures are lowered, the pressoreceptors become less active, and sympathetic nerves to the kidney are released from their "braking" action. As a result, the kidney is showered with sympathetic impulses, which stimulate renin release.

A second important regulatory system for renin secretion is provided by the direct action of *pressure in the afferent arterioles* of the kidney itself. When this pressure rises, renin secretion is inhibited; when it falls (as in our example), secretion is enhanced. This arteriolar mechanism is independent of nerves. When they are cut, the response persists.

The third regulatory system is found in the *juxtaglomerular apparatus*. This composite structure consists of the secretory cells in the afferent arteriole and specialized cells of the distal tubule, called macula densa, which are in close contact with the secretory cells. A decrease in fluid delivery within the nephron to the macula densa results in a stimulation of the secretory cells, and more renin is released into the circulation. The decrease in fluid delivery occurs when the glomerular filtration rate is lowered, and this can occur in response to the lowered arterial pressure, especially if sympathetic nerve impulses constrict the afferent arterioles. (Note that the reduced glomerular filtration by itself will help compensate for fluid depletion because it reduces fluid excretion.) The mechanism secretes renin into the systemic circulation, where it catalyzes the formation of angiotensins II and III, and these stimulate release of aldosterone, etc. The relation of this regulatory system to the mechanism described in plate 58, which utilizes the same juxtaglomerular apparatus for matching the glomerular filtration rate of each nephron to its tubular reabsorptive capacity, is not understood at present.

CN: Use the same colors as the previous page for water (A), NaCl solute (B), ADH (H), and aldosterone (G). Use red for blood vessel (C).
1. Begin with the figure in the upper left showing extracellular volume depletion. Note the use of gray in coloring the symbols of increase and decrease in the chain of events leading to the release of renin (D) by the cells of the juxtaglomerulus (which receive blood vessel color in the enlarged view in center of page).
2. Color renin's (D) role in hormonal regulation, (in the material under the enlargement), going from the liver on the left to the adrenals on the right.
3. Color the effects of aldosterone (G) in the lower right corner by following the numbered sequence which leads to the actions of ADH on the left.

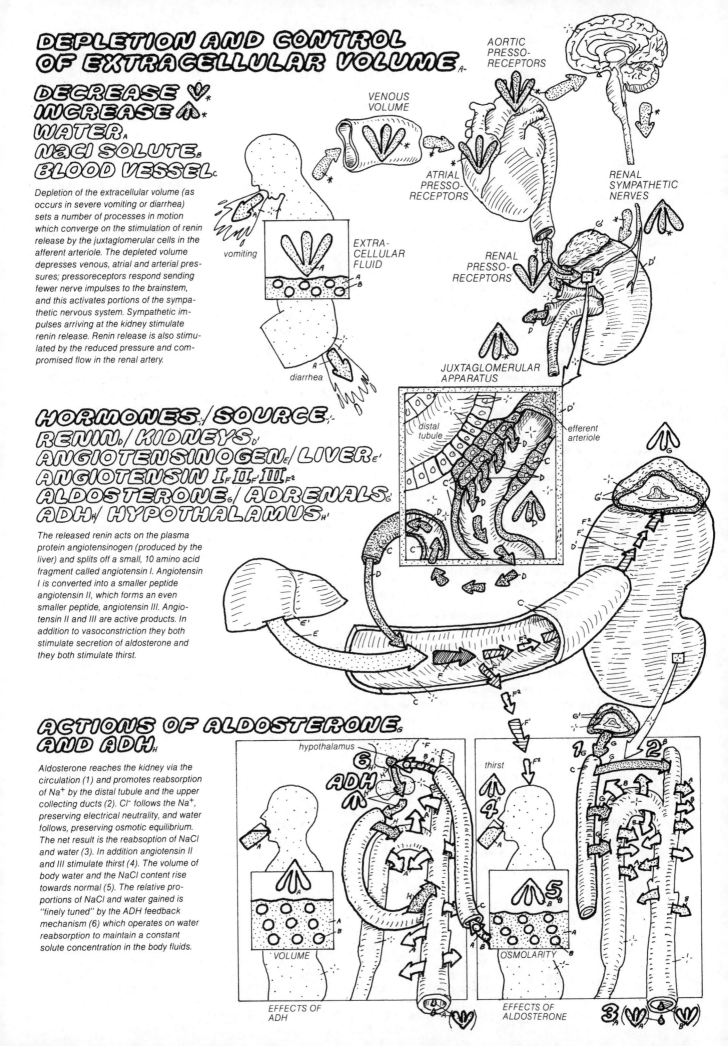

DEPLETION AND CONTROL OF EXTRACELLULAR VOLUME A

DECREASE ∨ *
INCREASE ∧ *
WATER A
NaCl SOLUTE B
BLOOD VESSEL C

Depletion of the extracellular volume (as occurs in severe vomiting or diarrhea) sets a number of processes in motion which converge on the stimulation of renin release by the juxtaglomerular cells in the afferent arteriole. The depleted volume depresses venous, atrial and arterial pressures; pressoreceptors respond sending fewer nerve impulses to the brainstem, and this activates portions of the sympathetic nervous system. Sympathetic impulses arriving at the kidney stimulate renin release. Renin release is also stimulated by the reduced pressure and compromised flow in the renal artery.

AORTIC PRESSORECEPTORS

VENOUS VOLUME

ATRIAL PRESSORECEPTORS

RENAL SYMPATHETIC NERVES

RENAL PRESSORECEPTORS

vomiting

EXTRACELLULAR FLUID

diarrhea

HORMONES / SOURCE
RENIN D / KIDNEYS D'
ANGIOTENSINOGEN E / LIVER E'
ANGIOTENSIN I F II F' III F²
ALDOSTERONE G / ADRENALS G'
ADH H / HYPOTHALAMUS H'

The released renin acts on the plasma protein angiotensinogen (produced by the liver) and splits off a small, 10 amino acid fragment called angiotensin I. Angiotensin I is converted into a smaller peptide angiotensin II, which forms an even smaller peptide, angiotensin III. Angiotensin II and III are active products. In addition to vasoconstriction they both stimulate secretion of aldosterone and they both stimulate thirst.

JUXTAGLOMERULAR APPARATUS

distal tubule

efferent arteriole

ACTIONS OF ALDOSTERONE G AND ADH H

Aldosterone reaches the kidney via the circulation (1) and promotes reabsorption of Na$^+$ by the distal tubule and the upper collecting ducts (2). Cl$^-$ follows the Na$^+$, preserving electrical neutrality, and water follows, preserving osmotic equilibrium. The net result is the reabsoption of NaCl and water (3). In addition angiotensin II and III stimulate thirst (4). The volume of body water and the NaCl content rise towards normal (5). The relative proportions of NaCl and water gained is "finely tuned" by the ADH feedback mechanism (6) which operates on water reabsorption to maintain a constant solute concentration in the body fluids.

hypothalamus

ADH

thirst

VOLUME

OSMOLARITY

EFFECTS OF ADH

EFFECTS OF ALDOSTERONE

ORGANIZATION & FUNCTIONS OF THE DIGESTIVE SYSTEM

The *digestive system* (also called the *digestive tract* or *tube*) is basically a tube open at both ends, making the lumen of the digestive system really an extension of the environment. The food enters from one end, the lining of the tube absorbs the usable substances, and the waste products leave from the other end. This design is already present in the simpler forms of animals. With evolution, only the complexity of the system increases; the structure of the tract is modified to adapt to new needs as animals' food habits change.

In mammals, including humans, the digestive system ingests food, which is usually in forms completely unsuitable for use by the body cells and thus must be transformed to smaller and simpler forms. This is accomplished by two kinds of digestive activities: *mechanical* and *chemical*. In mechanical digestion, solid food masses are torn apart, ground and vigorously shaken, and mixed with the various juices from the *digestive glands* to dissolve the food as much as possible as well as make it suitable for chemical digestion. The steps of mechanical digestion occur at several stages, aided by a variety of mechanical activities generated by the muscular walls of the digestive system. As a result, a rich soupy juice is formed. This soup is not necessarily in the final form from which food substances can be absorbed.

To transform the soup that results from mechanical digestion, chemical digestion must occur. In this operation, mainly accomplished by various *hydrolytic enzymes*, the larger and/or more complex food molecules are broken down gradually to the smallest components, which can be absorbed and delivered to body cells for consumption. The last function of the digestive system is to eliminate the unused waste materials without interfering with the processing of the incoming food.

In the human *mouth*, *salivary glands* secrete *saliva*, a mucus, aiding in mechanical digestion and dissolving the food. The throat (*pharynx*) and the *esophagus* transport the food into the *stomach*, which acts as a reservoir to receive all the food at once but delivers it to the *intestine* in intervals. In the stomach, food is subjected to vigorous mechanical movements that mix it with the *gastric juices*. Gastric juices, containing *mucus*, *acid*, and *enzymes*, are secreted by the stomach glands and surface cells. A small amount of chemical digestion, but no absorption of any significance, occurs in the stomach.

In the *small intestine*, the dissolved food particles (the *chyme*) are subjected to further mechanical shaking, mixing, and movements that mix them with the *intestinal juice* and propel them forward. Intestinal juices contain secretions of the glands of the small intestine and of the large accessory digestive glands (the *pancreas* and *liver*). The pancreatic secretion is an alkaline juice rich in hydrolytic enzymes that chemically digest essentially all the food substances. The liver secretes *bile*, which contains substances facilitating fat digestion. The small intestine is also the only place where the chemically digested food is absorbed. *Absorption* occurs across the inner lining of the intestine. Most of the absorbed food is delivered to the *intestinal-hepatic portal* venous system, which takes it to the liver; from there the nutrients move, via the bloodstream, to the rest of the body. The absorbed fatty foods bypass the liver and are delivered to the blood via the lymphatic circulation.

The *large intestine* (colon) is where the waste products of digestion are accumulated, dehydrated, and prepared for *excretion*. The water, salts (sodium), and some vitamins of bacterial origin are also absorbed in the colon. The *rectum* and the *anus* expel the feces (*defecation*), which in adult humans occurs usually once or twice a day.

Humans consume foods from a variety of animal and plant sources. In the fresh form, all these foods contain different amounts of the main classes of nutrients: *proteins*, *carbohydrates*, and *fats*. An apple contains mostly carbohydrates, some protein, and a very small amount of fat; meats contain a lot of protein, some fat, and a very small amount of carbohydrate. During chemical digestion, proteins are broken down first into *oligopeptides*, which are digested into smaller *peptides* and finally into *amino acids*, the building blocks of all peptides and proteins. Free amino acids are then in the form suitable for absorption by the intestinal mucosa and delivery to the liver and other body cells.

Dietary sources of carbohydrates are starches (polysaccharides) and *disaccharides* (e.g., table sugar [sucrose] and milk sugar [lactose]). Polysaccharides are broken down to *oligosaccharides* and finally to di- and *monosaccharides*; the disaccharides are broken down to monosaccharides directly. Monosaccharides or simple sugars like glucose, fructose, and galactose are the forms in which the body can absorb carbohydrates.

Dietary fats are provided mainly as *triglycerides*, which are broken down in the intestine into their constituents, *glycerol* and *fatty acids*. Occasionally, *mono-* or *diglycerides* are also produced. These simpler fats are then absorbed across the mucosa. Before entry into the blood, triglycerides are resynthesized and incorporated into lipoprotein particles called *chylomicrons*, which are then transported via the *lymphatic system* to the blood.

CN: Use blue for L and a light gray (or a single light color) for structures H-K. Notice the use of overlapping colors in the stomach region of the central illustration to suggest the presence of one organ in front of another.
1. Color the same structure in both the anatomic and the functional diagrams, before going on to the next structure. Color the titles along the right edge of the page. Color the inner edge of the doughnut, which is intended to demonstrate that the digestive tract, from the mouth to the anus, is essentially outside the body.

DIGESTIVE SYSTEM.

DIGESTIVE TRACT.
ORAL CAVITY A
PHARYNX B
ESOPHAGUS C
STOMACH D
SMALL INTESTINE E
LARGE INTESTINE F
RECTUM G

DIGESTIVE GLANDS.
SALIVARY GLANDS H*
LIVER I*
GALL BLADDER J*
PANCREAS K*

The function of the digestive system is to ingest, digest and absorb food substances into the bloodstream and to eliminate remaining wastes. Of the digestive system structures, those in the mouth and stomach act primarily in the mechanical and chemical digestion of foods. The parts of the small intestine act in both the chemical digestion and absorption of food substances. The large intestine (colon) absorbs remaining water and excretes waste products of digestion through its exit end, the rectum and anus. To facilitate digestion, numerous exocrine glands secrete a variety of juices containing water, enzymes, and mucus into the digestive lumen. Some of these glands, such as those in the stomach and intestinal walls, are internal; others, such as the pancreas, liver, and salivary glands, are independent organs (external glands). During the absorption process, the products of the breakdown of proteins and carbohydrates, as well as water, minerals, and water-soluble vitamins, are transported directly across the mucosal wall of the small intestine into the portal circulation. Fats and fat-soluble vitamins are absorbed via the lymphatics.

ANATOMIC ORGANIZATION.

FUNCTIONAL ORGANIZATION.

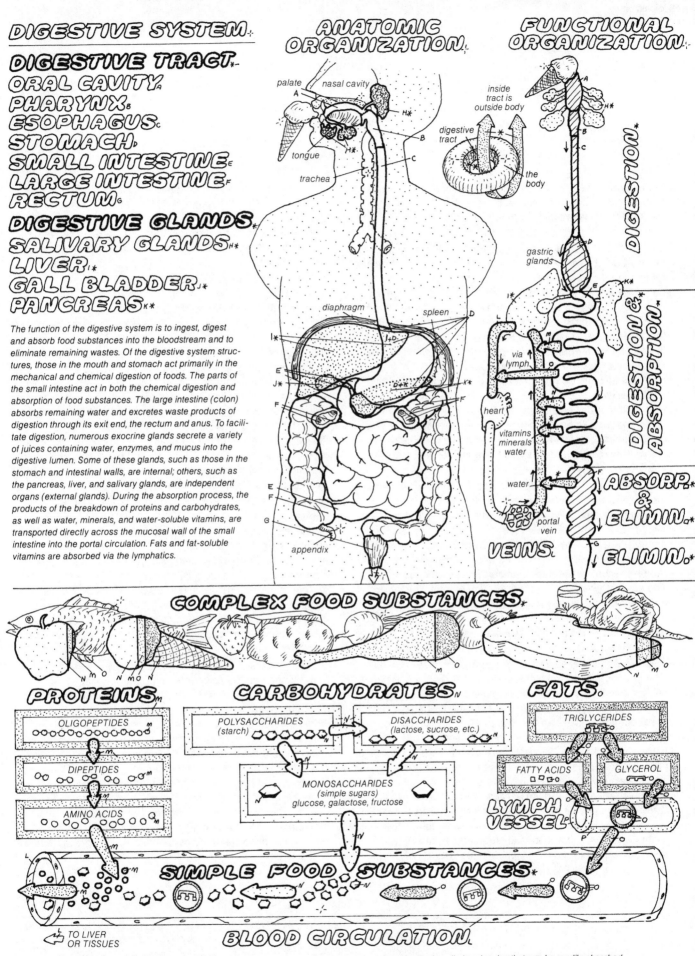

DIGESTION*

DIGESTION & ABSORPTION*

ABSORP. & ELIMIN.*

ELIMIN.*

VEINS.

COMPLEX FOOD SUBSTANCES*

PROTEINS M

OLIGOPEPTIDES

DIPEPTIDES

AMINO ACIDS

CARBOHYDRATES N

POLYSACCHARIDES (starch)

DISACCHARIDES (lactose, sucrose, etc.)

MONOSACCHARIDES (simple sugars) glucose, galactose, fructose

FATS O

TRIGLYCERIDES

FATTY ACIDS

GLYCEROL

LYMPH VESSEL

SIMPLE FOOD SUBSTANCES*

TO LIVER OR TISSUES

BLOOD CIRCULATION.

Foods such as meats, fruits, dairy products, bread, and vegetables are rarely found in a readily absorbable form. The complex food substances in the diet are proteins (meat, eggs, beans), carbohydrates (bread, rice, potatoes), and fats (milk, egg, butter, oils). Digestive enzymes secreted by the different digestive glands chemically break down these substances into simpler (smaller) molecules that can be readily absorbed by the intestinal mucosal cells into the bloodstream. Proteins are digested into amino acids, complex carbohydrates (polysaccharides), into simple sugars (monosaccharides) such as glucose, and fatty triglycerides into fatty acids and glycerol.

DIGESTION IN THE MOUTH: CHEWING, SALIVA, & SWALLOWING

In the *mouth*, the first station in digestion, food is exposed to both mechanical and chemical processes designed to turn the solid food pieces into a shape that can be easily swallowed.

MECHANICAL EVENTS. Several structures in the mouth aid in ingestion and mechanical digestion of the food: the *lips*, the *teeth*, the *tongue*, and the muscles of the *cheeks*. Adult humans have 32 teeth arranged in two sets attached to the upper and lower jaw bones. Human teeth are adapted to an *omnivorous* diet; the 8 front *incisors* are designed for cutting; the 4 *canines*, for tearing; the 8 *premolars*, for crushing; and the 12 *molars*, for grinding. Chewing (*mastication*) involves not only the movements of the jaws and the action of the teeth but also the coordinated movement of the tongue and other muscles of the *oral* (mouth) *cavity*. The activities of the masticatory muscles and the tongue are controlled by both voluntary and involuntary nervous control mechanisms. The mere placing of food in the mouth can activate some of the involuntary reflex mechanisms, the centers of which are in the *brain stem*.

SOURCE AND FUNCTIONS OF SALIVA. The chewing and mechanical actions of the mouth would be extremely difficult without the aid of *saliva*, a *mucus*-containing juice secreted by the salivary glands. There are three pairs of *salivary glands*: the *parotid* in the cheeks secrete a watery (*serous*) juice; the *submandibular* (under the lower jaw) and *sublingual* (under the tongue) secrete both serous and mucous saliva. The salivary glands are *acinar exocrine* glands. The *serous acini* secrete the watery saliva, and the *mucous acini* secrete a more viscous fluid containing the glycoprotein substance *mucin*, which gives the saliva its characteristic sticky and viscous texture.

The three glands secrete from 1 to 2 L of saliva each day. Of this, 25% is secreted by the parotid, 70% by the submandibular, and 5% by the sublingual glands. The serous saliva, containing more than 90% water, keeps the mouth wet, aids in speech, helps dissolve the food particles, and helps form a wetter mold from which the food *bolus* is produced. The dissolving of food particles is also necessary for activation of the *taste buds*, because the taste receptors respond only to dissolved substances. Serous saliva contains the salivary digestive enzyme *ptyalin*, an *amylase* that breaks down the starches. Another salivary enzyme is *lysozyme*, an antibacterial enzyme presumably secreted as a disinfectant; lysozyme destroys the bacteria in the food and mouth by lysing their cell wall. This is one reason animals instinctively lick their wounds. The saliva contains *sodium* and certain other *minerals* as well. The mucous saliva, containing mucin, functions principally as a lubricant and glue while the bolus is formed in the mouth and transported along the throat and esophagus. Without saliva, chewing and swallowing become very difficult tasks.

Saliva formation and secretion are under autonomic nervous control (see plate 25). *Parasympathetic* nerves originating in the *salivary nuclei* of the brain stem stimulate both serous and mucous salivary secretion; *sympathetic* nerves inhibit the secretion of serous saliva. This explains why the mouth becomes dry during fear and excitement (a sympathetic condition) and salivary juice flows profusely during relaxation or expectation of food and pleasure. During oral digestion, the presence of food, particularly dry or sour foods, in the mouth serves as a strong stimulus, which is communicated by sensory nerves to the brain stem salivary centers. These in turn activate the parasympathetic nerves to the salivary glands, increasing their production of saliva. Similarly, food odors acting through the olfactory (smell) senses and even thoughts of food, can increase salivary flow.

SWALLOWING AND BOLUS TRANSPORT IN THE ESOPHAGUS. After the bolus is appropriately formed in the mouth, the movements of the tongue gradually push it backward. Presence of the bolus on the back of the tongue activates the swallowing (*deglutition*) reflexes, which are centered in the brain *medulla*. When the tongue moves back to force the bolus into the throat (*pharynx*), the *soft palate* closes the nasal passages, and the *epiglottis* moves over the *glottis* to close the *larynx* and *trachea*. These protective reflexes prevent the bolus from entering the upper and lower respiratory passages.

When the bolus arrives in the pharynx, other reflexes transport it to the *esophagus*, a tubular organ connecting the throat with the *stomach*. The muscular wall of the esophagus contains layers of *circular* and *longitudinal smooth muscles* whose coordinated movements give rise to a special wavy contractile movement called *peristalsis*, which begins in the upper esophagus and travels toward the stomach. As a result, the bolus is propelled from throat to stomach. Although gravity may aid bolus transport in the human esophagus under normal circumstances, it is not a necessary condition; food can be swallowed in a supine position as well. Indeed, in ruminants (think of a grazing animal), food and water are usually propelled along the esophagus against gravity with little difficulty.

CN: Use red for C and a dark color for Q. Note that M is colored gray.
1. Color the structure of the mouth including the 3 salivary glands, but not the arrows indicating the amount of saliva production, nor the cell diagrams below them. Color the diagrams of teeth function.

2. Color the salivary material previously omitted, in addition to the rest of the salivary section along the right side of the page.
3. Color the panel on swallowing. First complete the diagram (1) on the left showing the mouth and throat structure, prior to swallowing.

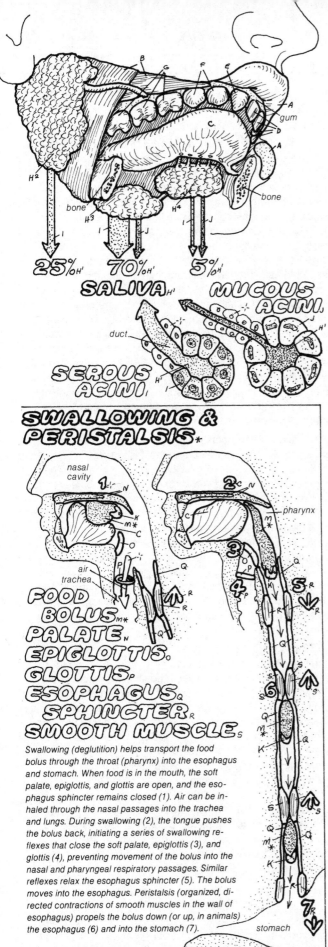

25%
70%
5%

SALIVA

MUCOUS ACINI

duct

SEROUS ACINI

SWALLOWING & PERISTALSIS *

nasal cavity

1

2

pharynx

air
trachea

FOOD
BOLUS
PALATE
EPIGLOTTIS
GLOTTIS
ESOPHAGUS
SPHINCTER
SMOOTH MUSCLE

stomach

Swallowing (deglutition) helps transport the food bolus through the throat (pharynx) into the esophagus and stomach. When food is in the mouth, the soft palate, epiglottis, and glottis are open, and the esophagus sphincter remains closed (1). Air can be inhaled through the nasal passages into the trachea and lungs. During swallowing (2), the tongue pushes the bolus back, initiating a series of swallowing reflexes that close the soft palate, epiglottis (3), and glottis (4), preventing movement of the bolus into the nasal and pharyngeal respiratory passages. Similar reflexes relax the esophagus sphincter (5). The bolus moves into the esophagus. Peristalsis (organized, directed contractions of smooth muscles in the wall of esophagus) propels the bolus down (or up, in animals) the esophagus (6) and into the stomach (7).

MECHANICAL EVENTS *

LIPS
MUSCLES/CHEEKS
TONGUE
TEETH *
INCISORS 8
CANINES 4
PREMOLARS 8
MOLARS 12

The chewing action (mastication) of the structures in the mouth (lips, tongue, cheeks, teeth, jaw) breaks the food materials into smaller pieces and forms a bolus for swallowing.

CUT   TEAR   CRUSH   GRIND

In the adult human, a total of 32 permanent teeth act to cut, tear, crush, and grind the food materials. Teeth are absent in the newborn. Deciduous (temporary) teeth (20) from between 6 to 24 months as the infant begins to use solid foods. Permanent teeth appear from 6 to 21 years.

CHEMICAL EVENTS *
SALIVARY GLANDS → SALIVA
PAROTID
SEROUS SALIVA
SUBMANDIBULAR:
SEROUS S., MUCOUS S.
SUBLINGUAL:
MUCOUS S., SEROUS S.

DAILY
1 - 2
LITERS

Each day about 1.5 liters of saliva are secreted by the 3 pairs of salivary glands: parotid, sublingual, and submandibular (submaxillary). Saliva is a mixture of water, mucus (containing mucin, a glycoprotein), minerals (sodium), and enzymes (ptyalin or amylase) for starch digestion and lysozyme (an antibacterial enzyme). Salivary glands contain cells arranged as serous or mucous acini, which secrete a watery or thick mucous solution, respectively. Parasympathetic nerves with centers in the brain stem stimulate salivary secretion.

CONTENT & FUNCTION OF SALIVA *

99½% WATER: DISSOLVES BOLUS FOR TASTE

The water in saliva helps dissolve food particles, facilitating taste sensation. Dry foods and sour (acid) juices induce copious salivary secretion.

MUCUS: LUBRICATION FOR BOLUS

The glycoprotein mucin secreted by the mucous acini gives saliva its sticky and lubricating property. Without saliva, bolus formation is very difficult, and swallowing is painful.

ENZYME: AMYLASE BEGINS STARCH DIGESTION

Salivary amylase begins the chemical breakdown of starches, forming oligo-, tri-, and the disaccharide maltose. Amylase action is important for the taste sensation of sweetness.

MINERALS & LYSOZYME: ANTI-BACTERIAL ACTION

The mineral sodium and the organic enzyme lysozyme in the salvia act as disinfectants, killing bacteria and microorganisms in food.

PHYSIOLOGY OF THE STOMACH

The *stomach* is a large muscular sac connected at its opening to the esophagus and at its end to the *duodenum* of the small intestine. Two *sphincters*, the *cardiac* and the *pyloric*, act as unidirectional flow valves permitting food to move into and out of the stomach. The stomach functions as a reservoir, receiving the ingested food in one portion. It disinfects the food, mixes the bolus with the gastric juice, and partially digests the ingested proteins. Finally, the stomach delivers a well-mixed, soupy *chyme* to the small intestine, in regular intervals, for further processing.

STOMACH SECRETIONS. Numerous exocrine *gastric glands* (*pits*) secrete *mucus*, *acid*, and *enzymes* into the stomach lumen. Each gland contains three types of cells, which together produce the bulk of *gastric juice*. The cells near the gland's neck (*mucous cells*) secrete the gastric mucus. (Mucus is also secreted by the cells lining the stomach's inner surface.) In the gland's deeper zone, there are two other cell types: the *chief* cells secrete the proenzyme *pepsinogen*, which is later converted to the gastric enzyme *pepsin* in the stomach's lumen; the *parietal* cells (also called the *oxyntic* cells) secrete a concentrated solution of *hydrochloric acid* (H^+Cl^-). Other, rarer cell types (endocrine or paracrine) present in the glands secrete *hormones* into the blood capillaries or tissue spaces.

ACTIONS OF STOMACH SECRETIONS. Stomach acid has several functions. The acidic gastric juice acts as a superior solvent, dissolving foodstuffs not soluble in water. Acid is necessary to activate the gastric enzyme pepsin (see below). Acid is a strong disinfectant, killing bacteria and other microorganisms in the ingested food. Finally, acid has a regulatory function: it stimulates the duodenum to secrete hormones to release bile and pancreatic juices (plate 70).

Pepsin is the only digestive enzyme of any significance produced in the stomach. It cleaves food *proteins*, forming small *peptides*. This action is probably not crucial for protein digestion because one of the proteases of the pancreatic juice (chymotrypsin) performs a similar function later in the small intestine. Pepsin may serve a regulatory function: the small peptides produced stimulate the sensory receptors in the gastric mucosa to initiate hormonal and nervous signals aimed to increase stomach motility and secretion (see plates 70, 71). When secreted by the chief (*zymogen*) cells, pepsin is in its inactive form, a larger protein called pepsinogen. Acid in the lumen promotes conversion of pepsinogen to pepsin. Pepsin, once formed, also attacks pepsinogen, producing more pepsin molecules (*autocatalysis*).

The stomach *mucus*, in addition to providing similar functions as the salivary mucus, forms a thick protective coat covering the inner linings of the stomach in order to protect it from mechanical damage and, perhaps, from the corrosive actions of the acid in the gastric juice. The breakdown of this coat is one of the causes of *ulcers*.

CELL PHYSIOLOGY OF ACID SECRETION. Stomach glands secrete a concentrated solution of *hydrochloric acid* that may reach a pH value near 1. If placed on the skin, this acid would cause serious burns and tissue damage. Gastric wall cells' impermeability to acid, as well as the protective action of the alkaline stomach mucus, prevents this damage from occurring in healthy individuals. Parietal cells secrete acid by directly pumping *hydrogen ions* from inside the cell out into the gland lumen, using an *active transport* mechanism. The pump obtains hydrogen ions from the dissociation of intracellular water ($H_2O \rightarrow H^+ + OH^-$). The hydrogen ions are pumped out in exchange for K^+ ions, which are pumped in. Parietal cells contain many mitochrondria, which utilize oxygen heavily and produce much ATP. The pumping mechanism, which consists of enzymes associated with intracellular *canaliculi* membranes, use the ATP. The parietal cell canaliculi are modified endoplasmic reticulum. Upon hormonal or nervous stimulation, the active transport mechanism is activated, resulting in hydrogen ions being secreted into the canaliculi, which converge and open into the gland lumen.

Parietal cells also contain large amounts of *carbonic anhydrase*, an enzyme that promotes carbon dioxide hydration: ($CO_2 + H_2O \rightarrow [H_2CO_3] \rightarrow H^+ + HCO_3^-$). The hydrogen ions produced in this reaction will combine with the hydroxyl ions left from water dissociation to form a new water molecule, replacing the one utilized by the pump. The parietal cell at the serosal (blood side) border then exchanges *bicarbonate* (HCO_3^-) ions produced in the above reaction with *chloride ions*; the chloride ions move in, and bicarbonate ions move out of the cell. The chloride-bicarbonate exchange is also an active transport mechanism, involving pumping and ATP utilization. The chloride ions are then transported across the cell and out into the stomach gland's lumen, where they combine with the hydrogen ions to form hydrochloric acid.

GASTRIC MOTILITY. Shortly after food enters the stomach, when sufficient gastric juice has been produced, special weak contractions (*mixing waves*) begin in stomach *fundus* and spread to *pylorus*. These waves (occurring every 20 sec.) help mix the food with the gastric juice. Later on, less frequent but much stronger *peristaltic* waves occur and force the *chyme* against the closed *pyloric sphincter*, resulting in chyme back flow. This movement vigorously mixes food with gastric juice, forming a soupy solution (chyme), which can now be processed by the intestinal enzymes. Gradually, the pyloric sphincter opens a little, allowing, with each peristaltic wave, delivery of some chyme into the *duodenum*. The rate of this process depends on the food content: carbohydrates empty rapidly; fats slowly; protein-rich foods, at an intermediate rate. This differential rate is regulated by hormones and nerves (see plates 70 and 71).

CN: Use dark colors for A, E, L, S, U.
1. Color the stomach in the upper right corner, notice the different secretory cells adjacent to the body and antrum portions, indicating their location. Color the stomach wall enlargement at the top.
2. Before coloring the gastric gland illustration in the center of the page, color the material on the four types of cells that surround the gland. Then color their location along the length of the gastric pit (L^1).
3. Color the gastric motility panel, coloring each figure completely before going on to the next.
4. Color the four situations that determine whether or not the stomach will empty its contents into the duodenum.

STOMACH *
CARDIAC SPHINCTER A
FUNDUS B
BODY C
ANTRUM D
PYLORIC SPHINCTER E
LONGITUDINAL MUS. F
CIRCULAR MUS. G
OBLIQUE MUS. H
RUGAE (FOLDS) I

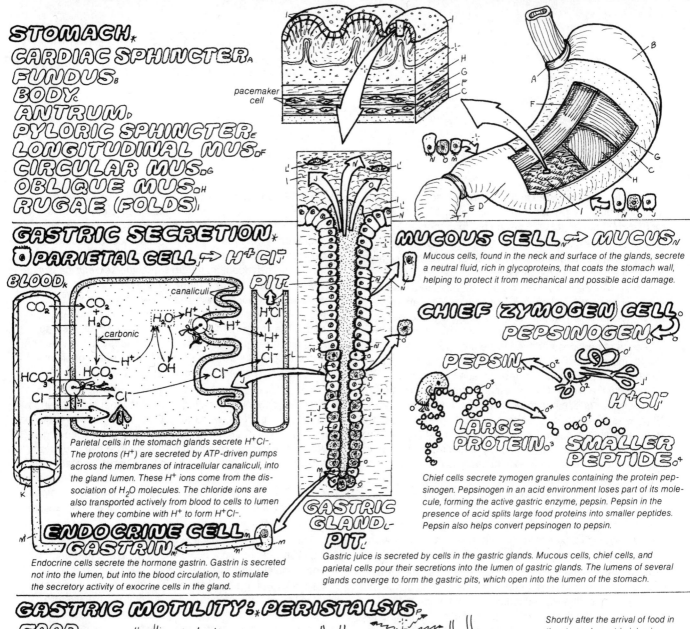

pacemaker cell

GASTRIC SECRETION *
❶ PARIETAL CELL → H^+Cl^- J

BLOOD K

PIT

canaliculi

CO_2 + H_2O

carbonic

H_2O → H^+

H^+

H^+

Cl^-

OH

HCO_3^-

Cl^- → Cl^-

Cl^-

H^+Cl^-

H^+

H^+

Cl^-

Parietal cells in the stomach glands secrete H^+Cl^-. The protons (H^+) are secreted by ATP-driven pumps across the membranes of intracellular canaliculi, into the gland lumen. These H^+ ions come from the dissociation of H_2O molecules. The chloride ions are also transported actively from blood to cells to lumen where they combine with H^+ to form H^+Cl^-.

ENDOCRINE CELL GASTRIN M

Endocrine cells secrete the hormone gastrin. Gastrin is secreted not into the lumen, but into the blood circulation, to stimulate the secretory activity of exocrine cells in the gland.

MUCOUS CELL → MUCUS N

Mucous cells, found in the neck and surface of the glands, secrete a neutral fluid, rich in glycoproteins, that coats the stomach wall, helping to protect it from mechanical and possible acid damage.

CHIEF (ZYMOGEN) CELL O
PEPSINOGEN O1

PEPSIN O2

H^+Cl^-

LARGE PROTEIN O3

SMALLER PEPTIDE O4

Chief cells secrete zymogen granules containing the protein pepsinogen. Pepsinogen in an acid environment loses part of its molecule, forming the active gastric enzyme, pepsin. Pepsin in the presence of acid splits large food proteins into smaller peptides. Pepsin also helps convert pepsinogen to pepsin.

GASTRIC GLAND PIT L

Gastric juice is secreted by cells in the gastric glands. Mucous cells, chief cells, and parietal cells pour their secretions into the lumen of gastric glands. The lumens of several glands converge to form the gastric pits, which open into the lumen of the stomach.

GASTRIC MOTILITY: PERISTALSIS P

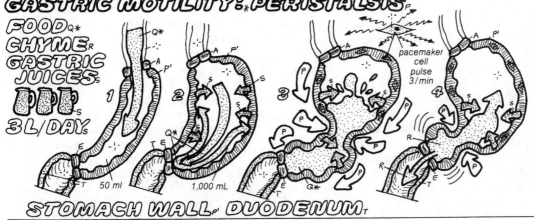

FOOD Q *
CHYME R
GASTRIC JUICES S

3 L/DAY S

50 ml   1,000 mL

pacemaker cell pulse 3/min

STOMACH WALL P' DUODENUM T

Shortly after the arrival of food in the stomach, gastric juice is copiously secreted to mix with and digest the food. To enhance mixing and digestion, the muscular wall of the stomach begins a series of regular contractions. These peristaltic contractions are generated by a discharge of pacemaker cells in the muscular wall and travel from fundus to antrum, being strongest in the antrum. The hormone gastrin and the parasympathetic nerve (vagus) regulate the strength of these contractions.

GASTRIC EMPTYING *

LIQUIDITY OF CHYME R
CHYME IN DUODENUM R
HIGH ACIDITY IN DUODENUM J'
FATS IN DUODENUM T

Another function of gastric peristalsis is to deliver at regular intervals the mixed and partly digested chyme into the small intestine (gastric emptying). Highly liquid chyme (with carbohydrate foods) increases the rate of emptying. Both highly acid chyme (with protein food) and fatty chyme decrease the rate. The reduction is to allow more time for digestion of the chyme in the intestine.

HORMONAL REGULATION OF DIGESTIVE ACTIVITIES

The motility and secretory activities of the digestive system are under both neural and hormonal control. This plate focuses on the control of digestion by *hormones*.

GASTRIN: STOMACH STIMULATORY HORMONE. *Gastrin* is a single chain peptide hormone secreted by certain isolated endocrine cells scattered in the walls of *stomach glands*, particularly in the *antrum* region. It is secreted into the blood in response to stimulation by *small peptides* present in the ingested food. The gastric mucosa contains sensory *chemoreceptor* and *stretch receptor* cells which detect the presence of peptides and food in the stomach lumen. Acting via the intrinsic nerve connections in the *stomach wall*, the receptor neurons signal the *gastrin-secreting cells* to release gastrin into the blood. The released gastrin returns to the stomach's *fundus* (main body) by way of the *bloodstream*, stimulates the stomach's glands to secrete *gastric juice*, and stimulates the stomach's muscular wall to increase *motility*. The action of gastrin is one reason that secretion and motility can continue even if all the external nerves to the stomach are cut (stomach denervation).

Gastrin is also very important clinically because excessive amounts of it are related to *ulcer* formation. Occasionally, certain *tumors* in the *pancreatic islets* secrete large quantities of gastrin, leading to excessive acid secretion in the stomach. This condition often results in gastric ulcers and bleeding.

DUODENAL HORMONES. Like the stomach antrum, the *duodenum* is also partly an endocrine organ. The duodenal wall contains scattered endocrine cells that secrete three hormones: *choleocystokinin* (CCK), *secretin*, and *gastric inhibitory peptide* (GIP), all of which are peptides. Secretin has an important place in the history of endocrinology because it was the first hormone to be discovered. In 1902 English physiologists Bayliss and Starling noted that when extracts of the duodenum were injected into the blood of fasting dogs (in which all the nerves to the pancreas had been cut), the secretion of pancreatic juice was markedly augmented. Based on this observation, Bayliss and Starling correctly postulated that, under normal conditions, the duodenum secretes into the blood a substance that, upon reaching the pancreas, stimulates the secretion of pancreatic juice (hence the name "secretin"). The term "hormone" was then adopted for such blood-borne humoral messengers. At the time of this discovery, all physiological regulations, including those of digestive activities, were thought to occur by the actions of nerves and the nervous system.

Secretin's target appears to be the cells lining the *ducts* of the *pancreatic acini* (aggregates of exocrine cells surrounding a cavity with a duct outlet) because secretion augments mainly the secretion of *bicarbonate*-rich juice, which is known to be produced by the *duct cells*. The signal for secretin secretion is the presence of acid in the duodenal lumen. This acid acts on the sensory chemoreceptors in the duodenal mucosa. The receptors in turn signal the *secretin-producing cells* to release their hormone. The secretion of the highly *alkaline*, bicarbonate-rich pancreatic juice helps neutralize the acid in the duodenal chyme, an important function because the small intestine wall is not as well protected against acid hazards as is the stomach's and because the intestinal and pancreatic enzymes work best in a neutral or slightly alkaline environment (see plate 72).

A third digestive hormone is choleocystokinin (CCK), a peptide hormone originating in the duodenal mucosa endocrine cells that has two targets. One is the *gallbladder*. Upon stimulation by CCK, the gallbladder contracts, releasing its stored *bile* into the duodenum. The alkaline bile neutralizes the acid and *emulsifies* the *fat* in the chyme, facilitating its chemical digestion by the pancreatic enzyme *lipase* (see plates 72, 73). The stimulus for CCK release into the blood is the arrival of fat- or acid-rich chyme from the stomach into the duodenum.

The pancreas is the second target organ for CCK. Here, the CCK acts on the *acinar cells* of the exocrine pancreas and stimulates their production and release of *pancreatic enzymes*, which are extremely important for the chemical digestion of various foodstuffs (see plate 72). It was previously believed that this hormonal action was expressed by another hormone called pancreozymin. Now it is believed that pancreozymin and CCK are the same hormone.

The gastrointestinal hormones discussed so far, all stimulate digestive activities. Recently, an inhibitory hormone (*gastric inhibitory peptide*, GIP), originating from the duodenal mucosa, has been discovered. GIP *inhibits* the stomach glands and muscles, decreasing their secretion and motility. This hormone's physiological role may be both to protect the duodenum against excessive acid and to regulate the rate of *gastric emptying*. Thus, high fat or acid content in the chyme causes release of this inhibitory peptide into the bloodstream, from which it is transported to the stomach, where it exerts its inhibitory actions. If the food is fatty, the reduced motility of the stomach results in slower chyme delivery to the duodenum, permitting increased time for digestion of what is already there. If the chyme is too acidic, GIP action again reduces acid secretion, thus diminishing the chances of acid damage to the duodenum.

CN: Use red for E and dark colors for C, D, I, K, M and Q.
1. Begin with the gastrin panel and follow the numbered sequence. Go on to the upper right panel, then the lower left, and lower right.

FOOD.ᴬ* CHYME.ᴬ'*
STOMACH WALL.ʙ
STRETCH.& CHEMORECEPTOR.ᶜ'
SMOOTH MUSCLE.ᴅ

HEART. BLOOD CIRC.ᴇ.ᴇ'
PROTEIN (PEPTIDE).ꜰ
GALLBLADDER. BILE.ɢ.ɢ'
PANCREAS.ʜ

GASTRIN SECRETING CELL.ɪ'

The bulk (1) of the food as well as its peptides (2) stimulate the stretch and chemoreceptors in the stomach wall (3). These
receptors, acting via local hormones or
nerve reflexes, stimulate the endocrine
cells (4) in the wall of the stomach to
secrete the hormone gastrin into
the blood. Gastrin circulates,
arriving back to the stomach to
stimulate secretion of acid
and enzyme (5) by the
gland cells, as well as to
increase motility by
acting on the smooth
muscles (6).

PARIETAL CELL:ᴊ H⁺Cl⁻

GASTRIC-INHIBITORY PEPTIDE (GIP).ᴋ SECRETING CELL.ᴋ'

Arrival of acid chyme (1) in the duodenum
stimulates the chemoreceptors in the duodenal
wall (2). These act on the endocrine cells in
the duodenal wall (3) to stimulate secretion
of the hormone gastric inhibitory peptide
(GIP) into the blood. GIP acts on the
smooth muscle cells (4) and the
glands (5) in the stomach, decreasing their activity. This
prolongs the delivery of chyme
and allows more time for the
small intestine to digest
the chyme.

DUODENUM.ᴸ

CHOLECYSTOKININ (CCK).ᴹ SECRETING CELL.ᴹ'

Delivery of acid and fatty chyme (1) to the duodenum
stimulates special chemoreceptors (2) in the duodenal
wall, which in turn stimulate the secretion of the hormone
cholecystokinin (CCK) from the endocrine cells (3) in the
duodenal wall. CCK stimulates contraction of the gall
bladder (4) and emptying of the bile into the duodenum
(5) to facilitate fat digestion. CCK also acts on the acinar
cells of the pancreas (6) to stimulate secretion of digestive enzymes into the duodenum (7).

FATS.ᴺ

ACINAR CELLS: ENZYMES.ᴼ'

SECRETIN.ᑫ SECRETING CELL.ᑫ'

Delivery of acid chyme (1) in the duodenum stimulates
the chemoreceptors in the duodenal wall (2), which in turn
stimulate the endocrine cells (3) of the duodenal wall to
secrete a hormone, secretin. Secretin acts on the duct
cells (4) of the exocrine pancreas, stimulating the secretion of a watery alkaline juice rich in bicarbonate. This
juice neutralizes the acid (5) and enhances the activity of
pancreatic enzymes in the duodenum.

pH⁻

DUCTILE CELLS: BICARBONATE.ᴿ'

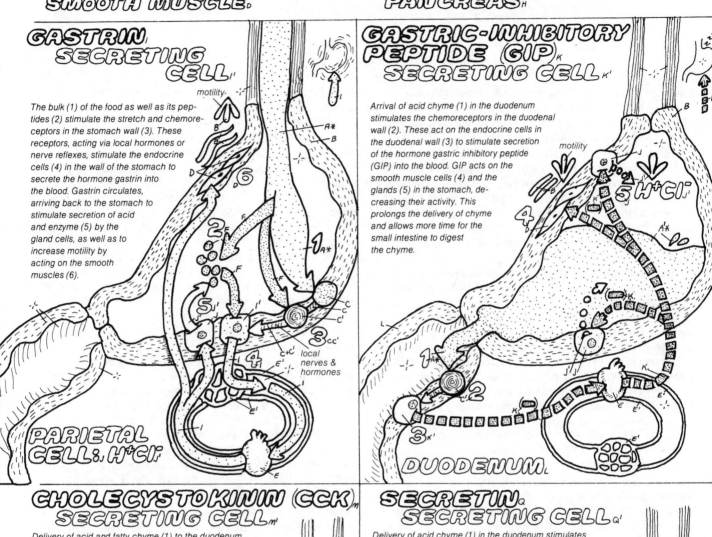
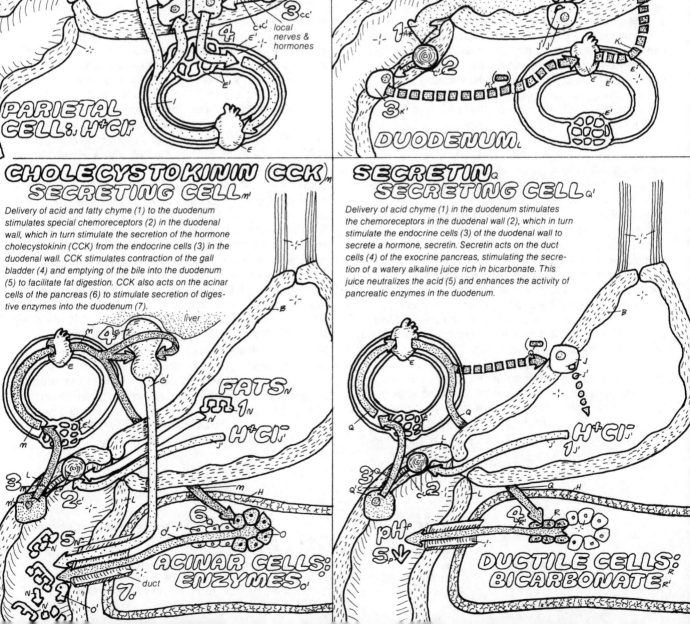

NEURAL REGULATION OF DIGESTION

Our knowledge of *autonomic nervous system* control of digestive activities precedes even that of hormonal control. Pavlov, the Russian physiologist and Nobel laureate, made many discoveries in this area.

AUTONOMIC CONTROL OF THE DIGESTIVE SYSTEM. The digestive system is innervated profusely with the nerve fibers of both the *sympathetic* and *parasympathetic* divisions, but the parasympathetic division's regulatory role, carried out primarily by the *vagus nerve*, seems to be paramount. In general, the parasympathetic system *increases* gastro-intestinal activity (secretion and motility), and the sympathetic system has a net *inhibitory* effect.

The parasympathetic vagus nerve contains both motor and sensory fibers. The motor fibers enhance digestive activities by stimulating local neurons of the *intrinsic nervous system*, located in the *gut wall*. The smaller intrinsic neurons in turn stimulate the *smooth muscles* and *gland cells*. Although the sympathetic fibers directly influence the smooth muscle and secretory cells in certain instances, the sympathetic system's general inhibitory effects on digestion are caused indirectly, by *constricting* the *blood vessels* in the digestive tract. The reduction in blood flow diminishes both secretory and contractile activity. The numerous *afferent sensory* fibers in the vagus nerve inform the *brain* about the condition of the gut and its content.

INTRINSIC (ENTERIC) NERVOUS SYSTEM. The intrinsic nervous system consists of two sets of *ganglia* or *plexi*: the superficial *submucosal* plexus mainly regulates the *digestive glands*, and the *myenteric* plexus, located deeper within the muscle layers, is primarily concerned with gut *motility*. The plexi, function in part, as the peripheral ganglia of the para-sympathetic system within the gut (see plate 25).

The plexi contain local sensory and motor neurons as well as interneurons. *Sensory neurons* are connected to the sensory *chemoreceptors*, which detect different substances in the gut lumen, and *stretch receptors*, which respond to the tension in the gut wall caused by the food and chyme bulk. The short effector *motor neurons* increase digestive gland activity or induce smooth muscle contraction. The myenteric and submucosal plexi in the same region communicate with each other, as well as with plexi farther in the gut, through *interneurons*. The vast numbers of neurons and neuronal connections in the plexi constitute the enteric nervous system, which carries out many digestive reflexes independently, in addition to mediating brain influence on digestive functions.

PHASES IN NEURAL REGULATION OF DIGESTION. Nervous system regulation of digestive activities is traditionally divided into three consecutive phases: *cephalic* (brain, mental), *gastric* (stomach), and *intestinal*.

CEPHALIC PHASE. When one is hungry, odors or even thoughts of foods commonly evoke *salivary* secretion (mouth watering). Experiments have shown that this anticipatory response also involves the secretion of a small amount of *gastric juice*. When food is placed in the mouth, gastric juice production is substantially increased, as is salivary secretion. There is also an increase (albeit a small one) in the secretion of *pancreatic juice*. These gastric and pancreatic secretions during the cephalic phase prepare the gut to receive food. One function of this step may be regulatory: the presence of some acid and pepsin in the stomach will help form peptides, which stimulate more juice production when food arrives in the stomach.

These anticipatory and reflex activations of digestion, particularly the stomach activity, have been labeled the "cephalic" (brain) phase because both the higher and the digestive centers of the brain play essential roles here. The main brain centers regulating digestive functions are in the *medulla oblongata*, where the taste fibers also have their primary centers and where the cell bodies of the vagus and salivary nerves are located. The higher cortical and olfactory centers influence these medullary motor centers in order to regulate digestion. All the cephalic responses, including those of the vagus and nerves to the salivary glands, are conducted by the parasympathetic outflow.

GASTRIC PHASE. When food enters the stomach, the mechanical stretch receptors sense the increase in bulk, and the chemoreceptors detect the presence of peptides in the food. These sensors signal the information to two targets: (1) the effector neurons in the local enteric plexi and (2) the brain medullary centers for digestion. Both these targets reflexly increase the stomach's secretion and motility over that occurring during the preceding cephalic phase (80% of gastric juice secretion compared to 10%), because this secretion deals with the bulk of the stomach's digestive functions. Also during this *gastric* phase, *gastrin* becomes active.

INTESTINAL PHASE. The arrival of the chyme in the duodenum initiates the *intestinal* phase of nervous control, during which gastric secretion and motility are at first increased to promote further digestion and emptying. As the small intestine becomes filled with acidic and fatty chyme, inhibitory signals (mostly hormonal) decrease stomach activity to prolong emptying and allow time for intestinal digestion.

CN: Use dark colors fo F & J.
1. Color the diagram of the sympathetic and para-sympathetic nervous system in the upper right corner in order to familiarize yourself with their effect on the digestive process. Notice the presence of the parasympathetic ganglia in the organs themselves. These have been deleted from the other diagrams for purposes of simplification.
2. Color the three phases of digestion (in the enclosed sections of the page).
3. Color the diagram of the intrinsic nervous system in the lower left corner.

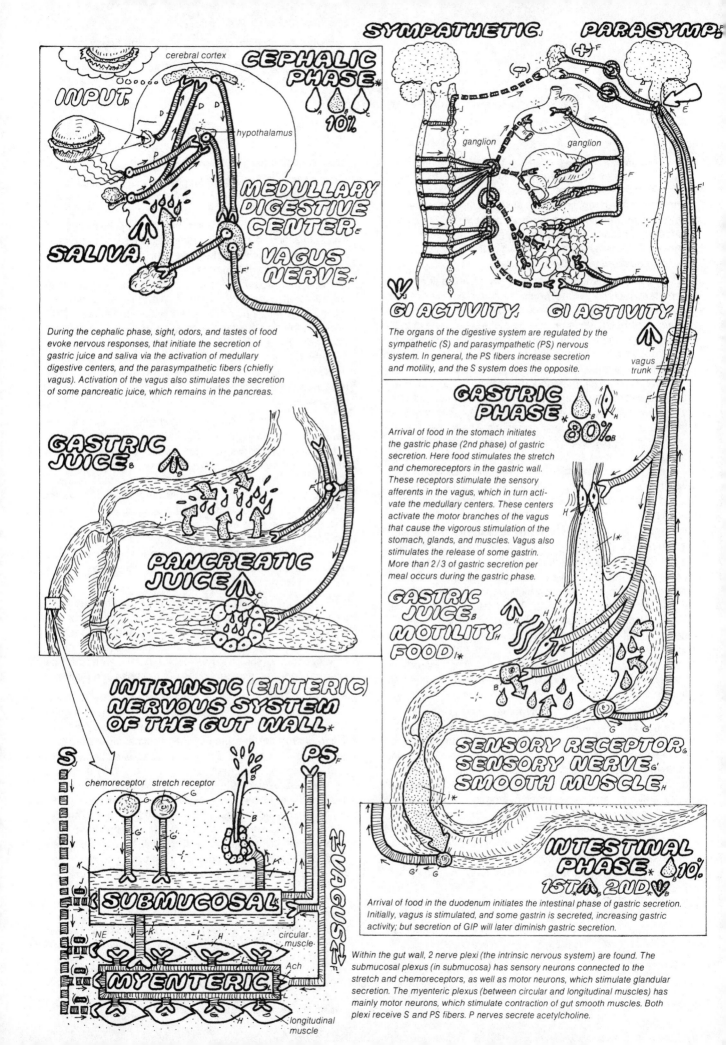

CEPHALIC PHASE *

INPUT

cerebral cortex

hypothalamus

10%

MEDULLARY DIGESTIVE CENTER

SALIVA

VAGUS NERVE

During the cephalic phase, sight, odors, and tastes of food evoke nervous responses, that initiate the secretion of gastric juice and saliva via the activation of medullary digestive centers, and the parasympathetic fibers (chiefly vagus). Activation of the vagus also stimulates the secretion of some pancreatic juice, which remains in the pancreas.

GASTRIC JUICE

PANCREATIC JUICE

INTRINSIC (ENTERIC) NERVOUS SYSTEM OF THE GUT WALL *

S   **PS**

chemoreceptor   stretch receptor

SUBMUCOSAL

NE   circular muscle

MYENTERIC

Ach

longitudinal muscle

SYMPATHETIC.   PARASYMP.

ganglion   ganglion

GI ACTIVITY.   GI ACTIVITY.

vagus trunk

The organs of the digestive system are regulated by the sympathetic (S) and parasympathetic (PS) nervous system. In general, the PS fibers increase secretion and motility, and the S system does the opposite.

GASTRIC PHASE *   80%

Arrival of food in the stomach initiates the gastric phase (2nd phase) of gastric secretion. Here food stimulates the stretch and chemoreceptors in the gastric wall. These receptors stimulate the sensory afferents in the vagus, which in turn activate the medullary centers. These centers activate the motor branches of the vagus that cause the vigorous stimulation of the stomach, glands, and muscles. Vagus also stimulates the release of some gastrin. More than 2/3 of gastric secretion per meal occurs during the gastric phase.

GASTRIC JUICE MOTILITY FOOD

SENSORY RECEPTOR SENSORY NERVE SMOOTH MUSCLE

INTESTINAL PHASE * 10%
1ST   2ND *

Arrival of food in the duodenum initiates the intestinal phase of gastric secretion. Initially, vagus is stimulated, and some gastrin is secreted, increasing gastric activity; but secretion of GIP will later diminish gastric secretion.

Within the gut wall, 2 nerve plexi (the intrinsic nervous system) are found. The submucosal plexus (in submucosa) has sensory neurons connected to the stretch and chemoreceptors, as well as motor neurons, which stimulate glandular secretion. The myenteric plexus (between circular and longitudinal muscles) has mainly motor neurons, which stimulate contraction of gut smooth muscles. Both plexi receive S and PS fibers. P nerves secrete acetylcholine.

ROLE OF THE PANCREAS IN DIGESTION

The *pancreas* is a large gland located underneath the stomach that has both endocrine and exocrine functions. The hormones of the *pancreatic islets*, *insulin* and *glucagon*, and their roles in regulating carbohydrate metabolism and blood sugar are discussed in plate 116. Here we focus on the digestive functions of the pancreas, namely, the production of pancreatic juice by the exocrine part of the gland, which constitutes more than 98% of its bulk.

The exocrine pancreas produces two physiologically important secretions. One, produced by the pancreatic *acini*, consists of a nearly complete set of *hydrolytic enzymes* for chemical breakdown of most large molecules found in the diet. The second is a watery secretion rich in *sodium bicarbonate*. This alkaline solution helps neutralize the gastric acid in the duodenum and provides a suitable chemical environment for the function of pancreatic enzymes. The exocrine pancreas consists of numerous acini, each comprised of a single layer of epithelial cells surrounding a cavity into which the secretory cells pour their secretions. The acinar cells secrete the digestive enzymes. The cavity opens into a duct through which the secretions of the acinar cells flow out. The ducts of pancreatic acini are lined with the *ductile* cells, which secrete the bicarbonate-rich solution. The smaller ducts all coalesce and converge, finally connecting to the main *pancreatic duct*, which joins the *duodenal lumen*.

FORMATION, COMPOSITION, AND FUNCTIONS OF THE BICARBONATE SOLUTION. The *active transport* mechanism of bicarbonate secretion by the duct cells is not well understood. The duct cells contain high amounts of the enzyme *carbonic anhydrase*, which may be involved in the active secretion of bicarbonate. To secrete bicarbonate, the duct cells possibly operate like the turned-around parietal cells of the stomach (see plate 69), which secrete acid into the stomach lumen and bicarbonate into the blood. The pancreatic duct cells do the opposite, secreting bicarbonate ions (along with a lot of sodium ions) into the duct lumen and acid into the blood.

The presence of sodium bicarbonate in the *pancreatic juice* gives this fluid an alkaline pH of about 8, enabling it to neutralize the acid chyme delivered from the stomach. Upon entry into the duodenum, the sodium bicarbonate reacts with the hydrochloric acid (H^+Cl^-) producing sodium chloride and carbonic acid. The latter acid is unstable and dissociates into carbon dioxide and water, so the hydrogen ions are gradually and effectively eliminated from the chyme in the duodenum. The reduction in duodenal acidity has two positive effects: (1) It reduces the noxious effects of acid on the duodenal mucosa, which is without much protection. (2) It makes the duodenal environment suitably alkaline for activation of pancreatic and intestinal digestive enzymes.

PANCREATIC ENZYMES. The physiological stimulus for acinar cell secretion of pancreatic enzymes is the presence of *fat* and *protein* in the duodenum; these stimuli trigger secretion of the duodenal hormone choleocystokinin (CCK). The stimulation of the vagus nerve also increases enzyme production. The acinar cells of the pancreas produce a viscous secretion rich in protein (enzymes), which are secreted in *zymogen granules*. Initially, most of the enzymes are secreted in their *inactive* forms (i.e., as larger *proenzyme* molecules). This is an advantage because the pancreatic enzymes are so powerful that they could digest the pancreas in a short time if they were not inhibited during their transport from the acinar cavity to the intestinal lumen. In the disease, *acute pancreatitis*, these enzymes are activated before reaching the intestine; thus, they digest the pancreas, causing death within days.

A key pancreatic proenzyme is *trypsinogen*, which is activated upon arrival in the duodenal lumen by the hydrolytic action of *enterokinase*, an enzyme secreted by the *duodenal mucosa*. The activation produces *trypsin*, a well-known all-purpose *protease* that can attack and hydrolyze many kinds of *proteins*. Among the targets of trypsin attack are the other inactive proenzymes secreted by the pancreas, particularly the proteases and lipases. In certain individuals, the intestinal mucosa is deficient in enterokinase. As a result, trypsin is not formed, other proteases are not activated, and dietary proteins remain undigested, causing protein deficiency and disease.

Some of the pancreatic enzymes, such as *amylase*, are secreted from the acinar cells in an already active form. Presumably, these enzymes do not pose any danger to the pancreatic tissue. Pancreatic amylase attacks the large dietary *polysaccharides* such as those found in the starches, forming smaller *oligo-* and *disaccharides* like dextrose and maltose (glucose-glucose).

Further digestion of disaccharides into such monosaccharides as glucose, fructose, and galactose occurs by the action of enzymes secreted by the intestinal mucosa (e.g., maltase and lactase). *Lipases* of the pancreas attack *triglycerides*, decomposing them into *glycerol* and *fatty acids* or to *monoglycerides* and fatty acids. The type of conversion depends on the type of lipase. Pancreatic proteases attack peptide bonds located between different but specific amino acids. As a result, the pancreatic proteases convert all the dietary *proteins* into *dipeptides*. The final hydrolysis of dipeptides to free *amino acids* occurs by the action of other proteases secreted from the *intestinal mucosa*.

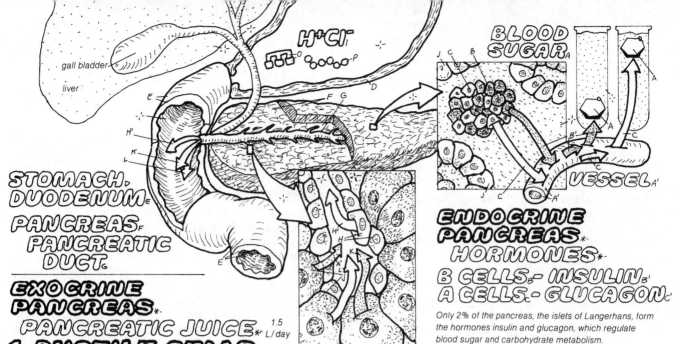

gall bladder

liver

STOMACH, DUODENUM, PANCREAS, PANCREATIC DUCT.

EXOCRINE PANCREAS *

PANCREATIC JUICE *

1.5 L/day

1. DUCTILE CELLS

$$\Downarrow$$

BICARBONATE (NaHCO$_3^-$)

$$+$$

H$^+$Cl$^-$

$$\Downarrow$$

NaCl + [H$_2$CO$_3$]

$$\Downarrow$$

H$_2$O, CO$_2$

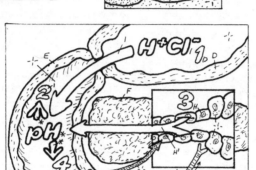

secretin

vagus nerve

BLOOD SUGAR

ENDOCRINE PANCREAS *

HORMONES *

B CELLS - INSULIN,

A CELLS - GLUCAGON,

Only 2% of the pancreas, the islets of Langerhans, form the hormones insulin and glucagon, which regulate blood sugar and carbohydrate metabolism.

VESSEL

More than 98% of the pancreas is devoted to its exocrine function: the secretion of pancreatic juice by the pancreatic acini and their duct cells. The enzymes and bicarbonate in the juice facilitate digestion in the small intestine. The duct cells of the pancreatic acini secrete a watery fluid rich in bicarbonate, designed to neutralize the gastric acid in the duodenum. This protects the duodenal wall and helps activate the intestinal and pancreatic enzymes. Not shown is the process by which bicarbonate is formed within the duct cells and is actively transported to the lumen. In the intestinal lumen, sodium bicarbonate reacts with the hydrochloric acid, forming salt and carbonic acid, which dissociates into carbon dioxide and water. In this way, gastric acid is neutralized, and the duodenal pH is increased to neutral levels.

2. ACINAR CELLS,

TRYPSINOGEN, & INACTIVE PROENZYMES,

ENTEROKINASE,

TRYPSIN,

ACTIVE DIGESTIVE ENZYMES *

FAT, PROTEIN,

Acinar cells of the pancreas secrete a variety of digestive enzymes to break down food substances into smaller absorbable molecules. Some, such as amylase, are secreted in an active form. But the proteases and the lipases are secreted as inactive proenzymes, to be activated once they reach the intestinal lumen. Trypsin, a powerful protease, plays a key part.

It is secreted in its inactive form (trypsinogen), which is activated by the intestinal protease, enterokinase, into trypsin. Trypsin then attacks other inactive proteases, converting them to active forms. Trypsin also activates the pancreatic lipase. Pancreatic enzyme secretion is stimulated by the vagus nerve and the hormone CCK.

PANCREATIC AMYLASE,

OLIGO- & DI-SACCHARIDES,

Pancreatic amylase attacks large polysaccharides (starches), converting them into smaller oligo- and disaccharides.

PANCREATIC LIPASE,

TRIGLYCERIDES,

F.A., GLYC., MONOGLYC.,

Pancreatic lipase attacks triglycerides, forming mostly fatty acids and monoglycerides along with some glycerol.

PANCREATIC PROTEASE,

PEPTIDES & AMINO ACIDS,

Pancreatic proteases (trypsin, chymotrypsin, elastase, carboxypeptidase) attack various proteins. Each, hydrolyzing peptide bonds between different amino acids, forming smaller peptides and amino acids.

THE LIVER AND BILE IN DIGESTION

The *liver* has many functions in controlling metabolism and deactivating hormones, drugs, and toxins. In addition, through *bile* formation, it plays a very important role in digesting *fats*.

LIVER STRUCTURE IN RELATION TO BILE FORMATION. The liver is the largest gland in the body. The formation of bile (nearly 0.5 L per day) is its major exocrine function. To understand how bile is formed, we must understand the structure of the basic anatomical-functional unit of the liver: the *liver lobule*. Each liver lobule is part of a hexagon in which the lobules are connected peripherally to the incoming blood and centrally to a vein that drains the blood.

The liver receives blood from two sources, the *hepatic artery* and the *portal vein*, bringing blood from the heart and the intestines, respectively. Blood flows out of the liver and into the heart via the *hepatic vein*. The liver thus has the unique ability to receive and sample the absorbed food substances before they reach the general circulation.

The *liver cells* (*hepatocytes*) are packed in walls (slabs) of cells, which are separated by blood *sinusoids* (a highly porous type of capillary). The incoming arterial and portal blood are mixed as they flow into these sinusoids. After the hepatocytes extract their oxygen and nutrient needs from this pool of blood, it flows into the centrally located branch of the hepatic vein. The hepatocytes form bile and secrete it into small *canaliculi*, which coalesce to form first the smaller and then the larger *bile ducts*. In these ducts, bile flows in the opposite direction of blood, preventing their mixing.

The various bile ducts finally coalesce to form the *hepatic duct*, which emerges from the liver. The hepatic duct bifurcates to form the *cystic duct*, which leads to the *gallbladder*, and the *common bile duct*, which, together with the *pancreatic duct*, connects with the *duodenum*. The *sphincter of Oddi* regulates the bile outflow from the common bile duct into the duodenum. When this sphincter is closed, the bile accumulates in the common bile duct, flowing back into the cystic duct and the gallbladder, where it is temporarily stored. After meals, the gallbladder contracts, releasing bile into the duodenum.

BILE COMPOSITION. Besides water (97%), bile contains two major organic constituents, *bile salts* and *bile pigments*, as well as such inorganic salts as *sodium chloride* and *sodium bicarbonate*. All of these are produced by the liver cells. Bile salts (also called *bile acids*) such as *cholic acid* and *deoxycholic acid* are formed from *cholesterol* within the liver cells (hepatocytes). To form bile pigments, *bilirubin*, the metabolite of heme formed during *hemoglobin* catabolism (*red blood cell* destruction), is taken up from blood and conjugated to glucuronic acid to form the golden yellow (bile color) bilirubin-glucuronide, which is more water-soluble than bilirubin and thus is excreted in the bile. About 4 g of bile salts and 1.5 g of

bile pigments are secreted every day in the bile. Most of the bile salts are reabsorbed by the intestines and delivered back to the liver; the bile pigments are mostly excreted with the feces. How liver cells secrete sodium, chloride, and bicarbonate ions into bile is poorly understood.

BILE FUNCTION. Of the major bile constituents, only the salts play a physiologically important role. Bile pigments are basically excretory products. The typical bile salts cholate and deoxycholate are *fat solubilizing agents*. They have both a fat-soluble *hydrocarbon ring* and several *charged groups*, enabling them to mix with fat and water, respectively. Thus, the addition of bile salts to a fat and water mixture increases fat's solubility. In the presence of bile salts, large fat droplets in the chyme become dispersed, forming smaller fat particles, a process called *emulsification*. In the emulsified form, fats can be much more easily and efficiently digested by the water-soluble enzyme *lipase* from the pancreas (see plate 72). The products of lipase digestion (glycerides and fatty acids) form special fatty aggregates called *micelles* (see plate 7), which the intestinal mucosal cells can readily absorb. In the absence of bile, fat digestion diminishes markedly even though the enzyme lipase is present.

GALLBLADDER FUNCTION. The gallbladder is a storage sac for bile, which is produced continuously by the liver but is delivered to the duodenum only after meals, particularly meals containing fats. Before meals, the sphincter of Oddi, located at the opening of the common bile duct into the duodenum, is closed, causing the flowing bile to back up, filling the gallbladder. While the bile is stored in the gallbladder, the bladder wall absorbs some of its water, concentrating the bile. The arrival of fatty food in the duodenum stimulates the release of the duodenal hormone CCK (see plate 70), which acts on the gallbladder, causing it to contract. The bile is then released into the duodenum to act.

Two major problems and diseases are associated with abnormalities of the gallbladder function. One is *gallstones*. In certain individuals, excess amounts of *cholesterol* (usually a minor bile constituent) in the bile precipitate, perhaps as a result of excessive removal of water, forming gallstones. Gallstones in the bile duct, may cause severe abdominal pain, requiring surgery. If the enlarged stones obstruct the common bile duct, bile flows back into the liver and eventually leaks into the blood, causing *jaundice*, a disorder characterized by a yellowish color of the skin and eyes due to deposition of bilirubin and related bile pigments in the capillaries and tissue spaces. Jaundice may also occur due to excessive *hemolysis of the red cells*, a condition that occurs in certain diseases and produces unusually large amounts of bilirubin. Liver damage such as occurs after certain viral infections (hepatitis) also causes jaundice.

CN: Use red for M, blue for L, golden yellow for F, and a very light color for B.
1. Begin with the upper drawing of the various organs involved.
2. Color the enlargement of a liver section, beginning with blood from the digestive system entering from the portal vein (L). Include the liver cell diagram, noting that some of the cholesterol form-ing the bile salts is made within the liver cell. Color the diagram on the far right which provides an overview of the flow of blood and bile.
3. Color the box showing the function of bile.
4. Color the illustrations depicting gall bladder function. Follow the numbered sequence. Note that no. 8 is found in the material below.
5. Complete the material at the bottom.

STOMACH A
LIVER B
HEPATIC DUCT C
GALLBLADDER D
CYSTIC DUCT E
BILE F COMMON BILE DUCT F'
PANCREATIC DUCT G
PANCREATIC LIPASE G'
DUODENUM H
SPHINCTER OF ODDI I
H⁺Cl⁻ J FATS K

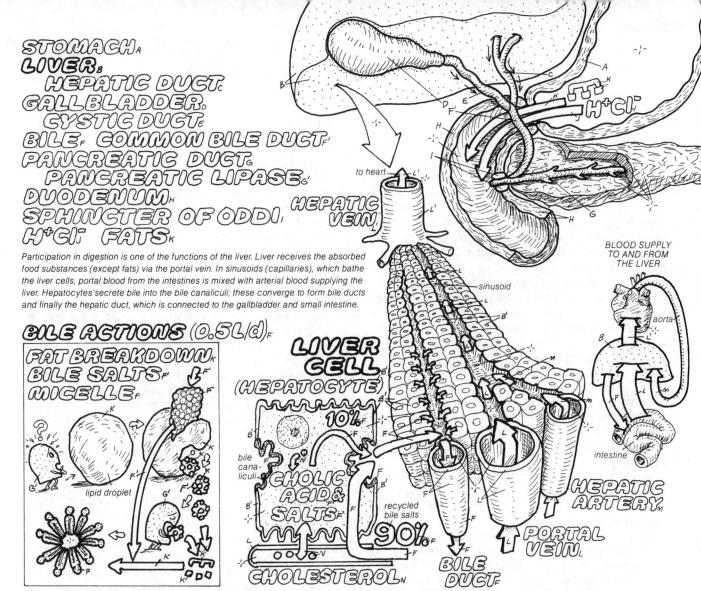

BLOOD SUPPLY TO AND FROM THE LIVER

to heart

HEPATIC VEIN J

sinusoid

aorta

intestine

HEPATIC ARTERY M

PORTAL VEIN L

BILE DUCT F

Participation in digestion is one of the functions of the liver. Liver receives the absorbed food substances (except fats) via the portal vein. In sinusoids (capillaries), which bathe the liver cells, portal blood from the intestines is mixed with arterial blood supplying the liver. Hepatocytes secrete bile into the bile canaliculi; these converge to form bile ducts and finally the hepatic duct, which is connected to the gallbladder and small intestine.

BILE ACTIONS (0.5L/d) F

FAT BREAKDOWN K
BILE SALTS F'
MICELLE P

lipid droplet

LIVER CELL (HEPATOCYTE)

bile canaliculi

CHOLIC ACID & SALTS

10%

90%

recycled bile salts

CHOLESTEROL N

Bile salts emulsify large fat droplets, breaking them down into smaller particles on which pancreatic lipase can act more efficiently. Bile salts also combine with products of lipase digestion (monoglycerides and fatty acids), forming lipid micelles, which are readily absorbed by the intestinal mucosa.

Bile contains bile salts and bile pigments. Bile salts (salts of cholic acid) are formed from cholesterol in the liver and secreted in the bile to aid in the digestion of fats. 90% of bile salts is reabsorbed in the blood and recycled via the entero-hepatic circulation.

GALL BLADDER FUNCTION D

Bile is continuously secreted from the liver (1). Before meals, the sphincter of Oddi is closed (2), directing the bile flow into the gallbladder for storage (3). Mucosal cells of the gallbladder actively reabsorb sodium, chloride, and water, concentrating the bile (4). After meals, the fat in the chyme (5) dilates the sphincter of Oddi and stimulates the release of the hormone CCK. CCK induces the contraction of the gallbladder (6), releasing bile into the duodenum (7). Increased cholesterol or decreased water, in the bile, favors formation of gallstones. Gallstones may obstruct bile flow, resulting in jaundice (8).

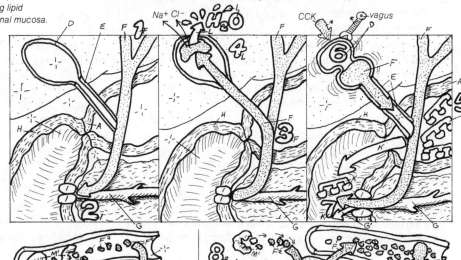

Na⁺ Cl⁻   H₂O

CCK   vagus

BILE PIGMENTS (BILIRUBIN) F²
RED BLOOD CELL M

The liver cells also secrete for the purpose of excretion certain derivatives of bilirubin, a yellowish pigment that is a metabolite of hemoglobin. Some of the bile pigments are excreted in the feces; others are reabsorbed in the blood for excretion in the kidneys. Jaundice, a disease caused by elevated levels of bilirubin in the blood, tissues, and skin, can occur when biliary flow is obstructed due to gallstones. Liver damage or excessive hemolysis of the red blood cells may also cause jaundice.

NORMAL

GALL STONES N'

JAUNDICE F³

H₂O in bile
cholesterol in bile

bilirubin in blood and tissues

STRUCTURE AND MOTILITY OF THE SMALL INTESTINE

The *small intestine* is a long, convoluted tube specialized both for *completion of digestion* and *absorption* of nutrients. The length of an animal's small intestine depends on its *dietary habits*. It is shorter in meat eaters (*carnivores*) and longer in grass eaters (*herbivores*). In the omnivorous human, the small intestine is of medium length (about 3 m, or 10 ft), although it is two to three times longer in cadavers due to loss of muscle tone.

SEGMENTS OF SMALL INTESTINE. At its beginning, the small intestine is connected to the stomach where the *pylorus sphincter* controls *chyme* inflow into the *duodenum*, the first intestinal segment. The *jejunum* and *ileum* are the second and third segments. The ileum joins with the *cecum* of the large intestine by the *ileocecal valve*, which controls outflow from the small intestine. The different segments vary in their functions. The duodenum is highly secretory (mucus, enzymes, and hormones); the jejunum and ileum are specialized for nutrient absorption. Though absorption occurs across the entire surface of the jejunum and ileum, various substances are selectively absorbed in different segments (see plate 75).

STRUCTURE OF THE INTESTINAL WALL. The structure of the intestinal wall broadly resembles that of other parts of the digestive tract, but there are histological variations suitably adapted for its particular absorptive functions. The most superficial (near lumen) is the *mucosa*, which contains the *absorptive cells*. Beneath the mucosa is the *submucosa*, containing the *glands* and small blood vessels. Two layers of smooth muscles, the *circular* and *longitudinal*, are found deep under the submucosa. These are responsible for intestinal motility. Groupings of nerves, the *submucosal and myenteric plexi*, are located within these layers (see plate 71). A supportive layer of connective tissues, the *serosa*, forms the intestine's outer cover.

The inner wall of the small intestine is extensively folded (*plicae circularis*), tripling the surface area. Each fold in turn contains numerous microscopic structures called *villi* (fingers). The villi (about 30 million, 30/mm^2) expand the absorptive surface area another ten times.

HISTOPHYSIOLOGY OF VILLI. Each villus consists of a single layer of *surface epithelial cells* covering an inner core of very small blood and lymph vessels, *autonomic nerve fibers*, and *smooth muscle cells*. Most of the surface epithelial cells of the villi are *absorptive*, but some are *secretory*. The absorptive cells are glued tightly together by *desmosomes* and *tight junctions* (see plate 2) so that nutrients can pass only across, not between, the cells. The absorptive cells, occurring mostly on the hills of the villi, contain *microvilli* (brush border) on their luminal surface; these effectively increase each cell's absorptive surface by twenty times. Altogether, the plicae circularis, the villi, and the microvilli increase the absorptive surface by six hundred-fold, providing an area the size of a tennis court for nutrient absorption. The microvilli contain actin and are capable of contractile movements, which aid in absorption.

During absorption, nutrients are transported across the brush border into the absorptive cells and out into the villus core. Here the water-soluble nutrients are delivered to the *blood capillaries*. These capillaries coalesce to form venules, which leave the villus to form small veins leading finally into the *portal vein*, which carries nutrients to the liver. The fatty nutrients are taken up by the small blind lymph capillaries (*lacteals*), which coalesce to form the lymph vessels. These join with the larger lymph vessels, which ascend the trunk and connect with the large veins, where the fatty nutrients are delivered to the blood circulation. The nerve fibers and smooth muscles of the villus control blood flow and contraction of the villus as a whole. The contractile movements aid the flow of blood and absorbed nutrients in the villus.

The secretory cells, another type of villus epithelial cells, are located deep in the intervilli valleys (*intestinal crypts*), where they may form glands, secreting mucus or other substances into the crypts. The entire population of surface epithelial cells continuously turns over, being replaced every few days. New cells are formed deep in the crypts and migrate up toward the tips of the villi, where they are shed into the intestinal lumen. The shedding and destruction of these cells (20 million cells/day in humans) provides one source for intestinal enzymes (e.g., entero-kinase).

INTESTINAL MOVEMENTS. The small intestine shows two important movements: *segmentation* and *peristalsis*. Segmentation movements are achieved by sustained contractions of the circular muscles. Superimposed on these sustained contractions, which tend to entrap the chyme within a small segment of the intestine, are other *mixing movements* that shake the chyme, mix it with the intestinal juice, and promote its absorption. The peristaltic movements generated by the coordinated contractions of the circular and longitudinal muscles usually occur toward the large intestine, propelling the intestinal chyme down the tract. It takes several hours for the chyme to move from the duodenum to the end of the ileum. Intestinal movements are generated by the *intrinsic nerve plexi* of the intestinal wall (see plate 71) and do not require the autonomic nerves for their generation, but these nerves can regulate the intensity of the contractions.

CN: Use red for I, purple for J, blue for K, and a very light color for G.
1. Begin with the diagram in the upper left corner and follow the enlargments down the page. Note that the internal structures of the three villi have been segregated (lacteal in the left, blood vessels in the center, nerves and muscles on the right) for purposes of clarity.
2. Color the tiny diagram of a tennis court in the lower left corner, demonstrating the total absorptive area of the small intestine.
3. Color the motility panel. Note that only the portion of the intestinal wall which is contracted is to be colored.

SMALL INTESTINE *
DUODENUM. 10"/25 CM$_A$
JEJUNUM. 4'/1.1 M$_B$
ILEUM. 6'/1.7 M$_C$

The small intestine, located between the stomach and large intestine, is where chemical digestion is completed and absorption of essentially all the food substances occurs. The small intestine consists of three segments: duodenum, jejunum, and ileum.

INTESTINAL FOLDS$_{G'}$
(PLICAE
CIRCULARES) **3X SURFACE AREA**

INTESTINAL WALL *
SEROSA$_{B'}$
LONGITUDINAL MUS.$_D$
CIRCULAR MUSCLE$_E$
SUBMUCOSA$_F$
MUCOSA$_G$

The inner lining of the small intestine is greatly folded to increase the surface area for absorption. The folding creates first the visible plicae circulares and then the microscopic villi. The epithelial cells covering the villus contain microvilli. The microvilli, villi, and foldings together expand the absorptive surface 600x.

LACTEAL$_H$
ARTERY, CAP., VEIN$_K$
SMOOTH MUSCLE$_L$
AUTONOMIC NERVES$_M$
SUBMUCOSAL PLEXUS$_N$
MYENTERIC PLEXUS$_O$
CELL FORMATION$_P$
ABSORPTIVE CELLS$_{G^2}$
SECRETORY CELLS$_Q$

Each villus is an absorptive unit. Its numerous epithelial cells transport substances into the inner villus space, where blood capillaries, blind lymph ducts (lacteal), and nerves and smooth muscle are found. Between the villi, are found the intestinal crypts, which contain secretory glands. Fats and fat-soluble vitamins move into the lacteals; all other absorbed substances move into the capillaries. Muscles and nerves regulate the movement of the villi, enhancing absorption and transport.

The contractions of the small intestine (seen in box below) result in a segmentation movement that traps and mixes the chyme, allowing time for absorption, as well as a peristalsis movement that gradually moves the chyme down the tract. These movements are regulated by nerve and muscle layers of the intestinal wall.

VILLI (10X)$_{G^2}$

absorption   absorption

Crypt of Lieberkuhn (intestinal gland)

secretion

MICROVILLI (20X)$_R$
ABSORPTION *

TOTAL SMALL INTESTINE SURFACE AREA
(600X) *

INTESTINAL MOTILITY *
CHYME$_S$
SEGMENTATION$_{E'}$
CONTRACTIONS$_{E'}$
PERISTALSIS$_E$

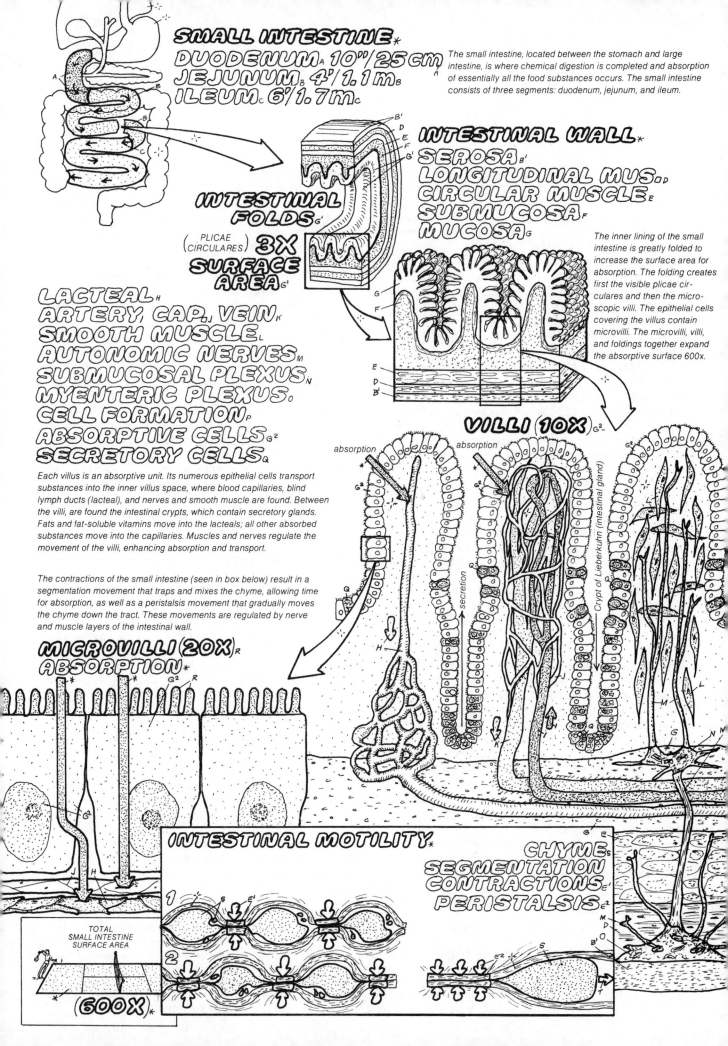

ABSORPTION IN THE SMALL INTESTINE

After complete chemical digestion, the *absorptive cells* of the intestinal epithelium absorb nutrients and deliver them to the bloodstream so they can be made available to the body cells for use and consumption. Intestinal absorption of nutrients utilizes many of the same transport mechanisms occurring in other body cells, including both physical (e.g., diffusion) and physiological (e.g., active transport) mechanisms (plates 8, 9). In addition, some transport mechanisms are unique to the intestinal absorptive cells. After passing across the mucosal epithelium, the water-soluble nutrients flow into the *blood capillaries* of the villi, and the fatty and fat-soluble nutrients flow into the *lacteals* and the lymphatic vessels before entering the bloodstream.

MINERAL AND SALT ABSORPTION. Some substances pass across the mucosa by simple physical diffusion and osmosis. For example, *potassium* passes primarily by diffusion, and *water* transport occurs by *osmosis*, following the transport of salts and other osmotically active substances such as glucose. The transport of other minerals and salts (e.g., iron, calcium, and sodium) requires the operation of more complicated physiological systems.

Iron is transported from the intestinal lumen, across the mucosa, and into the plasma by an iron-binding protein called *transferrin*. When iron is available in excess, it is combined with *ferritin*, a ubiquitous iron-binding protein, and stored within the mucosal cells. In dietary iron deficiency, iron is released from ferritin and delivered to the blood. *Calcium* is actively taken up by a mechanism in the brush border (microvilli) and transported across the cell by a *calcium-binding protein* for delivery to the blood. This protein is made in the mucosal cells under *vitamin D_3* stimulation (see plate 114).

Sodium, the body's major extracellular electrolyte, is transported by an energy (ATP)-dependent (active) mechanism. Sodium passes across the brush border bound with carriers. Inside the cell, a plasma membrane mechanism, probably the membrane *Na-K-ATPase*, located on the *basolateral* borders of the absorptive cells, pumps the sodium out into the intercellular space, from which it diffuses into the blood. The operation of this pump keeps the intracellular concentration of sodium low, enabling inward diffusion from the intestinal lumen.

GLUCOSE AND AMINO ACID ABSORPTION. The active transport of sodium is of particular importance because the transport of several other substances (such as *glucose* and some *amino acids*) occurs mainly in conjunction with it. Glucose and amino acids are believed to be transported across the brush border by sodium-dependent carriers, building high intracellular concentrations of these substances, which permit them to pass across the basal membranes by diffusion or facilitated diffusion. If the sodium pump is inhibited, glucose and amino acid transport are diminished because the intracellular sodium concentration rises, preventing the diffusion of sodium from the lumen. Several amino acids are initially taken up at the brush border as *dipeptides*. At this time, or during mucosal transport, various *dipeptidase enzymes* present in the brush border attack these peptides, releasing free amino acids, which either diffuse or are actively transported.

FAT ABSORPTION. Products of *fat* digestion are *monoglycerides*, *glycerol*, and *fatty acids*. Fatty acids are either short chained or long chained. *Short chain* fatty acids pass across the mucosa by diffusion and, being fairly water-soluble, enter the blood capillaries along with other water-soluble nutrients. However, *long-chain* fatty acids and other fatty nutrients, including *cholesterol*, receive special treatment during absorption. These fatty products diffuse across the brush border. Within the mucosal cells, the *triglycerides* are resynthesized on the smooth endoplasmic reticulum and packed with cholesterol and other fatty substances within certain *lipoprotein particles* called *chylomicrons*. The packaging process occurs within the Golgi apparatus (see plate 1).

Chylomicrons, like other members of the family of lipoprotein particles, contain a coat of protein and a core of fat (see plate 129), allowing large amounts of fat to float in the bloodstream without coalescing. After formation, chylomicrons are extruded by exocytosis from the mucosal cells into the lacteals, move into the larger lymph vessels, and finally pour into the veins in the upper trunk near the neck.

VITAMIN ABSORPTION. Vitamins are divided into two categories, *water-soluble* and *fat-soluble*. The water-soluble vitamins, such as the B family and C, pass across the mucosa by diffusion and also by association with specialized membrane carriers. *Vitamin B_{12}* (cyanocobalamine) is the largest of the vitamins, and its transport utilizes yet another mechanism.

The secretory cells in the stomach wall normally produce a specific transport mucoprotein (proteins containing special polysaccharides) called *intrinsic factor*. In the chyme, intrinsic factor binds with vitamin B_{12}; the absorptive cells use endocytosis to take up this complex. Vesicles are released at the basal surface by exocytosis. Diseases of the stomach (e.g., gastritis) reduce vitamin B_{12} absorption because they deplete intrinsic factor, causing pernicious anemia (see plate 136). Fat-soluble vitamins such as vitamins D, A, and K are absorbed in the chylomicrons along with the fatty nutrients.

CN: Use the same colors as were used on the previous page for lacteal (A), capillary (B), absorptive cell (C), and microvilli (D).
1. Color the four bold titles, beginning with lacteal (A), and the structures they refer to, before you color anything else. Then begin coloring the substances absorbed from the lumen of the intestine, starting with iron (Fe^{++}). Follow each substance as it enters the cell, moves through, and out again either into a capillary or lacteal. When you come across a transport mechanism that is displayed at the top of the page, color both examples: in the cell membrane below and in the membrane at the top of of the page.

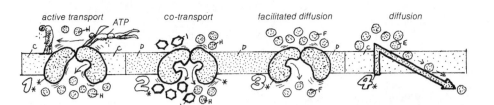

Nutrients, vitamins, minerals, and water are transported from the intestinal lumen to the blood and lymph by the absorptive cells of intestinal mucosa using a variety of cellular and membrane transport mechanisms discussed earlier in the book.

active transport   ATP   co-transport   facilitated diffusion   diffusion

1*   2*   3*   4*

ACTEAL. CAPIL. ABSORPTIVE CELL. MICROVILLI.

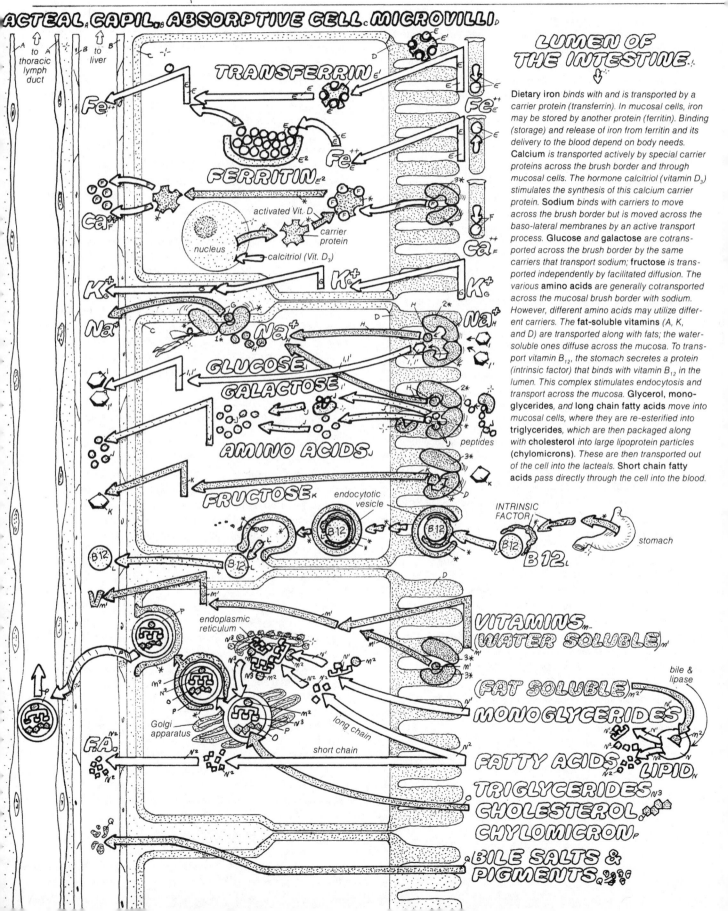

LUMEN OF THE INTESTINE.

Dietary iron binds with and is transported by a carrier protein (transferrin). In mucosal cells, iron may be stored by another protein (ferritin). Binding (storage) and release of iron from ferritin and its delivery to the blood depend on body needs. **Calcium** is transported actively by special carrier proteins across the brush border and through mucosal cells. The hormone calcitriol (vitamin D_3) stimulates the synthesis of this calcium carrier protein. **Sodium** binds with carriers to move across the brush border but is moved across the baso-lateral membranes by an active transport process. **Glucose** and **galactose** are cotransported across the brush border by the same carriers that transport sodium; **fructose** is transported independently by facilitated diffusion. The various **amino acids** are generally cotransported across the mucosal brush border with sodium. However, different amino acids may utilize different carriers. The **fat-soluble vitamins** (A, K, and D) are transported along with fats; the water-soluble ones diffuse across the mucosa. To transport vitamin B_{12}, the stomach secretes a protein (intrinsic factor) that binds with vitamin B_{12} in the lumen. This complex stimulates endocytosis and transport across the mucosa. **Glycerol, monoglycerides**, and **long chain fatty acids** move into mucosal cells, where they are re-esterified into **triglycerides**, which are then packaged along with **cholesterol** into large lipoprotein particles (chylomicrons). These are then transported out of the cell into the lacteals. **Short chain fatty acids** pass directly through the cell into the blood.

to thoracic lymph duct

to liver

TRANSFERRIN

Fe   Fe⁺⁺

FERRITIN

Ca   Ca⁺⁺

activated Vit. D

carrier protein

nucleus   calcitriol (Vit. D_3)

K⁺   K⁺   K⁺

Na⁺   Na⁺   Na⁺

GLUCOSE

GALACTOSE

peptides

AMINO ACIDS

FRUCTOSE

endocytotic vesicle

B12   B12   B12

INTRINSIC FACTOR

B12   B12   stomach

V

VITAMINS (WATER SOLUBLE)

endoplasmic reticulum

bile & lipase

(FAT SOLUBLE)

MONOGLYCERIDES

Golgi apparatus

long chain

short chain

FA   FATTY ACIDS   LIPID

TRIGLYCERIDES

CHOLESTEROL

CHYLOMICRON

BILE SALTS & PIGMENTS

FUNCTION OF THE LARGE INTESTINE

FUNCTIONAL ANATOMY. The *large intestine* (colon) is a wide (6 cm, or 2.5 in) and short (120 cm, or 4 ft) tube extending between the small intestine and the *rectum*. The large intestine processes the remaining undigested chyme into *feces* (stools), a relatively solid and bulky material that can be excreted at intervals. In doing so, the colon absorbs water and conducts specific movements, some of which enable fecal excretion in appropriate intervals. The colon wall contains many exocrine glands that secrete a viscous *mucus* to help mold the feces and protect the colon wall from mechanical damage that could result from the flow of the solid contents.

The unidirectional *ileocecal valve*, which connects the ileum of the small intestine with the *cecum* of the large intestine, performs two functions: (1) it permits discontinuous delivery of chyme to the large intestine, allowing time for the colon to perform its functions, and (2) it prevents *bacteria* from penetrating the normally clean small intestine. The cecum and its vestigial extension, the *appendix*, contain a high concentration of bacteria. A *gastroileal reflex* controls the ileocecal valve. Increase in gut motility occurring after meals relaxes the valve; colon *distension* inhibits it. Occasional peristaltic waves are responsible for the valve's periodic opening.

After the cecum, the colon consists of the *ascending, transverse, descending,* and *sigmoid* segments. The ascending and transverse colons are the sites of *absorption* and *secretory activities*. Absorption of water dehydrates the colon content, facilitating the formation of solid fecal matter. The descending and sigmoid colons are the sites of *storage* of fecal matter.

The sigmoid colon joins with the *rectum*, a muscular cavity functioning in short-term storage of feces and in stimulation of *defecation* (fecal excretion, bowel movement). The *anus* is the end organ of the digestive tract. It is a *sphincter system*, consisting of an internal smooth muscle sphincter and an external striated muscle sphincter, designed for both involuntary and voluntary control of defecation.

ABSORPTION OF SODIUM, WATER, AND VITAMINS; SECRETION OF POTASSIUM. To form solid feces, the remaining chyme entering the colon must be dehydrated. This is achieved by *absorption of water* across the colonic surface epithelium. The absorption of this water is important in the body's water economy because about 2 L of water are absorbed daily in the colon. Water absorption occurs in an *obligatory* manner by *osmosis* following the active absorption of sodium. *Potassium*, however, is secreted in the large intestine, creating a major problem of potassium depletion during severe diarrhea (see plate 77).

Nutrients such as glucose and amino acids are not absorbed in the colon, but certain vitamins and drugs can be efficiently absorbed (hence drug administration by rectal suppositories). The slow rate of feces movement in the colon permits the bacteria inhabiting the large intestine to digest the unused *cellulose* and other fibers, grow, and proliferate. Therefore, as the colon content moves toward the rectum, the mass of the solids of dietary origin gradually diminishes, while the bacterial debris gradually increases. As a result, nearly one-third of the stool's solid mass is of bacterial origin. Bacterial contribution is also one reason why diets rich in pectin and cellulose fibers (e.g., fruits and raw vegetables) result in higher stool mass. The metabolism and death of the colon bacteria provide a useful source for several vitamins, such as the B family and K. This source becomes very important during dietary vitamin deficiency.

LARGE INTESTINE MOTILITY. Three types of movements characterize large intestine motility: segmentation, peristalsis, and mass movement. The *segmentation movements* entrap the colon contents within a small segment; the contraction of muscle layers then turns and churns the content, exposing it to the epithelial cells for sodium and water absorption. The *peristalsis movements* occur in regular intervals, passing along as waves of contractions down the colon. Thus, the gradually dehydrating feces move toward the descending colon for storage. The descending colon exhibits another type of movement called *mass movement*. Here, and in the sigmoid colon, a strong peristaltic wave forces large "masses" of feces into the rectum at once. Such contractions occur a few times daily, usually after meals.

NEURAL CONTROL OF COLON MOVEMENTS AND DEFECATION. Hormones do not play any role in colon movements. Instead, both the intrinsic nerve plexi and the extrinsic parasympathetic nerves (vagus in the upper colon and sacral nerves in the lower colon, rectum, and anus) regulate colon motility and the defecation reflex. The nervous control of segmentation and slow peristalsis is basically similar to that of the small intestine (i.e., under enteric plexi control but influenced in intensity by parasympathetic nerves). Although the mass movements can be generated intrinsically by the plexi, the brain and extrinsic nerves play major roles in regulating them. Thus, anxiety (exams) and the presence of coffee and food in the mouth can activate the colon and the urge for a bowel movement.

DEFECATION REFLEX. Mass movements force fecal matter into the rectum, distending this organ. Rectal distension triggers the defecation reflex, which involves contraction of the sigmoid colon and rectum to force the feces out and relaxation of the normally closed anal sphincters to permit outflow. This occurs in infants, in whom voluntary control of defecation has not developed. In adults, rectal distension also signals the brain, creating the urge for bowel movement. The external anal sphincter, a striated muscle that develops voluntary control by the end of infancy, is voluntarily relaxed, permitting fecal outflow. Other voluntary mechanisms such as pressure from the abdominal and respiratory muscles (diaphragm) also aid in defecation. In adults, the defecation reflex may be inhibited voluntarily, postponing the bowel movement.

CN: Use red for the blood capillary (T) and light blue for Q. Use dark colors for K, L, and O.
1. Begin with the upper left illustration, noting that some of the titles refer to structures found in other locations.
2. Color the enlargement of the cecum (below the upper illustration). Note that the circular muscle

(O) refers to the three motility diagrams which will be colored next.
3. Color the absorption and secretion process in the lower left, noting that the title, absorption, is divided into two colors.
4. Color the steps involved in the defecation reflex, following the numbered sequence.

LARGE INTESTINE *

CECUM A
ILEOCECAL VALVE B
APPENDIX C
ASCENDING COLON D
TRANSVERSE COLON E
DESCENDING COLON F
SIGMOID COLON G
RECTUM H
ANAL CANAL I, ANUS J
INTERNAL SPHINCTER
(SMOOTH MUSCLE - INVOL.) K
EXTERNAL SPHINCTER
(SKELETAL MUSCLE - VOL.) L

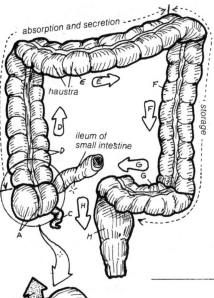

absorption and secretion
haustra
ileum of small intestine
storage

The large intestine (colon) is that part of the digestive tract located between the ileum of the small intestine and the rectum. It consists of the cecum, the ascending, transverse, descending, and sigmoid segments. It functions to remove the undigested chyme from the small intestine, to absorb the remaining water and electrolytes, and to store the solid wastes for elimination at prolonged intervals. Most of the absorption occurs at the proximal half; the storage occurs at the distal half.

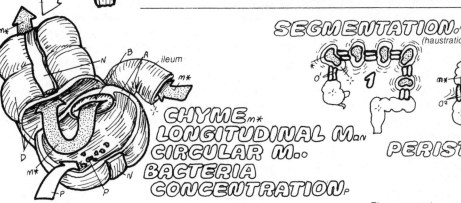

CHYME m*
LONGITUDINAL M. N
CIRCULAR M. O
BACTERIA
CONCENTRATION P

SEGMENTATION O
(haustration)

MOTILITY *

PERISTALSIS +

MASS MOVEMENT S

The ileocecal valve controls the flow of chyme from ileum into cecum. It also prevents the back flow of fecal matter into the small intestine. Arrival of the peristaltic waves of the ileum relaxes the ileocecal sphincter, permitting regular delivery of chyme into the cecum.

The segmentation movements of the large intestine mix and turn the fecal matter to facilitate the absorption of water and electrolytes. Peristaltic movements promote the gradual transport of the feces toward the rectum. The segmentation and the peristaltic movements are directed by the local intrinsic nervous system. The mass movements, controlled in part by the parasympathetic nerves, cause rapid propulsion of large bulks of feces from the intestine into the rectum.

ABSORPTION & SECRETION

H_2O, Na^+_R, K^+_S
CAPILLARY T

Sodium is actively absorbed in the large intestine; potassium is secreted. Water follows sodium by obligatory osmosis, causing dehydration of the feces and compacting of the solid waste.

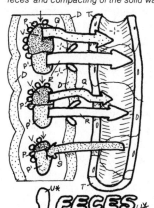

FECES U*
DIETARY ORIGIN V
NON DIETARY P (bacteria debris)

Bacteria in the large intestine digest the cellulose, decreasing the dietary sources of the feces. Their debris add to the nondietary portion.

DEFECATION REFLEX

ABDOMINAL MUSCLES Z

COLON MOTILITY

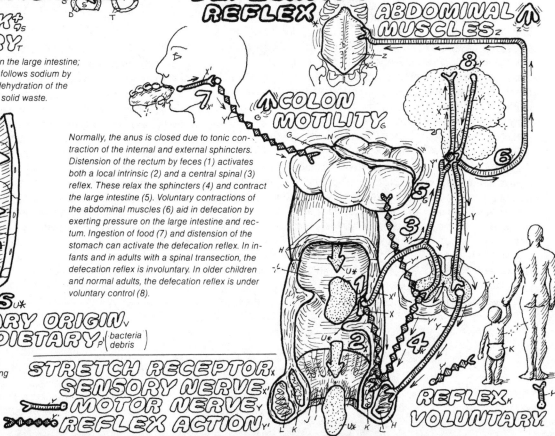

Normally, the anus is closed due to tonic contraction of the internal and external sphincters. Distension of the rectum by feces (1) activates both a local intrinsic (2) and a central spinal (3) reflex. These relax the sphincters (4) and contract the large intestine (5). Voluntary contractions of the abdominal muscles (6) aid in defecation by exerting pressure on the large intestine and rectum. Ingestion of food (7) and distension of the stomach can activate the defecation reflex. In infants and in adults with a spinal transection, the defecation reflex is involuntary. In older children and normal adults, the defecation reflex is under voluntary control (8).

STRETCH RECEPTOR X
SENSORY NERVE X
MOTOR NERVE Y
REFLEX ACTION Y

REFLEX K
VOLUNTARY

DIGESTIVE DISORDERS AND DISEASES

Digestive disorders and diseases are among the most common problems in the body. Some of these, such as vomiting, are normal responses to ingesting toxins or excess food; others, such as ulcers, may have complicated causes, including stress.

VOMITING. *Vomiting* is a useful physiological defense response. The *vomiting reflex* aids in the rejection of undesirable food from the stomach by expelling it out through the esophagus and mouth. The vomiting reflex is initiated by activation of *sensory receptors* in the stomach wall. Both chemoreceptors and stretch receptors may be involved. Among the stimuli normally activating these receptors is the presence of too much food, which causes excessive stretching of the stomach. Poisons and microbial toxins initiate vomiting by acting on the chemoreceptors of gastric mucosa. These sensory signals are communicated by sensory fibers in the *vagus* to a *vomiting center* in the *brain medulla*. This center also responds to certain toxic substances in the blood.

Activation of the vomiting center results in a complex of reflex responses: the *glottis* closes to keep the vomit from entering the respiratory passages; the *cardiac sphincter* in the lower esophagus opens; massive contractions of the *abdominal* and *respiratory* muscles occur to exert external pressure on the stomach; the *vagus nerve* stimulates the stomach vigorously; and, finally, a strong wave of *reverse peristalsis* moves from the pylorus to the cardia. As a result, the stomach contents are expelled, eliminating the source of toxicity and discomfort.

GASTROINTESTINAL ULCERS. *Ulcers* are wounds occurring in the inner lining of the stomach and small intestine, particularly in the duodenum. In fact, although stomach ulcers are better known, only 10% of ulcers occur in the stomach. The rest are associated with the duodenum. Stomach ulcers are, however, more dangerous. There is little doubt that ulcers are caused by the corrosive and noxious effects of acid on the gut wall. At first, these ulcers are superficial. If exposure to acid continues unchecked, the wound deepens, reaching the vascular layers deep in the wall, and bleeding occurs. The bleeding is worsened by digestion of food which increases both acid secretion and stomach motility. This bleeding, which makes ulcers painful and dangerous, can be detected by the presence of fresh blood clots in the feces.

Several factors and conditions may contribute to ulcers. One is the excessive production of acid, which may result from increased activity of the vagus nerve or gastrin secretion. Indeed, in many serious ulcer cases, cutting of the vagus nerve to the stomach (vagatomy) markedly ameliorates the condition. The cause of the increased vagal activity is not known. Gastrin-producing tumors of the pancreas also cause ulcers. Another cause may have to do with too little resistance to acid by the gut wall.

It is widely believed, though not fully established, that stress causes ulcers, at least in certain individuals. In rats, a few hours of exposure to stressful situations such as immobilization can cause widespread gastric and duodenal ulcers. These effects may be due to the catabolic effects of high levels of corticosteroids secreted by the adrenal cortex during stress, which weaken the gut wall's resistance to acid. Other stress hormones, such as adrenalin, are less likely to cause ulcers.

DIARRHEA. *Diarrhea* is characterized by excessive and frequent discharge of watery feces. The condition is sometimes caused by an increase in intestinal motility, delivering great quantities of watery chyme to the large intestine. The colon's inability to absorb the excess water causes watery and frequent fecal discharges. Different factors may be responsible for the increased motility. Thus, certain fruits, such as prunes, contain substances that naturally increase intestinal motility. Diarrhea can also be caused by the actions of certain toxins on the epithelial cells of intestinal glands. For example, *cholera toxin* causes the intestinal glands to secrete large quantities of electrolytes (sodium, chloride, bicarbonate) into the lumen. Water follows by osmosis. A cholera victim can lose about 10 L of water per day, a lethal condition if not treated.

Certain diarrheas are caused by enzyme deficiency in the small intestine. For example, certain individuals lack the enzyme *lactase*, produced by mucosal cells. Therefore, they cannot digest lactose, the sugar in milk and dairy products. The undigested lactose increases the lumen osmolarity, resulting in decreased water absorption in the small intestine and increased chyme delivery to the colon, causing diarrhea. Diarrhea may also be of *nervous* (psychogenic) origin. For example, anxiety increases parasympathetic activity to the lower bowels, increasing intestinal motility, which in turn decreases absorption time, leading to diarrhea.

CONSTIPATION. Reduced intestinal motility, particularly of the large intestine, is responsible for *constipation*, a common digestive disorder. The reduced motility increases the storage time, which in turn increases the amount of water absorbed from the feces. Dried feces are less bulky and therefore less likely to initiate movements. Causes of constipation are not well understood. *Learning to inhibit* the defecation reflex during childhood may be one cause. *Dietary habits* may be another. Increased *fiber content* (raw vegetables, fruits) in the diet improves fecal bulk, which in turn stimulates colon motility and defecation. Although the average adult defecates once daily, many healthy people have less frequent bowel movements. Indeed, mild and occasional constipation does not pose any physiological problems, but prolonged constipation is accompanied by abdominal discomfort, headaches, loss of appetite, and even depression. Sudden prolonged constipation, however, may be due to diseases of the colon.

CN: Use red for E and dark colors for A and F.
1. Begin with the upper panel, completing the diarrhea segement first.
2. Color the belching material, noting that the symbol for fermentation (J) is that of an undigested carbohydrate. The bean shape for (L) is self-explanatory.
3. Color lactose intolerance and then vomiting.
4. Color peptic ulcers. The diagram in the lower left corner represents a bleeding ulcer.

DIARRHEA*

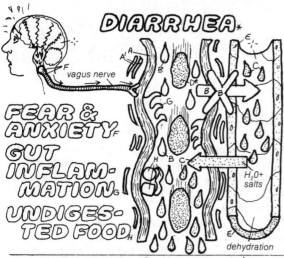

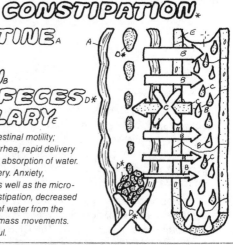

vagus nerve

FEAR &
ANXIETY F
GUT
INFLAM-
MATION G
UNDIGES-
TED FOOD H

H₂O+
salts

dehydration

CONSTIPATION*

LARGE INTESTINE A
MOTILITY A'
ABSORPTION B
SECRETION C FECES D*
BLOOD CAPILLARY E

Diarrhea is often caused by increased intestinal motility; constipation, by decreased motility. In diarrhea, rapid delivery of chyme to the colon leaves little time for absorption of water. Defecation is frequent, and feces are watery. Anxiety, ingestion of certain foods (e.g., prunes), as well as the microbial infections can cause diarrhea. In constipation, decreased colon motility causes excessive removal of water from the feces, as well as delaying peristalsis and mass movements. Defecation is infrequent and can be painful.

BELCHING & FLATULENCE*

SOURCES OF
EXPELLED GAS: i-
SWALLOWED i°
FERMENTATION J
PUTREFACTION K
CERTAIN FOODS L
BACTERIA M

The gas in the gastrointestinal tract is derived from air trapped in swallowed food, from fermentation of some dietary materials by the intestinal bacteria (mainly in the cecum and colon), and from putrefaction of foods. Gas may be expelled from the mouth by belching or from the anus by flatulence. Excessive gas is a usual source of pain and discomfort.

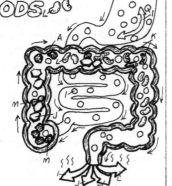

LACTOSE INTOLERANCE: DIARRHEA & FLATULENCE*

MILK J' UNDIGESTED LACTOSE J²
BACTERIA M
GAS N

In some individuals, deficiency in the digestive enzyme lactase prevents absorption of lactose, the disaccharide sugar in milk and milk products. Accumulation of lactose in the lumen of the small intestine increases osmotic pressure, reducing water absorption and promoting diarrhea. The undigested lactose will be utilized by the bacteria in the cecum and colon, forming gas and causing discomfort and flatulence. These individuals must avoid milk products.

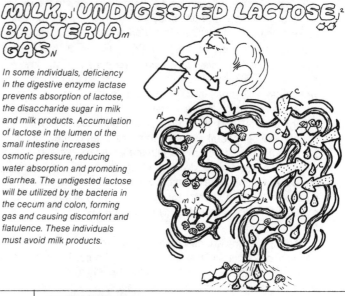

VOMITING*

INPUT FROM SENSORS: o
Stomach or intestinal toxins.
Extreme stomach distention.
Emotional distress.
Severe headache.
Nausea-producing smells,
tastes, sights and motion.

VOMITING CENTER P (in medulla)

VOMITING REFLEX: F
GLOTTIS CLOSES Q
CARDIAC SPHINCTER
OPENS R
DIAPHRAGM S ABDOM-
INAL MUS. CONTRACT T
REVERSE STOMACH
PERISTALSIS &
PRESSURE INCREASE
FOOD EXPELLED V

vagus

spinal nerves

Ingestion of excessive or poisonous food irritates the stomach mucosa, activating sensory fibers to the vomiting center in the brain medulla. Motor signals from this center evoke the vomiting reflex. Thus, saliva flows, glottis closes, and muscles are contracted, increasing pressure on the stomach. Reverse peristalsis in the stomach, aided by increased intra-abdominal pressure, expels the food out through the relaxed cardiac sphincter and esophagus, pharynx, and the mouth.

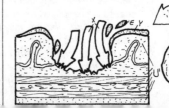

PEPTIC ULCERS*

10% GASTRIC U
90% DUODENAL W
ACID X
MUCUS Y
BLOOD E'

anxiety
stress
deadlines

Normally, the stomach and intestinal mucosa are protected against the eroding action of gastric hydrochloric acid. A special mucosal coat may play a role in this protection. Certain disorders of the wall or excess acid secretion, tend to erode the wall, creating wounds (ulcers), which, if deep enough, will reach the vascular layers and cause bleeding. Only 10% of ulcers are formed in the stomach, the rest occur in the duodenum. Endogenous abnormalities resulting in excessive activity of the vagus nerve, or increased secretion of gastrin, as well as psychophysiological factors stemming from anxiety and stress are believed to be involved in the causation of ulcers.

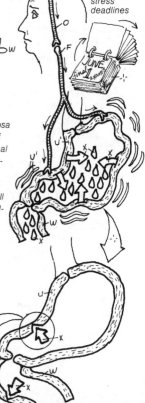

FUNCTIONAL ORGANIZATION OF THE NERVOUS SYSTEM

The nervous system is responsible for sensory and motor activities, for behavior (instinctive and learned), and for regulating activities of the internal organs and systems. To appreciate its importance, consider the problems faced by persons who have just become blind or deaf or the difficulties encountered by victims of brain stroke or spinal injury.

The nervous system as a whole may be divided into two systems, the *central nervous system* (CNS) and the *peripheral nervous system* (PNS). The CNS, consisting of the *brain* and *spinal cord*, processes sensory information and integrates it with past experience to produce appropriate motor commands. The PNS consists of the *sensory receptors* (organs), which are specialized to detect changes in the external environment or in the body interior and to communicate these signals to the CNS via the *afferent sensory nerves*.

Another part of the PNS is the *motor effectors*. These consist of the *voluntary skeletal muscles*, responsible for body and limb movements, and the *smooth muscles* and *glands*, which effect changes in visceral organ motility and secretions. *Efferent motor nerves* extending from the CNS to these organs are also part of the PNS. Based on these different targets, the peripheral motor system has been divided into a *somatic* division, which deals with the voluntary skeletal muscles, and an *autonomic* division, which deals with the visceral effectors. Although the autonomic and somatic systems are distinct in terms of their motor output nerves and targets, they may share both peripheral sensors and certain central nervous centers.

CNS operations are carried out by the *sensory*, *motor*, and *association* centers in the brain and spinal cord. The different zones of the CNS are devoted wholly or partly to any one of these functions. In this regard, three points must be noted. First, nerve centers are organized in a hierarchy (i.e., the sensory centers are divided into *lower* and *higher* ones, as are the motor and association centers). Only the lower centers are in direct contact with the peripheral neural structures. In order for the higher centers to communicate with the periphery, they must go through the lower centers and vice versa.

Second, nerve centers, whether sensory, motor, or association, consist of many *nerve cells* (neurons) and *synapses* (see plates 16, 17) that connect these cells to one another, to those in other centers, or to neurons in the periphery. Although the morphology (shapes) of the nerve cells may vary in different parts, the connections of the nerve cells mainly determine their function.

Third, all nerve centers operate on the basic principles of *excitation* and *inhibition* (the function being determined by the type of synapses between them). (See plate 82.)

To understand the operations of different sensory, motor, and associative processes, consider the human reaction to a loud and strange sound. The sound is detected by the ear, which transduces the waves into nerve signals and sends them, via the afferent auditory nerve, to the lower hearing center in the brain stem. Here signals are initially processed and then sent to the lower motor centers in the brain stem and spinal cord to activate the startle or head-turning reflexes. At the same time, activation of the autonomic centers results in increased heart and breathing rates in preparation for eventual running (fleeing).

In addition, the lower hearing centers communicate nerve signals to the higher hearing centers in the cortex, where other qualities of the sound are evaluated, with the results communicated to the cortical association centers. Here the sound is examined in relation to other sensory stimuli (e.g., visual) converging simultaneously. If further motor actions, particularly voluntary actions such as running or fleeing from the site, become necessary, appropriate commands are issued to the higher motor centers, which in turn signal the lower motor centers to activate the appropriate muscle groups. The signal from the brain stem also activates the reticular formation of the brain, which excites the cortex globally, increasing general awareness and vigilance. The higher centers can also enhance the behavioral drives and autonomic responses necessary for carrying out these motor tasks.

Division of the CNS into the higher and lower centers stems from the evolutionary path of the nervous system and in practice offers certain advantages for defense and survival. The earliest nerve centers may have resembled the rudimentary operations of the spinal cord (i.e., direct contact between the lower sensory and motor components), enabling very fast spinal reflexes such as limb withdrawal, which occur without the influence of the higher centers. These defensive reflexes maximize survival. Thus, spinal cord structure in this respect remains fairly uniform throughout evolution.

With brain evolution, new higher sensory and motor centers emerged above the older lower centers and thus controlled their activity. This arrangement improved the nervous abilities of selectivity, skillfulness, and adaptiveness of the responses. Indeed, in the lower vertebrate brain, the cerebral cortex — the site of highest and finest analysis of sensory and motor integration, learning, and skilled tasks — is rudimentary, emerging with the evolution of the mammalian brain and reaching prominence in primates. In humans, the association areas have enlarged, occupying most of the brain cortex. This trend has enabled such adaptive capacities as learning, introspection and planning, speech and language to develop.

CN: Use a dark color for the A structures and a very light color for E.
1. Begin with the anatomical illustration at the top of the page. Because the peripheral nerves (B) are so numerous and so small in this drawing, you may wish to color over the many lines representing nerves.
2. Color the organizational diagram across the middle of the page, starting with the borders of the CNS (A) and the PNS (B). Then follow the numbered sequence beginning with the

body's various sensory receptors (C¹) receiving stimuli from the environment. The ascending sensory pathways and the descending motor pathways within the CNS receive the same labels as the centers from which they orginate.
3. Color the bottom material, working from left to right. In the section of spinal cord, note the presence of sensory areas within the overall motor portions of the anterior half as well as motor areas within the predominantly sensory posterior half.

CENTRAL NERVOUS SYSTEM.
BRAIN.
SPINAL CORD.
PERIPHERAL N. S.
SENSORY
AFFERENT
NERVES.
MOTOR
EFFERENT
NERVES.
SOMATIC
(VOLUNTARY).
AUTONOMIC
(INVOLUNTARY).

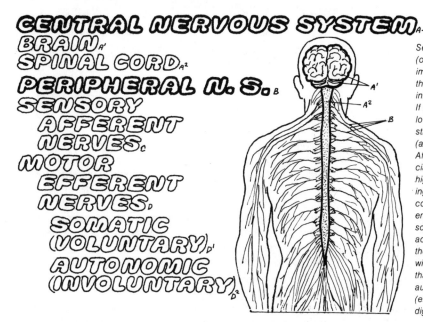

Sensory stimuli excite the peripheral sensory receptors (1) (organs), evoking nerve impulses in the sensory cells (2). These impulses are conveyed along the sensory (afferent) nerves (3) to the CNS (brain and spinal cord). Here the sensory signals are initially analyzed and integrated in the lower sensory centers (4). If necessary, reflex responses are generated by activation of lower association (5) and motor systems. To deal with complex stimuli, sensory signals are conveyed, via the central sensory (ascending) pathways (6) to the higher sensory structures (7). After analysis and integration within the higher sensory and association centers (8), appropriate signals are transmitted to the higher motor centers (9) and then via the central motor (descending) pathways (10) to the lower motor centers (11). Final motor commands are sent out, via the lower motor neurons and peripheral motor (efferent) nerves (12), to the peripheral effectors. In the somatic NS; these effectors are the skeletal muscles (13). Their activation generates bodily movements. Central commands for the visceral effectors (smooth muscles and glands) are generated within special structures in the brain (limbic system, hypothalamus, and medulla oblongata) (9A) and are sent out via the autonomic (sympathetic and parasympathetic) nerve fibers (efferents) (12A) to regulate the activity of blood vessels, heart, digestive system, etc. (13A).

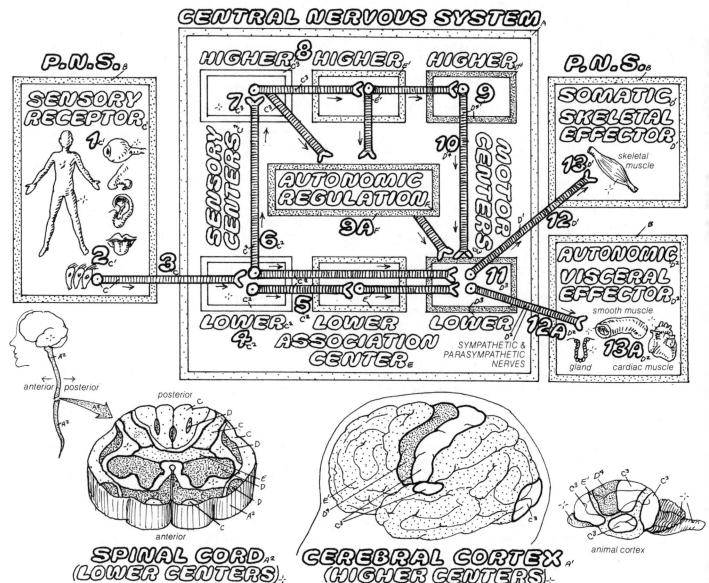

Sensory, motor, and association functions are carried out by different parts of the CNS. In the spinal cord (the oldest part of CNS), the structures in the anterior (ventral) half carry out motor functions; those in the posterior (dorsal) half deal with sensory functions. The middle part of the cord is concerned with association functions, connecting sensory with motor as well as the right half with the left. In the brain cortex (phylogenetically, the newest part of the CNS), sensory functions are carried out by parts located mainly in the posterior half. Motor functions are dealt with by areas in the anterior (frontal) half. Note that, in the spinal cord, the relative size of the association areas is small compared to that of the sensory and motor areas. In the cortex, the size of the association areas far exceeds that of the sensory and motor. Note also the marked increase in the size of the association areas with evolution, indicating the importance of these areas in the higher functions of the nervous system (learning, perception, language).

BRAIN STRUCTURES & GENERAL FUNCTIONS

Housed in the skull, the *brain* consists of all the parts of the central nervous sytem (CNS) above the spinal cord. Anatomically, it may be divided into two major parts, a lower *brain stem* and a higher *forebrain*. The brain stem is situated directly above and has extensive connections with the spinal cord. The brain stem, the most primitive part of the brain, consists of several parts, including the medulla, pons, cerebellum, midbrain, hypothalamus, and thalamus.

Brain stem structures carry out many vital somatic, autonomic, and reflexive functions that deal with vegetative functions for body maintenance and survival. The centers for control of respiration and cardiovascular and digestive functions are located in the *medulla*, the most primitive (the "lowest") of brain stem structures. The *pons* has structures involved in the function of cerebellum and motor control, in addition to other inhibitory control centers for respiration. Other nuclei in the reticular core of the pons and medulla (reticular formation) are involved in sleep and in the generalized control of excitation of higher forebrain structures.

The *cerebellum*, located in the back of the brain stem and attached to the midbrain, is a major motor structure involved in movement coordination. Somatic motor centers (nuclei) in the *midbrain* are involved in regulation of walking and posture and of reflexes for head and eye movements. The *hypothalamus* contains numerous centers (nuclei, areas) for regulating the internal environment (homeostasis), including those for controlling body temperature, blood sugar, hunger and satiety, sexual behavior, and hormones. The *thalamus* is a complex structure involved in integrating sensory signals and relaying them to the higher forebrain structures, particularly to the *cerebral cortex*. The thalamus also participates in motor control and in regulating cortex excitation. Brain stem organization is fairly similar in the different vertebrates, particularly among mammals. Several pathways connect the brainstem to the lower sensory and motor centers in the spinal cord and the higher ones in the forebrain.

The brain stem capacities and importance in behavior and nervous regulation may be shown by observing the motor and behavioral abilities of anencephalic (no brain) infants, who are born without the forebrain. Such infants usually do not survive for long, but during their short lives, they are capable of many behaviors. They can find the nipple and suckle milk; they can smile, frown, cry, and make other infant sounds; they can move the head and limbs in a manner similar to normal newborns.

The human forebrain consists of two nearly symmetrical *cerebral* (brain) *hemispheres*, each comprised of the cerebral cortex, the *basal ganglia*, and the *limbic system*. The two hemispheres are connected by a massive bundle of fibers called the *corpus callosum*. The cerebral cortex is a sheet of nerve cells (gray matter) about 5 mm thick that covers the surface of the hemispheres (cortex = bark). The large area of the cortex and the need to fit this sheet within the skull produces the folds and convolutions (sulcus = furrow; gyrus = convolution). The cortex and the associated large mass of nerve fibers (white matter) make up the bulk of the cerebral hemispheres. In humans, the cerebral cortex is extremely well developed both in size and nerve cell organization, enabling it to be the site of the highest and most intricate analysis of sensory and motor information (see plate 105).

Each hemisphere's cortex is divided into four lobes. The *frontal lobe* extends from the anterior tip of the hemisphere back to the *central sulcus* (fissure of Rolando). The posterior areas of the frontal lobe are specialized for motor functions, including those for language (see plates 90, 105); the anterior areas are involved in learning, planning, and other higher psychological processes. The *occipital lobe*, located in the back of the hemisphere, is involved mainly in visual operations (see plate 94). The dorsal (top) and lateral areas between the frontal and occipital lobes are called the *parietal lobe*. It is specialized for somatic sensory functions (e.g., skin senses) and the related association roles (see plate 87). Certain areas in the parietal lobe are also very important in cognitive and intellectual processes. The *temporal lobe* comprises the hearing centers and related association areas, including some speech centers. Other areas of the temporal lobe are important in memory (see plate 103). The anterior and basal areas of the temporal lobe are involved in the sense of smell and in functions related to the limbic system.

Another brain system found in the forebrain is the *basal ganglia*, which consist of structures mainly involved in motor processes. In lower animals, the basal ganglia are the only higher motor structures. In humans, the structures of the basal ganglia work in conjunction with the motor areas of the cortex and cerebellum for planning and coordinating gross voluntary movements. The third forebrain system of structures is the *limbic system*. Also called the *limbic lobe*, the components of this system are intimately involved in the expression of instinctive behaviors, emotions, and drives. the overall size and organization of the limbic system do not change significantly during the course of mammalian evolution, indicating this system's involvement with basic behaviors common to all species of mammals (see plates 91, 102).

Even though many functions, particularly motor and sensory, are fairly well localized to distinct areas and parts of the brain, different brain sections are well connected. Particularly in regard to such brain activities as learning, memory, and consciousness (global functions), the brain probably works as a whole.

CN: Use dark colors for B, C, E, F, and G.
1. Begin in the upper right corner by coloring the cortex of the two cerebral hemispheres (A¹) and the list of titles, without coloring the structures to which the titles refer. Then start with the material in the upper left corner and work your way down to the limbic system and across to the basal ganglia.
2. Color both views at the bottom simultaneously. The vertical broken line in the mid-sagittal view shows the location of the coronal section.

FOREBRAIN & BRAIN STEM

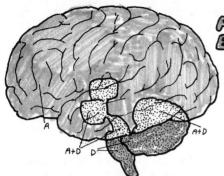

Anatomically, the brain may be divided into a forebrain (cerebrum) subserving higher nervous functions (perceptions, voluntary motor control, emotion, cognition, and language), and a brain stem, serving the regulation of internal bodily functions, involuntary reflexes, and also serving as relay stations for signal transmission to and from the forebrain.

THE BRAIN*
FOREBRAIN_A
CEREBRAL CORTEX_A'
LIMBIC SYSTEM_B
BASAL GANGLIA_C
BRAIN STEM_D
THALAMUS_E
HYPOTHALAMUS_F
MIDBRAIN_G
PONS_H
MEDULLA_I
CEREBELLUM_J

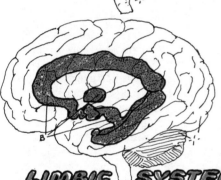

gyrus
sulcus
central sulcus
K
L
sylvian fissure
N
m

LOBES OF THE CEREBRAL CORTEX*

FRONTAL_K
PARIETAL_L
OCCIPITAL_M
TEMPORAL_N

The forebrain consists of two hemispheres, each divided into 4 lobes: frontal, parietal, occipital, and temporal. The lobes are covered by the cortex, a wide, thin (3-5 mm) sheet of gray matter that must fold extensively to fit in the skull (hence, the convolutions — sulci and gyri). The occipital lobe performs higher visual functions; the temporal lobes house the auditory and associated language and cognitive areas; the parietal lobes host the somatic sensory and related association areas; the frontal lobes contain the higher motor and association motor areas. Deep within the lobes are the white matter (fibers) and other forebrain structures (limbic system, basal ganglia).

LIMBIC SYSTEM_B

The limbic system consists of several forebrain structures that deal with the integration and expression of emotions, feelings, and drives. In lower animals, the limbic system is intimately connected with the olfactory sense. In higher animals, it is well connected with the cortex of the frontal lobes.

BASAL GANGLIA_C

The structures of the basal ganglia are higher motor centers, functioning in harmony with the motor cortex in the control of higher voluntary motor activity. In birds and lower animals that lack a true cortex (neocortex), the structures of the basal ganglia are the highest centers of motor control.

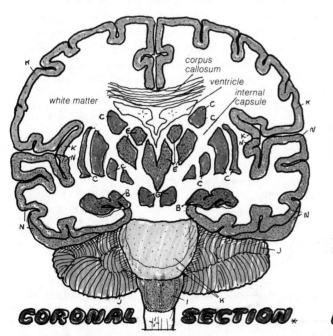

corpus callosum
ventricle
internal capsule
white matter

CORONAL SECTION*

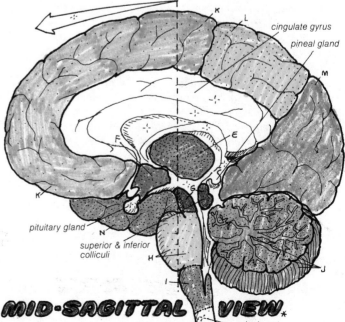

cingulate gyrus
pineal gland
pituitary gland
superior & inferior colliculi
spinal cord

MID-SAGITTAL VIEW*

To view the inner structures of the brain, either a coronal (cross) section or a sagittal section can be used. The coronal section shown (left diagram) depicts the relationship between the cortex and the underlying white matter and nerve centers (basal ganglia, thalamus). Note the corpus callosum connecting the two hemispheres. The sagittal section cuts along the medial plane, exposing the hidden medial cortical structures as well as many of the brain stem structures.

The *spinal cord* (SC) is one of the two main parts of the central nervous system (CNS). The SC is about 40-45 cm (16-18 in.) long, extending within the inner cavities of the vertebral column from the neck to the loin. Practically all the voluntary skeletal muscles in the neck, trunk, and limbs receive their supply of motor nerves from the SC. All the sympathetic and some of the parasympathetic motor outputs to the skin and visceral organs also emerge from the SC. All sensory signals from the peripheral receptors of the skin, muscles, and joints in the trunk and limbs are sent to the SC.

The spinal cord performs two basic functions. First, it can act as a nerve center, integrating the incoming sensory signals and activating the motor output directly, without any brain intervention. This function is manifested in the operation of spinal reflexes, which are extremely important in defending against noxious stimuli and maintaining body support. Second, the SC is the intermediate nerve center (station) between the periphery and the brain. Thus, all voluntary and involuntary motor commands from the brain to the body musculature must first be communicated to the spinal motor centers, which process these signals appropriately before passing them to the muscles. Similarly, sensory signals from the peripheral receptors to the brain centers are first communicated to the SC sensory centers, where they are partly processed and integrated before delivery to the brain sensory centers. The SC fiber pathways serve in such two-way communication between the brain and the cord.

SC structural organization can best be studied by observing a cross section of the cord. Throughout its length, an outer band of *white matter* surrounds an inner core of *gray matter*. This pattern appears fairly uniform throughout the SC length, the right and left halves of which are symmetrical. The white matter consists mostly of myelinated nerve fibers (axons) grouped in bundles. The cell bodies of these fibers are either in the brain or in the SC. The gray matter consists of nerve cells (neurons), their processes, and the numerous synapses between the nerve cells.

The SC gray matter is shaped like the letter H (or more like a butterfly, whose wings are called horns). The gray matter may be divided into three functional zones: the *dorsal (posterior) horns* are sensory; the *ventral (anterior) horns* are involved in motor functions; and the middle zone carries out, in part, association functions between the sensory and motor zones.

The SC gray matter is populated by large and small neurons. The large neurons are either motor or sensory. The motor neurons, located in the ventral horns, are output neurons that send their motor fibers to the voluntary skeletal muscles via the ventral (motor) roots. Motor neurons are grouped in clusters, each cluster serving a different muscle. In the thoracic, lumbar, and sacral segments of the SC, *autonomic motor neurons* are clustered separately.

The peripheral sensory input to the spinal cord arrives in the dorsal horns via the *primary sensory neurons*, the cell bodies of which are located outside the SC gray matter, in the dorsal (sensory) root ganglia. These primary sensory cells have a bifurcating axon, the central branch of which emerges from the sensory root entering the dorsal horn to synapse with the sensory relay cells and the interneurons. Located in the dorsal horns, the large sensory relay cells give rise to fibers that cross over to the opposite side and ascend the SC white matter to communicate the incoming peripheral sensory signals to the higher brain centers.

The *interneurons* (association neurons), which are small, provide excitatory and inhibitory connections between the primary sensory neurons and the motor neurons in the ventral horns of the same or the opposite side. Some of the dorsal root sensory fibers continue uninterrupted to enter in the ventral horn of the same side, where they synapse directly with the motor neurons. These and other interneuron-mediated local connections between sensory and motor neurons provide the structural basis (i.e., nerve circuits) for the operation of spinal reflexes (see plate 89). Motor neurons receive input not only from the sensory neurons and the interneurons, but also from the neurons in the higher brain centers (see plate 90). Because communication between all brain neurons and the voluntary skeletal muscles is possible only through the spinal motor neurons, they are called the "final common path."

The SC white matter is divided into bundles (funiculi), each containing nerve fibers (axons) traversing between the SC and the brain. These bundles form the SC *pathways*, which are either *ascending* or *descending*. Generally, the ascending pathways are sensory, taking signals from the cord to the brain, and the descending pathways are motor, bringing commands from the brain to the cord. The fibers of the motor and sensory pathways are segregated in functionally distinct bundles. For example, signals relating to fine touch, pressure, and proprioception ascend in the dorsal pathways, pain and temperature signals ascend in the lateral pathways, voluntary motor signals descend in the dorsolateral pathways, and involuntary motor signals descend in the ventral pathways.

CN: Use a dark color for F.
1. Begin in the upper left corner.
2. Color the organization of the spinal cord. Color the gray matter gray (D) and leave the white matter (E) uncolored.
3. Color the organization of gray matter, represented by the left half of the spinal cord section shown below. Note that only the borders of the three zones of gray matter are colored gray. In the right half of the illustration color the various tracts of the white matter. The entire gray matter portion of this half is colored gray.

FUNCTIONS OF SPINAL CORD A
SENSORY SIGNALS B
PATHWAYS TO & FROM BRAIN C
MOTOR SIGNALS C

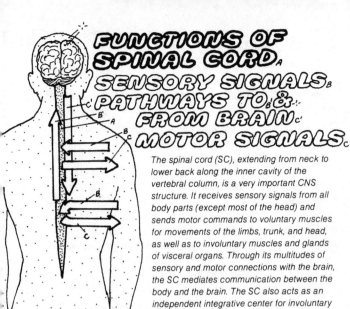

The spinal cord (SC), extending from neck to lower back along the inner cavity of the vertebral column, is a very important CNS structure. It receives sensory signals from all body parts (except most of the head) and sends motor commands to voluntary muscles for movements of the limbs, trunk, and head, as well as to involuntary muscles and glands of visceral organs. Through its multitudes of sensory and motor connections with the brain, the SC mediates communication between the body and the brain. The SC also acts as an independent integrative center for involuntary (spinal) reflexes.

ORGANIZATION OF SPINAL CORD A
GRAY MATTER D*
WHITE MATTER E
SENSORY NEURON B
INTERNEURON (ASSOCIATION) F
MOTOR NEURON C
ASCENDING TRACT (SENSORY RELAY N.) B'
DESCENDING TRACT (UPPER MOTOR N.) C'

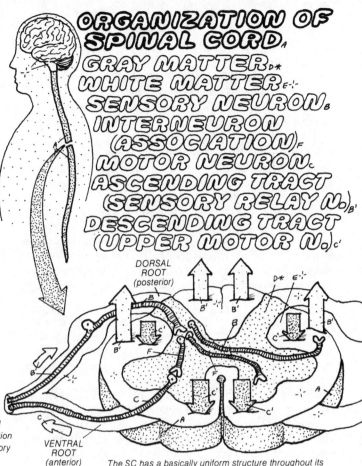

DORSAL ROOT (posterior)

VENTRAL ROOT (anterior)

The SC has a basically uniform structure throughout its length. It is arranged into an inner mass of gray matter (GM) surrounded by an outer band of white matter (WM). In cross section, GM of SC is shaped like an H or butterfly. GM consists of cell bodies of neurons, their dendrites, short axons, and synapses, making it the site of neural (synaptic) analysis, integration, and transmission. GM is connected with the dorsal and ventral roots through which the SC communicates with the periphery. WM consists of ascending (sensory) and descending (motor) fibers (pathways) connecting the SC with the brain. The fatty myelin sheath around the fibers gives WM its name.

ORGANIZATION OF GRAY MATTER D*

The GM of the SC is roughly organized into dorsal and ventral horns (DH, VH). DH carries out sensory functions and VH, motor. A middle zone is involved in association functions between DH and VH of the same and opposite sides. DH receives sensory signals, which arrive via the dorsal roots. Sensory afferents conveying various modalities (pain, touch, etc.) terminate in different laminae of DH. DH analyze, integrate, and transmit these signals to association and motor neurons in the SC or to relay neurons going to the brain. VH contains cell bodies of spinal motor neurons, the fibers of which leave the SC through the ventral (motor) roots, innervating voluntary muscles. Within each VH, motor neurons are grouped in discrete nuclei, each dealing with a separate muscle. The middle association zone contains inhibitory and excitatory interneurons. The short axons of these interneurons make specific connections between the sensory and motor elements of the DH and VH of the same and other segments. These connections underlie spinal integration and spinal reflexes.

DORSAL HORN (SENSORY) D' B
MIDDLE ZONE (ASSOCIATION) D² F
VENTRAL HORN (MOTOR NUCLEI) D³ C²

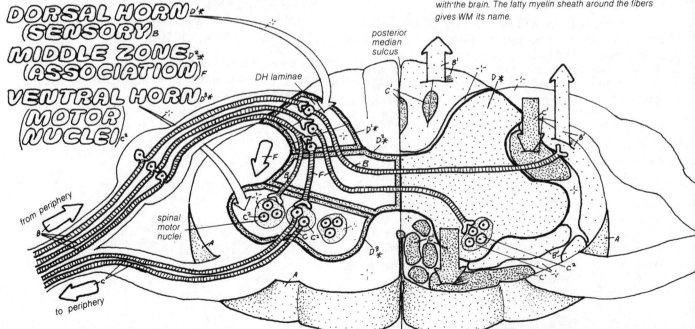

DH laminae

posterior median sulcus

from periphery

spinal motor nuclei

to periphery

WM of the SC is segregated into bundles (funiculus, column, tract) of descending and ascending fibers (axon of large neurons). Ascending fibers are generally sensory; descending fibers are motor; some descending fibers are for regulation of sensory input. Major ascending sensory pathways connect the SC with the medulla, brainstem reticular formation, and thalamus. Major descending tracts connect the forebrain voluntary motor areas as well as midbrain involuntary motor centers with the SC.

ORGANIZATION OF WHITE MATTER E

THE PERIPHERAL NERVOUS SYSTEM

The *peripheral nervous system* (PNS) was defined in plate 78. Here we focus on PNS organization, describing particularly the *peripheral nerves*, which connect the CNS with the sensory receptor and with the motor effectors (muscles and glands).

PERIPHERAL NERVE STRUCTURE. A typical peripheral nerve trunk consists of thousands of nerve fibers grouped in bundles and sheathed by connective tissue coats. The bundles are usually segregated by function (*sensory* or *motor*) or by target (arm, muscle groups, skin zone). Each nerve fiber is the axon of either a peripheral sensory, motor, or autonomic neuron. The nerve fibers are of varying diameters. Some have a myelin sheath (are myelinated). Large-diameter myelinated fibers (Type A) conduct rapidly (up to 120 m/sec., nearly 250 mi/hr.); small-diameter fibers (Type C, autonomic, pain) are unmyelinated and conduct slowly (< 1 m/sec.).

CRANIAL AND SPINAL NERVES. Peripheral nerves are associated with both brain and spinal cord. The 12 pairs of brain nerves (*cranial nerves*) are referred to by names or roman numerals (see table). Cranial nerves emerge from different brain sites. The *spinal nerves*, however, have a more uniform orientation. Each of the 31 pairs of spinal nerves is associated with one of the spinal vertebrae and is formed from a union of fiber bundles emerging from the two *spinal roots*. The *dorsal roots* carry sensory fibers; the *ventral roots* carry motor (somatic and autonomic) fibers. Therefore, each spinal nerve is a mixed nerve containing somatic motor, somatic sensory, and autonomic fibers. Like their corresponding vertebrae, the spinal nerves are divided into 8 cervical (neck), 12 thoracic (chest), 5 lumbar (loin, lower back), and 5 sacral (sacrum bone). There is also one coccygeal nerve. In general, the cervical nerves innervate targets in the neck, shoulders, and arms; the thoracic nerves, the trunk; the lumbar nerves, the legs; and the sacral-coccygeal nerves, the genitalia, pelvic, and groin areas.

AUTONOMIC INNERVATION AND FUNCTION. The various somatic sensory and motor nerves are mentioned in the plates where the system they serve is described. The autonomic nervous system is introduced in plate 25. Autonomic motor nerves regulate motility and secretion in the skin, blood vessels, and visceral organs by stimulating smooth muscles and exocrine glands. Autonomic regulation is carried out by two types of nerves: *sympathetic and parasympathetic*.

SYMPATHETIC NERVES. The sympathetic motor outflow is via the spinal nerves emerging from the thoracic and lumbar segments of the spinal cord, innervating targets in the visceral core and the body periphery (skin and vessels of muscles). Even the targets in the head (e.g., the eyes [iris]) receive sympathetic innervation via the spinal nerves. The sympathetic nerves found within the spinal nerve trunks are postganglionic fibers (i.e., their cell bodies are in the chain of *sympathetic ganglia*, located on both sides along the vertebral column). The postganglionic sympathetic neurons are driven by the shorter myelinated *preganglionic sympathetic*

neurons, which are located in the lateral horns of the spinal cord and send their axons into the sympathetic ganglia.

The neurons of the sympathetic chains are connected by interneurons; this particular organization underlies the generalized discharge characteristic of the sympathetic nervous system. Other additional sympathetic ganglia are found in the viscera, which are associated with the autonomic splanchnic nerves, innervating targets like the stomach and the adrenal medulla. In accord with the nonselective and diffuse function of the sympathetic system, the sympathetic fibers innervate practically every visceral and peripheral organ in the body. In many cases, they innervate the blood vessels in these organs, thus controlling the blood flow.

PARASYMPATHETIC NERVES. The parasympathetic nerves are associated with only certain cranial nerves, such as the III, V, and X, as well as with nerves emerging from the sacral spinal cord. The most prominent parasympathetic nerve is the vagus (wanderer; X cranial nerve), which innervates many visceral organs, including the lungs, heart, and digestive tract. The fibers found within the parasympathetic nerves are basically preganglionic, and their cell bodies are located in the motor nuclei of the brain stem or in the sacral spinal cord. The postganglionic neuron is short and emerges from a peripheral ganglia located near or in the target organ. The parasympathetic innervation of visceral organs is discrete and selective. Some targets, like the heart and digestive system, receive profuse innervation; others, like the kidney, have sparse innervation.

CENTRAL AUTONOMIC CONTROL. The peripheral autonomic fibers are controlled by nerve centers in the brain stem, particularly the medulla and the hypothalamus. The medullary centers deal with the routine automatic control of such internal systems as the cardiovascular, respiratory, and digestive systems. The sympathetic hypothalamic centers are involved in controlling body temperature and bodily responses to emotional states such as fear and excitement, fight and flight. To exert these controls, the hypothalamic and medullary centers send neurons to stimulate the preganglionic autonomic neurons in the spinal cord, which in turn stimulate the postganglionic neurons going to the peripheral effectors.

CRANIAL NERVES AND THEIR FUNCTIONS.

CRANIAL NERVE	NAME	FUNCTIONS
I	olfactory	sensory, smell
II	optic	sensory, vision
III	oculomotor	motor, autonomic, eye movement
IV	trochlear	motor, eye movement
V	trigeminal	sensory, motor, autonomic
VI	abducent	motor, eye movement
VII	facial	motor, facial movement
VIII	vestibular	sensory, hearing/balance
IX	glossopharyngeal	somatic/autonomic motor to tongue, pharnyx
X	vagus	autonomic motor/sensory
XI	accessory	motor, neck and shoulder
XII	hypoglossal	motor, tongue movement

CN: Use dark colors for C and E.
1. After reading the introductory paragraph in the upper left corner, begin with the diagram of the peripheral nervous system in upper right corner. Note that the 12 cranial nerves contain various sensory, motor and autonomic nerves. Color them accordingly. The spinal nerves contain all three, and this is shown in the cut nerve to which all the spinal nerves (A) are pointing. Color all 31 spinal nerves.

2. Color the large diagram describing the organization of spinal nerves. Begin with the anatomical portion to the left and include the directional arrows. Then color the cutaway drawing on the right side. Note that the title, autonomic efferent motor, refers to both the sympathetic and parasympathetic nerves, and is colored gray. In this large illustration, only the sympathetic system (F) is shown, whereas the parasympathetic (G) is included in the bottom diagram.

The peripheral nervous system (PNS) consists of peripheral nervous structures and nerves that serve in the somatic and autonomic divisions. In the somatic division, sensory nerves connect the special (e.g., ear) and general (e.g., skin) sensory receptors to the spinal cord (SC) and brain, and motor nerves connect the central nervous system (CNS) to the skeletal muscles. In the autonomic division, visceral sensory fibers (inflow) and the sympathetic (S) and parasympathetic (PS) fibers (motor outflow) connect the visceral organs and effectors to the S and PS ganglia as well as to the SC and brain. Fibers within a nerve trunk vary in size and speed of conduction.

SPINAL NERVES (31 PAIR)ₐ
DORSAL ROOT & GANGLION_B
SOMATIC & VISCERAL AFFERENT SENSORY N._C
VENTRAL ROOT,
SOMATIC, EFFERENT MOTOR NERVE_E
AUTONOMIC EFFERENT MOTOR:*
SYMPATHETIC: PREGANGLIONIC_F
POSTGANGLIONIC_F'
SYMPATHETIC CHAIN,_F² GANGLION_F³
PARASYMPATHETIC: PREGANG._G
POSTGANGLIONIC_G' GANGLION_G²

PERIPHERAL NERVES

Of the total of 43 pairs of peripheral nerves, 12 are associated with the brain (cranial nerves) and 31 with the SC (spinal nerves). Cranial nerves emerge directly from the brain, but the spinal nerves form by the merger of the dorsal and ventral roots of the SC. The cranial nerves are identified by names or roman numerals, spinal nerves by name and number of corresponding vertebrae. Some cranial nerves are purely sensory, others are motor or mixed. Some contain partly autonomic fibers; others are largely autonomic. Spinal nerves are generally mixed, containing sensory, motor, and autonomic fibers.

SENSORY_C
MOTOR_E
AUTONOMIC*

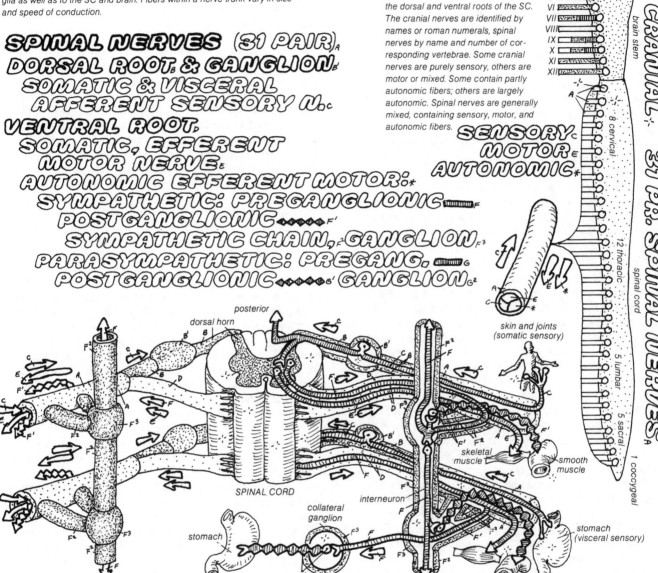

12 PR. CRANIAL + 31 PR. SPINAL NERVESₐ

brain stem
8 cervical
spinal cord
12 thoracic
5 lumbar
5 sacral
1 coccygeal

skin and joints (somatic sensory)

skeletal muscle

smooth muscle

posterior
dorsal horn
SPINAL CORD
interneuron
collateral ganglion
stomach
stomach (visceral sensory)

The sensory afferents from visceral organs enter the CNS via cranial or spinal nerves. Motor outflow to visceral effectors is via the S and PS motor fibers. PS fibers leave via the cranial and sacral nerves; the S outflow is via the thoracic and lumbar spinal nerves. Between the CNS and its visceral targets, both S and PS outflow consist of two neurons and a ganglion (G). The first neuron (pre-G) beginning in the brain or the SC, synapses with the second neuron (post-G) inside G. The G of the S system are located parallel to the SC, forming a chain. The G of the PS system are located near the target organs. Some targets (e.g., the stomach) contain complicated nervous networks (plexus) of their own that are innervated by the post-G fibers. The pre-G sympathetic fibers begin in the intermediate motor horns of the SC and terminate in the S ganglia. The post-G neurons leave the G and course along a spinal nerve to serve blood vessels and sweat glands of the somatic area served by that spinal nerve. Other post-G fibers in the S system leave the G to reach their target via an independent visceral nerve. The neurons in the S ganglia are connected by interneurons, allowing for simultaneous and generalized discharge from several S ganglia, even when only one G is activated from the brain or periphery. In contrast, the proximity of PS ganglia with their targets and lack of inter-G connections allow for discrete activation of specific targets by the PS system. Both S and PS pre-G neurons within the brain or SC are controlled by descending fibers from higher centers in the hypothalamus and medulla, enabling the hypothalamus and medullary centers to exert their control over internal bodily functions (digestion, blood flow, etc.). Also, via its connections with the limbic system, the hypothalamus controls internal bodily responses during arousal and emotional states.

SOMATIC & AUTONOMIC* MOTOR RESPONSE

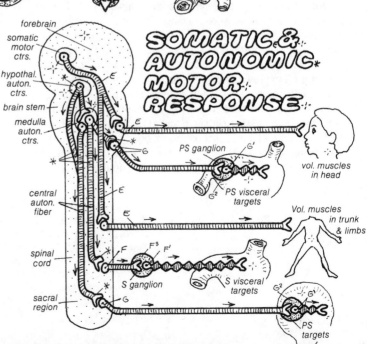

forebrain
somatic motor ctrs.
hypothal. auton. ctrs.
brain stem
medulla auton. ctrs.
central auton. fiber
spinal cord
sacral region

PS ganglion
PS visceral targets
vol. muscles in head
Vol. muscles in trunk & limbs
S ganglion
S visceral targets
PS targets

MECHANISMS OF EXCITATION & INHIBITION

SYNAPSE IMPORTANCE IN NEURAL FUNCTION. Although neurons are the excitable cells of the nervous tissue, functionally they are not the true units of the *nervous system*. Single neurons are unable to carry out the typical actions of this system, be they simple reflexes or complex thought processes. These actions are carried out by nerve nets and circuits, which are the true functional units of the nervous system. Nerve nets and circuits are made up of two or more (sometimes millions of) neurons that interact with one another through *excitatory* and *inhibitory synapses*. Therefore, the synapses, by providing controllable functional connections between neurons, are responsible for the *integrative* functions of the CNS. Without the trillions of synapses, reflexes would not operate, communication between the periphery and the CNS and within the CNS would cease, and the brain's integrative operations would fail. Plates 16 and 17 discuss the cellular mechanisms of signal transmission in the excitatory and inhibitory synapses. Here we focus on their integrative and regulatory functions.

EXCITATORY/INHIBITORY SYNAPTIC INTERACTIONS. As an example of synaptic interaction in a nerve net, take a typical spinal motor neuron, with its cell body and dendrites (its receptive zone) in the ventral horn of the spinal cord. To make a muscle fiber contract, this motor neuron must be excited to its threshold level; it will then discharge nerve impulses along its axon to excite the muscle. To prevent a muscle from contracting or to relax it, the motor neuron must be suppressed (inhibited). The cell body and dendrites of this motor neuron are the site of thousands of synapses made by the ending of sensory neurons, interneurons, or neurons originating in the brain. Some of these synapses are excitatory (E); others are inhibitory (I). An E and an I synapse may be side by side. Although any neuron in the nervous system (the motor neuron, in this case) can receive both E and I synapses at once, a neuron can contribute only either the E or I type of synapse; furthermore, all the contributing terminals would be the same (E or I). Thus, all motor neurons contribute only E terminals to their targets. Many of the descending neurons originating in the brain also have E terminals.

Neurons that provide E synaptic terminals are called *excitatory* (E) *neurons*. Most of the large neurons (macroneurons) that connect the CNS with the periphery or communicate between the major parts of the CNS are of the E type. However, the I terminals, which are crucial for central synaptic integration, are often provided by the small *inhibitory neurons* (= short axon neurons, interneurons, microneurons). Thus, if a sensory fiber from the periphery or a descending motor fiber from the brain needs to inhibit a spinal motor neuron, it must first excite the I type interneurons, which in turn can inhibit the motor neuron via their I synapses. In a mature animal or human, all the E and I terminals to a motor neuron are per-

manently in place; only the pattern of nervous activity or the degree of sensory or descending motor input determines which terminals (E or I) will be used.

Activation of each synaptic terminal produces a slow, weak, and graded *synaptic potential*; *EPSP* in the E terminal and *IPSP* in the I terminal (see captions in the illustration and plates 16, 17 for details of the electrical and ionic aspects of EPSP and IPSP). Because these terminals are impinging on a simple neuron, the EPSPs and IPSPs can add up (summation, see below) algebraically. When E synapses are more active than I synapses or more E synapses are active, excitation will prevail; otherwise, inhibition will prevail. If the two types of synapses are equally active, their effects will be cancelled out. Summation of synaptic interaction is one basis for neuronal integration (i.e., the balance of the excitatory and inhibitory input to the receptive zone of a postsynaptic neuron determines if the axon will fire nerve impulses).

SPATIAL AND TEMPORAL SUMMATION. In the neuromuscular junction, a single action potential causes the release of enough neurotransmitter (acetylcholine) to cause a full endplate potential and a consequent muscle twitch (plate 17). In the central synapses, however, the energy of a single EPSP or IPSP is usually insufficient to activate the postsynaptic neuron. To increase efficiency, the level of excitation or inhibition at the postsynaptic surface must increase. The algebraic addition of synaptic potentials at the receptive surface of a postsynaptic neuron is called *synaptic summation*. There are two types of summation: *spatial* and *temporal*. Spatial summation occurs when the presynaptic input is summated across the different synaptic sites, impinging on the same postsynaptic neuron. Spatial summation may involve both E and I types of synapses, from one, two, or more presynaptic neurons. Temporal summation (i.e., that of individual synaptic potentials in *time*) may involve a single synapse. Here an increase in the *frequency* of discharge (number of impulses/unit of time) will heighten the effectiveness of the synapse. In the E synapse, this increases the probability of discharge; in the I synapse, it decreases it. Opportunities for spatial summation are created by increasing either the number of terminal connections of a single presynaptic neuron to the same postsynaptic one (as in the central terminal branchings of the sensory afferents) or the number of active presynaptic neurons (recruitment). In temporal summation, high-frequency stimulation of the presynaptic neuron results in the accumulation of enough E or I currents to cause excitation or inhibition of the target neuron when single-impulse or low-frequency stimulation is insufficient. *Convergence* of several neurons of the same type on a single postsynaptic neuron is another device to create opportunities for spatial and temporal summation.

CN: Use very light colors for A and B.
1. Begin with the excitatory neuron (A) and color its fiber from the brain down the spinal cord in the upper left drawing. Then color its enlargement along with an inhibitory neuron (B). The afferent (sensory) and efferent (motor) fibers are colored (A) since they are also excitatory. Color the typical excitatory and inhibitory synaptic ending of each.

neuron in the large illustration. Include the synaptic potential charts (on each side of illustration).
2. Color convergence and divergence next.
3. Color the spatial and temporal summation with the accompanying synaptic potential chart. Note that in both cases the summations are shown for excitatory potentials. They can also occur in inhibitory systems.

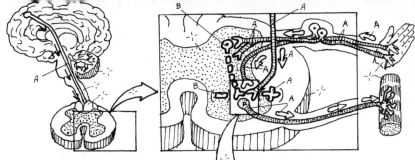

Neurons interact with other neurons through excitatory (E) and inhibitory (I) connections (synapses). E synapses activate and I synapses deactivate the postsynaptic neuron. The action of each neuron (exerted by its output synapses) is either E or I. However, each neuron can receive both E and I synapses (input) from other neurons. The large neurons, projecting from one CNS area to another, are generally of the E type. I neurons are smaller and have short axons. I neurons often function as "interneurons" because they provide local inhibitory connections between the large E neurons.

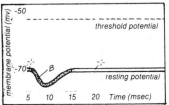

INHIBITORY NEURON SYNAPSE (IPSP)
(Inhibitory Postsynaptic Potential)

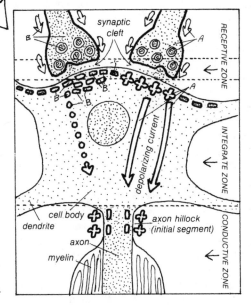

The release of the I neurotransmitter at the I synapse causes inflow of Cl⁻ or outflow of K⁺, creating a localized "hyperpolarization" of the postsynaptic membrane (i.e., an IPSP). The increased negative charges (determined by the strength of the IPSPs) "prevent" flow of depolarizing current toward the axon, decreasing the probability of axon depolarization and discharge.

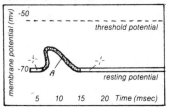

EXCITATORY NEURON SYNAPSE (EPSP)
(Excitatory Postsynaptic Potential)

The release of E neurotransmitter at the E synapse causes inflow of positively charged ions (Na⁺), creating localized "depolarization" (i.e., EPSP) of the postsynaptic membrane. Current flows from this area to the resting (polarized) initial segment of the axon. The strength of this current depends on the strength of EPSPs. The axon will discharge nerve impulses if the depolarized current is at or above the axon firing threshold.

CONVERGENCE

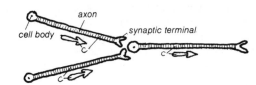

Convergence (C) occurs when a neuron receives synaptic input from several other neurons. Divergence (D) occurs when the synaptic output from one neuron is distributed to more than one neuron (by branching of the axon). C and D may occur in both E and I neurons. C and D are fundamental in the physiology of neuronal networks, because they provide the structural basis of such important synaptic phenomena as spatial summation (see below), facilitation, and occulsion.

DIVERGENCE

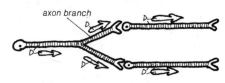

SUMMATION OF SYNAPTIC POTENTIALS
SPATIAL

ONE UNIT ACTIVE (weak signal) SEVERAL UNITS ACTIVE (strong signal)

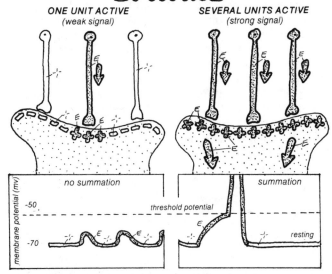

A single EPSP is usually too weak to cause activation of a neuron. One solution is to increase the number of input sites (i.e., activation of more presynaptic units). The EPSPs from all the active sites of the postsynaptic neuron now add up (spatial summation), creating the threshold current for axon discharge. Spatial summation occurs in inhibitory synapses as well, with the opposite effect.

TEMPORAL

LOW IMPULSE FREQUENCY (weak signal) HIGH IMPULSE FREQUENCY (strong signal)

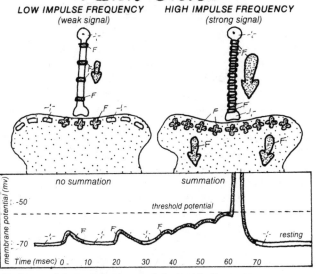

Another way to increase the strength of signals at the E synapses is by summating them in "time", (i.e., by increasing the frequency [number/sec.] of impulses arriving at the same synapse). Thus, in "temporal" summation, successive input impulses cause a "rapid" buildup of positive charges, leading to threshold excitation of the axon. Temporal and spatial summation can occur together as well. Temporal summation also occurs at the I synapses, with the opposite effect.

TYPES OF SENSORY RECEPTORS

The *sensory receptors* are highly specialized cells or organs by which the nervous system detects the presence of and changes in the different *forms of energy* in the external and internal environments. The sensory receptors transform these various forms of energy into a unitary language (i.e., *action potentials*, which are then sent to the CNS). Each sensory receptor is equipped with parts that confer its ability to detect the stimulus and to *transduce* (translate) the physical energy into nerve signals. For example, the skin's Pacinian corpuscle is sensitive to indentation in the skin, which is detected by the corpuscle's capsule and transmitted as waves of mechanical deformation to the nerve ending in the corpuscle core. The nerve ending transduces the pressure into an electrical depolarization, which activates the attached sensory nerve.

The physical world around us contains numerous forms of energy, not all of which we are able to detect. The detectable ones are classified into the categories of *mechanical, chemical*, thermal, and *photic* (light) energies. The body may have one or more types of sensory receptors to deal with any one kind of these energy forms. Some receptors, like those in the skin, are modified dendrites. In the eye's photoreceptors, much of the cell is modified for detection and transduction of light rays. Some receptor cells (such as the skin and smell receptors) act as independent single sensory units. In other cases, the receptors are housed as organized masses of cells within a sensory organ (such as the eye's retina). The structural integrity of the retina as a whole is essential for form perception and other spatial functions.

Based on the energy form to which they respond, sensory receptors are classified as *mechanoreceptors, chemoreceptors, thermoreceptors, nociceptors,* and *photoreceptors.* Most of these are described further in the appropriate plates along with the sense they serve. Here a general description of the different functional categories will be given.

MECHANORECEPTORS. The *mechanoreceptors* make up the most diverse group of sensory receptors. Found in the skin, muscles, joints and the visceral organs, they are sensitive to mechanical deformation of the tissue and cell membranes. This deformation can arise in various ways, including indentation, stretch, and hair movement. The *skin receptors* include the largest variety of mechanoreceptors. Many of the sensory nerve endings encapsulated in a fibrous (connective tissue) covering are believed to be mechanoreceptors.

Opinions differ regarding the types of cutaneous (skin) mechanoreceptors that carry out the various modalities of skin sensation. Apparently, *light* (fine) *touch* is detected by the superficially located receptors, such as the *Meissner's corpuscle, Merkel's disk*, and the nerve plexi found around the roots of skin hairs, *hair root plexi. Crude touch* and *pressure* may be detected by the deeper receptors, such as the *Krause's endbulb, Ruffini's ending*, and the *Pacinian corpuscle.*

Changes in the *muscle length/tension* are detected by stretch receptors in the *muscle spindle*; changes in the *tendon length/tension* are detected by the *Golgi tendon organ.* Specialized receptors in the joints signal changes in the joint or limb *displacement* and *position* in space (*joint receptors, kinesthetic receptors*).

More specialized mechanical receptors containing *hair cells* (cells with a modified cilia) are found in the inner ear. Movement of the ciliary hair deforms the cell membrane, activating the hair cell. These hair cells are found in the *cochlea* (hearing organ), where they respond to mechanical waves generated by sounds, and in the *vestibular apparatus* (balance organ), where they respond to fluid mechanical waves caused by head movements. Walls of many visceral organs contain stretch receptors that signal distension. The *baroreceptors* in the walls of certain arteries (carotid and aorta) are well known examples. These are sensitive to changes in the distension of the arterial wall caused by changes in blood pressure.

THERMORECEPTORS AND NOCICEPTORS. The sensations of *warmth* and *cold* are conveyed by thermoreceptors, which are probably *free nerve endings* in the skin. Other specialized types of free nerve endings (*nociceptors*) respond to stimuli that cause *pain*. Certain neurons in the brain *hypothalamus* are also sensitive to changes in *blood temperature.*

CHEMORECEPTORS. Numerous sensory stimuli of a chemical nature are detected by a variety of *chemoreceptors.* Thus, *olfactory receptor cells* in the nose detect environmental odors. *Taste receptor cells* in the tongue's taste buds detect certain substances in food that may be advantageous (sweets, salts) or harmful (bitter substances) for the body. Other types of internal chemoreceptors detect changes in the physiologically important blood substances. For example, sensor cells in the *carotid* and *aortic bodies* detect oxygen, certain hypothalamic *osmoreceptors* regulate blood *osmolarity* by detecting blood *sodium* levels, and other hypothalamic receptors (*glucoreceptors*) detect blood *glucose* levels.

PHOTORECEPTORS. The retina, the nervous part of the eye, contains *photoreceptor* cells (rods and cones) that can detect *light* energy. The visible light rays make up a specific band in the spectrum of electromagnetic wave energy *Rods*, being more abundant and more sensitive, serve in peripheral and night vision; *cones* work only in daylight and can detect red, blue, and green colors (specific wavelengths of light).

CN: Because of the many colors needed on this plate, you may have to use the same color for different letter labels if you run short.
1. Color the three upper panels from left to right.
2. Color the various mechanoreceptors, most of which appear in the block of skin to the right. The illustrations and titles depicting the form of stimulus (on the left) are not to be colored. Do color, where present, the nerve carrying the action potential (C).
3. Color the categories of receptors. In some situations, the receptor is so small that an arrow pointing to it is colored instead.

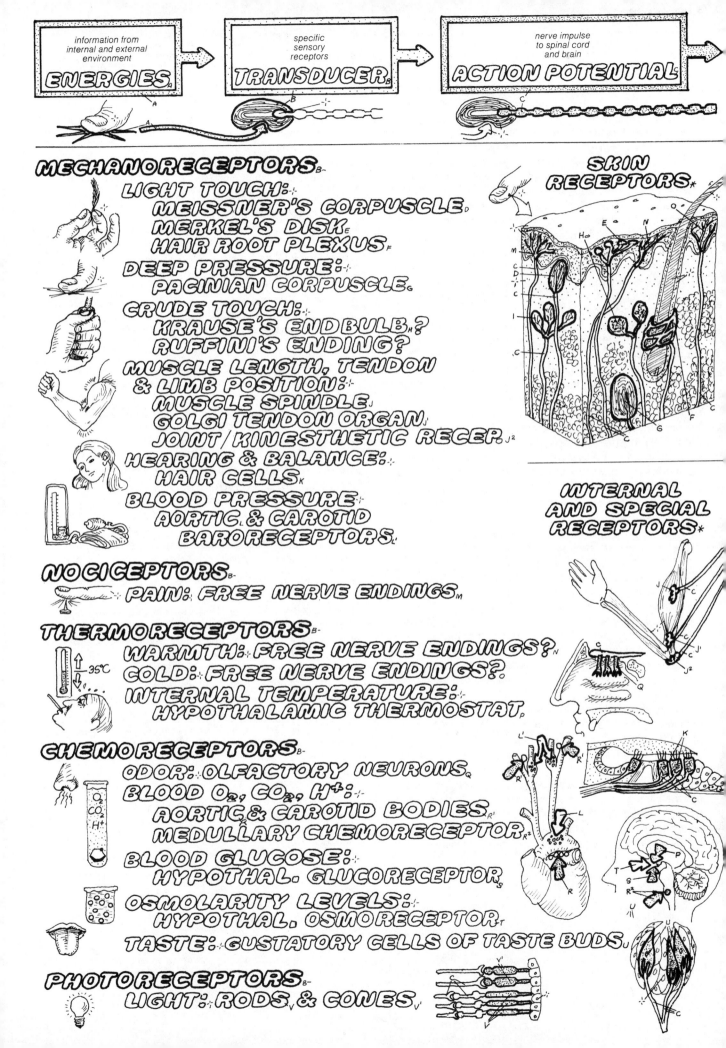

ENERGIES — information from internal and external environment

TRANSDUCER — specific sensory receptors

ACTION POTENTIAL — nerve impulse to spinal cord and brain

MECHANORECEPTORS

LIGHT TOUCH:
MEISSNER'S CORPUSCLE
MERKEL'S DISK
HAIR ROOT PLEXUS

DEEP PRESSURE:
PACINIAN CORPUSCLE

CRUDE TOUCH:
KRAUSE'S END BULB?
RUFFINI'S ENDING?

MUSCLE LENGTH, TENDON & LIMB POSITION:
MUSCLE SPINDLE
GOLGI TENDON ORGAN
JOINT/KINESTHETIC RECEP.

HEARING & BALANCE:
HAIR CELLS

BLOOD PRESSURE
AORTIC & CAROTID BARORECEPTORS

NOCICEPTORS
PAIN: FREE NERVE ENDINGS

THERMORECEPTORS
WARMTH: FREE NERVE ENDINGS?
COLD: FREE NERVE ENDINGS?
INTERNAL TEMPERATURE:
HYPOTHALAMIC THERMOSTAT

CHEMORECEPTORS
ODOR: OLFACTORY NEURONS
BLOOD O_2, CO_2, H^+:
AORTIC & CAROTID BODIES
MEDULLARY CHEMORECEPTOR
BLOOD GLUCOSE:
HYPOTHAL. GLUCORECEPTOR
OSMOLARITY LEVELS:
HYPOTHAL. OSMORECEPTOR
TASTE: GUSTATORY CELLS OF TASTE BUDS

PHOTORECEPTORS
LIGHT: RODS & CONES

SKIN RECEPTORS

INTERNAL AND SPECIAL RECEPTORS

RECEPTORS AND SENSORY TRANSDUCTION

One of the basic questions in neurophysiology is how the sensory receptors transduce (translate) the energy in a stimulus (be it physical, like pressure, or chemical, like odors) into the common language of nervous system communication (i.e., the nerve impulse). This phenomenon is called *sensory transduction*. In this plate, we consider the mechanism for this ability as well as certain other functional properties of sensory receptors. We focus on the skin receptors, which have been extensively studied.

SENSORY TRANSDUCTION. The Pacinian corpuscles are sensory receptor organs found in the deep layers of the skin. They are believed to be mechanical receptors sensitive to such stimuli as *pressure*. The mechanism by which mechanical pressure may produce nerve signals in the sensory fiber has been extensively studied in these receptors. Each Pacinian corpuscle consists of the ending of a sensory fiber branch wrapped in a fibrous connective tissue capsule. The nerve ending has the structural and functional properties of neuronal dendrites and is the true transducer in this system, as will become clear below. The nerve ending, as it emerges from the capsule, is continuous with a myelinated fiber containing nodes of Ranvier (see plate 15) and having the functional properties of axons. The connective tissue capsule is not absolutely necessary for transduction by the receptor, but it helps by adequately spreading the mechanical stimulus and contributing to receptor adaptation (see below).

When a mechanical pressure is applied to the skin surface, the skin deforms, resulting in stimulation of the underlying Pacinian corpuscles. Because the corpuscles are laid deep in the skin, only very strong deformations (stimuli) can activate them. Thus, the Pacinian corpuscles respond to "pressure" type of stimuli of high enough strength. The numerous layers of connective tissue fibers making up the capsule act as cushions, transmitting and spreading the mechanical waves to all parts of the nerve ending in the corpuscle core. Remember that the ending membrane, like all neuronal membranes, shows a resting membrane potential, with the inside being negatively charged, and that sodium ions are present at rest on the outside in high concentrations (see plates 11, 13). In response to the deformation wave, the nerve ending membrane structure is temporarily altered. This increases the membrane permeability to the positively charged sodium ions, which move into the ending interior, depolarizing the membrane. This depolarization is called the *receptor potential*.

The receptor potential, also called the *generator potential* because it generates the nerve impulse in the adjoining axon, belongs to a family of membrane potentials called *graded potentials*. In contrast to the *action potentials*, which occur only in axons and obey the all-or-none law (see plate 14), the receptor potentials, like all graded potentials, do not show spikes, and their amplitude varies directly with the strength of the stimulus and the amount of sodium entering the ending. Thus, the amplitude (strength) of the receptor potential increases directly with the increase in the stimulus strength, up to a limit, of course.

An electrotonic current will flow between the nerve ending and the adjacent first node of Ranvier for as long as the receptor potential lasts. This is because the inside of the activated nerve ending acts as a positive pole while the inside of the first node, being at rest, acts as negative pole. The strength of this current is also proportional to the amplitude of the receptor potential. Once the current reaches the node's excitation threshold, the node produces an action potential (nerve impulse) that is conducted along the sensory fiber.

FREQUENCY CODING OF STIMULUS INTENSITY. The first node continues to fire action potentials as long as the receptor potential lasts. The amplitude of the receptor potential, being proportional to the *stimulus strength*, also determines the *frequency* (number/sec.) of the nerve impulse. This relationship forms the basis of the frequency coding of sensory messages (i.e., the stronger the stimulus, the higher the impulse frequency in the sensory nerve). This is how the brain detects the changes in stimulus intensity. If the stimulus intensity continues to increase, causing more extensive skin indentation, but the maximal discharge frequency of the nerve branch is reached, then the adjacent corpuscles are activated, first those belonging to the same sensory neuron and then those belonging to the neighboring sensory neurons. This is called *receptor recruitment*.

RECEPTOR ADAPTATION. Many receptors decrease their firing rate while the stimulus continues to be applied at the same strength. These receptors are called *rapidly adapting*. In contrast, the *slowly adapting* receptors continue to fire throughout the stimulus duration at the same or somewhat lower rates. The receptors for fine touch and pressure (e.g., hair root plexi and Pacinian corpuscles) are rapidly adapting types; the joint and muscle mechanoreceptors, serving the kinesthetic and proprioceptive senses, are slowly adapting types. Pain receptors, for obvious survival-related reasons, show little adaptation. Thanks to receptor adaptation, we are normally less aware of the presence of clothing on our skin, although we feel it when we first put it on and again briefly every time we move. The mechanism of receptor adaptation is poorly understood. In the encapsulated receptors, the elastic properties of the connective tissue fibers engulfing the nerve ending (e.g., in the capsule of the Pacinian corpuscle) may be partly responsible for the phenomenon of receptor adaptation.

CN: Letter label A should receive the same color as was given the transducer (B) on the previous page. Action potential (C) on both plates should get the same color. Use dark colors for D and E.
1. Begin with the upper diagrams and the three stages of sensory transduction. Note that, in stage (1), the symbols of positive and negative charges are colored over with the background colors, whereas in steps 2 and 3, they may or may not be colored. If they are, the positive charge receives the Na+ permeability color (D).
2. Color the lower chart, color the vertical column of short bars on the far left that represents the frequency of action potentials, and note that, in the five lengthy arrows representing different receptors, light lines have been drawn to suggest the frequency of action potentials (you will have to color over these lines when you fill in the arrows).

TRANSDUCTION & RECEPTOR POTENTIAL *
PACINIAN CORPUSCLE A
NAKED (FREE) NERVE ENDING (DENDRITE) B
MYELINATED SENSORY NERVE FIBER (AXON) C

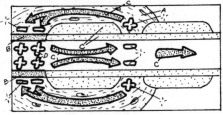

Pacinian corpuscles, found in deep layers of skin, transduce (convert) pressure (mechanical) stimuli to electrical nerve signals. Each corpuscle consists of a sensory nerve ending attached to a myelinated axon and surrounded by a connective tissue capsule. Only the nerve ending acting as the dendrite (receptive element) of the sensory neuron is the real transducer, converting mechanical energy to nerve signals.

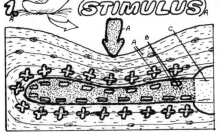

1. PRESSURE (STIMULUS) A

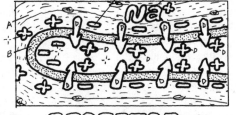

2. Na⁺ PERMEABILITY D
RECEPTOR POTENTIAL B
at the nerve ending

3. ACTION POTENTIAL C
at the first node
NERVE IMPULSE C
conducted to CNS

Pressure to the skin stretches the fiber layers of the capsule, deforming the membranes of the nerve ending (1). The perturbation increases membrane permeability to Na^+; Na^+ flows in, depolarizing the membrane (2), creating a graded potential (receptor potential or generator potential) across the membrane of the nerve ending. This potential creates a current between the nerve ending region and the first node of Ranvier in the axon (3). When strong enough (i.e., at threshold), the current will evoke an action potential (nerve impulse) in the node to be conducted to the CNS.

STIMULUS STRENGTH A AND FREQUENCY OF NERVE IMPULSE C

AMPLITUDE OF RECEPTOR POTENTIAL B
THRESHOLD POTENTIAL E

membrane potential (mv)

+30
-60
-90

The amplitude (size, strength) of the receptor potential (in millivolts) increases in proportion to the stimulus strength. At threshold intensity, the axon will fire an action potential. Further increases in amplitude will increase the frequency of discharge of nerve impulses by axon (sensory fiber). Thus, the function of Pacinian corpuscles is to convert a continuum of mechanical stimuli of differing energies into a modular frequency code. Signal transmission by a frequency code is a fundamental property of nerve fibers in the sensory and motor systems.

SENSORY RECEPTOR ADAPTATION *

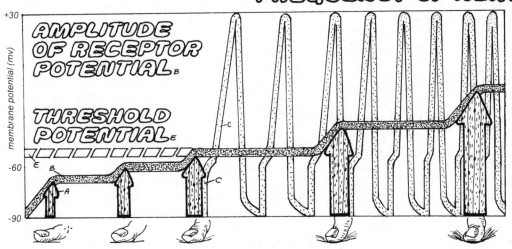

frequency of action potentials

PAIN
JOINT POSITION
MUSCLE STRETCH
SLOW ADAPTING

PRESSURE F
TOUCH G
RAPIDLY ADAPTING *

Seconds 1 2 3 4 5 6 7 8

Receptor adaptation is the phenomenon whereby sensory receptors decrease or stop impulse production even when the stimulus is sustained. Some receptors adapt rapidly, such as the fine touch and pressure receptors (Pacinian corpuscles, Merkel's disks, hair root plexus). This property helps to increase tactile discrimination. Others adapt slowly, such as the pain receptors (nociceptors) in the skin and the proprioceptors of the muscle and joints. Pain usually signals discomfort and danger and should be attended to, hence the need for a persistent signal.

SUSTAINED STIMULUS A

SENSORY UNITS, RECEPTOR FIELDS, & TACTILE DISCRIMINATION

Once sensory stimuli are transduced into nerve signals, they are communicated to the CNS along the *afferent* nerves by the fibers of the *primary sensory neuron*. The *cell bodies* of these neurons are located in the *dorsal root ganglion* near the spinal cord. The sensory neurons are pseudounipolar cells with a bifurcated nerve fiber. In the past, the peripheral segment of the nerve fiber bringing sensory signals toward the cell body was called the dendrite and the centrally directed segment, the axon. However, modern views hold that the sensory neuron's true dendrite is only the nerve ending (i.e., the *sensory receptor*). The remaining fiber segments of the sensory neuron all show the structural and functional properties of an *axon*.

A primary sensory neuron, its fibers and all the peripheral *end branchings*, and the synaptic central terminals constitute a *sensory unit*. Thus, the body periphery is served by numerous independent sensory units bringing in the somatic sensory messages to the CNS.

RECEPTIVE FIELD. The area of the skin (or any other body part) served by a sensory unit is called the *receptive field* of that unit. Because the peripheral branchings of sensory fibers have a radial orientation, the receptive fields of the sensory units have a circular shape. When the peripheral end branches of two neighboring sensory units are far apart, the stimuli impinging on the receptive field of one unit will not activate the neighboring unit. If the branches of two neighboring units commonly innervate some areas of a target, the receptive fields will overlap. Overlapping receptive fields provide a basis for certain neural analysis and integration of the sensory input because the stimuli impinging on the receptive field of one unit will also elicit impulses in the neighboring units, albeit to different degrees. The CNS neurons sense this differential activation, forming the basis of tactile discriminaton.

TACTILE SENSITIVITY AND DISCRIMINATION. The human skin is endowed with remarkable *tactile* abilities. However, not all of its parts show equal capacities. The *fingertips*, the *lips*, the *genitalia*, and the *tongue tip* are the most sensitive areas. The tactile sensory skills may be divided into two categories, *intensity discrimination* and *spatial discrimination*. Intensity discrimination (i.e., sensitivity) refers to the ability to judge *stimulus strength*; spatial discrimination involves the ability to differentiate between the *locations* of *point stimuli*.

To test intensity discrimination, a pointed probe is pushed gradually onto the skin surface until the subject reports sensing it. This is the point of *tactile threshold*. At this point, the depth of the skin dip formed by the probe is measured using the probe's scale. This depth gives a quantitative reading of tactile sensitivity. The tongue tip is the most

sensitive body area in this regard, followed by the fingertips, in which a mere 6 micron dip can be detected. In the palms, the threshold is four times higher; on the back of the hands, trunk, and legs, it is ten to twenty times higher. Note that high threshold means low sensitivity. Therefore, the highest tactile sensitivities are associated with such body parts as the fingertips and tip of the tongue, which are actively used to sample and investigate the environment.

The neural basis of differential tactile sensitivity lies in the number of sensory branches and sensory units per unit area of the skin. The fingertips contain many more units than the back. Therefore, stimuli of similar strength (causing the same amount of skin indentation) will activate more sensory fibers or units on the fingertip than on the back. The *convergence* (see plate 82) of the primary afferents from the fingertip units onto the spinal sensory relay cells is also higher, so brain cells can be activated by relatively weak stimuli applied to the fingertip, but stimuli of similar strength applied on the back will be below the brain's detection level.

The fingertips also show the highest *spatial discrimination* ability. This is assessed by the *two-point discriminiation test* in which the tips of caliper arms are placed on the skin and the distance between them is reduced until the subject reports sensing only one point. This minimum distance is an index of spatial discrimination: the less the distance, the higher the discrimination. This *minimum separable distance* is narrowest in the fingertips (1-2 mm) and widest in the back (up to 70 mm).

The neural basis for spatial discrimination lies in the size and degree of receptive fields' overlap. In the fingertips, the receptive fields are small and the degree of overlap high; the opposite is true for the back or legs. Therefore, in the fingertips, even two closely applied stimuli are likely to activate two different sensory units as one point impinges on the receptive field of one unit while the second point impinges on another. So long as the central neuron receives messages from two separate units, the two points can be discriminated from each other. If both point stimuli fall in the receptive field of one unit, only one point will be deciphered. The high receptive field overlap in the fingertips also permits other discriminative abilities. For example, if a tactile stimulus impinges on the receptive field center of one unit, it activates that unit maximally. But due to receptive field overlap, the same stimulus activates the receptive field periphery of the neighboring unit. The activation of the second unit is, however, weaker. The differential rate of activity between the two neighboring units forms the basis of *lateral inhibition* (discussed in the bottom diagram on this plate), a phenomenon serving to sharpen contrast and enhance discrimination.

CN: Use dark colors for A, G, and I.
1. Begin with the sensory unit at the top of the plate.
2. Color the illustrations of receptive fields. Note the many vertical bars (C¹) along the axon. These represent the number of impulses traveling along the axon. The cell bodies have been omitted for simplification.

3. Color the examples representing the two forms of tactile discrimination: spatial and intensity. Color the points of the calipers used to make the measurements in these demonstrations.
4. Color the explanation of lateral inhibition. Note the broken line of the inhibitory neuron (I) in the spinal cord and the dotted lines representing inhibited ascending sensory pathways.

SOMATIC (AFFERENT) SENSORY UNIT (SU)

SENSORY RECEPTORA
END BRANCHINGB
SENSORY FIBER (AXON)C
CELL BODYD
CENTRAL TERMINALE

Primary somatic afferent (sensory) neurons send a single long peripheral process that traverses within a peripheral nerve and branches out near or within its target organ (skin, joints, muscle spindles, tendons, teeth, tongue, etc.). The central process enters the spinal cord or brain stem to synapse with central neurons. A single sensory neuron with all its peripheral end-branchings and central terminals is called a sensory unit (SU). The specific peripheral area served by a SU is called the receptive field (RF).

RECEPTIVE FIELDS (RF)*
STIMULUSG
NERVE IMPULSEC'

Receptive fields (RF) of neighboring SUs in the skin may be separate or overlapping. Stimulation within each separate RF evokes impulse activity only in the corresponding SU; stimuli falling within the over-lapping RFs will evoke activity in all the participating units. However, as shown at right, when a tactile stimulus activates more peripheral branches of one unit than another, then impulse activity in that unit will be correspondingly higher than activity in the neighboring units.

PATHWAYF

ascending sensory pathway

spinal cord

firing

quiet

SEPARATE RF*

OVERLAPPING RF*

low activity

high activity

low activity

TACTILE DISCRIMINATION

The tactile sensitivity of a particular part of skin is directly pro-portional to the number of SUs innervating that area as well as to the degree of overlap within the RFs of these SUs. Sensitivity is usually inversely related to the size of the RF. Thus, the fingertips, receiving a high number of SUs, with small overlapping RFs, are far more more sensitive than the skin on the back.

Spatial discrimination: In the two-point discrimination test, the spatial discriminative ability of the skin is determined by measuring the minimum separable distance between two tactile point stimuli. The tongue, lips, and fingertips rank high in this ability (1-3 mm). The back of the hands, the back and legs rank low (50-100 mm).

Intensity discrimination: Sensitive areas are also better able to discriminate differences in the intensity of tactile stimuli. Thus, an indentation of 6 microns on the fingertip is sufficient to evoke a sensation. This threshold is 4 times higher in the palm.

SPATIAL (2-POINT) DISCRIMIN.H

*1-2 mm*H

*30-70 mm*H

FINGER (MANY UNITS)H

units overlap

BACK (FEW UNITS)H

no overlapping

INTENSITY DISCRIMIN.G'

6μ

24μ

ONE-POINT DISCRIMINATION	TWO-POINT DISCRIMINATION	ONE-POINT DISCRIMINATION	TWO-POINT DISCRIMINATION
	less than 1 mm	*more than 1 mm*	*less than 30 mm*	*more than 70 mm*

LATERAL INHIBITION*
INHIBITORY NEURONI

In the SUs with overlapping RF, when activity in one SU is higher than in the neighboring units (e.g., by recruitment of more peri-pheral branches of the main SU), signal transmission from the less active neighboring units to CNS neurons are suppressed by inhibiting their synapses. This phenomenon is called lateral inhibition and serves to sharpen contrast discrimination.

spinal cord

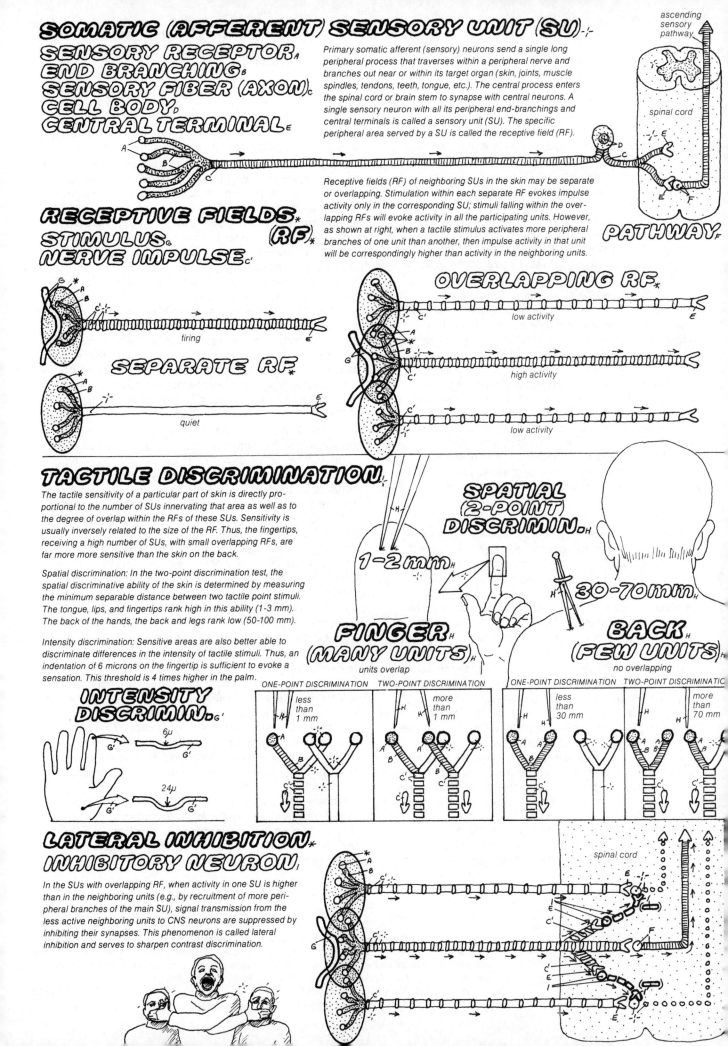

SOMATIC SENSORY PATHWAYS

Although we know that each sensory receptor type is designed to respond mainly to one type of stimulus, the question remains as to how the brain differentiates between these various modalities (i.e., how cold is sensed apart from heat and pain or pressure from touch). This is particularly problematic because all sensory nerve fibers communicate with the brain using the same language, that of nerve impulses. To answer, we consider here the functional segregation of somatic sensory modalities and the channeling of sensory signals along the various pathways and synaptic stations of the spinal cord and brain.

FUNCTIONAL DIFFERENTIATION IN THE SENSORY NERVE. If we dissect a single nerve fiber from the thousands found in a sensory nerve and record its activity, we find that the largest increase in its activity occurs when a particular type of stimulus is applied. This "appropriate" stimulus may be touch, pain, thermal, etc. If we were to stimulate that fiber electrically the subject would probably report only the sensation associated with that particular type of stimulus. The generalization that the fibers of a sensory nerve each conduct signals concerning only one type of sensory modality (presumably because it is connected to only one particular type of receptor) has been named the "doctrine of specific nerve energies." This and the specificity of sensory receptors may provide clues about how sensory modalities are differentiated in the nervous system.

THE SPINOTHALAMIC PATHWAY. Generally, crude tactile, pain (nociceptive), and temperature (thermal) signals are conveyed by unmyelinated, small-diameter fibers (type C). The cell bodies of these fibers are small, and their central terminals in the spinal cord release peptide transmitters (e.g., substance-P by the pain fibers) (see also plate 88). Signals relaying fine touch and pressure as well as proprioceptive modalities (from joints and muscles) are carried by fast-conducting, large, myelinated fibers (type A) having large cell bodies.

Upon entry to the spinal cord, the various sensory fibers become segregated into two categories. Thin fibers carrying pain, temperature, and crude tactile sensations, collectively referred to as the nondiscriminative modalities, terminate in the dorsal horn of the spinal cord, where they synapse with the secondary relay cells, the fibers of which decussate (cross over) and enter the white matter to ascend in the spinothalamic pathways toward the brain. This pathway has two clear divisions: the pain and temperature modalities are segregated in a lateral division, and the crude tactile fibers are bundled in an anterior (ventral) division. The spinothalamic fibers terminate in the thalamus, the most important subcortical sensory relay/integration center. The spinothalamic pathway and its related modalities represent a basic, primitive somatic sensory system seen in all vertebrates.

DORSAL COLUMNS AND DISCRIMINATIVE MODALITIES. Phylogenetically, a more recent somatic sensory system, well developed in primates and humans, is represented by the pathway taken by the large myelinated fibers carrying the modalities of fine touch and pressure and proprioception (discriminative tactile). These fibers enter the spinal cord but do not terminate or synapse in the dorsal horn. Instead they ascend, without crossing, up the sensory pathways of the posterior (dorsal) columns (funiculi), to end in the medulla, where they make their first synapse. The axons of the secondary sensory cell arise here, cross over (decussation), and ascend in the medial lemniscus to end in the same area of the thalamus where the spinothalamic fibers end. The dorsal column-lemniscal system is called the discriminative pathway because such important sensory capacities as precise localization, two-point discrimination, fine touch, vibration, stereognosis (object recognition by manipulation), and limb/body position in space are all conveyed by this system.

THALAMIC RADIATION TO SENSORY CORTEX. From the thalamus arise the third-order neurons, forming the fibers of somatic radiation, which project to the sensory cortex (primary somatic sensory cortex) located in the postcentral gyrus. In all the relay stations, the dorsal horn, the medulla, and the thalamus, the sensory impulses are filtered and integrated so that the messages arriving in the sensory cortex have already undergone certain fine tuning. What happens in the cortex is the subject of plate 87. This fine tuning is in part controlled by the sensory cortex, which sends certain descending sensory control fibers to the subcortical relay stations to regulate the quality and quantity of the messages arriving in the cortex (feedback control circuits).

INPUT TO LOWER MOTOR CENTERS AND RETICULAR FORMATION. A major function of the somatic sensory afferents from the skin, joints, and muscles is to activate the spinal reflexes (plate 89). This is accomplished by collaterals or main branches of the primary afferents as they enter the spinal cord. These branches synapse with the spinal interneurons, which will in turn synapse with the spinal motor neurons to complete the motor reflex circuits. The nociceptive fibers carrying pain signals are of course the most important here due to their protective functions, but information from other modalities is also necessary for appropriate adjustment of the reflexes.

On their way to the brain, the ascending fibers send collateral branches to the midbrain motor centers to influence involuntary motor activity and to the centers in the reticular formation to influence sleep and wakefulness (plate 100), arousal and attention, and central inhibition of pain (plate 88).

CN: Use dark colors for F, G, H, and J.
1. Before coloring the nerves and pathways, color structures A-E along with their location on the human figure. Then begin with the block of skin in the lower left corner and color the various receptors and their sensory neurons (F-H) and follow them to the various ascending pathways (F¹-H¹)

beginning in the spinal cord. Color the rectangular enlargement of the dorsal horn.
2. After completing the pathways, color the decussation title (J) in the center, and the two arrows identifying the decussation sites.
3. Color the material on descending controls at the top of the page.

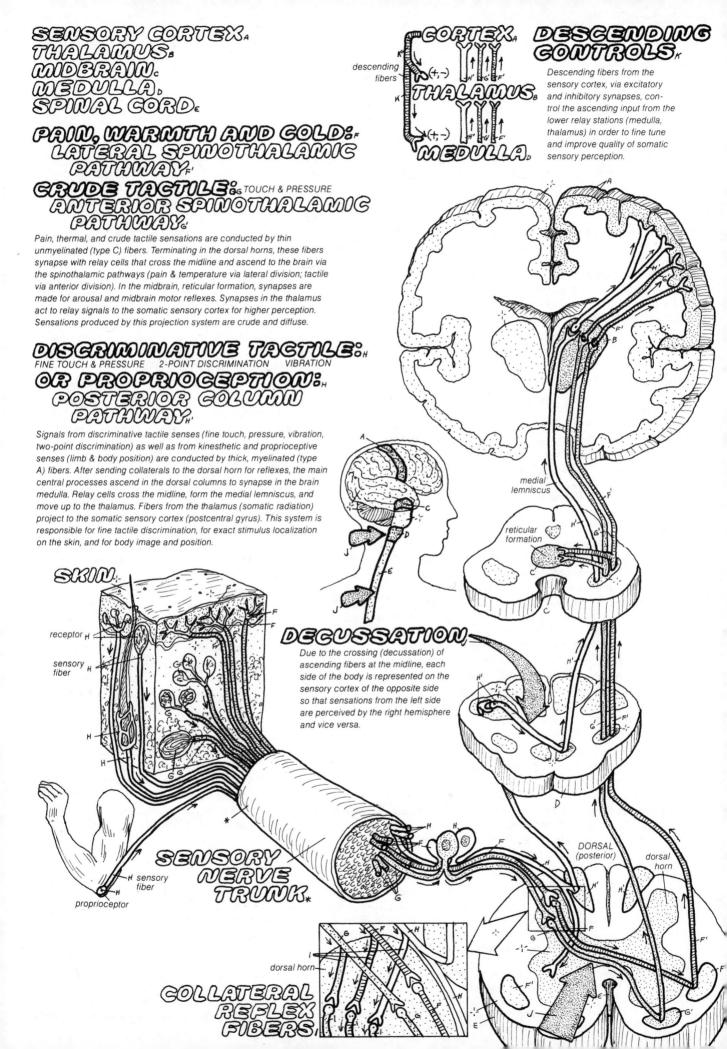

SENSORY CORTEX ᴀ
THALAMUS ʙ
MIDBRAIN ᴄ
MEDULLA ᴅ
SPINAL CORD ᴇ

PAIN, WARMTH AND COLD: ꜰ
LATERAL SPINOTHALAMIC PATHWAY ꜰ'
CRUDE TACTILE: ɢ TOUCH & PRESSURE
ANTERIOR SPINOTHALAMIC PATHWAY ɢ'

Pain, thermal, and crude tactile sensations are conducted by thin unmyelinated (type C) fibers. Terminating in the dorsal horns, these fibers synapse with relay cells that cross the midline and ascend to the brain via the spinothalamic pathways (pain & temperature via lateral division; tactile via anterior division). In the midbrain, reticular formation, synapses are made for arousal and midbrain motor reflexes. Synapses in the thalamus act to relay signals to the somatic sensory cortex for higher perception. Sensations produced by this projection system are crude and diffuse.

DISCRIMINATIVE TACTILE: ʜ
FINE TOUCH & PRESSURE   2-POINT DISCRIMINATION   VIBRATION
OR PROPRIOCEPTION: ʜ
POSTERIOR COLUMN PATHWAY ʜ'

Signals from discriminative tactile senses (fine touch, pressure, vibration, two-point discrimination) as well as from kinesthetic and proprioceptive senses (limb & body position) are conducted by thick, myelinated (type A) fibers. After sending collaterals to the dorsal horn for reflexes, the main central processes ascend in the dorsal columns to synapse in the brain medulla. Relay cells cross the midline, form the medial lemniscus, and move up to the thalamus. Fibers from the thalamus (somatic radiation) project to the somatic sensory cortex (postcentral gyrus). This system is responsible for fine tactile discrimination, for exact stimulus localization on the skin, and for body image and position.

DESCENDING CONTROLS ᴋ

Descending fibers from the sensory cortex, via excitatory and inhibitory synapses, control the ascending input from the lower relay stations (medulla, thalamus) in order to fine tune and improve quality of somatic sensory perception.

CORTEX ᴀ
descending fibers
THALAMUS ʙ
MEDULLA ᴅ

SKIN

receptor ʜ
sensory fiber ʜ

medial lemniscus
reticular formation

DECUSSATION

Due to the crossing (decussation) of ascending fibers at the midline, each side of the body is represented on the sensory cortex of the opposite side so that sensations from the left side are perceived by the right hemisphere and vice versa.

SENSORY NERVE TRUNK *

sensory fiber ʜ
proprioceptor

DORSAL (posterior)
dorsal horn

COLLATERAL REFLEX FIBERS

dorsal horn

ORGANIZATION & FUNCTIONS OF THE SENSORY CORTEX

In plate 86, we learned that different sensory pathways convey functionally distinct sensory modalities and that sensory pathways converge onto a part of the *thalamus* from which sensory signals radiate to the *primary somatic sensory cortex*. Here we consider how the sensory cortex detects the source and qualities of the various sensory stimuli. People who have suffered damage to their sensory cortex (e.g., from gunshot wounds or strokes) may be aware of the sensory stimuli (especially of pain and temperature), but they are very poor in discriminating the intensity of tactile stimuli and their exact source (spatial discrimination). Indeed, these individuals become completely disabled in *stereognosis* — the ability to recognize the shapes of objects by manipulation alone.

SOMATOTOPIC ORGANIZATION AND THE SENSORY HOMUNCULUS. The projection pathways anatomists studied revealed that each point on the skin surface is connected to a point on the sensory cortex. In addition, early neurophysiological studies showed that if one bends single hairs on an animal's skin and determines the point on the sensory cortex where the stimulus evokes a potential change and then connects these cortical points, one obtains a rather faithful representation of the body surfaces/parts (somatotopic map). This finding gave a clue as to how the brain can localize the stimulus source in the body. A similar knowledge in the human was not available.

During the 1940s, Wilder Penfield, the great Canadian neurosurgeon, began his functional exploration of the human cortex, including the sensory cortex. Working with conscious patients undergoing brain surgery (the brain has no pain receptors, and electrical stimulation of the cortex does not cause pain), he stimulated the sensory cortex with a mild current. The patients reported various tactile sensations in a body part (e.g., the toes, the fingers, or the back). Connecting these points, Penfield noted that the sensory cortex contained a representation of the body, which he termed the *sensory homunculus* (= little man). The legs are represented on the hidden medial portion of the postcentral gyrus, the trunk on the top, the arms, hands, and head on the larger exposed lateral surface. Due to the crossing of ascending projections, the body's left side is represented on the right hemisphere and the right side on the left hemisphere.

In contrast to the animals' somatotopic sensory map, which closely resembled the animal's figure, the human homunculus shows important distortions in two respects. First, the area for the hands is interposed between the areas for the head and the trunk. Second, the representation is not proportional to the size of the body part. Thus, the hands and the face have large representations, and the trunk and legs have

small ones. Within the hand area, each finger and the thumb has an independent representation, the largest being for the index finger. Within the face, the lips have large representation. Indeed, the extent of sensory cortex devoted to a body part is proportional to the part's density of innervation, tactile sensitivity, and prowess, not to its size.

CORTICAL STRUCTURE AND COLUMNAR ORGANIZATION. The sensory cortex consists of 6 layers parallel to the cortical surface and numbered from top to bottom. These layers are populated by small and large pyramidal neurons as well as stellate and fusiform neurons. Layer IV neurons receive the afferent sensory input; those in layers V and VI are output neurons projecting relay and feedback control signals to other CNS areas. Smaller neurons of layers II and III serve as local association neurons connecting neighboring cortical areas.

In the late 1950s, it was discovered that when a microelectrode for recording the activities of single neurons was inserted in the cortex perpendicular to the cortical surface, all the neurons along the electrode's path had a uniform receptive field and responded to the same tactile modality. When the electrode was inserted obliquely, however, different groups of neurons were encountered. At first they responded to different modalities in the same receptive fields; then if the electrode was moved far enough, the receptive field changed as well. It was therefore concluded that functionally the cortex neurons are organized in cylinders or *columns*. Each column is about 3-5 mm long, less than 1 mm wide, and contains about 100,000 neurons. Each group of columns deals with a particular part in the periphery (i.e., has similar receptive fields), and each single column deals with only one modality. Thus, in a group of columns that corresponds to an area in a finger, one column deals with proprioception, the next with touch, and another with pressure. There are no separate columns for temperature and pain, these modalities being served by a few cells in some tactile columns.

Within each column, some cells imitate the behavior of the sensory receptors (e.g., increasing their firing rate with increasing stimulus intensity); these cells are called *simple cells*. Other cells increase their activity only when a stimulus moves across the skin in a particular direction; these cells are called *complex cells*. The response pattern of the neurons in the cortical columns is called *feature detection*. Through feature detection in the primary and association sensory cortex, the complex world of sensory stimuli is molded into a perception pattern. The columnar organization has been extensively studied in the visual cortex (see plate 94).

CN: Use a dark color for W (the neurons in the cortex). Ideally, you should have 28 colors available for this plate (A-Z, 1, 2); in all likelihood you won't and will have to use the same color for more than one letter label.
1. Begin with material at the top, noting that the two structures representing the thalamus (C) are buried deep below the thin cortex layer (A).
2. Color the representation of body parts on the sensory cortex. Also color the homunculus which is a proportional drawing based upon the amount of cortex devoted to a

particular part. Note that the last two sections of the cortex are colored gray because they represent the pharynx and intra-abdominal structures which are not shown.
3. Color the organization of the sensory cortex. Note that the left three columns deal with three specific modalities. The right column shows neuron connections within a modality.
4. Color the examples of convergence and divergence in the lower right corner, starting with the many receptor fields (1) of convergence and working up to the cortex. Do the same for divergence.

SOMATIC SENSORY CORTEX, ASSOCIATION SENSORY CORTEX, THALAMUS.

The primary somatic sensory cortex (somesthetic cortex, S) is the area just behind the central fissure (postcentral gyrus). Receiving the specific somatic radiation from the thalamus, this area is the main cortical area involved in analysis and integration of sensory input from the skin, joints, etc., somesthetic perception, and sensations related to body position and movement. Lesions of this area do not abolish tactile sensations, but diminish discriminatory tactile sensations (fine localization, strength of stimuli) and sensation of bodily movements. The association sensory cortex is located posterior to the primary sensory cortex from which it receives input. Loss or removal of this area results in disorders of body image and complex somatic-visual sensations.

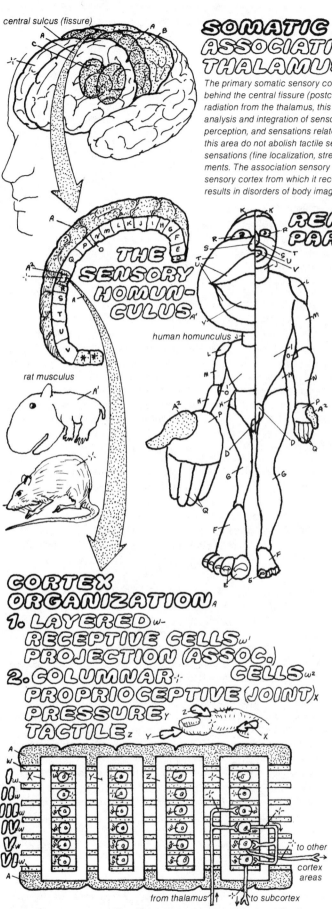

central sulcus (fissure)

THE SENSORY HOMUNCULUS.

rat musculus

human homunculus

REPRESENTATION OF BODY PARTS ON THE CORTEX.

Stimulation of discrete areas of the somesthetic cortex (S) in conscious patients undergoing brain surgery evokes tingling sensations in discrete body areas. Careful mapping of these referred (projected) sensations has revealed that the body is represented on the S cortex in an orderly manner (sensory homunculus), with the trunk on the top, the hands at the middle, and the face at the bottom of the postcentral gyrus. The human sensory homunculus is distorted, so the areas of the body that are most sensitive and capable of fine tactile discrimination (hands, fingers, and lips) have large representations and the trunk and legs have relatively small representations. In lower animals (rat) this representation is more proportional; still the muzzle and vibrissae have relatively larger representation than other parts of the body.

TRANSMISSION OF STIMULI TO NEURONS IN CORTEX. CONVERGENCE.

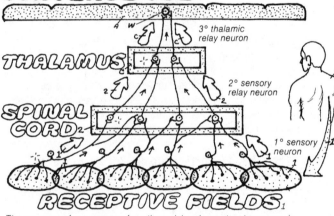

THALAMUS
3° thalamic relay neuron

SPINAL CORD
2° sensory relay neuron

1° sensory neuron

RECEPTIVE FIELDS

The presence of convergence from the peripheral receptors to neurons in the S cortex allows a single cortical cell to respond to sensory signals from large areas of the body (sum of the receptive fields of all the primary sensory afferents). This arrangement is observed generally in the spinothalamic projection system serving crude tactile sensation. In contrast, the presence of divergence, wherein a single peripheral afferent sends signals to many cortical cells, allows for more refined discrimination of tactile signals. This arrangement is seen in the dorsal column-lemniscal projection system, serving discriminatory tactile sensations. The divergence pattern is so organized that with point stimulation of a skin part, some cortical cells are activated more than others, providing for contrast discrimination.

DIVERGENCE.

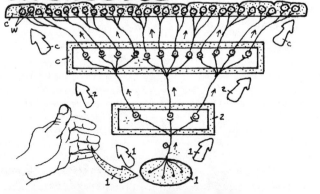

CORTEX ORGANIZATION
1. LAYERED RECEPTIVE CELLS PROJECTION (ASSOC.) CELLS
2. COLUMNAR PROPRIOCEPTIVE (JOINT) PRESSURE TACTILE

to other cortex areas

from thalamus   to subcortex

The S cortex, like other areas of the neocortex, is organized both horizontally and vertically. Horizontally, it is organized into six layers of neurons, with cells of each layer involved in a different transmission function. Layer IV cells receive the specific somatic radiation from the thalamus. Vertically the S cortex is organized into columns. Each column is about 1 mm wide and a few mm long. There are thousands of columns in the S cortex. All cells in a particular column respond to a signal relating to a single modality from a distinct body area. Thus a column sensitive to tactile stimuli from a finger is located next to another column responsive to proprioceptive sensation from the joints in that finger, etc.

PHYSIOLOGY OF PAIN

The sense of *pain* is complex because it involves not only a sensation but feelings and emotions as well. For this reason, the neurophysiology of pain involves structures not normally considered as part of the sensory nervous system. Furthermore, classically, the ascending sensory (excitatory) aspects of pain signals have been emphasized. The intrinsic capacity of CNS structures to suppress pain signals has recently become the focus of much attention and research.

PAIN RECEPTION. The sense of pain is served by *free nerve endings* located in the skin and certain visceral tissues. Pain can be caused by stimuli of different natures. For example, strong *mechanical* stimuli (intense pressure), very hot and very cold *thermal* stimuli, and certain *chemical* stimuli such as acidic substances all can cause pain. It is important to note that the pain receptors generally have a high threshold of stimulation, so they are usually activated when stimulus strength is very high. Because such strong stimuli are usually noxious, pain sensation is also called *nociception*, and the pain receptors activated by nociceptive stimuli are called *nociceptors*. One view holds that all nociceptive stimuli cause *tissue damage*, the extent of which may vary from the slight effects of a simple pinch to the severe consequences of burns. Tissue damage results in the local release of certain internal *nociceptive substances* such as *serotonin*, *substance-P*, *histamine*, and *kinin* peptides (bradykinin, etc.) in the injured tissue. These substances then act on the free nerve endings, activating pain signals.

TWO PAIN SYSTEMS. There appear to be two systems of pain transmission to the CNS, which are associated with two distinct types of pain experience. When one steps on a thumbtack, one feels a *sharp* sensation, followed a while later by a more *dull* pain sensation. In addition to arriving earlier, the sharp and *prickling* sensation is *short lasting*, and its *source* can be accurately *localized*. The dull sensation is *long lasting* and *diffuse*; it *hurts* and *aches*, but the ache source cannot be pinpointed and generally is ascribed to a larger body part.

It is now believed that the sharp pain is conveyed by thin but myelinated, relatively *fast*, nerve fibers (*type A-delta*), and the dull, aching, and hurting pain by unmyelinated *slow* conducting *type C* fibers. Conduction velocity in the A-delta fibers is about 10 times faster than in the C fibers. Both types of fibers terminate in the *dorsal horn* and ascend by the *spinothalamic* pathway. Whereas the slow/aching pain signals make a major input into the brain stem *reticular formation* and essentially terminate in the *thalamus*, the sharp/fast pain signals ascend more directly to the thalamus and up to the *sensory cortex*. The cortical component gives the fine localization capacity to the sharp/fast pain system, whereas the heavy subcortical projection of the dull/slow pain system to the reticular formation and the structures of the *limbic system* is associated with the aching/hurting component. Patients with damage to the sensory cortex can still feel pain and are hurt by it, but they are unable to accurately localize the source.

CENTRAL, DESCENDING PAIN INHIBITION. It has recently been shown that electrical stimulation of certain neuronal groups in the brain stem reticular formation makes the conscious animal completely oblivious to pain stimuli. Further research has indicated that, from the reticular formation, *descending* control fibers project to the dorsal horn of the spinal cord, where they *suppress* the relay of pain signals to the brain. This system is believed to help animals and humans cope with the debilitating hurtful consequences of pain arising during physical stress and fighting. It is presumably the active training of this *descending inhibition* that gives the Yogis of India their great tolerance of pain and athletes and soldiers their ability to continue struggling in the face of bodily hurts and trauma.

ENDORPHINS. One mechanism by which higher reticular centers inhibit pain is beginning to be understood. Descending fibers activate certain *inhibitory interneurons* in the dorsal horn, which release a peptide neurotransmitter called *enkephalin* (one of the *endorphins*). Enkephalin suppresses the transmission of pain signals by binding with particular receptor molecules (*opiate receptors*) present in the synapses of cells in the dorsal horn. The binding either decreases the amount of the neurotransmitter *substance-P* released from the *type C pain afferents* or induces *postsynaptic inhibition* of the relay cells. Morphine and other opiate analgesics (pain killers) act in the same way as endorphins to relieve pain.

AFFERENT PAIN INHIBITION. The interneurons of the dorsal horn may also be involved in a different type of pain inhibition. It has long been known that skin rubbing relieves the dull/hurtful pain sensation originating from that or a nearby area. Rubbing activates the large, fast-conducting *tactile fibers* (type A-alpha) while pain is conveyed by C fibers. In the dorsal horn, branches of touch fibers activate inhibitory interneurons, which in turn inhibit the synaptic transmission of pain signals. This is called the *gate theory of afferent inhibition*. Presumably, the more powerful tactile signals limit the transmission gates in the dorsal horn to their own, suppressing and excluding access for the weaker pain signal. The gate theory of afferent inhibition as well as central inhibition of pain by way of endorphins may have implications for the phenomenon of *acupuncture analgesia*.

REFERRED PAIN. The afferent pain fibers originating from the same area show extensive convergence onto the dorsal horn relay cells. In certain cases, the convergence may take place by fibers from different areas, causing the relay cell to be activated by pain originating in different body parts. Usually, one part is a visceral area or organ. This mechanism may underlie the phenomenon of *referred pain*. For example, pain originating in the heart is often felt as coming from the inner aspects of the left arm. Physicians make extensive use of referred pain, for which maps have been constructed, as means of diagnosing problems in the visceral organs (e.g., heart conditions).

CN: Use dark colors for E and H.
1. Begin on the left side of the upper panel and work your way up to the brain cortex in the upper right corner. Note that only the dorsal horn regions of the spinal cord gray matter is colored gray.
2. Color the panel on pain inhibition and follow the numbered sequence. The process described is an elaboration of the function of the descending pathway (J) in the diagram above. Note the reduction in frequency of nerve impulses (represented by the vertical bars that receive the color H) in the inhibited relay fiber (5). The terminal of the incoming nerve receives the color of the pain producing substance-P (M), which is itself being inhibited (3).
3. Color the gate theory panel.
4. Color the referred pain panel and then the phantom limb pain material.

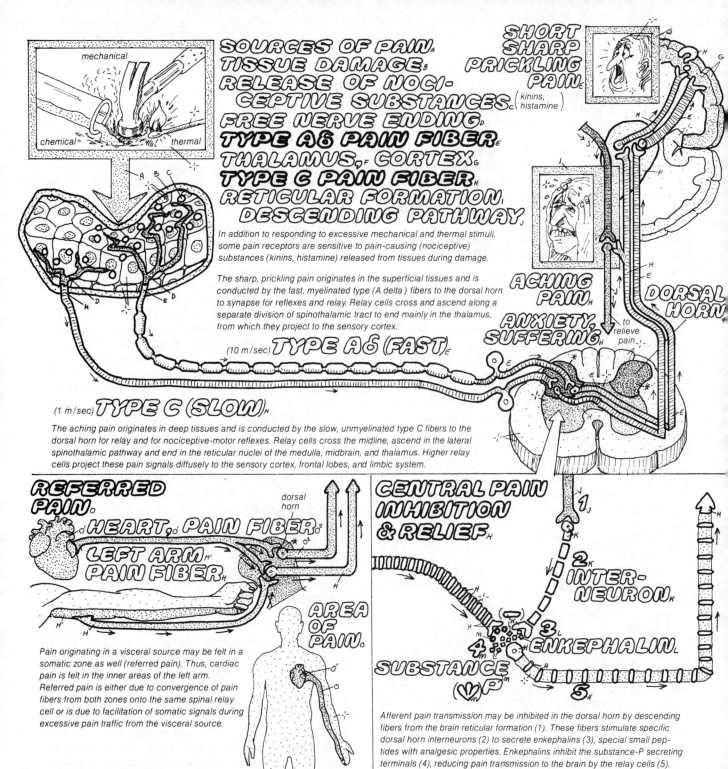

SOURCES OF PAIN.
TISSUE DAMAGE:
RELEASE OF NOCI-
CEPTIVE SUBSTANCES (kinins, histamine)
FREE NERVE ENDING.
TYPE Aδ PAIN FIBER.
THALAMUS, CORTEX.
TYPE C PAIN FIBER.
RETICULAR FORMATION.
DESCENDING PATHWAY.

mechanical / chemical / thermal

In addition to responding to excessive mechanical and thermal stimuli, some pain receptors are sensitive to pain-causing (nociceptive) substances (kinins, histamine) released from tissues during damage.

The sharp, prickling pain originates in the superficial tissues and is conducted by the fast, myelinated type (A delta) fibers to the dorsal horn to synapse for reflexes and relay. Relay cells cross and ascend along a separate division of spinothalamic tract to end mainly in the thalamus, from which they project to the sensory cortex.

(10 m/sec) **TYPE Aδ (FAST)**

SHORT SHARP PRICKLING PAIN

ACHING PAIN
ANXIETY, SUFFERING
to relieve pain

DORSAL HORN

(1 m/sec) **TYPE C (SLOW)**

The aching pain originates in deep tissues and is conducted by the slow, unmyelinated type C fibers to the dorsal horn for relay and for nociceptive-motor reflexes. Relay cells cross the midline, ascend in the lateral spinothalamic pathway and end in the reticular nuclei of the medulla, midbrain, and thalamus. Higher relay cells project these pain signals diffusely to the sensory cortex, frontal lobes, and limbic system.

REFERRED PAIN.
HEART, PAIN FIBER.
LEFT ARM PAIN FIBER.

dorsal horn

AREA OF PAIN.

Pain originating in a visceral source may be felt in a somatic zone as well (referred pain). Thus, cardiac pain is felt in the inner areas of the left arm. Referred pain is either due to convergence of pain fibers from both zones onto the same spinal relay cell or is due to facilitation of somatic signals during excessive pain traffic from the visceral source.

CENTRAL PAIN INHIBITION & RELIEF.
INTER-NEURON.
ENKEPHALIN.
SUBSTANCE-P

Afferent pain transmission may be inhibited in the dorsal horn by descending fibers from the brain reticular formation (1). These fibers stimulate specific dorsal horn interneurons (2) to secrete enkephalins (3), special small peptides with analgesic properties. Enkephalins inhibit the substance-P secreting terminals (4), reducing pain transmission to the brain by the relay cells (5).

PHANTOM LIMB PAIN.

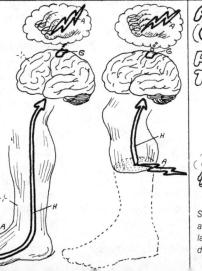

Phantom pains seemingly orginate in the amputated limbs. Irritation of severed endings of pain fibers at the sites of amputation signal pain to the same areas of the sensory cortex. These signals are "projected" to their original source in the limb or body, creating a phantom sensation.

AFFERENT INHIBITION (GATE THEORY).
PAIN.
TOUCH.

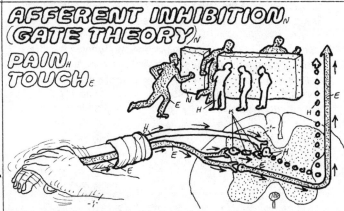

Strong tactile stimulation of skin (rubbing) diminishes pain originating from that area. This is due to the effect of afferent inhibition: Tactile signals conducted by large type A fibers inhibit transmission of pain conducted by type C fibers in the dorsal horn by blocking the synaptic "gates" normally used by the smaller fibers.

Reflexes are programmed, stereotyped, predictable motor responses to certain specific sensory stimuli. They are therefore the most elementary form of nervous action. Reflexive responses make up most of the behavior of simpler animals, as well as that of the newborns of higher animals. Although the *spinal reflexes* (i.e., those associated with *spinal cord* control of trunk and limb muscles) are better known, many *brain reflexes* also exist. These regulate such things as eye movements, head turning, chewing, and sneezing. Reflexes associated with the somatic nervous system and voluntary muscles are better known, but numerous *autonomic reflexes* associated with visceral effectors (heart, smooth muscles, and glands) also exist.

The operation of any reflex requires the active participation of all components of the *reflex arc*: (1) the *sensory receptors*, which detect the *stimulus*; (2) the *afferent nerve*, which conveys the sensory signal centrally; (3) an integrative *synaptic center*, which analyzes the sensory input; (4) the *efferent nerve*, which conducts the motor output to the periphery; and (5) a *motor effector* (e.g., skeletal muscle, smooth muscle, glands) which carries out the response. The complexity of a reflex response corresponds mainly to the complexity of the reflex center; this in turn depends on the number of interneurons and synapses involved.

MUSCLE SPINDLE AND THE STRETCH REFLEX. The most extensively studied reflex is the *stretch reflex*, also the simplest known reflex because there is only one synapse in the path of its arc (*monosynaptic reflex arc*). Large skeletal muscles contain several spindle shaped organs (muscle spindles), which are sensory organs detecting changes in the length or tension of muscle fibers. Muscle spindles contain the sensory receptors of the stretch reflex.

Each spindle contains modified muscle fibers called intrafusal fibers (L. fusus = spindle). At its middle, each intrafusal fiber has a mechanical *stretch receptor* connected to a sensory nerve. Stretching the muscle activates the spindle receptor, firing nerve signals to the spinal cord. In the spinal cord, the terminals of the spindle sensory fiber make direct excitatory synaptic contact with *alpha motor neurons* serving the same muscle. Alpha motor neurons are the large neurons that innervate ordinary muscle fibers (extrafusal). Activation of alpha motor neurons by the spindle sensory fibers and the resultant contraction lead to shortening and return of the muscle fiber to its original length.

The stretch reflex continuously monitors the length or tension of muscle fibers and keeps them constant during rest. This helps give the muscles tonus (tone), maintaining their readiness for action. Spindle fibers also have a *contractile segment* located on the sides of the stretch receptor. Contraction of these motor segments stretches the spindle sensory segment, activating the stretch reflex. The spindle motor segments are innervated by smaller spinal motor neurons called *gamma motor neurons* (gamma efferents), which are in turn driven by neurons from the higher brain

centers, especially from the brain stem. Whenever the higher centers need to increase the readiness of skeletal muscles, they activate the gamma efferents, causing the stretch reflex.

THE KNEE JERK REFLEX AS A POLYSYNAPTIC SPINAL REFLEX. The stretch reflex is simple and monosynaptic. Most spinal reflexes are *polysynaptic* (i.e., the reflex arc and center involve one or more *interneurons* and higher numbers of synaptic connections). A well-known example is the *knee jerk reflex*, in which the stretch reflex plays a part. The person whose knee jerk reflex is being tested is placed on a high chair with his or her legs dangling. When the patellar tendon, connecting the thigh extensors to the tibia, is tapped below the knee, the lower leg shows a fast reflex extension (knee jerk) due to contraction of the thigh *extensor* muscles. This is basically a stretch reflex: tapping the tendon pulls on the tendon fibers, which in turn stretch the muscle and the spindle fibers, activating the stretch reflex. However, execution of a proper knee jerk reflex requires not only the activation of extensor muscles but also the relaxation of the opposing *flexor* muscles. Because all lower motor neurons are excitatory, the only way to obtain flexor relaxation is to inhibit their motor neurons. This is achieved by inhibitory interneurons that are activated by a branch of the spindle sensory fiber.

IMPORTANCE OF INTERNEURONS. The associative interneurons of the spinal cord, particularly the inhibitory ones, underlie the operation of all complex spinal reflexes. For example, in the *withdrawal reflex* (a limb *flexor reflex*), noxious (sharp or hot) stimuli activate the pain fibers, the collaterals of which stimulate interneurons that, in turn, excite motor neurons going to the flexor muscles on the same side, causing ipsilateral limb withdrawal (as occurs after you touch a very hot object). Even in a simple withdrawal reflex, the ipsilateral extensors must be simultaneously relaxed. The withdrawal of one leg in the standing posture often throws the body's weight onto the other leg. Here the extensors of the contralateral leg must be stimulated while the flexors are inhibited (*crossed extensor reflex*).

The excitatory and inhibitory circuitry for activation of most of the spinal reflexes is already present at birth. The activation of any particular circuit (reflex) depends mainly on the stimulus type and location, on the skin, tendon, etc.

SPINAL SHOCK AND INDEPENDENCE OF SPINAL REFLEXES. Spinal reflexes do not require the participation of the brain. They can be seen in animals that have undergone *spinal transection* (severing of the connection between the brain and the cord). Similarly, these reflexes can be observed, although somewhat abnormally, in quadriplegic humans who have suffered damage to various connections between the brain and the spinal cord. However, for a short period after spinal transection, spinal reflexes disappear. This period of *spinal shock* is short in lower animals (minutes in frogs) and long in higher animals (weeks to months in humans), presumably because a higher animal's brain exerts more control over its spinal cord, (encephalization).

CN: Use dark colors for B, D, and O.
1. Begin with the upper panel. Note that the title for letter label O, interneuron, is found in the knee jerk reflex panel.
2. Color the stretch reflex panel, beginning with the overview in the small rectangle on the left. Then follow the numbered sequence.
3. Color the knee jerk reflex panel, noting that a

muscle spindle (K & L) has been greatly enlarged for purposes of illustration. Note that the flexor muscle that is inactivated by an inhibited efferent motor nerve (D) is left uncovered. In the withdrawal reflex panel, leave the extensor muscle uncolored.
4. Finish up the cross-extensor reflex, once again leaving uncolored the relaxed muscles in the two examples.

REFLEX ARC *
RECEPTOR A
AFFERENT (SENSORY NERVE) B
SPINAL CORD OR BRAIN C (INTEGRATING SYNAPTIC CENTER)
EFFERENT (MOTOR) NERVE D
EFFECTOR E

SPINAL NERVE F
GANGLION G
DORSAL ROOT H
VENTRAL ROOT

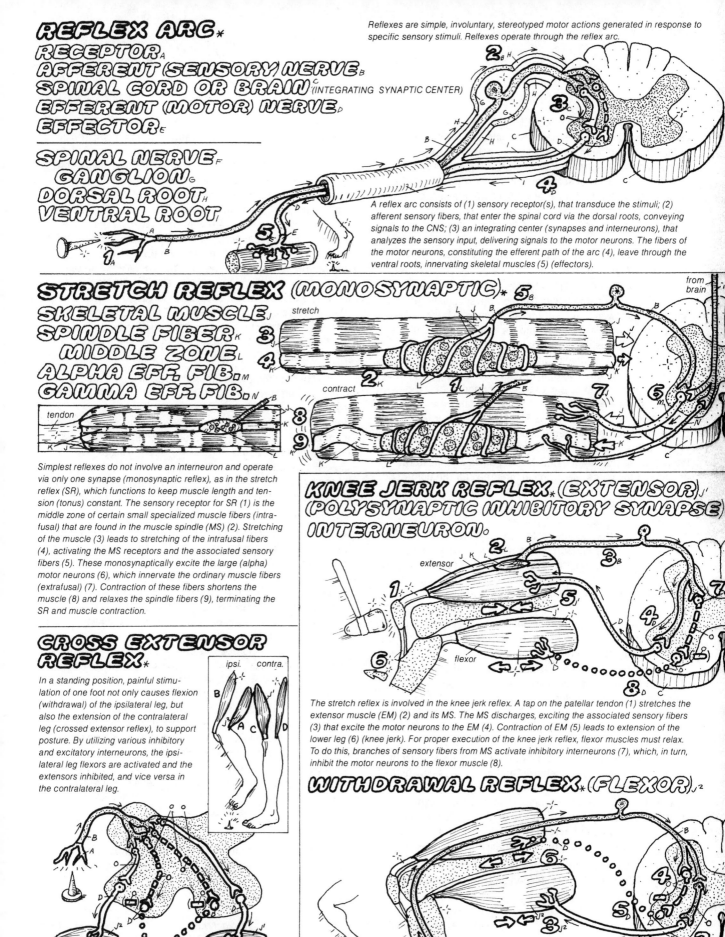

Reflexes are simple, involuntary, stereotyped motor actions generated in response to specific sensory stimuli. Relfexes operate through the reflex arc.

A reflex arc consists of (1) sensory receptor(s), that transduce the stimuli; (2) afferent sensory fibers, that enter the spinal cord via the dorsal roots, conveying signals to the CNS; (3) an integrating center (synapses and interneurons), that analyzes the sensory input, delivering signals to the motor neurons. The fibers of the motor neurons, constituting the efferent path of the arc (4), leave through the ventral roots, innervating skeletal muscles (5) (effectors).

STRETCH REFLEX (MONOSYNAPTIC) *
SKELETAL MUSCLE J
SPINDLE FIBER K
MIDDLE ZONE L
ALPHA EFF. FIB. M
GAMMA EFF. FIB. N

Simplest reflexes do not involve an interneuron and operate via only one synapse (monosynaptic reflex), as in the stretch reflex (SR), which functions to keep muscle length and tension (tonus) constant. The sensory receptor for SR (1) is the middle zone of certain small specialized muscle fibers (intrafusal) that are found in the muscle spindle (MS) (2). Stretching of the muscle (3) leads to stretching of the intrafusal fibers (4), activating the MS receptors and the associated sensory fibers (5). These monosynaptically excite the large (alpha) motor neurons (6), which innervate the ordinary muscle fibers (extrafusal) (7). Contraction of these fibers shortens the muscle (8) and relaxes the spindle fibers (9), terminating the SR and muscle contraction.

KNEE JERK REFLEX * (EXTENSOR) J¹
(POLYSYNAPTIC INHIBITORY SYNAPSE) INTERNEURON.

The stretch reflex is involved in the knee jerk reflex. A tap on the patellar tendon (1) stretches the extensor muscle (EM) (2) and its MS. The MS discharges, exciting the associated sensory fibers (3) that excite the motor neurons to the EM (4). Contraction of EM (5) leads to extension of the lower leg (6) (knee jerk). For proper execution of the knee jerk reflex, flexor muscles must relax. To do this, branches of sensory fibers from MS activate inhibitory interneurons (7), which, in turn, inhibit the motor neurons to the flexor muscle (8).

CROSS EXTENSOR REFLEX *

In a standing position, painful stimulation of one foot not only causes flexion (withdrawal) of the ipsilateral leg, but also the extension of the contralateral leg (crossed extensor reflex), to support posture. By utilizing various inhibitory and excitatory interneurons, the ipsilateral leg flexors are activated and the extensors inhibited, and vice versa in the contralateral leg.

ipsi. contra.

A FLEXOR C EXTENSOR
B EXTENSOR D FLEXOR
ipsilateral contralateral

WITHDRAWAL REFLEX * (FLEXOR) J²

The withdrawal reflex is a defensive flexor reflex elicited in response to noxious (painful) stimulation of the foot. Sensory pain signals (1) excite motor neurons to the flexors (2), eliciting flexion and withdrawal of the leg (3). Simultaneously, via inhibitory interneurons (4), motor neurons to the extensor muscles are inhibited (5) to relax the extensors of the same leg.

VOLUNTARY MOTOR CONTROL

BRAIN AND VOLUNTARY MOVEMENT. That the centers for *voluntary control* of movement are in the brain is easily demonstrated in people who have had accidents that severed their spinal cord (spinal transection). Such persons lose voluntary control of muscles innervated by segments of the spinal cord below the transection level. When this occurs at the neck level (cervical transection), the result is quadriplegia. The quadriplegic is incapable of voluntary movement in any trunk and limb muscles, although head, tongue, and eye movements, controlled by the cranial nerves and brainstem centers, are still present.

MOTOR CORTEX AND THE MOTOR HOMUNCULUS. Late in the nineteenth century, two German physiologists, Fritsch and Hitzig, showed that electrical stimulation of certain areas of the frontal cortex in experimental animals resulted in contraction of muscles and gross limb movements. In the twentieth century, Penfield, a Canadian neurosurgeon, noted that electrical stimulation of the human cortex in an area just in front of the central sulcus evoked contraction of body muscles. This area, located on the percentral gyrus, is called the *primary motor cortex* (MC). The MC, like its sensory counterpart across the central sulcus, shows a *somatotopic* organization. The areas controlling the legs and trunks are on the top of the precental gyrus. Moving down on the lateral aspects of the gyrus, we come to the areas for the control of hand muscles, followed by areas for the control of head, tongue, and other muscles involved in speech. Thus, a *motor homunculus* is present in the primary motor cortex. The representation is contralateteral for the trunk and limb muscles but bilateral for the speech muscles. As with the sensory homunculus (plate 87), the representation is not proporational to the size of the body part (muscles) but to the degree of skilled movement and motor capacities of the part. The hands and digits, which are capable of great versatility of movements, have very large representations, as do the tongue and speech muscles. But the massive leg muscles have relatively small cortical represenation.

Animal brain stimulation studies indicate that when deeper layers of the cortex are stimulated with weak currents, contractions of single muscles or discrete muscle groups can be evoked. Surface stimulation with strong currents causes contractions of complex muscle groups, perhaps because the current spreads to wider cortical areas (see below).

Like the sensory cortex, the motor cortex has six horizontal layers (although the MC is thicker and contains mainly pyramidal neurons), as well as a vertical columnar organization. Neurons from the deeper layers appear to be the output neurons. The Betz cells, the very large pyramidal neurons once thought to be the cortical substrate of voluntary movement, constitute a small proportation of these neurons.

PYRAMIDAL TRACT AS THE OUTPUT PATHWAY FOR MOTOR CORTEX. A major descending motor pathway, the *pyramidal tract*, originates in the MC. The pyramidal tract is present only in higher animals (such as mammals) and is especially well developed in primates and humans. The fibers of the pyramidal tract descend in the brain and spinal cord white matter to terminate, some directly and some indirectly (via interneurons), on the motor neurons of the spinal cord and brain stem. The motor neurons from the brainstem and spinal cord in turn innervate voluntary muscles of the head and trunk/limbs, respectively. The portion of pyramidal tract terminating in the brain stem is called the *corticobulbar tract*, and the remaining part is the *corticospinal tract*. The term *upper motor neuron* is applied to the neurons of the pyramidal tract. The efferent motor neurons innervating the muscles are referred to as *lower motor neurons*. The *pyramidal system* consists of the primary motor cortex and the pyramidal tract. It is the chief executor of voluntary motor commands in general and of skilled movements in particular.

The fibers of the pyramidal tracts, particularly the corticospinal portion, decussate before reaching their destination in the spinal cord. In humans, more than 80% of the fibers decussate at the level of the medulla (pyramidal decussation) to innervate the spinal motor neurons on the opposite side; most of the rest decussate at lower levels. This almost complete decussation is the basis of contralateral motor control by the higher brain centers. Thus, the motor areas in the right hemisphere control muscles on the left side and vice versa.

THE PREMOTOR CORTEX AND MOVEMENT PATTERNS. Detailed stimulation studies have revealed another cortical motor area in front of the MC called the premotor area (premotor cortex). This area acts as the association area for MC, generating movement patterns that are then communicated to specific areas in the MC. Thus, MC neurons are controlled by the premotor cortex neurons. When a person is asked to move an arm, the premotor neurons increase activity before the MC neurons. Stimulation of the premotor association cortex causes whole, purposive movments of muscle groups. The association motor cortex receives association fibers from other cortical areas, especially from the sensory association area. Lesions (damage) or ablation (removal) of the primary motor cortex or pyramidal tract may cause marked paralysis (inability to initiate voluntary movement) or at least paresis (weakness in voluntary movement), but damage to the premotor cortex results in poverty of skilled movements. When damage is restricted to areas in front of the hand area, skilled manipulation is impaired. When damage is restricted to the area in front of the speech muscles area, articulation is impaired (see also plate 105).

The illustration on the lower right corner of this plate represents diagrammatically the relationship between the cortical motor and sensory areas in a cartoon of a real-life situation.

CN: Use the same color for the spinal cord (C) as you used on the previous page. Use dark colors for F, G, and H.
1. Begin with the upper left diagram. Note the three-dimensional spinal cord (C) section at the bottom of the diagram with the corticospinal tracts (F) also being represented three-dimensionally.
2. Color the representation of body muscles on the motor cortex (the motor homunculus) in the lower left corner. When you color these drawings, you will probably have to use some colors more than once.
3. Color the diagrams on motor and premotor cortex in the upper right region.
4. Color the material in the lower right, beginning with the anatomical description of the structures involved in the numbered sequence.

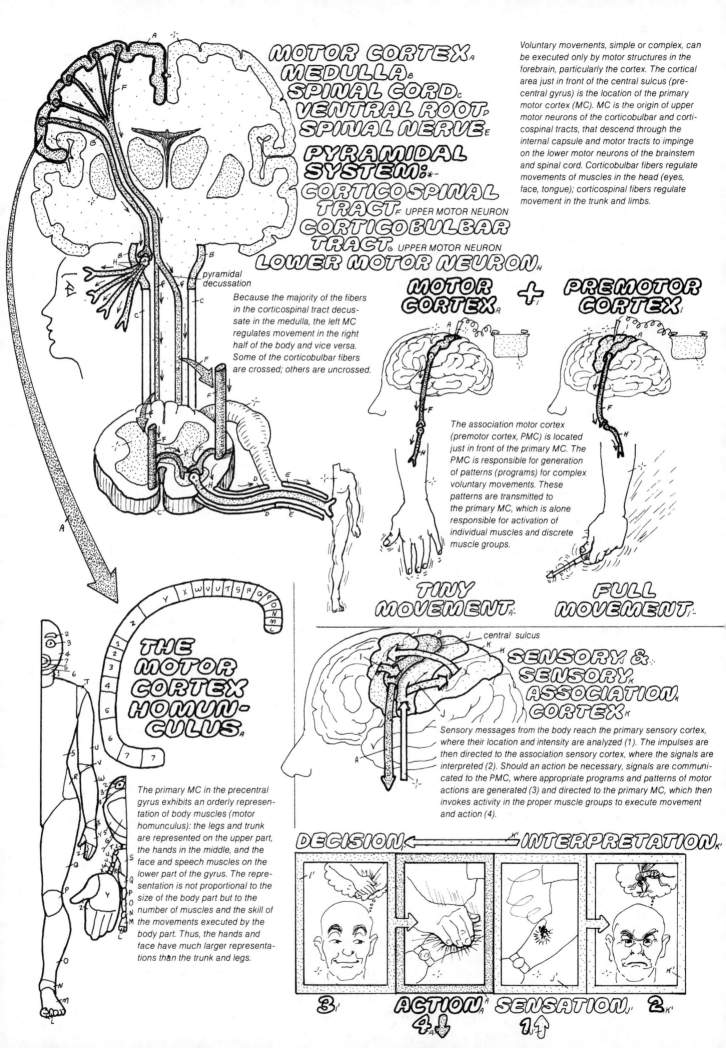

MOTOR CORTEX.A
MEDULLA.B
SPINAL CORD.C
VENTRAL ROOT.D
SPINAL NERVE.E

PYRAMIDAL SYSTEM:*
CORTICOSPINAL TRACT.F UPPER MOTOR NEURON
CORTICOBULBAR TRACT.G UPPER MOTOR NEURON
LOWER MOTOR NEURON.H

pyramidal decussation

Voluntary movements, simple or complex, can be executed only by motor structures in the forebrain, particularly the cortex. The cortical area just in front of the central sulcus (precentral gyrus) is the location of the primary motor cortex (MC). MC is the origin of upper motor neurons of the corticobulbar and corticospinal tracts, that descend through the internal capsule and motor tracts to impinge on the lower motor neurons of the brainstem and spinal cord. Corticobulbar fibers regulate movements of muscles in the head (eyes, face, tongue); corticospinal fibers regulate movement in the trunk and limbs.

Because the majority of the fibers in the corticospinal tract decussate in the medulla, the left MC regulates movement in the right half of the body and vice versa. Some of the corticobulbar fibers are crossed; others are uncrossed.

MOTOR CORTEX.A ＋ PREMOTOR CORTEX

The association motor cortex (premotor cortex, PMC) is located just in front of the primary MC. The PMC is responsible for generation of patterns (programs) for complex voluntary movements. These patterns are transmitted to the primary MC, which is alone responsible for activation of individual muscles and discrete muscle groups.

TINY MOVEMENT.A   FULL MOVEMENT.I

THE MOTOR CORTEX HOMUNCULUS.A

The primary MC in the precentral gyrus exhibits an orderly representation of body muscles (motor homunculus): the legs and trunk are represented on the upper part, the hands in the middle, and the face and speech muscles on the lower part of the gyrus. The representation is not proportional to the size of the body part but to the number of muscles and the skill of the movements executed by the body part. Thus, the hands and face have much larger representations than the trunk and legs.

central sulcus

SENSORY & SENSORY ASSOCIATION CORTEX.K

Sensory messages from the body reach the primary sensory cortex, where their location and intensity are analyzed (1). The impulses are then directed to the association sensory cortex, where the signals are interpreted (2). Should an action be necessary, signals are communicated to the PMC, where appropriate programs and patterns of motor actions are generated (3) and directed to the primary MC, which then invokes activity in the proper muscle groups to execute movement and action (4).

DECISION.I' ⟵ INTERPRETATION.K'

ACTION.A 4↓   SENSATION.J 1↑   2.K'

3.I'

BASAL GANGLIA & CEREBELLUM IN MOTOR CONTROL

In addition to the cortical motor areas and the pyramidal system (discussed in plate 90), important for voluntary motor function control, the brain has other motor control systems served by structures such as the *basal ganglia* (BG) and *cerebellum* (CB).

BASAL GANGLIA. The BG consist of some forebrain structures (*caudate, putamen, globus pallidus*) and some midbrain structures (*substantia nigra, red nucleus, subthalamus*). In birds and lower vertebrates that lack the neocortex, the BG are the major brain structures for higher motor control. In the human and higher animals, the BG were historically thought to be part of a separate motor system (hence called the *extrapyramidal system*) functioning in control of gross and unskilled (yet voluntary) motor activities such as those involved in postural control and locomotion. Recent findings of extensive two-way connections between the forebrain BG and the motor cortex have emphasized the interaction between the two motor systems and redefined the pyramidal and extrapyramidal systems as a unified motor brain.

BG FUNCTIONS. The knowledge of BG involvement in motor functions comes mainly from the striking motor abnormalities associated with damage and degeneration in these structures. These well-known neurological diseases are illustrated in the lower panel of the plate. Unfortunately, little is known of BG's role in normal motor control. In general, two types of functions are assumed for it. One involves control of voluntary gross movements (e.g., postural control during locomotion or ballistic limb movements). These controls are exerted by changing the levels of muscle tension and kinesthetic feedback activity. Other, recently established BG function relates to the initiation of voluntary motor commands.

BG OUTPUT CONNECTIONS. BG output pathways are consistent with these two functions. The first is performed by BG midbrain motor components (red nucleus, substantia nigra) via such descending motor pathways as the rubrospinal and reticulospinal tracts (*extrapyramidal tracts*). These tracts control the activity of *gamma motor neurons* of the spinal cord (see plate 78) and hence control muscle tension, the stretch reflex, and proprioceptive and kinesthetic activities. BG influence on initiating voluntary motor activities is exerted directly on the pyramidal system via extensive output from the forebrain BG to the *premotor cortex*, which then signals the motor cortex and the pyramidal tract. The descending pyramidal fibers activate the *alpha motor neurons*, causing muscle contraction. Indeed, even before any voluntary movement is executed, forebrain BG neurons increase their firing rate.

CEREBELLUM AND MOTOR COORDINATION. The cerebellum (CB) is one of the most primitive brain structures. Located outside the cerebrum, it consists of an overlaying and highly folded *cerebellar cortex* and deeply situated *cerebellar nuclei*. Much of our knowledge about CB function comes from lesion studies in animals and humans (see the lower panel of the plate). These studies indicated that the CB is essential for proper and smooth *coordination* of voluntary motor activity but not for its *initiation*.

FUNCTIONAL DIVISIONS OF CB. Although CB structure appears fairly uniform, its different parts (lobes) are involved in various types of coordinative control. The *flocculo-nodular lobe*, the most primitive part, located in the base of the CB, works in conjunction with the vestibular system, controlling equilibrium, balance, and head / eye coordination. The *vermis* (snake), located over the midline, is less primitive than the flocculo-nodular lobe and coordinates gait and posture during locomotion. The vermis receives extensive projections from proprioceptors of joints and muscles. The most advanced parts of the CB are the *cerebellar hemispheres*, extensively developed in primates and humans. The CB hemispheres work closely with the premotor cortex to coordinate fast, skilled motor activities.

CEREBELLUM AS THE COMPARATOR COMPUTER. The CB hemispheres have two-way communication channels with both the motor cortex in the brain and the voluntary muscles in the periphery to coordinate motor performance. Remember that whenever a voluntary movement is desired (e.g., picking up a glass), the *premotor cortex* generates the movement patterns, sending them to the *primary motor cortex*, which then activates the muscles via the descending upper and lower motor neurons (see plate 90). Each time the premotor cortex sends these signals, it also sends a copy to the CB via the cerebellar relay pathways in the *pons*. The CB matches these commands with muscle performance and signals the motor cortex via relay centers in the *thalamus*, informing it of any wrong commands or needs for adjustments.

At the same time, the CB, acting through the red nucleus and its descending connections with the gamma motor neurons, modifies muscle tension and the stretch reflex to bring the muscles in line with motor cortex commands. Thus, the CB oversees the ongoing communication between the cortex and the muscles, controlling motor performance.

CEREBELLAR INPUT AND OUTPUT. The input from the motor cortex and the muscles arrives via brain stem relay centers to the CB cortex, where the CB circuits analyze it. The outcome is relayed by the prominent Purkinje cells in the CB cortex to the deep CB nuclei. Interestingly, the Purkinje cells are inhibitory neurons, releasing GABA as the neurotransmitter. However, the neurons of CB nuclei, being the real CB output neurons (they innervate CB targets in the midbrain: red nucleus and thalamus) are excitatory. How do the CB cortex circuits make decisions? How much is preprogrammed genetically, and how much learning is involved? These are questions under investigation.

BASAL GANGLIA:
CAUDATE NUCLEUS
PUTAMEN
GLOBUS PALLIDUS
SUBTHALAMUS
SUBSTANTIA NIGRA

THALAMUS
PREMOTOR CORTEX
MOTOR CORTEX
RED NUCLEUS
EXTRAPYRAMIDAL TRACTS
MUSC. SENS. SIGNAL
PYRAMIDAL TRACT

Neurons in the BG increase activity before the execution of voluntary movements, implying that voluntary movements are planned and initiated in the BG (1). Impulses are then sent to the PMC (2) via the thalamus. PMC then activates the primary MC (3) to execute movement (4) and CB (5) to coordinate (6) it. BG, via separate descending motor pathways (7) (extrapyramidal tracts), directly influence muscle spindle tension (8), facilitating the execution of all kinds of movement.

The basal ganglia (BG) are motor structures in the forebrain (caudate, putamen, and globus pallidus) and midbrain (subthalamus, substantia nigra, and red nucleus) involved in higher regulation of both voluntary and complex involuntary movements. Lesions of BG are associated with dramatic motor disorders (see below). As part of the extrapyramidal system, the BG not only have extensive two-way connections with cortical motor areas and cerebellum (CB) but function together with these structures in movement regulation.

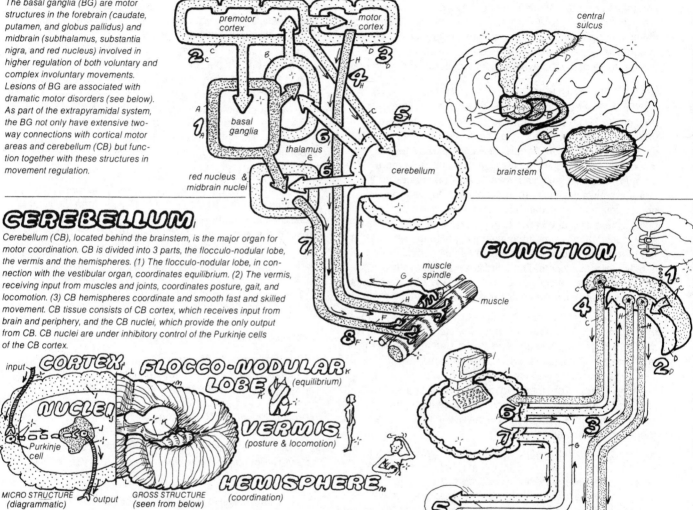

premotor cortex · motor cortex · basal ganglia · thalamus · red nucleus & midbrain nuclei · cerebellum · central sulcus · brain stem · muscle spindle · muscle

CEREBELLUM

Cerebellum (CB), located behind the brainstem, is the major organ for motor coordination. CB is divided into 3 parts, the flocculo-nodular lobe, the vermis and the hemispheres. (1) The flocculo-nodular lobe, in connection with the vestibular organ, coordinates equilibrium. (2) The vermis, receiving input from muscles and joints, coordinates posture, gait, and locomotion. (3) CB hemispheres coordinate and smooth fast and skilled movement. CB tissue consists of CB cortex, which receives input from brain and periphery, and the CB nuclei, which provide the only output from CB. CB nuclei are under inhibitory control of the Purkinje cells of the CB cortex.

CORTEX
NUCLEI
FLOCCO-NODULAR LOBE (equilibrium)
VERMIS (posture & locomotion)
HEMISPHERE (coordination)

FUNCTION

input · Purkinje cell · MICRO STRUCTURE (diagrammatic) · output · GROSS STRUCTURE (seen from below)

CB has been likened to a computer that monitors the commands of the cortical motor centers and the performance of motor effectors. It will then act to improve motor performances. For example, to hold a glass steady, PMC (1) signals the appropriate motor patterns to MC (2). MC transmits the signals to muscles in the arms and hands (3). At the same time PMC, sends similar signals to CB (4). Sensory receptors in muscles and joints will then inform CB of their performance and position (5). CB matches performance of effectors with the motor commands from the PMC and sends signals back to PMC (6) to adjust its future commands and correct any errors. CB also signals the muscles to modify their tension and readiness-tonus state in an appropriate manner.

DISORDERS OF BASAL GANGLIA & CEREBELLUM

SUBSTANTIA NIGRA:
PARKINSON'S DISEASE

Degeneration of dopamine-releasing neurons in substantia nigra results in Parkinson's disease. Occurring mainly in the elderly, it is characterized by bradykinesia (poverty of movement), rigidity and tremor (shakes).

CAUDATE & PUTAMEN:
CHOREA

Degeneration of neurons in striatum (caudate-putamen) causes "choreas," diseases in which orderly progression of voluntary movements, as in walking, is replaced by rapid, involuntary dancing movements (St. Vitus' dance).

CEREBELLUM:

Damage to CB results in poor posture, equilibrium, and ataxia (disorders of movement coordination). Gait is wide based and drunkenlike. Postural (extensor) muscles are weak (hypotonia). Speech is slurred. Voluntary movements are accompanied by "intention tremor" and dysmetria (past-pointing).

GLOBUS PALLIDUS:

Degeneration of globus pallidus results in "athetosis," involuntary writhing (twisting and turning) movements of the limb.

ATHETOSIS

ATAXIA

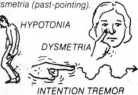

HYPOTONIA
DYSMETRIA
INTENTION TREMOR

THE EYE'S OPTICAL FUNCTIONS

The *eye* is a complex sensory organ designed to perform both *optical* functions for image formation and *nervous* functions for photic transduction, image analysis, and image transmission to the brain. The eye's optical apparatus forms and maintains sharp focus of an object's image on the retina, the eye's nervous part. Photoreceptors on the retina convert the incident photons into nerve signals and transmit them via integrative neural elements to the brain's visual centers. In this plate, we study the eye's optical structures and functions.

IMAGE FORMATION BY THE EYE'S COMPOUND LENS SYSTEM. The eye's optical function is to refract (bend) light rays emitted from objects to form sharp images of them onto the retina. The objects may be as simple as a point light source (a small distant candle) or as complex as two- or three-dimensional objects (lines, circles, cubes, or more complex bodies such as a bird in flight). The images on the retina are always smaller than the objects. Indeed, in the human fovea (a small patch of retina in the eye's posterior pole), where spatial vision is highly developed, image size is always less than 1 mm! An image falling onto the retinal surface produces neural signals on a mosaic of photoreceptors in different retinal spots. The retina then sends this two-dimensional map of signals to the brain, where they are used to reconstruct three-dimensional images that we perceive.

Light rays pass through several transparent media in the eye before they stimulate the retina's photoreceptors. These media help refract and converge the rays so that the image falling on the retinal surface is smaller than the real object. The first of these media is the *cornea*, which because of its higher density (compared to air) and its curved surface, refracts the rays inward. Next the rays pass through the *aqueous humor*, a viscous fluid in the eye's anterior chamber (between the lens and cornea).

The eye's crystalline *lens* acts like a biconvex glass lens. Parallel rays (i.e., from more than 6 m [20 ft]) entering the lens periphery are refracted inward, converging on a *focal point* behind the lens and along its *optical axis*, a straight line passing through the lens center. The *focal distance* (i.e., the distance between the focal point and the lens) is fixed in a glass lens. For the normal human lens, this distance varies (see below), being about 16 mm at rest. Behind the lens there is one last transparent medium, the gel-like *vitreous humor*, which helps keep the eyeball's spherical shape. Because of the lens's biconvexity, the image formed on the retina is inversed. The brain inverts this image so that our mental images are right side up.

ABNORMAL IMAGE FORMATION. The study of the eye's abnormal optical performance is helpful for a better understanding of the normal focusing mechanisms. Two kinds of abnormalities are usually encountered: those caused by the eyeball's structural deformation and those caused by the lens' rigidity. If the eyeball is too long and elipsoid (*myopia*), the focal point falls in front of the retina, causing visual images to appear blurred. In order to see clearly, the viewer needs to bring the object nearer to the eye (*nearsightedness*). This condition can be corrected by placing biconcave lenses in front of the eye. They diverge the light rays before they enter the eye, in effect bringing the object closer. If the eyeball is too short (hyperopia), the focal point falls behind the retina. People with this affliction see distant objects better (*farsightedness*). This condition is corrected by using a biconvex lens that converges the light rays before they pass through the eye, in effect moving the object farther away. Another type of optical abnormality occurring with aging that seriously interferes with near-vision is related to the loss of lens elastically (see below).

OCULAR RESPONSE IN NEAR-VISION. The lens is held by ligaments attached to the *ciliary muscles*. When the ciliary muscles contract, the ligaments loosen, releasing tension on the lens; the lens, being elastic, relaxes, assuming a more spherical shape. This decreases the lens' focal distance. Relaxation of the ciliary muscles pulls on the ligaments, which in turn increases tension on the lens, making it flatter and increasing the lens focal distance. Therefore, in order to form sharp images of distant objects on the retina, the ciliary muscle relax to flatten the lens; for near objects, the muscles contract to increase the lens' curvature.

The lens' ability to change curvature for sharp focusing is called *accommodation*. During childhood and early maturity, the lens is elastic and accommodates well. With aging, the crystalline lens hardens, decreasing its elasticity and accommodation power. Indeed, by their early fifties, both men and women have totally lost their capacity for accommodation (presbyopia), a condition requiring the use of biconvex corrective lenses for near-vision actions such as reading.

Accommodation also involves changes in the *pupil* size. The circular pupil is a hole formed by a ring of smooth muscles called the *iris*. Contraction of the iris *sphincter* muscles constricts the pupil; that of the *dilator* muscles widens it. The pupil has two functions. One is in the *light reflex*. When exposed to bright light, the pupil constricts to permit less light to enter the eye. In the dark, the pupil dilates to allow in more light. Another pupillary function occurs during near-vision responses. When the eyes shift from looking at a far object to a near one, the pupil constricts. This *pupillary constriction* response increases the depth of visual field, like a pinhole, permitting sharp focus and clear vision of near objects. Because the two eyes converge during the near-vision response, the pupillary constriction occurring in this is also called the *convergence response*.

CN: Use your lightest colors for A, B, C, and G. Use dark colors for E and H.
1. Begin with the structure of the eye and its comparison to the camera.
2. Color the panel on light refraction.
3. Color the defects and correction of image formation.
4. Color the three examples of near-vision adjustments, noting that presbyopia on the right is a problem with example one, accommodation.
5. Follow the numbered sequence of steps in the control of near-vision reflexes.

STRUCTURE OF THE EYE *

CORNEA ₐ
AQUEOUS HUMOR ᵦ
LENS c
CILIARY MUSCLE ᴅ
LIGAMENTS ᴇ
IRIS ꜰ
VITREOUS HUMOR ɢ
RETINA ʜ
CHOROID ɪ
SCLERA ᴊ
OPTIC NERVE ᴋ

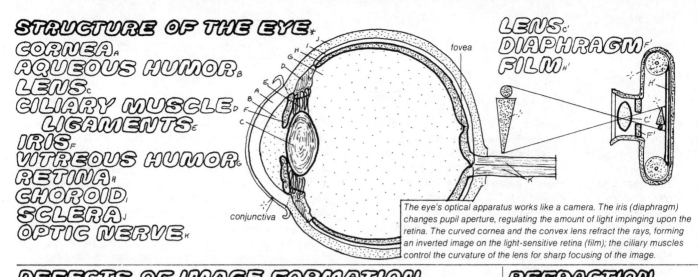

fovea

LENS c'
DIAPHRAGM ꜰ'
FILM ʜ'

conjunctiva ᴊ

The eye's optical apparatus works like a camera. The iris (diaphragm) changes pupil aperture, regulating the amount of light impinging upon the retina. The curved cornea and the convex lens refract the rays, forming an inverted image on the light-sensitive retina (film); the ciliary muscles control the curvature of the lens for sharp focusing of the image.

DEFECTS OF IMAGE FORMATION *

DEFECT

too short

too long

CORRECTED

convex lens

concave lens

HYPEROPIA ʟ¹
(farsighted)

The eyeball is considered normal (emmetropic) if, at rest, it can focus parallel rays from distant objects sharply on the retina. Short eyeballs focus images behind the retina (far-sighted; hyperopia), a defect corrected by using convex lenses (glasses). Long eye-balls, focus images in front of the retina (nearsighted, myopia), a defect corrected by using concave lenses (glasses).

MYOPIA ʟ²
(nearsighted)

REFRACTION *
LIGHT RAY *

Light rays passing through the transparent media of different densities are bent (refraction). The denseness and curvature of the medium determine the degree of refraction. Refraction is necessary for forming a small image on the retina. The eye's refractive media (cornea, humors, lens) together act as a single convex lens system (reduced eye) enabling the formation of small inverted images of distant objects behind the lens on the retina.

NEAR VISION ADJUSTMENTS *

1. ACCOMMODATION *

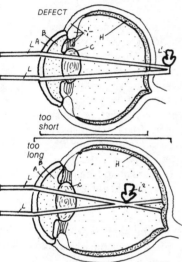

distant viewing

near

As a distant object moves closer, the image moves behind the retina. To keep the image sharply on the retina, the lens must accommodate: ciliary muscles contract, lens ligaments relax, and the lens becomes rounder. This moves the focal point closer to the lens, keeping the image sharply on the retina. With aging, the lens hardens and is less able to accommodate. After 55, accommodation is impossible (presby-opia), and corrective lenses (glasses) for reading, etc., are needed.

PRESBYOPIA c'

degree of accommodation

YRS 10   20   30   40   50   60   70

2. PUPIL CONSTRICTION *

distant ꜰ   near ꜰ

During accommodation, the iris also constricts to narrow the pupil, permitting increased depth of focus. For very close objects, external eye muscles move the eyeballs inward to keep sharp focus.

3. CONVERGENCE *

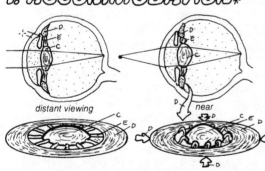

MUSCLE ᴍ

Rays from an approaching object (1) form an image behind the retina (2). The blurred image signals (3) are sensed by the brain visual centers (4), which activate midbrain motor centers (5) to send corrective motor signals for accommodation. Parasympathetic fibers (6) constrict the ciliary muscles (7) to relax the lens for sharp focus-ing and stimulate the iris to close the pupil (8). Sympathetic fibers stimulate the iris to dilate the pupil, (9).

CONTROL OF
NEAR VISION REFLEXES *
CONSTRICTS ← PARASYMPATHETIC N. ₒ

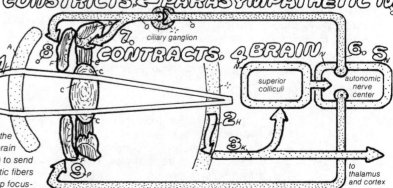

ciliary ganglion

CONTRACTS ᴅ   BRAIN ɴ

superior colliculi

autonomic nerve center

to thalamus and cortex

DILATES ᴘ ← SYMPATHETIC N. ᴘ

THE RETINA & PHOTORECEPTION

STRUCTURE AND NEURONS OF THE RETINA. The *retina* is the eye's *neural* part. It consists of five cell types, three of which compose the retina's three main layers: the outermost *photoreceptor cell* (PR-cell) layer, the middle *bipolar cell* (BP-cell) layer, and the innermost *ganglion cell* (G-cell) layer. Two other cell types, the *horizontal cells* and the *amacrine cells*, are found adjacent to the PR-cell and G-cell layers, respectively. The retina covers most of the eye's inner surface, and its structure is fairly uniform throughout except in two small but very important areas; the *blind spot* and the *fovea* (see below).

The only cells sensitive to light, photoreceptors are located in the outermost zone of the retina so that the light rays must pass through all retinal layers before striking them. Externally, the PR-cells are apposed to the nonneural *pigment cells*, which contain the black pigment *melanin*. Melanin makes the interior of the eye black (as in a camera), preventing back reflection of light. PR-cells transduce the light energy into conductance changes in their membranes. Depolarization of the membrane excites postsynaptic BP-cells, which in turn synaptically influence the G-cells. When G-cells fire action potentials, PR-cell states are communicated to the brain via G-cell axons. These axons congregate toward the *optic disk*, a point in the back of the eye, where they coalesce to form the *optic nerve*, which leaves the eye toward the brain. A lack of photoreceptors in the optic disk makes it insensitive to light (hence, the blind spot).

Horizontal cells both modulate the activity of neighboring PR-cells and influence the interaction of PR-cells with BP-cells. Amacrine cells modulate the activity of the neighboring G-cells and their interaction with BP-cells. Interestingly, retinal neurons, except the G-cells, do not fire action potentials (nerve impulses); instead, they produce slow, graded potentials (nerve signals) that electrotonically stimulate or inhibit other cells.

RODS AND CONES. There are two kinds of PR-cells, *rods* and *cones*. Rods are much more abundant than cones, are very sensitive to light, and function in dim light (*night vision*). The eyes of nocturnal animals contain mainly rods. Cones require more light for activation, are sensitive to *colors*, and function best in *day vision*. Different parts of the retina contain different concentrations of rods and cones. Rods are found mainly in the retinal periphery; cones are heavily concentrated in the fovea. There are also more G-cells per unit area of fovea; these provide direct private channels (nearly one cone to one G-cell) for communication between cones and brain cells. In the retinal periphery, the receptor-neuron ratio is very high (100 rods to 1 G-cell) to increase G-cell light sensitivity. Because of these structural adaptations, the fovea is used for day vision, color vision, and vision requiring great *visual acuity*, such as reading small print. Indeed, when inspecting an object carefully, the eyes move such that the fovea is placed directly along the eye's optical axis. In contrast, the retinal periphery is ideal for night vision, being so sensitive that candlelight can be seen 10 miles away.

MOLECULAR PHYSIOLOGY OF PHOTORECEPTION. Rods are the only sensory receptors that are *not* depolarized by sensory stimulation; instead, they are *hyperpolarized*. We will see the function of this hyperpolarization shortly. How does light cause hyperpolarization? In the dark, sodium ions continuously enter the rods via sodium channels. The influx of sodium ions maintains the cell in a depolarized state. Light causes the sodium channels to close, making the inside less depolarized (i.e., hyperpolarized). How does light close the sodium channels?

In their outer zone, rods contain numerous membranous *disks*, each including millions of molecules of *rhodopsin*, the photoreceptor molecule (visual purple). Rhodopsin is a membrane-bound protein consisting of a protein, *opsin*, and a light-sensitive pigment, *retinal* (retinaldehyde, retinine). Retinal has a hydrocarbon chain that, in its "11-cis" position, enables the retinal to bind with opsin. Light photons cause the chain to switch to the "trans" position. This event, called the *light reaction*, cause the breakdown of rhodopsin into opsin and retinal. This separation activates another part of rhodopsin, which works as an enzyme, stimulating a cascade of reactions leading to a decrease in the concentration of an intracellular messenger, cyclic-GMP (a relative of cyclic AMP). This decrease in messenger concentration closes the sodium channels, hyperpolarizing the rod membrane. The phototransduction mechanism is so sensitive and effective that the human visual system can detect, under proper conditions, a single quantum of light impinging on the retina!

DARK ADAPTATION. Looking at a highly illuminated white sheet for some time markedly but temporarily reduces vision. closing the eyes regains vision. The excess light decomposes rhodopsin, diminishing its supply and reducing vision. In the dark, rhodopsin slowly reforms by recombination of opsin with *vitamin A*, the oxidized form of retinal. During this *dark adaptation*, retinal sensitivity gradually but markedly increases (100,000 times in 30 min.). Vitamin A deficiency can lead to night blindness (inability to see in dim light).

INHIBITION AND EXCITATION AMONG CELLS IN THE RETINA. Objects in the visual fields are a collection of light and dark points, each of which correspondingly forms light and dark point images on the retina. How does the retina inform the brain of this mosaic of light and dark spots? A highly simplified scheme may be as follows: In the dark, rods are depolarized; this activates the inhibitory BP-cells, which in turn inhibit the G-cells. Inhibited G-cells send no messages to the brain, indicating darkness. In the light, hyperpolarized rods no longer activate the inhibitory BP-cells. The G-cells, relieved from inhibition, send nerve impulses to the brain, indicating light.

CN: Use the same colors as were used on the previous page for light ray (A), retina (C), and optic nerve (D).
1. Begin in the upper left corner with the small diagram of the eye. Continue on the peripheral portion of the retina (C¹) and then the fovea area (C²) on the left. For each section, work from the pigment epithelium (E) upwards.
2. Color the lower panel, beginning with the enlarged rod cell (G) on the far left. Note that only the cell membrane (G¹) is colored, along with the disks (M¹) in the outer zone. Color the summary of light and dark reactions in the upper right section. The protein opsin (M), being part of the disk membrane (M¹), receives the same color. Color the detailed explanation of the effect of the absence of light, including the diagram to its right. Do the same with the light reaction.

PERIPHERY

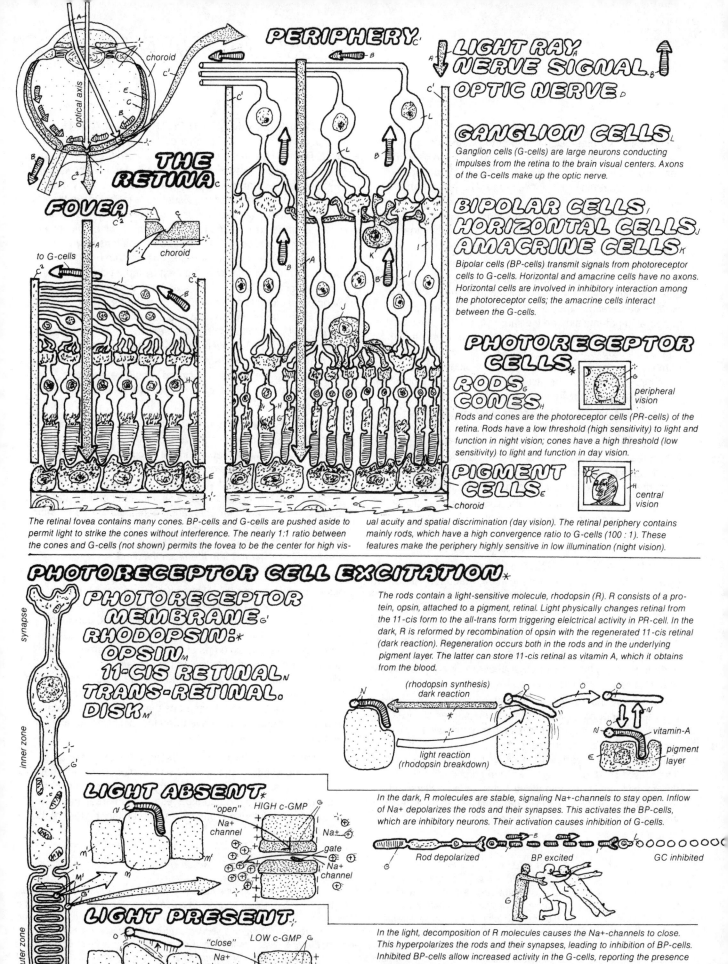

THE RETINA

choroid

optical axis

FOVEA

to G-cells

choroid

LIGHT RAY
NERVE SIGNAL
OPTIC NERVE

GANGLION CELLS

Ganglion cells (G-cells) are large neurons conducting impulses from the retina to the brain visual centers. Axons of the G-cells make up the optic nerve.

BIPOLAR CELLS,
HORIZONTAL CELLS,
AMACRINE CELLS

Bipolar cells (BP-cells) transmit signals from photoreceptor cells to G-cells. Horizontal and amacrine cells have no axons. Horizontal cells are involved in inhibitory interaction among the photoreceptor cells; the amacrine cells interact between the G-cells.

PHOTORECEPTOR CELLS

RODS
CONES

peripheral vision

Rods and cones are the photoreceptor cells (PR-cells) of the retina. Rods have a low threshold (high sensitivity) to light and function in night vision; cones have a high threshold (low sensitivity) to light and function in day vision.

PIGMENT CELLS

central vision

choroid

The retinal fovea contains many cones. BP-cells and G-cells are pushed aside to permit light to strike the cones without interference. The nearly 1:1 ratio between the cones and G-cells (not shown) permits the fovea to be the center for high visual acuity and spatial discrimination (day vision). The retinal periphery contains mainly rods, which have a high convergence ratio to G-cells (100 : 1). These features make the periphery highly sensitive in low illumination (night vision).

PHOTORECEPTOR CELL EXCITATION

PHOTORECEPTOR MEMBRANE
RHODOPSIN:
OPSIN
11-CIS RETINAL
TRANS-RETINAL
DISK

synapse

inner zone

outer zone

pigment cell

The rods contain a light-sensitive molecule, rhodopsin (R). R consists of a protein, opsin, attached to a pigment, retinal. Light physically changes retinal from the 11-cis form to the all-trans form triggering elelctrical activity in PR-cell. In the dark, R is reformed by recombination of opsin with the regenerated 11-cis retinal (dark reaction). Regeneration occurs both in the rods and in the underlying pigment layer. The latter can store 11-cis retinal as vitamin A, which it obtains from the blood.

(rhodopsin synthesis) dark reaction

light reaction (rhodopsin breakdown)

vitamin-A

pigment layer

LIGHT ABSENT

"open"
Na+ channel

HIGH c-GMP

gate

Na+ channel

In the dark, R molecules are stable, signaling Na+-channels to stay open. Inflow of Na+ depolarizes the rods and their synapses. This activates the BP-cells, which are inhibitory neurons. Their activation causes inhibition of G-cells.

Rod depolarized   BP excited   GC inhibited

LIGHT PRESENT

"close"
Na+ channel

LOW c-GMP

inside   outside

In the light, decomposition of R molecules causes the Na+-channels to close. This hyperpolarizes the rods and their synapses, leading to inhibition of BP-cells. Inhibited BP-cells allow increased activity in the G-cells, reporting the presence of light to the brain.

Rod hyperpolarized   BP inhibited   GC excited

BRAIN AND VISION

From the light and dark spots on its surfaces, the *retina* generates a two-dimensional map of the visual world. The receptive field of a ganglion cell is circular, so each ganglion cell reports on the presence or absence of light and its intensity in a circular spot on the retina. The brain then uses structures of its visual system, chiefly the visual cortical areas, to synthesize a three-dimensional picture of the visual field from the information it receives from the two retinas.

SUBCORTICAL VISUAL PATHWAYS AND CENTERS. Upon exiting from the eye, the axons of ganglion cells form the *optic nerve*. The optic nerves of the two eyes converge at the *optic chiasm*, where the fibers coming from the nasal segment of each retina cross over to the opposite side, and the temporal segments stay on the same side. After crossing, the optic nerve is called the *optic tract*. Due to the crossing, the right and left optic tracts carry signals relating to the left and right *visual fields*, respectively. Thus, neurologists can use tests of deficiencies in the visual fields to determine possible locations of lesions in the visual pathway.

A portion of optic tract fibers enters the midbrain and terminates in the *superior colliculus* and *reticular formation*. The superior colliculus helps coordinate accommodation and light reflexes involving the lens and the pupil as well as some eye and head movement reflexes. The reticular formation functions in cortical arousal, excitability, and in sleep.

The bulk of the optic tract fibers, concerned with visual perception, end in the visual centers of the thalamus, the *lateral geniculate body*. Here the first synapse is made, and considerable integration of impulses takes place.

PRIMARY VISUAL CORTEX. The lateral geniculate body gives rise to the fibers of *optic radiation*, which fan out toward a portion of the posterior lobe of the cerebral hemispheres called the *primary visual cortex*. The entire retina is represented on the cortex, point to point, in a precise and organized way so that each retinal half and each quadrant can be clearly demarcated. People with a gunshot wound or other circumscribed lesion in their primary visual cortex become blind with respect to a limited and specific area of the visual field.

Although retinal representation in the visual cortex is very precise, it is not equal with respect to the shares of *fovea* and *periphery*. The fovea, which is less than 1 mm wide, has a very large representation; the remaining, larger part of the retina (the periphery) has a relatively smaller representation. The reason for this inequality lies in the higher density of the ganglion cells emerging from the fovea as well as in the essential role of the fovea in visual acuity, spatial perception, and color vision, functions that demand more neuronal units and brain area.

COLUMNAR ORGANIZATION OF THE VISUAL CORTEX. Visual cortex neurons, like other cortical areas, are arranged in six layers. The stellate cells in layer IV are the only ones receiving specific visual input from the thalamus. Three major discoveries were made recently about the physiology and functional organization of the visual cortex. First, as in the somatic sensory cortex, neurons positioned perpendicularly to the surface form a functional "column" in which all cells have the same receptive field. Second, columns responding to the right and left eyes alternate, one column being excitable by input from one eye, the next by input from the other eye (*ocular dominance columns*). Third, cells of layer IV, the principal recipients of visual input from the thalamus, do not have a circular receptive field and do not respond to point sources of light. Instead, they are sensitive to such stimuli as *lines* and *edges*. Moreover, a line stimulus must have a particular *orientation* to activate a cortical cell. These orientation-specific neurons, called *simple cells*, will not respond if the line with their preferred orientation is displaced in the receptive field.

There are, however, neurons that respond to the same stimulus in any position within the receptive field as long as the same orientation is maintained. These *complex cells* are located within the same column above and below the simple cells, in layers II, III, and V, and receive signals from the simple cells. The hierarchy of all simple and complex cells dealing with a particular orientation constitutes an *orientation column*. In a way, the complex cells may also be construed as movement detectors because they continue to fire during the stimulus movement in the receptive field.

VISUAL ASSOCIATION AREAS. Surrounding the primary visual cortex are the *visual association areas*, which are well developed in primates. The visual association areas receive their input mainly from the primary visual cortex, although some input is also provided from the subcortical visual centers. The cells and columns of the association areas further analyze the processed output of the primary visual cortex (lines, orientations, movements) to encode more complex and detailed visual patterns. Indeed, in the monkey's higher visual association areas, certain neurons that respond to an intact model of a monkey face have been found. Distorted face models will not evoke activity!

In the brain cortex, the visual association areas are adjacent to the auditory and somatic association areas. Moreover, certain *higher-order association areas* interposed between the visual, somatic, and auditory association areas integrate the output of all these areas to create a unitary picture of the sensory world.

CN: Use darker colors for D-I.
1. Begin with the large illustration in the upper left. The two colors of the visual field (A, B) are transmitted along the visual pathways to the primary visual cortex in the back of the brain (which in this example is represented by the reverse placement of those two colors). Throughout the rest of the page, the primary visual cortex will receive just the A[1] color. Note that in this example the various structures making up the visual pathways are identified by a fat arrow pointing to their location on the pathways. These arrows receive the colors of the visual field (A, B).
2. Color the example of retinal representation on the primary visual cortex, which in this case receives the colors of the periphery (J) and fovea (K).
3. Color the route of visual signals in the cerebral cortex.
4. Color the organization of cells, starting with the cells on the left, shown responding to the orientation of a pitcher. Then do a section of the cortex in which an example of neural connectivity is displayed. Next color the example on the right demonstrating the two aspects of cell column organization: by orientation (which includes both simple and complex cells) and by ocular dominance (an alternation of right and left eye). For purposes of simplification, cells are shown only in layers II, III, and V.

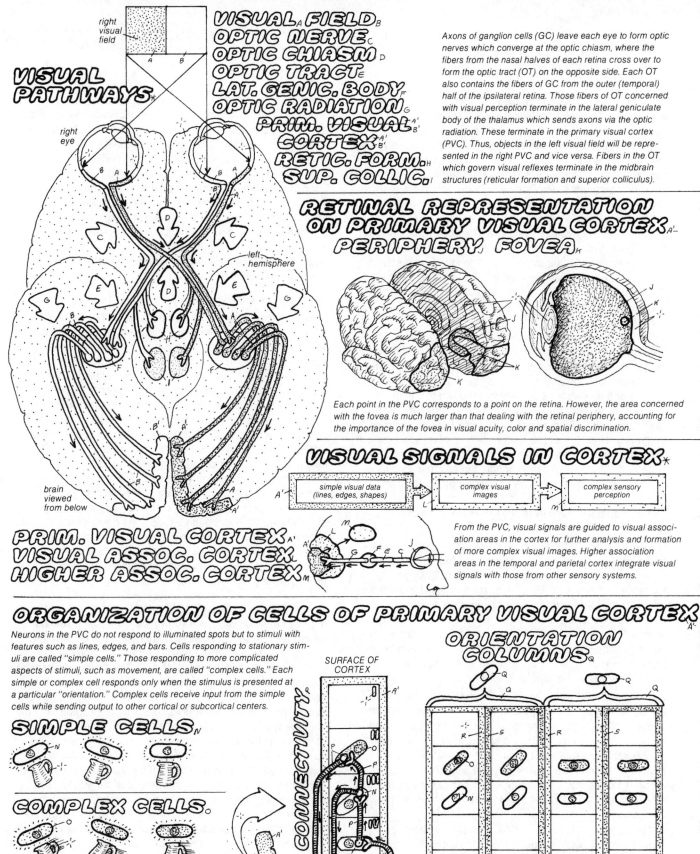

VISUAL PATHWAYS *

right visual field

right eye

left hemisphere

brain viewed from below

VISUAL FIELD A B
OPTIC NERVE C
OPTIC CHIASM D
OPTIC TRACT E
LAT. GENIC. BODY F
OPTIC RADIATION G
PRIM. VISUAL CORTEX A' B'
RETIC. FORM. H
SUP. COLLIC. I

Axons of ganglion cells (GC) leave each eye to form optic nerves which converge at the optic chiasm, where the fibers from the nasal halves of each retina cross over to form the optic tract (OT) on the opposite side. Each OT also contains the fibers of GC from the outer (temporal) half of the ipsilateral retina. Those fibers of OT concerned with visual perception terminate in the lateral geniculate body of the thalamus which sends axons via the optic radiation. These terminate in the primary visual cortex (PVC). Thus, objects in the left visual field will be represented in the right PVC and vice versa. Fibers in the OT which govern visual reflexes terminate in the midbrain structures (reticular formation and superior colliculus).

RETINAL REPRESENTATION ON PRIMARY VISUAL CORTEX A'
PERIPHERY FOVEA K

Each point in the PVC corresponds to a point on the retina. However, the area concerned with the fovea is much larger than that dealing with the retinal periphery, accounting for the importance of the fovea in visual acuity, color and spatial discrimination.

VISUAL SIGNALS IN CORTEX *

| simple visual data (lines, edges, shapes) | ⟹ | complex visual images | ⟹ | complex sensory perception |

From the PVC, visual signals are guided to visual association areas in the cortex for further analysis and formation of more complex visual images. Higher association areas in the temporal and parietal cortex integrate visual signals with those from other sensory systems.

PRIM. VISUAL CORTEX A'
VISUAL ASSOC. CORTEX L
HIGHER ASSOC. CORTEX M

ORGANIZATION OF CELLS OF PRIMARY VISUAL CORTEX A'

Neurons in the PVC do not respond to illuminated spots but to stimuli with features such as lines, edges, and bars. Cells responding to stationary stimuli are called "simple cells." Those responding to more complicated aspects of stimuli, such as movement, are called "complex cells." Each simple or complex cell responds only when the stimulus is presented at a particular "orientation." Complex cells receive input from the simple cells while sending output to other cortical or subcortical centers.

SIMPLE CELLS N

COMPLEX CELLS O

NEURAL CONNECTIVITY P

SURFACE OF CORTEX

SUB-CORTEX
output   input

ORIENTATION COLUMNS Q

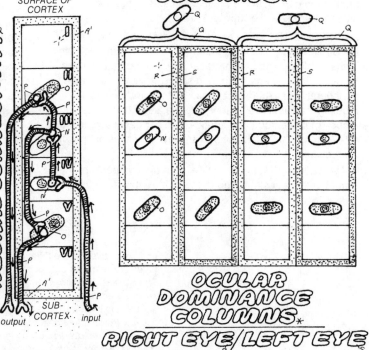

Simple and complex cells corresponding to a particular orientation are arranged in columns running perpendicular to the cortical surface (orientation columns). Simple cells are found mainly in layers III and IV of the cortex; complex cells are in layers II, III, and V. These "orientation columns" are placed side by side across the cortex in such a way that those receiving input from one eye alternate with those connected to the other eye (ocular dominance columns).

OCULAR DOMINANCE COLUMNS *
RIGHT EYE / LEFT EYE
R   S

SOUNDS AND THE EAR

SOUND WAVES. The *ear* transduces *sound wave* energy into electrical nerves impulses that are relayed to the *central auditory system*, where they are perceived as meaningful sounds. Sound is produced by particles vibrating in a physical medium (air, water, or solids). Sounds, which, unlike photons, cannot travel in a vacuum, are characterized in terms of two physical parameters: *amplitude* and *frequency*. A measure of sound energy is the peak amplitude, or the height of the sinusoidal oscillation. To describe the degree of *loudness* as perceived by our auditory system, we express sound intensity as a logarithmic unit, *decibels* (dB), relative to a standard sound pressure (0.0002 dyne/cm^2). The loudness range of common sounds varies between 20 dB (a whisper from 1 m [3 ft] away) to 130 dB (jet engines). Each 20 dB increase in loudness corresponds to a 10-fold increase in the actual intensity of the sound.

Frequency refers to the number of oscillation cycles per second or Hertz (1 Hertz [Hz] = 1 cycle/sec.). Sound waves range in frequency from 1 to over 100,000 Hz. The perceptual counterpart of sound frequency is *pitch*. The dominant frequency of the newborn's cry is high, resulting in a high-pitched sound. After puberty, men develop larger larynxes (voice box) with thicker vocal cords, which lower their sound frequency. Thus, the sound frequency (Hz) of an adult male is lower than that of an adult female, resulting in the difference in pitch. A third perceptual component of sound is its quality, or *timbre*, a property not easily explicable by the physical characteristics of sound waves. Identical musical tones of the same loudness emitted from two different instruments are thus perceived quite differently.

HEARING ABILITIES. The human ear at the peak of its performance (during childhood and the early teens) can detect sound waves in the frequency range of 20 to 20,000 Hz. However, the sensitivity (i.e., the hearing threshold) is not the same for the various frequencies. Highest sensitivity is observed for sounds in the range of 1 to 4 KHz. Different animals' hearing abilities vary. Dogs can hear sounds up to 50 KHz (hence the dog trainer's whistle is often inaudible to the human ear). Bats are actually capable of hearing ultrasound waves (i.e., above 20 KHz).

EAR STRUCTURE. The ear in all mammals has three parts: the outer, middle, and inner ear. The *outer ear* consists of the *pinna* (earlobe) and the *ear canal*, which together act like a funnel, collecting and chanelling sound waves from a large peripheral field into the ear canal. In dogs and rabbits, the mobile pinna acts like a radar antenna, searching for and focusing on the source of sounds. The ear canal acts like a resonance chamber, helping to amplify the waves of specific frequencies.

The ear canal terminates in the *eardrum* (tympanic membrane), a thin but strong round sheet of elastic connective tissue that vibrates in response to sound wave pressure. The eardrum is the first component of the *middle ear*, which also contains three distinct *ossicles* (three small bones, each the size of a matchhead). They form a lever system with one end connected to the eardrum and the other end to the inner ear's *oval window*. The eardrum vibrates in unison with the first bone, the *malleus* (hammer), and the second bone, the *incus* (anvil), which is attached almost perpendicularly to the malleus, forming a lever. The incus displaces the third bone, the *stapes* (stirrup), which in turn vibrates the oval window of the inner ear. These mechanical systems and properties of the inner ear increase the effective pressure on the oval window 20 times.

THE INNER EAR. The vibrations of the oval window set into motion various mechanical components of the *cochlea*, the hearing organ of the inner ear. (The inner ear also houses the vestibular apparatus for balance [see plate 97].) The cochlea's function is to transform the sound waves into electrical signals for transmission to the brain. The cochlea (snail) is a complicated coiled tube consisting of several components, some of which play purely mechanical roles and others of which are important in mechanoelectrical transduction. The coiling serves to maximize the tube's length while keeping its total size minimal. A *basilar membrane* running along the length of the cochlea separates it into upper (scala vestibuli) and lower (scala tympani) chambers. The two chambers communicate at the cochlea's apex, the helicotrema. A thinner membrane (Reissner's membrane) running through the scala vestibuli along the basilar membrane creates a third middle chamber, the *scala media*, which is filled with the fluid *endolymph*. Other chambers contain *perilymph*, which has a different ionic composition. These variations are important in the physiology of the cochlea.

HAIR CELLS AND AUDITORY TRANSDUCTION. Movements of the oval window form waves in the scala vestibuli's perilymph. These waves are transmitted to the endolymph and then to the underlying basilar membrane, forcing it to vibrate (see plate 96 for more details). The basilar membrane supports the *hair cells* of the *organ of Corti*, the true *mechanoelectrical transducers*. The transduction elements, being specialized ciliary structures (*stereocilia*), are located at the apical side of the hair cells. The tips of the stereocilia are in contact with the *tectorial membrane* (roof membrane), and the cell bodies of the hair cells communicate via chemical synapses with the end branches of the *auditory nerve fibers*. Basilar membrane vibration displaces the stereocilia, producing a receptor potential that is synaptically transmitted to the endings of the auditory nerve fibers. Here *nerve impulses* (action potentials) are produced and conducted by the primary auditory afferents to the brain auditory centers.

CN: Use dark colors for A, J, K, P, and R.
1. Begin with the upper panel, coloring all numbers and titles. Note that the three frequencies (C) illustrated are in the low amplitude range (whisper). The one example of high amplitude (B) (loud sound) is on the far right.
2. Color the structures and functions of the ear.
3. Color the structure of the cochlea (L) next, beginning with the overall view on the far left, then the enlarged section taken from near the cochlea's base, and then the horizontal diagram representing the entire cochlea with the spirals straightened out. Note that the scala vestibuli (N) and the scala tympani (N^2) receive the same color because they communicate and contain the same fluids (perilymph) (P^1). In this horizontal illustration, the tectorial membrane (R) has been deleted for purposes of simplification. Color the bottom cartoon depicting sequence of events involved in the transmission of sound waves.

CHARACTERISTICS OF SOUND WAVES

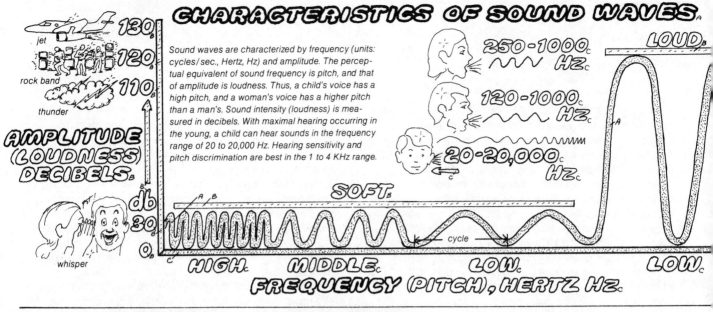

AMPLITUDE (LOUDNESS) DECIBELS

Sound waves are characterized by frequency (units: cycles/sec., Hertz, Hz) and amplitude. The perceptual equivalent of sound frequency is pitch, and that of amplitude is loudness. Thus, a child's voice has a high pitch, and a woman's voice has a higher pitch than a man's. Sound intensity (loudness) is measured in decibels. With maximal hearing occurring in the young, a child can hear sounds in the frequency range of 20 to 20,000 Hz. Hearing sensitivity and pitch discrimination are best in the 1 to 4 KHz range.

jet — 130
rock band — 120
thunder — 110
db 30
0

whisper

250-1000 Hz
120-1000 Hz
20-20,000 Hz

LOUD

SOFT

cycle

HIGH   MIDDLE   LOW   LOW
FREQUENCY (PITCH), HERTZ Hz

FUNCTIONS OF THE EAR

OUTER EAR (funneling)
PINNA
AUDITORY CANAL
TYMPANIC MEMBRANE

MIDDLE EAR (amplification)
MALLEUS
INCUS
STAPES

INNER EAR (transduction)
OVAL WINDOW
ROUND WIN.
COCHLEA
AUDITORY N.

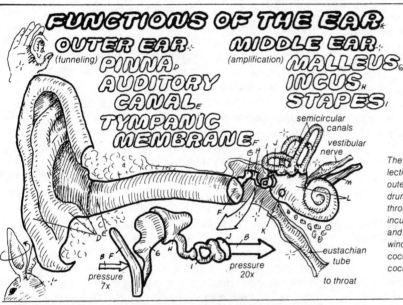

semicircular canals
vestibular nerve
eustachian tube
to throat
pressure 7x
pressure 20x

The ear's outer, middle, and inner parts perform the functions of collecting, amplifying, and transducing sound energy, respectively. The outer ear's pinna and ear canal funnel sound waves onto the eardrum, causing it to vibrate. The eardrum vibration is amplified 20-fold through the lever action of the middle ear's three ossicles (malleus, incus, and stapes) and differential vibrating surfaces of the eardrum and the inner ear's oval window. Stapes movement displaces the oval window and subsequently the basilar membrane of the inner ear's cochlea, generating frequency-dependent traveling waves in the cochlea which transduces sound waves into nerve impulses.

SOUND TRANSMISSION IN THE COCHLEA

SCALA VESTIBULI/PERILYMPH
SCALA TYMPANI/PERILYMPH
SCALA MEDIA / ENDOLYMPH
BASILAR MEMBRANE
ORGAN OF CORTI, HAIR CELL
TECTORIAL MEMBRANE
AUDITORY N., GANGLION

The cochlea is a coiled, blind duct, wide at the base and narrow at the apex. The basilar membrane (BM) supporting the organ of Corti runs along the cochlea's length, dividing it into two chambers: an upper scala vestibuli and a lower scala tympani. A thin membrane (Reisner's) running above the organ of Corti creates a third middle chamber, the scala media. The scalae vestibuli and tympani are filled with perilymph fluid and communicate at the apex. The scala media contains endolymph, which bathes the organ of Corti and its hair cells. The oval and round windows at the two ends of the perilymph absorb and release the sound waves, respectively, and maintain the perilymph in a compressed state.

COCHLEA
base   apex

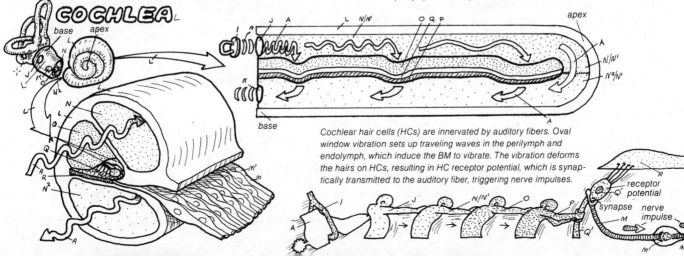

apex
base

Cochlear hair cells (HCs) are innervated by auditory fibers. Oval window vibration sets up traveling waves in the perilymph and endolymph, which induce the BM to vibrate. The vibration deforms the hairs on HCs, resulting in HC receptor potential, which is synaptically transmitted to the auditory fiber, triggering nerve impulses.

receptor potential
synapse   nerve impulse

AUDITORY DISCRIMINATION; AUDITORY BRAIN

FREQUENCY DISCRIMINATION BY THE COCHLEA. In the *cochlea*, two structures are functionally most important for transduction and frequency analaysis: the *hair cells* (HCs) and the *basilar membrane* (BM). Along the length of the cochlea, the HCs ($\approx$ 23,500 in humans) can be divided into two groups: a single row (3500) of *inner HCs* and three rows (20,000) of *outer HCs*. Most of the fibers in the *auditory nerve* (about 30,000) innervate the inner HCs while a smaller portion innervates the more numerous outer HCs. Thus, compared to the outer HCs, the inner HCs have more direct communication channels with the brain auditory centers. Although this arrangement is functionally important, the HCs by themselves could not account for the cochlea's ability to perform frequency analysis.

The BM, which supports the HCs, plays a more crucial role in frequency discrimination because it shows clear structural variation along the cochlea's length. The BM is narrow at the cochlea's base near the *oval window*, widening as it approaches the apex. Thus, the BM stiffness varies systematically along the cochlea. During the 1940s, Georg von Békésy (Nobel laureate) noted that vibrations of the oval window set up *traveling waves* in the *perilymph*. Depending on the sound frequency, each wave produces a peak amplitude at a *distinct* location along the cochlea: the higher the sound frequency, the nearer the peak amplitude to the oval window.

At its peak amplitude, the traveling wave generates the largest displacement in the BM, making the parts of the BM near the oval window more sensitive to *high-frequency* sounds and the apical parts of the BM more sensitive to *low-frequency* sounds. The respective regions in the middle portions of the BM detect frequencies between these extremes. This is called the *place principle* of frequency discrimination. Because each part of the BM supports a separate set of HCs and their nerve fibers, the *auditory nerve fibers* emerging from the base of the cochlea transmit messages about high-frequency sounds while those farther away from the base do so for increasingly lower frequency sounds.

LOUDNESS DISCRIMINATION. Regardless of their particular frequencies, sound waves of higher intensity would produce bigger BM displacement on the respective cochlear regions. This in turn creates larger receptor potentials in the HCs of that region and consequently a higher rate of firing of nerve impulses in the associated auditory nerve fibers. Thus, the brain auditory centers detect loudness by comparing the changes in firing rate of impulses in a particular auditory nerve fiber relative to the resting levels. Differences in sound frequency are detected by noting which fiber (or neighboring set of fibers) is active compared to other auditory nerve fibers.

AUDITORY CENTERS AND CONNECTIONS. Brain auditory centers and pathways are complicated and still not well understood. The primary auditory neurons have their cell bodies in the cochlea's *spiral ganglia* located inside the cochlea. The central axons of these bipolar neurons, after emerging from the cochlea, merge with the fibers from the vestibular organ (see plate 97) to form the VIII cranial (*vestibulo-acoustic*) nerve, which enters the *medulla*. The auditory fibers terminate in the *cochlear nuclei*. From this synaptic station, numerous connections are made with other brain centers: (1) the medullary auditory centers serving in sound localization, in auditory reflex functions (e.g., regulating middle and inner ear muscles), and in the startle reflex; (2) the *midbrain centers* of *inferior colliculus* and *reticular formation*, important for regulating auditory reflexes related to head and eye movements and for sound localization. Auditory input to the reticular formation has major influences in arousal, attention, and wakefulness; (3) from inferior colliculi, higher projections serving in auditory perception are relayed to the *medial geniculate nuclei* of the *thalamus*. Due to extensive cross-connections, auditory projections are bilateral, with the contralateral projections being more intensive.

AUDITORY CORTEX. From the thalamus, *auditory radiation fibers* project to the *primary auditory cortex* located in the temporal lobe (mostly hidden in the floor of the Sylvian gyrus). As with the lower auditory centers, the primary auditory cortex also has a representation map (a *tonotopic map*) of the cochlea so that neurons in certain parts of the auditory cortex detect higher frequency sounds while neurons in another part respond to lower frequency sounds.

We seldom encounter pure tones in everyday life; indeed, speech and most environmental sounds consist of complicated patterns of frequency and loudness, continuously varying in time. Cortical areas analyze and integrate these sounds to form meaningful auditory images in our brain. As in the visual and somatic cortex (see plates 87, 94), some auditory cortex neurons respond only to complex sound patterns. For example, some fire when the pitch is altered in time; others fire when loudness of a particular sound frequency is changed. Damage to the primary auditory cortex in one hemisphere is not debilitating, but bilateral damage causes severe deficits in discrimination and temporal ordering of complex sounds and appreciation of sound qualities. Discrimination of pure tones is relatively unaffected because this function is also carried out by lower hearing centers.

AUDITORY ASSOCIATION CORTEX. From the primary auditory cortex, fibers project to the neighboring *auditory association areas*, where auditory signals are further analyzed and matched with data from other senses. Humans comprehend words and speech sounds in these higher association centers as well as in the adjacent Wernicke's area, well known to be involved in sensory aspects of language and speech functions (see plate 105).

CN: Use the same colors as on the previous plate for sound waves (A), cochlea (B), tectorial membrane (C), hair cells (D), auditory nerve (E), and basilar membrane (H). Light colors for M and N.
1. Begin with the structures of the organ of Corti at the top, moving down to the diagram of frequency analysis in the basilar membrane. Follow this to the right, where the auditory pathways (I) and the
primary auditory cortex (M) are shown.
2. Color the small brain drawing with the enlarged auditory cortex areas below it.
3. Color the localization of sound drawing, noting that the arrows, identified as time and loudness differences, relate to the sound waves emanating from the turning fork.
4. Color the effects of age on hearing.

SOUND WAVE TRANSDUCTION & DISCRIMINATION.

"ORGAN OF CORTI"

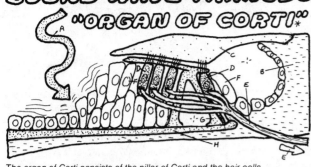

COCHLEA.
TECTORIAL MEMBRANE.
HAIR CELL.
AUDITORY NERVE.
SUPPORTING CELL.
PILLAR OF CORTI.
BASILAR MEMBRANE.

The organ of Corti consists of the pillar of Corti and the hair cells (HCs). HCs, the transducing elements of the ear, are located on the supportive cells above the basilar membrane (BM). There are three outer and one inner rows of HCs; each HC makes synaptic contact with auditory nerve fibers. Oval window vibration produces traveling waves in the choclea's perilymph, displacing the BM. The BM's displacement waves travel from the base to the apex of the cochlea. The traveling wave pattern varies with sound frequency: at high frequencies, the peak occurs at the base, where the BM is stiff; at lower Hz, the peak is shifted toward apical regions, where BM fibers are less stiff. Thus, the auditory nerve fibers emerging from the cochlea's base are concerned with sounds of high Hz, and those emerging toward the apex deal with sounds of increasingly lower Hz. An increase in the sound intensity (loudness) raises the amplitude of the traveling waves. This is ultimately reported to the brain by the increase in the nerve impulse number in the corresponding auditory fibers.

Primary auditory fibers converge to form the auditory nerve, which, along with the vestibular nerve (not shown), enter the medulla as the VIII cranial nerve. Auditory fibers synapse with neurons in the cochlear nuclei. From here fibers concerned with auditory perception ascend and terminate in the medial geniculate body (thalamus). In this path, some fibers cross over and other ascend uncrossed. From the thalamus, fibers of auditory radiation carry signals to neurons in the primary auditory cortex located on the superior aspects of the temporal lobe. Fibers concerned with auditory reflexes ascend (crossed and uncrossed) to midbrain centers (inferior colliculus and reticular formation).

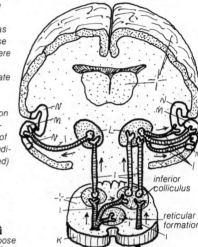

inferior colliculus

reticular formation

superior olive nuclei

cochlear nuclei

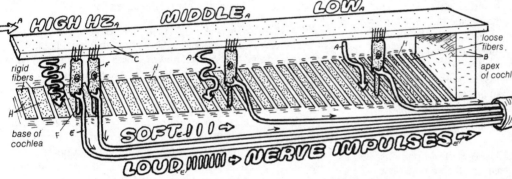

HIGH HZ. MIDDLE. LOW.

loose fibers

apex of cochlea

rigid fibers

base of cochlea

SOFT.

LOUD. NERVE IMPULSES.

AUDITORY PATHWAYS.

COCHLEAR REPRESENTATION ON PRIMARY AUDITORY CORTEX.

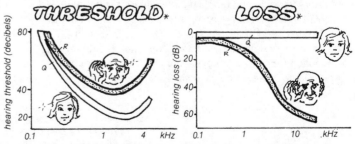

low Hz high

MEDULLA, MIDBRAIN, MEDIAL GENICULATE BODY, PRIMARY AUDITORY CORTEX, ASSOCIATION CORTEX.

Neurons of central auditory centers respond differentially to sounds of different frequencies. The cochlea has an orderly representation at the primary auditory cortex (PAC), so some PAC neurons respond best to high Hz sounds and others, to low Hz sounds. Neurons of PAC, like other cortical areas, are arranged in "columns" so that each column is specialized for a very narrow range of sound frequencies. Neurons of PAC columns provide the major input to the highest auditory association areas located adjacent to PAC. In humans, these association areas communicate with still higher cortical areas dealing with language and speech.

LOCALIZATION OF SOUND.
1. TIME DIFFERENCE.
2. LOUDNESS DIFFER.

(< 3,000 Hz)

(> 3,000 Hz)

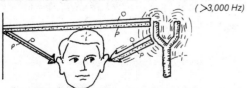

Localization of sound (which direction?) involves interaction of the two ears. If a sound comes from a source to the left, it will be relatively louder to the left ear and reach it faster. By comparing the signals arriving from the two ears, the brain determines the sound source. Comparison of the time difference works best for sounds below 3 kHz and that of loudness difference for sounds above 3 kHz.

AGE AND HEARING LOSS.

THRESHOLD*

hearing threshold (decibels)

80

40

20

0.1   1   4   kHz

LOSS*

hearing loss (dB)

0

20

40

60

0.1   1   10   .kHz

YOUNG
OLD

In old age, hearing becomes less sensitive (presbycusis), so sounds must be louder to be audible. This hearing loss is particularly severs in the high Hz range, possibly due to selective aging damage in the basal cochlea. This interferes with audibility of consonants. As one ages, hearing loss extends to lower Hz and eventually to the 1-4 KHz range used in normal speech.

THE SENSE OF BALANCE

Standing, postural control, and many normal movements often must occur against the earth's gravity. To counter and adapt to this force, several sensory and motor mechanisms have evolved, e.g., the general proprioceptive sense organs in joints, etc., and a special sense organ, the inner ear's *vestibular apparatus* (VA), which detects changes in the head's position in space. The VA is in fact one of the earliest of the special senses to arise in evolution. Through its connections with the brain and spinal cord, the VA sends signals to activate adaptive motor responses aimed at maintaining equilibrium. These responses involve the postural and support muscles of the limbs and trunk and the muscles that move the head and eyes.

VA FUNCTIONS. Knowledge of VA functions was gained by lesion studies in animals. Animals with bilateral removal of the VA are unable to stand upright, especially if blindfolded. When attempting to move forward or backward or to turn, they fall. If the VA is removed on only one side they exhibit postural and equilibrium deficits on that side.

The VA consists of two components: (1) the *semicircular canals* (SCCs) and (2) the *utricle* and *saccule*. Selective lesions of the utricle/saccule organs (see below) alone interfere with balance in the upright position or during translational (linear) acceleration (i.e., when the body is moved forward, backward, up, or down). Lesions of the SCCs interfere with balance during turning movements. In broad terms, the utricle/saccule organs (maculae) are often activated in response to alterations in *static equilibrium*; the SCCs are excited mostly in response to changes in *dynamic equilibrium*.

DETECTION OF LINEAR ACCELERATION BY THE MACULAR ORGANS. The saccule (little sac) and utricle (little bag) are two small swellings in the inner ear wall, each containing one *macula* (macular organ), which is bathed by *endolymph*. Each macula is a mechanoelectrical transducing receptor organ containing *hair cells* (HCs). Each HC contains numerous apical *stereocilia* and a single *kinocilium*. Branches of the *vestibular nerve* engulf the HCs basally. The cilia are embedded in an *otolithic membrane* that contains small (3-19 micron long) calcium carbonate crystals, called the *otoliths* (ear stones).

When the head moves (accelerates) linearly in any direction, the maculae move along. But the otoliths, being denser than the surrounding fluid, lag behind. This distorts the position of the stereocilla, resulting in the production of a receptor potential in the HCs. This potential synaptically triggers action potentials in the vestibular nerve fiber, which are then sent to the brain. Saccule and utricle orientation are such that their maculae inform the brain about any linear movement of the head and consequently of the body. Macular activation occurs, however, mainly during the onset (acceleration) and termination (deceleration) of the motion. Thus, in a moving car or elevator, we perceive motion only during the initial and terminal phases.

DETECTION OF ROTATIONAL ACCELERATION BY SCCs. The SCCs of the VA detect accelerations associated with rotational motion. The three fluid-filled canals are situated perpendicular to one another. Therefore, the rotational motion of the head in any direction stimulates at least one of the three SCCs. Each canal contains at each end one mechanoelectrical transducing sense organ called the *ampulla*. Like the macula, each ampulla contains HCs with similar ciliary structures. These cilia are, however, embedded in a gelatinous layer called the *cupula* (little cup), which extends across the canal's lumen and is attached to the canal's other wall. Rotational acceleration of the head moves the SCCs, displacing the attached ampulla/cupula along the same direction. But the endolymph fluid in the canal lags behind, due to its inertia. This differential movement of the fluid in relation to the cupula distorts the stereocilia, creating a receptor potential in the HCs. The receptor potentials trigger action potentials in the vestibular nerve fibers. The action potentials (nerve impulses) then inform the brain's vestibular centers about the particular rotational motion.

CENTRAL VESTIBULAR SYSTEM. The vestibular fibers leave the VA, join the auditory fibers to form the vestibulo-acoustic nerve (VIII cranial nerve), which enters the *medulla*, and terminate mainly in the *vestibular nuclei*. The vestibular nuclei act as integrating centers for vestibular information processing.

The vestibular nuclei motor output is chiefly directed at two targets. The first is the lower motor centers in the *spinal cord*. Here impulses activate mostly the gamma motor neurons, particularly to extensor muscles, increasing the tone and tension of the *body's muscles* (see plate 89). These responses are essential for postural control and balance. The second target is the upper *midbrain motor nuclei*, particularly those regulating *eye and head* movements. In addition, vestibular nuclei are connected reciprocally with the vestibular part of the *cerebellum* (the flocculo-nodular lobe), which controls appropriate execution of the motor commands (see plate 91). Vestibular nuclei also send fibers to the *higher somatic sensory systems* of the brain involved in the perception of head/body position (body image).

VESTIBULAR REFLEXES. The vestibular system's motor responses are executed through many *vestibular reflexes*. These are generally inborn, fast, and highly purposive. Thus, when we tilt to the left, our left leg extensor muscles are activated to ensure balance. Excessive tilting makes us lift the opposite leg and arm. If our head turns to the left while we view a stationary object to the right, *vestibulo-ocular reflexes* help maintain our gaze by moving our eyes to the right. In the absence of visual cues (as in the blind person), vestibular sensations and responses are the only cues available for orientation in space.

CN: Use dark colors for D and K.
1. Begin with the semicircular canals and detection of rotational acceleration. Color the diagram to the left of the semicircular canals. This demonstrates the three spatial planes occupied by the canals. Color the canals in the large drawing, and continue up through the various blow-ups. Color the directional arrows of the cartoons.

2. Color the detection of linear acceleration by the saccule and utricle, beginning with the section of the macula (upper right), then the blow-up and the diagram on top of the heads.
3. Color the vestibular reflex by beginning in the lower left corner and following the numbered sequence. You may wish to color the relevant parts of the diagram on the right simultaneously.

The inner ear's vestibular apparatus detects changes in the head's (and thus the body's) position in space, signaling the brain motor centers to adjust posture and maintain balance. The vestibular apparatus has two parts: 1. three semicircular canals (SCC), which respond to rotational acceleration in head movements (dynamic balance); 2. saccule and utricle (macular organs), which respond to head tilts and linear acceleration (static balance).

CRISTA

ROTATIONAL ACCELERATION*
(DYNAMIC BALANCE)

SEMICIRCULAR CANALS
POSTERIOR A
SUPERIOR B
LATERAL C
AMPULLA D
CUPULA E
SUPPORTING CELL F
HAIR CELL G
VESTIBULAR N. H

vestibulo-acoustic nerve (VIII cranial)

auditory nerve

vertigo
nystagmus

sea-sickness

cochlea

The 3 SCCs are fluid-filled tubes positioned at right angles to one another. Thus, during any rotational movement of the head, at least one canal will be active. Each canal has an ampulla, containing its sensory organ (crista). The crista contains hair cells (HCs) whose stereocilia are embedded in the gelatinous cupula. When the head begins to rotate in a particular direction, the canal fluid initially lags behind due to its inertia; this shears the cupula (which is moving with the canal wall), causing the stereocilia to bend in the opposite direction. This activates HC, which causes the vestibular nerve fiber to increase its signal to the brain to initiate motor reflexes in the opposite direction to keep balance. The fluid will catch up with the canal motion 20 seconds later, reducing the discharge from SCC.

MACULA

LINEAR ACCELERATION*
(STATIC BALANCE)

UTRICLE I SACCULE J
MACULA K
OTOLITH L
GELATINOUS LAYER M

posture

horizontal acceleration

vertical acceleration

The maculae are the sensory organs in both the saccule and utricle. Each macula consists of hair cells (HCs) covered by a gelatinous layer containing small calcite rocks (otoliths). The otoliths are covered by a fluid. Horizontal or vertical acceleration of the head in space exerts a stronger pull on the otoliths, which are denser than the surrounding fluid. This causes the otoliths to displace the stereocilia of the HCs and changes its receptor potential. The latter change will stimulate the fibers of the vestibular nerve (which are attached to the HCs) to increase their nerve impulses according to the intensity of displacement. At rest, these fibers discharge tonically.

VESTIBULOSPINAL REFLEXES*

EYE MUSCLES 7
MIDBRAIN MOTOR NUCLEI 6
HEAD & NECK MUSCLES

5

VESTIBULAR NERVE 2
VESTIBULAR NUCLEI 4
CERE-BELLUM 3

VESTIBULAR APPARATUS 1
8
SPINAL CORD
9
BODY MUSCLES

brain

spinal cord

The exerciser experiences linear and angular acceleration. His vestibular apparatus (1) senses these changes, informing the brain via the vestibular nerve (2). In the brain medulla, the vestibular nuclei (VN) (3) receive these signals and send them to the cerebellum (4) (flocculonodular lobe). The cerebellum issues appropriate directions to VN. Then VN signals the midbrain motor centers (5) to execute proper reflex movements of the head (6) and eye (7) muscles. These reflexes help maintain a steady image of the surrounding objects on the retina. At the same time, the vestibular nuclei signal the spinal motor nuclei (9) to activate appropriate postural muscles. These reflexes help maintain balance.

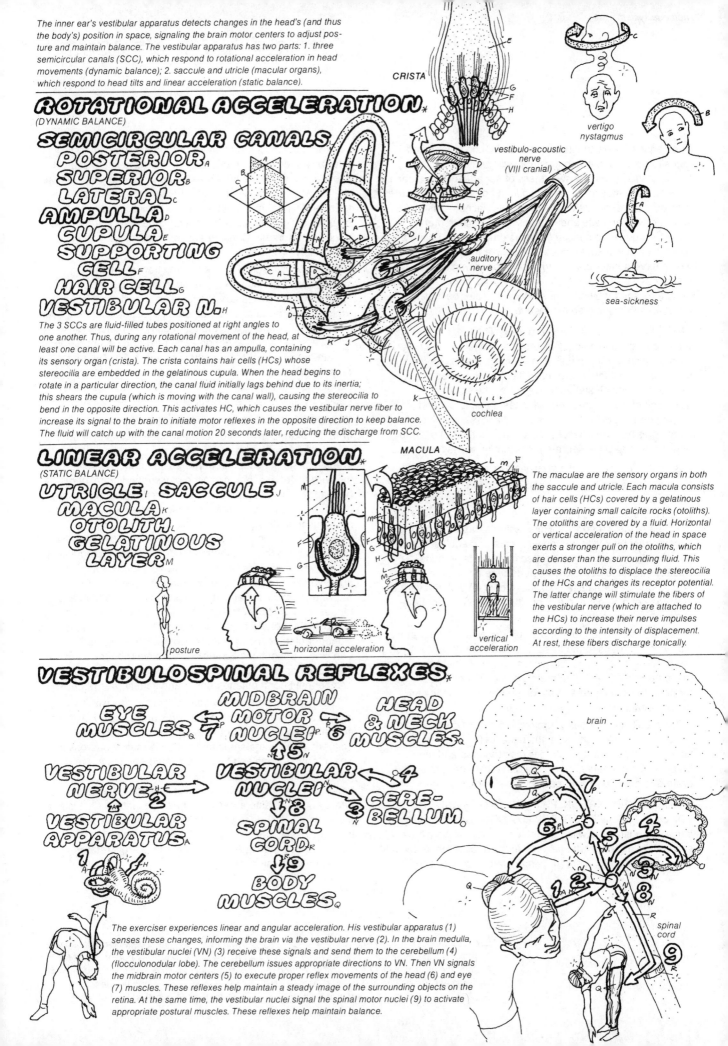

THE SENSE OF TASTE

The sense of *taste* (gustation) serves in food choice and ingestion, nutrition, energy metabolism, and electrolyte homeostasis. Thus, the sweetness of sugars and the pleasant saltiness (in small amounts) of sodium and calcium chlorides provide a drive for intake of these essential nutrients. Indeed, salt-deprived animals and humans show definite preference for salty foods. The sense of taste is crucial in detecting and rejecting harmful substances in food, like the plant alkaloids, which have a bitter taste. Early conditioned responses to food tastes have profound effects on future food habits. The sense of taste is served by special sensory receptor cells (*taste cells*) located mainly on the *tongue*.

TASTE PAPILLAE AND TASTE BUDS. The tongue contains numerous morphologically diverse *taste papillae*: the *fungiform* (mushroomlike) papillae in the front, the *vallate* (valley) in the back, and the *foliate* (flowerlike) on the sides. Each papilla contains clusters of *taste buds*. The smallest "organized" unit of taste sensation, each taste bud is a round aggregate of about 50 *taste cells* and a lesser number of *supporting cells* and *basal cells*. Glands at the bottom and between the papillae secrete fluids that rinse the taste buds.

TASTE CELLS AND GUSTATORY TRANSDUCTION. The taste cells are the sensory receptors and the transducing elements of gustation. The apical *microvilli* of the taste cells, which project into the taste pores (in the papillae canals), are the sites of taste reception. Here tastants (taste-producing substances) bind with the taste receptor molecules. Receptors for sugars have been isolated from the taste buds. Basally, the taste cell is connected to the endings of the *taste nerve fiber*. Several taste cells make contact with the various branches of a single nerve fiber. The sites of contact have synapse-like properties.

Tastant binding to the microvilli produces a *receptor potential*, which spreads to the basal end and synaptically activates the nerve ending, producing *action potentials* that travel along the taste nerve fiber to the brain's gustatory centers. Receptor potential amplitude and action potential frequency are proportional to tastant concentration.

Taste cells continuously turn over (every 10 days), with new taste cells regenerating from the basal cells. Taste nerve fibers are important for taste cell maintenance and regenration. Cutting a nerve fiber causes the taste cells to degenerate and nerve regrowth causes them to regenerate.

TASTE DISCRIMINATION: PRIMARY TASTES AND THE TONGUE'S REGIONAL SPECIALIZATION. Using mixtures of different substances with distinct tastes, researchers have shown that all tastes and flavors can be accounted for by any one or a combination of four primary tastes: *sweet, sour,*
salty, and *bitter*. Different areas of the tongue seem to be relatively specialized in this regard. The sweet sensation is evoked mainly on the tip. Areas posterior and lateral to the tip are especially sensitive to salty substances. Still more posterior and lateral areas of the tongue are responsive to the sour sensation. The bitter sensation is best detected on the back of the tongue. Considerable overlap exists, however, particularly on the tip.

Electrophysiological studies have shown that although some of the taste buds (e.g., those in the back) are uniquely responsive to bitter tastes, the taste buds in the anterior regions respond to a variety of taste sensations. Indeed, even single sensory fibers from the taste buds respond to all primary taste stimuli, although each fiber shows a preference (maximal activation) to one particular taste. The higher taste centers may discriminate the individual taste qualities by extracting and comparing the information relayed simultaneously by all the active taste fibers.

TASTE NERVES AND NERVE CENTERS. Taste sensation from the anterior two-thirds of the tongue is conveyed by the *chordae tympani* nerve, a branch of the *facial nerve* (VII cranial); taste sensation from the posterior third is served by the *lingual nerve*, a branch of the *glossopharyngeal nerve* (IX cranial). The various taste afferents terminate in the *gustatory nucleus* in the *medulla*, the first major center for integrating and routing taste signals. From this area, three groups of connections are made to other brain regions. First are the reflex connections to the neighboring medullary centers regulating the *salivary gland* and *stomach* activities, thus causing increased salivary flow and acid secretion (see plate 68). Saliva enhances gustation by helping to dissolve food subtances; only dissolved substances can activate the taste buds. The second group of connections is made with the higher centers in the *hypothalamus* and *limbic system*. The limbic projections serves the affective (hedonic, pleasant/unpleasant) repsonses to tastes, and the hypothalamus components contribute to hunger and satiety (see plate 101).

The third group of connections project, partly with the tongue's tactile fibers, to the *thalamus* and *cortex*. However, in contrast to the input from the tongue's tactile fibers, the taste input to the thalamus and cortex is mainly uncrossed. These connections and their associated centers serve in higher gustatory perception (i.e., recognition) and discrimination of taste modalities and flavors. Two separate but nearby cortical taste areas have been described. One is near the *tongue area* on the somatic sensory cortex, and another is in the *insula*, adjacent to the temporal lobe cortex (see plates 79, 87). Damage to these areas severely retards gustatory discrimination.

CN: Use dark colors for D and N.
1. Begin with the taste areas of the tongue.
2. Color the enlargement of the papillae (E) and the enlargement of the taste bud (G) and taste cell (H).
3. Color the section on taste pathways, while following the numbered sequence. Notice that a sagittal view of the medullary portion of the brain stem is shown in the upper square. The cerebral hemisphere is shown in a coronal view.
4. Color the two charts at the bottom.

TASTE AREAS OF THE TONGUE *

SWEET (A)
SALT (B)
SOUR (C)
BITTER (D)

Taste sensation is served by the taste buds (TB) on the tongue surface. Different areas of the tongue are sensitive to different taste modalities: the tip to sweets, the back to bitter, the lateral areas to sour, and the anterolateral areas to salty substances.

PAPILLAE (E)
EPITHELIUM (F)
TASTE BUD (G)
TASTE CELL (H)
MICROVILLI (I)
SUPPORTING CELL (J)
NERVE FIBER (K)

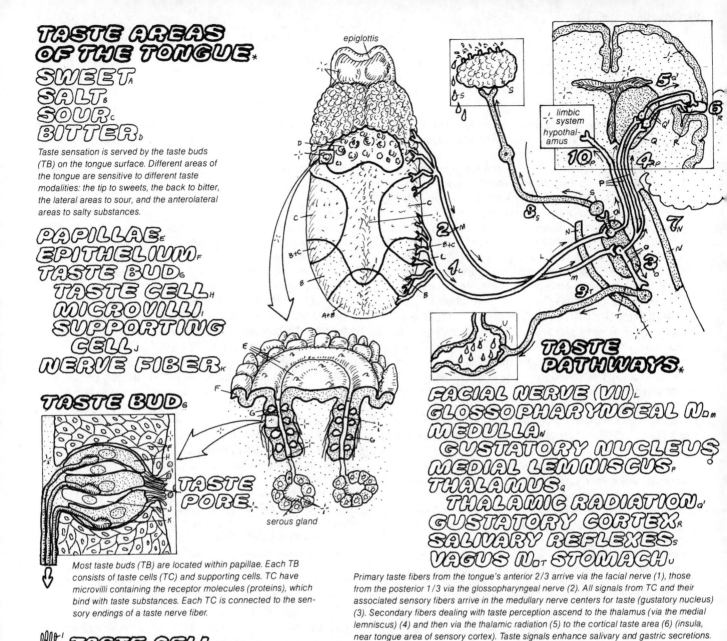

epiglottis

serous gland

TASTE BUD (G)

TASTE PORE (L)

Most taste buds (TB) are located within papillae. Each TB consists of taste cells (TC) and supporting cells. TC have microvilli containing the receptor molecules (proteins), which bind with taste substances. Each TC is connected to the sensory endings of a taste nerve fiber.

TASTE PATHWAYS *

FACIAL NERVE (VII) (L)
GLOSSOPHARYNGEAL N. (M)
MEDULLA (N)
GUSTATORY NUCLEUS (O)
MEDIAL LEMNISCUS (P)
THALAMUS (Q)
THALAMIC RADIATION (Q')
GUSTATORY CORTEX (R)
SALIVARY REFLEXES (S)
VAGUS N. (T) STOMACH (U)

Primary taste fibers from the tongue's anterior 2/3 arrive via the facial nerve (1), those from the posterior 1/3 via the glossopharyngeal nerve (2). All signals from TC and their associated sensory fibers arrive in the medullary nerve centers for taste (gustatory nucleus) (3). Secondary fibers dealing with taste perception ascend to the thalamus (via the medial lemniscus) (4) and then via the thalamic radiation (5) to the cortical taste area (6) (insula, near tongue area of sensory cortex). Taste signals enhance salivary and gastric secretions. Indeed, saliva helps dissolve substances before they can stimulate taste cells. The gustatory signals for these visceral reflexes are integrated at the level of the medullary digestive centers (7). Parasympathetic motor commands are sent to the salivary glands via the facial nerve (8) and to the stomach via the vagus nerve (9). Taste signals also go to the limbic system and hypothalamus (10) for affective responses.

TASTE CELL (H)
RECEPTOR POTENTIAL (H')
ACTION POTENTIAL (K')

The binding of taste molecules to TC microvilli generates a receptor potential in the basal aspects of TC, which, if sufficiently strong, will evoke a nerve impulse in the taste nerve fiber.

AFFECTIVE RESPONSE TO CONCENTRATION DIFFERENCES *

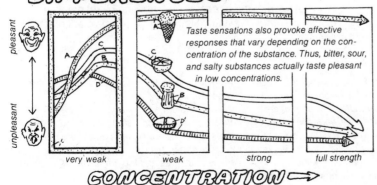

Taste sensations also provoke affective responses that vary depending on the concentration of the substance. Thus, bitter, sour, and salty substances actually taste pleasant in low concentrations.

pleasant

unpleasant

very weak   weak   strong   full strength

CONCENTRATION ➔

TASTE THRESHOLDS *

SACCHARIN (A')
SUCROSE (A²)
QUININE (D')
HCl (C')
NaCl (B')

Taste thresholds (minimum amount for evoking sensation) to different substances vary. Sweets show the highest thresholds (lowest sensitivity). Bitter substances show the lowest thresholds (highest sensitivity). Salty and sour substances fall in between. The thresholds among the sweet substances also vary. Thus, sucrose (table sugar) shows a much higher threshold than saccharin (artificial sweetener).

1   500

10   100   2,000   10,000

THE SENSE OF SMELL

The sense of *smell* (olfaction), one of the earliest distance senses to arise in evolution, is well developed in lower animals and nocturnal mammals. Olfaction is used to detect odors of foods, individuals, enemies, territory, the opposite sex, etc. In humans and animals alike, olfaction evokes, perhaps more than any other sense, emotional reactions resulting in strong approach/avoidance behaviors. The scent of a rose can bring pure pleasure; the smell of a rotten egg can cause nausea and vomiting. Although humans are not known for olfactory keenness, memories of odors carry deep and emotionally rich associations, particularly early experiences. The sense of smell is also important for regulating appetite and food intake. Congested sinuses usually causes loss of appetite.

THE OLFACTORY ORGAN AND RECEPTOR NEURONS. The *sensory receptors* for *odors* are found in the *olfactory mucosa* (olfactory organ) located in the superior nasal sinuses. The *olfactory receptor cells* are bipolar neurons with the cell body in the olfactory mucosa. A single dendrite projects superficially into the nasal cavity, terminating in several immotile *cilia*. The cilia, immersed in a fluid *mucus layer* (secreted by the *Bowman's glands* and the *supporting cells*), are the site of interaction between the odors and the receptor neuron (olfactory transduction).

The olfactory nerve consists of the axons of the olfactory neurons, which form small bundles to pass through the *cribriform plate* before entering the cranial cavity and the *olfactory bulb*. In the bulb, these axons synapse with the apical dendrites of the secondary relay cells (*mitral cells*) at a site called the *glomerulus*. The number of olfactory neurons is very high (millions), but the number of mitral cells is low (thousands), providing for a high convergence ratio of olfactory cells to mitral cells (1000:1). This arrangement is believed to be a basis for the great sensitivity of the olfactory system in detecting odors at low concentration.

OLFACTORY TRANSDUCTION. The odors, which are either gaseous or vapors of volatile substances, reach the nasal cavities through the nostrils. Sniffing (a reflex as well as a voluntary inspiratory action) increases the air flow through the upper sinuses. In the olfactory mucosa, the odors must first dissolve in the mucus fluid layer before they can activate the receptor cells. The membranes of the immotile cilia contain *receptor sites* (probably protein molecules) to which odors bind. The binding increases cyclic AMP levels (intracellular messenger) within the receptor cells, which serves to amplify the signal and eventually cause membrane depolarization by increasing permeability to a cation (*sodium* or *calcium*). The resulting receptor potential then induces action potentials in the axon, which are conducted to the olfactory bulb. The sum of receptor potentials of many olfactory neurons produces the electro-olfactogram, which may be recorded from the surface of the olfactory mucosa.

THE OLFACTORY BULB. The site of the first synaptic station for the olfactory impulses, the olfactory bulb is a layered and highly organized structure. There is evidence for some mapping of odors in the bulb: stimulation with a single odor (e.g., peppermint) enhances metabolic activity in a circumscribed region of the bulb (although olfactory nerve signals tend to stimulate much larger areas of the bulb). The olfactory messages are processed at the level of the glomeruli and the mitral cells of the bulb and then sent, via the *olfactory tract* (axons of mitral cells), to the higher olfactory areas of the brain. The olfactory bulb receives from the brain many *centrifugal fibers*, the major portion of which activates the *granule cells* of the olfactory bulb. These cells are local inhibitory interneurons serving to inhibit the mitral cells. This interaction between the bulb and brain acts as a negative feedback loop controlling the afferent input by regulating the mitral cell activity (output).

HIGHER OLFACTORY CENTERS. The olfactory output from the bulb to the brain has several targets, the main one being the *primary olfactory cortex* and the higher *olfactory association areas*. These areas are the sites of olfactory discrimination, perception, and memories. The second target is the *limbic system* structures (see plate 102) (e.g., *amygdala* and *septum*), where olfactory signals activate smell-related emotions and behaviors. The odors that provoke instinctive, stereotyped responses are called "pheromones." The intimate connections of olfaction and the limbic system in lower animals is the basis of the synonym *rhinencephalon* (nose brain) for the limbic system. Other targets of bulb output are the *hypothalamic centers* (see plate 101), regulating feeding, autonomic responses, and hormonal control (particularly the reproductive hormones), and the *reticular formation*, regulating arousal and attention. These signals are relayed indirectly through the limbic and cortical structures.

OLFACTORY DISCRIMINATION. Even humans, having only a modest sense of smell, can discriminate between thousands of odors. The mechanism of odor discrimination and the relative shares of periphery and central structures in this process are poorly understood. According to one theory, odors, like tastes, can be categorized into a finite set of primary odors (i.e., floral, peppermint, musky, camphor, putrid, pungent, ethereal). Some support for this theory comes also from studies of people with specific anosmias (inability to smell a particular class of odors); the defect is genetically transmitted, suggesting the possibility of a deficient receptor protein.

Single olfactory neurons can be activated by more than one odor type, although their responses are not uniform. To discriminate between the different odors, brain olfactory centers may extract the necessary information by reading the discharge *pattern* of many simultaneously active afferent fibers.

CN: Use dark colors for E and F; very light for A.
1. Color the upper left illustration showing the flow of odor molecules (E) through the interior of the nasal cavity. Then color the blowup of the olfactory bulb (B) and olfactory neurons (C), starting at the bottom with the interaction of the odor molecules (E) with the olfactory neuron cilia (F). Work your way upward, and complete the network of neural cells in the bulb. Then color the diagram on the right describing the regeneration cycle of olfactory receptor neurons.
2. Color the panel on olfactory pathways, noting that the three specific structures listed under limbic system (Q) are indeed parts of the system but receive separate colors for added emphasis. Color these structures and the frames of the small illustrations that relate to them.
3. Color the remaining panels.

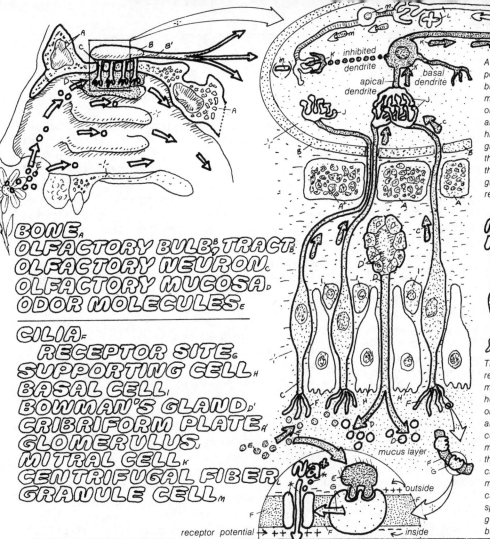

Axons of the olfactory neurons (ON) run through pores in the cribriform plate and enter the olfactory bulb (OB), where they excite the apical dendrites of mitral cells (MC) within the glomerulus. Here olfactory messages are segregated, refined, and amplified. Axons of the MC convey the excitation to higher brain centers via the olfactory tract. The granule cells of OB inhibit MC activity by acting on the basal dendrites of MC. Centrifugal fibers from the brain suppress MC activity by exciting the granule cells. This way the brain can check and regulate output of the OB.

BONE A
OLFACTORY BULB, TRACT B
OLFACTORY NEURON C
OLFACTORY MUCOSA D
ODOR MOLECULES E

CILIA F
RECEPTOR SITE G
SUPPORTING CELL H
BASAL CELL I
BOWMAN'S GLAND D'
CRIBRIFORM PLATE A'
GLOMERULUS J
MITRAL CELL K
CENTRIFUGAL FIBER L
GRANULE CELL M

NEURON REGENERATION C

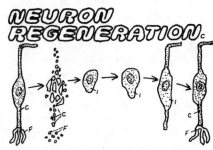

The sense of smell is served by millions of olfactory receptor neurons (ON) located in the olfactory mucosa/organ (OM). OM, one on each side of the head, are small areas in the roof and walls of superior nasal sinuses. OM also contains supporting cells and Bowman's glands, which secrete the mucus coat of OM. This mucus is essential, because odors must first dissolve in mucus before they can excite the ON. The dendrite of ON terminates in special cilia that are the site of olfactory transduction. Odor molecules bind with receptor proteins in the cilia, causing depolarization of ON. ON have a short life span (1-2 months); once degenerated, they will regenerate by division and differentiation of the resting basal cells.

ODOR THRESHOLDS *

Degree of sensitivity to odors, measured by odor thresholds (minimum amount needed for detection), varies depending on the odor. Threshold for mercaptan (sulfur compounds added to household natural gas) is 50,000 times lower than for peppermint oil; threshold for peppermint is 250 times lower than for ether.

MERCAPTAN N 0.000 000 4 N'
OIL OF PEPPERMINT * 0.02 *
ETHER * 5 * mg/L of air

SNIFFING REFLEX *

Sniffing increases the speed and volume of air flow in the upper nasal cavities, promoting access of air and odors to olfactory mucosa. Without sniffing, levels of many odors would be too low for detection.

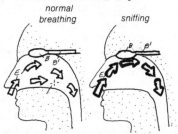

MASKING & ADAPTATION

A person can smell many different odors, but only one odor at a time because stronger odors tend to mask the effects of weaker ones (masking). Continuous exposure to one odor also diminishes sensitivity (adaptation). Odor concentration must be markedly increased to overcome odor adaptation.

OLFACTORY PATHWAYS *
BULB TRACT B, B' → OLFACTORY CORTEX O → ASSOC. CORTEX P

LIMBIC SYSTEM: Q*
AMYGDALA R
SEPTUM S
HYPOTHALAMUS T

From OB, signals are conducted along two pathways to higher brain structures: (1) to the primary olfactory cortex and higher olfactory association areas for perception and recognition of odors and their association with other sensory data; (2) to the limbic system structures (amygdala, septum) to activate instinctual behavior and emotions, as well as to the hypothalamus for regulation of drives and visceral activities (e.g., digestion). Olfactory signals also act on the reticular formation (not shown) to cause arousal.

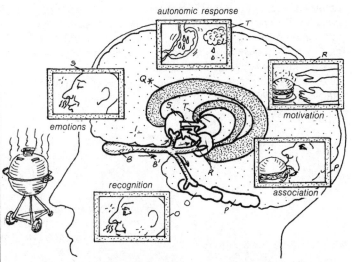

autonomic response

emotions

motivation

recognition

association

EEG, SLEEP/WAKEFULNESS, & RETICULAR FORMATION

BRAIN WAVES. In the 1930s, Hans Berger, a German psychiatrist, discovered that slow and weak waves of electrical activity could be recorded directly from the human scalp. This spontaneous electrical brain activity, present during complete rest or immobility and even in sleep, was called the *brain waves* or *electroencephalogram* (EEG). Although spontaneous, the waves' *amplitude* and *frequency* vary, depending on the state of mental (brain) activity and the recording site. When the subject is relaxed, with eyes closed, the EEG shows a moderate activity of about 8-14 c/s (*alpha rhythm*), which can best be recorded in the back of the head over the occipital (visual) area. The regular and synchronous alpha waves disappear (alpha block) when subjects open their eyes or when a light is flashed over the closed eyelids and are replaced by a faster, desynchronous wave pattern called the *beta waves* ($>$14 c/s). Beta waves, often present in the brain's frontal areas, are associated with arousal, alertness, and mental activity. Although EEG waves orginate in the cortex, their modulation in different states is mainly due to the influence of subcortical structures of the reticular formation.

RETICULAR FORMATION FUNCTIONS. In the brain's core, from medulla to midbrain and thalamus, loosely organized areas of gray matter form the *reticular formation* (RF). RF neurons show diffuse connections. Midbrain RF stimulation results in beta EEG, *awakening*, and *arousal;* midbrain RF lesions result in a coma-like state with persisting synchronized delta waves. Thus, the midbrain RF gives rise to the *ascending reticular activating system* (ARAS), a diffuse fiber projection that fans out to innervate many areas of the forebrain. The ARAS also includes the *reticular nuclei of the thalamus,* which also project diffusely to all areas of the *cerebral cortex.*

The nonspecific ARAS projections globally depolarize (excite) the cortex, in contrast to the *specific sensory* projections from the thalamus, which exert localized and functionally specific excitatory effects on the cortex. The general ARAS excitation facilitates cortical responsiveness to specific sensory signals from the thalamus. Normally, on their way to the cortex, the sensory signals from the *sensory afferents* stimulate the ARAS, via collateral branches of their axons. The excited ARAS in turn stimulates the cortex, facilitating sensory signal reception. When afferent systems are massively stimulated (loud noise, cold shower), the ARAS projections trigger generalized cortical activation and arousal.

SLEEP AND EEG. Before sleep onset, the EEG has the alpha pattern due to reduced motility and relaxation. As sleep progresses, the waves become slower and larger, occasionally interrupted by bursts of fast spindle-like activity. More detailed studies have shown that human sleep consists of two general *states* (kinds), which may be subdivided into specific *stages.* One state is the *orthodox* or *slow-wave sleep,* comprising four stages. After falling asleep, the individual goes sequentially

through stages 1, 2, 3, and 4. Each stage is characterized not only by specific EEG patterns but also by bodily or autonomic activities, such as changes in body position (turning over or to the sides), changes in heart and respiratory rates, bouts of sweating, and even a large burst of growth hormone secretion (see plate 112). *Stage 4,* showing synchronous, large, and slow *delta waves* (high amplitude, 3-4 c/s), is the deepest of the four stages. Interestingly, bed-wetting in children and sleepwalking (somnambulism) often occur during stage 4.

REM SLEEP. After stage 4, another, distinctly different state of sleep called the *paradoxical* or *REM* sleep commences. The EEG becomes betalike (fast and desynchronized), resembling wakefulness patterns, even though the person is extremely difficult to awaken (hence the term "paradoxical sleep"). During the REM sleep, muscle tone in the neck and support muscles is suppressed (hence the floppy limbs and head falling), but the eyes show rapid searching movements under the closed lids (hence the designation REM, for *rapid eye movement*). Dreams also occur mainly during REM sleep.

In adults, a complete sleep cycle (1, 2, 3, 4, REM) lasts about 90 min. The REM stage occupies about 20% of the 8 hrs. total of nightly sleep; the longest REM (and the dream) episodes occur early in the morning. In infants, total sleep time is more than 16 hrs. of which nearly 50% is spent in REM. In old age, the total amount of sleep diminishes by 1-2 hrs. and stage 4 disappears completely. REM sleep is essential. If subjects are deprived of it by being awakened each time REM sleep begins, they get an excess amount of REM sleep the next night, as though the body (brain) is compensating for the amount of REM lost. The causes and uses of REM sleep are not clearly understood.

SLEEP CENTERS. Sleep was once considered a passive phenomenon caused by reduced sensory stimulation. However, certain zones in the lower RF appear to be involved in active generation of sleep and regulation of its states. Thus, *serotonin* secreting neurons from the *raphe nuclei* in the pons' RF may augment the slow wave sleep. *Norepinephrine* releasing neurons originating in the *locus ceruleus* (a part of the medullary RF) may promote the REM sleep. Specific lesions of these areas selectively interfere with the two states of sleep. Certain areas of the *hypothalamus* also have sleep promoting activity. Drugs that lower or enhance the brain's norepinephrine and serotonin levels cause *insomnia* (sleeplessness) or sleep, respectively. In addition, certain sleep promoting proteins accumulate during wakefulness, particularly in the cerebrospinal fluid, and induce sleep when they reach high levels.

In short, the sleep promoting centers of the lower RF and the midbrain wakefulness centers (ARAS) of the upper RF interact to regulate not only the different stages of sleep but also the entire cycle of sleep and wakefulness.

CN: Use a gradation of colors from light to dark for the four stages of sleep (F-I), and a dark colors for J.
1. Color the afferent sensory (C) inputs from the special senses in the large head of the upper panel. Color the alpha and beta waves on the left and the remaining two diagrams.
2. Color the titles designating the stages of sleep
(not the brain waves) and the amount of time spent in each stage of the chart on sleep cycles. What has not been shown in this chart is that after each cycle the sleeper quickly goes back through stages 4-1 before starting the next cycle.
3. Color the sleep centers on the right and wakefulness vs. sleep at the bottom.

WAKEFULNESS *

RETICULAR FORMATION A
ASCENDING RETIC. ACTIVATING SYSTEM (ARAS) A'
THALAMUS RETICULAR NUCLEI B
DIFFUSE THALAMIC PROJECTION SYSTEM B'
SENSORY AFFERENT C CEREBRAL CORTEX D
ELECTRO ENCEPHALOGRAM (EEG) E

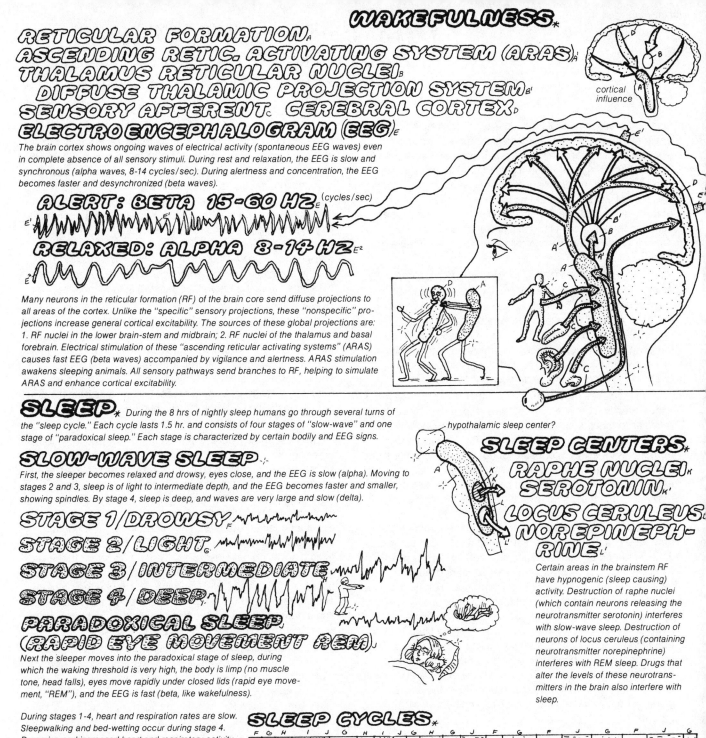

cortical influence

The brain cortex shows ongoing waves of electrical activity (spontaneous EEG waves) even in complete absence of all sensory stimuli. During rest and relaxation, the EEG is slow and synchronous (alpha waves, 8-14 cycles/sec). During alertness and concentration, the EEG becomes faster and desynchronized (beta waves).

ALERT: BETA 15-60 HZ E (cycles/sec)

RELAXED: ALPHA 8-14 HZ E²

Many neurons in the reticular formation (RF) of the brain core send diffuse projections to all areas of the cortex. Unlike the "specific" sensory projections, these "nonspecific" projections increase general cortical excitability. The sources of these global projections are: 1. RF nuclei in the lower brain-stem and midbrain; 2. RF nuclei of the thalamus and basal forebrain. Electrical stimulation of these "ascending reticular activating systems" (ARAS) causes fast EEG (beta waves) accompanied by vigilance and alertness. ARAS stimulation awakens sleeping animals. All sensory pathways send branches to RF, helping to simulate ARAS and enhance cortical excitability.

SLEEP *
During the 8 hrs of nightly sleep humans go through several turns of the "sleep cycle." Each cycle lasts 1.5 hr. and consists of four stages of "slow-wave" and one stage of "paradoxical sleep." Each stage is characterized by certain bodily and EEG signs.

SLOW-WAVE SLEEP
First, the sleeper becomes relaxed and drowsy, eyes close, and the EEG is slow (alpha). Moving to stages 2 and 3, sleep is of light to intermediate depth, and the EEG becomes faster and smaller, showing spindles. By stage 4, sleep is deep, and waves are very large and slow (delta).

STAGE 1/DROWSY F
STAGE 2/LIGHT G
STAGE 3/INTERMEDIATE H
STAGE 4/DEEP I
PARADOXICAL SLEEP (RAPID EYE MOVEMENT REM) J
Next the sleeper moves into the paradoxical stage of sleep, during which the waking threshold is very high, the body is limp (no muscle tone, head falls), eyes move rapidly under closed lids (rapid eye movement, "REM"), and the EEG is fast (beta, like wakefulness).

During stages 1-4, heart and respiration rates are slow. Sleepwalking and bed-wetting occur during stage 4. Dreaming and increased heart and respiratory activity occur during REM. As night progresses, the duration of REM sleep increases (more dreams near morning), and that of stage 4 decreases (more sleepwalking and bed-wetting early at night). Adults spend 20% of sleep in REM, infants, 50%. Old people have less sleep, less REM, and no stage 4.

hypothalamic sleep center?

SLEEP CENTERS *
RAPHE NUCLEI K
SEROTONIN K'
LOCUS CERULEUS L
NOREPINEPH-RINE L'

Certain areas in the brainstem RF have hypnogenic (sleep causing) activity. Destruction of raphe nuclei (which contain neurons releasing the neurotransmitter serotonin) interferes with slow-wave sleep. Destruction of neurons of locus ceruleus (containing neurotransmitter norepinephrine) interferes with REM sleep. Drugs that alter the levels of these neurotransmitters in the brain also interfere with sleep.

SLEEP CYCLES *

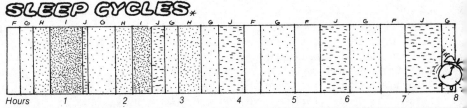

	F G H I	J	G H I	J	G H I	J G H	J G	J	F	G	F	J	F	J G
Hours	1		2		3		4	5		6		7			8

WAKEFULNESS VS. SLEEP *

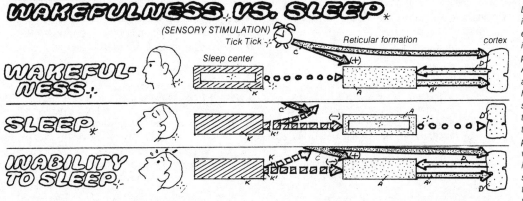

(SENSORY STIMULATION)
Tick Tick

Reticular formation   cortex

Sleep center

WAKEFUL-NESS

SLEEP *

INABILITY TO SLEEP *

During sleep, hypnogenic centers inhibit ARAS, decreasing cortical excitability. Reduction of sensory input (darkness, silence, supine position) helps diminish activity in ARAS. Reduced cortical excitability accounts for the absence of behavioral consciousness and responsiveness during sleep. Anxiety, thoughts, or sensory stimuli counter the effects of sleep centers on ARAS, promoting sleeplessness. During wakefulness, hypnogenic centers are inactive. ARAS is now able to stimulate the cortex. Vigilance is maintained by continuous excitatory input from sense organs.

THE HYPOTHALAMUS: ORGANIZATION & FUNCTIONS

One of the essential brain functions is to regulate the internal environment and to adaptively mediate the influences of the external environment on the internal organs and systems. The *hypothalamus* (H) is the highest brain structure directly concerned with the body's homeostasis and integration of internal activities. The strategic location of the H, in the base of the brain and above the pituitary gland, provides an ideal situation to exert control over the lower autonomic and endocrine systems and to be controlled by the higher forebrain centers.

STRUCTURE AND HYPOTHALAMIC INPUT/OUTPUT. Anatomically, the H can be divided into several relatively discrete nuclei. Occasionally, a particular function has been ascribed to a specific nucleus (e.g., the *suprachiasmatic nucleus* and the *diurnal rhythms*). In most cases, however, larger "areas" of the H, made up of more than one nucleus, are implicated in a function (e.g., the regulation of *body temperature* by areas in the *anterior* and *posterior* H).

The H receives input from all the major *sensory systems*, either directly, as in the case of the eye, or indirectly via the *reticular formation* (see plate 100). The sensory input informs the H about environmental conditions. For example, input from the eyes relays information about lighting conditions (the day's length), and input from the skin relays messages about environmental temperature. Sensory messages also come from the internal sensors (e.g., those in the mouth and the digestive system). A different kind of input coming from the *limbic system* (e.g., amygdala, septum) in the forebrain informs the H about the state of the animal's drives (hunger, thirst, sex) and emotions. A third category of input is provided by *hormones* and other blood-borne substances such as sodium ions and glucose, conveying messages from the body to the H about the salt, water and energy situation. Output from the H goes to the limbic system to interact with the structures controlling emotions and drives (see plate 102), to the *midbrain motor centers* for controlling somatic motor responses during emotional behaviors, to the *sympathetic* and *parasympathetic* autonomic centers in the *medulla* and *spinal cord* controlling visceral organs, and finally to the *pituitary gland* to control water, salt, metabolic, and hormonal parameters.

HYPOTHALAMIC FUNCTIONS. Localized electrical stimulation and discrete lesions in small areas of the H have been very successful in elucidating the functions of this structure. One of the most important functions of the H to be discovered by stimulation studies was control of the *autonomic nervous system*. The effects of hypothalamic stimulation on the *sympathetic nervous system* in particular were so marked that Charles Sherrington, the great English physiologist, called the H the "head ganglion of the sympathetic nervous system" (see also plates 25, 81). Here, the significance of H is in integrating autonomic responses with other brain area activities, with the animal's emotional state, and with the particular environmental conditions conveyed via the senses. Thus, the marked increase in cardiac activity and peripheral vasoconstriction, accompanied by vasodilation in the skeletal muscles, which occur after stimulation of the *lateral H*, are sympathetic responses that occur equally during generalized physical activation of the body (e.g., exercise, running, and fighting [fight/flight response], see plate 119). In a similar manner, the H integrates the autonomic and visceral responses exhibited during emotional stress (e.g., fear); here the H mediates the responses initiated in the emotional centers of the limbic system (see plate 102).

Another well-known function of the H is in regulating *body temperature*. Areas in the *anterior* and *posterior* H integrate the diverse sympathetic responses to cold, such as cutaneous vasoconstriction, piloerection, secretion of epinephrine and thyroxine, and the consequent increase in the metabolic rate (see plate 134). Certain areas in the H also help integrate parasympathetic nervous activity, but these effects are less marked than those of the sympathetic system.

The best-known studies implicating H in psychophysiological regulation of bodily homeostasis are those of *hunger/ satiety* and *feeding behavior*. Animals with discrete lesions in the *lateral* H show loss of appetite, reduced food intake (anorexia), and eventual wasting. In contrast, lesions in the ventromedial H lead to excessive eating (hyperphagia) and eventual obesity. Therefore, the lateral areas of H contain centers that enhance appetite and feeding, and the ventromedial areas contain the *satiety centers*. Together these two centers help control *feeding behavior*, *energy balance*, and possibly *body weight* while exerting inhibitory effects on one another (see plate 132). Other areas in *paraventricular nuclei* and the *dorsolateral* H have been implicated in the control of *water*, *thirst*, and *salt* (see plates 62, 110, 120).

When the *suprachiasmatic nuclei* of the H are lesioned, certain *diurnal rhythms* (daily 24-hr. rhythms), especially those of hormones and activity, are abolished. For example, secretion of *ACTH* and *cortisol* (see plate 121) is high in the morning and low in the evening; this diurnal cycle is abolished after lesions of the suprachiasmatic nuclei.

The H is one brain area demonstrating clear male-female differences. Not only does the H control the secretion of *sex hormones* in different ways in the two sexes (see plate 147), but the *preoptic areas* and the *anterior H* are intimately involved in the control and expression of sex-specific behaviors. Indeed, in several mammals, a small but discrete nucleus (the *sexually dimorphic nucleus of the H*) has been found. It is markedly larger in the male than in the female. The extremely important control functions of the H on the endocrine system are discussed elsewhere (see plates 109-111).

CN: Use a dark color for A.
1. Begin by coloring the hypothalamus (A) in the small cross-section of the brain. Then color the enlargement arrow pointing to the structures of the hypothalamus.
2. Color gray the title representing the source of input to the hypothalamus and of the hypothalamic outputs.
3. Color each title in the bottom panel and the structure to which it refers in the large drawing. Note that the paraventricular nucleus receives two colors to represent a dual role in H function. Also note that the lateral areas of H are not adequately represented in this sagittal diagram.

HYPOTHALAMUS

The hypothalamus (H) is the major brain center for regulation of internal body functions. Situated above the pituitary gland, underneath the thalamus (hypo-"thalamus"), H has numerous areas (nuclei), each involved in the regulation of some internal function. H has many connections to and from the forebrain limbic structures in addition to receiving input from the sensory organs, especially smell, taste, and the eyes. H, via its efferent (output) connections to the brainstem, spinal cord, and pituitary gland, controls somatic motor, autonomic motor, and hormonal secretions. H may be divided into lateral, medial, anterior, and posterior zones.

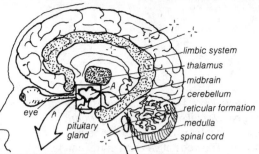

limbic system
thalamus
midbrain
cerebellum
reticular formation
medulla
spinal cord
eye
pituitary gland

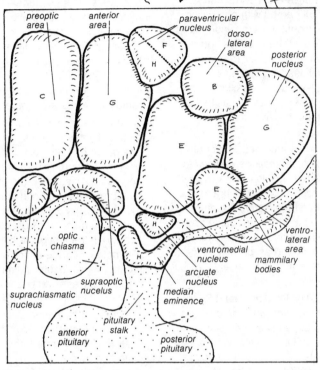

INPUT ⟹

SENSES*
RETICULAR FORMATION*
LIMBIC SYSTEM*
VISCERAL ORGANS*
HORMONES*
GLUCOSE, Na⁺*

⟹ OUTPUT*

MIDBRAIN (MOTOR)*
LIMBIC SYSTEM*
MEDULLA (P. SYMP.)* (SYMP)*
SPINAL CORD (SYMP.)*
PITUITARY (HORMONES)

preoptic area — anterior area — paraventricular nucleus — dorso-lateral area — posterior nucleus — ventro-lateral area — mammilary bodies — ventromedial nucleus — arcuate nucleus — median eminence — supraoptic nucleus — suprachiasmatic nucleus — optic chiasma — pituitary stalk — anterior pituitary — posterior pituitary

HYPOTHALAMIC FUNCTIONS

Stimulation of areas in the lateral H activates generalized sympathetic responses. These areas control the fight/flight reactions. Stimulation of smaller areas may induce adrenal medullary release of epinephrine, or vasodilation in the skeletal muscles. H influences the parasympathetic system also, via the parasympathetic centers in the brain medulla.

Stimulation or destruction of areas in the anterior H and preoptic areas has profound effects on the regulation of sex hormones (via the anterior pituitary) and sexual responses. A special nucleus, the sexually dimorphic nucleus, exists in this area, which is markedly larger (more active?) in some male mammals.

Some diurnal (daily) rhythms in bodily functions (e.g., for hormonal secretion or visceral activities) are regulated by the suprachiasmatic nuclei. Light and length of day stimulate this nucleus through direct retinal connections. This nucleus and the pineal gland (well known for its mediation of light effects/diurnal functions) also interact.

Stimulation of areas in the lateral H increases appetite and induces eating behavior (feeding center). Prolonged stimulation leads to overeating/overweight. Destruction of feeding center causes loss of appetite and wasting. Stimulation of ventromedial H causes cessation of eating (satiey center); lesioning leads to overeating/obesity. The satiety and feeding centers have reciprocal inhibitory interactions. Neurons in satiety centers are sensitive to blood glucose levels (hypothalamic "glucostat").

Stimulation of areas in dorsal/lateral H induces drinking behavior. Injection of acetylcholine or angiotensin II into these areas has similar effects; lesions interfere with water regulation and electrolyte balance. Neurons in this area are sensitive to blood osmolarity and sodium levels (hypothalamic "osmostat").

Body temperature is regulated by areas in H (at 37° C). Stimulation of the anterior H activates heat loss mechanisms (cooling center); stimulation of the posterior H activates heat conservation/production (heating center). These areas have reciprocal inhibitory interactions. Some of the neurons in this hypothalamic "thermostat" are sensitive to changes in blood or skin temperature.

Parts of H (e.g., the median eminence) act like an endocrine gland, secreting hormones. Those secreted via the posterior pituitary gland directly act on target organs (kidney, uterus, breast). Numerous other H hormones regulate activity of the anterior pituitary gland, which in turn regulates the activities of many organs and glands.

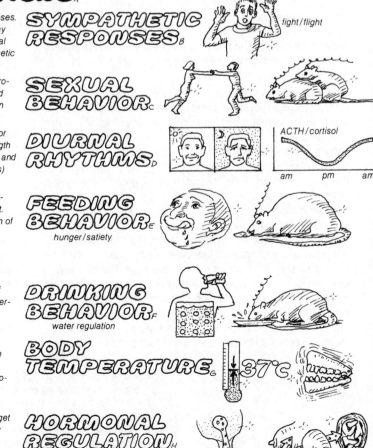

SYMPATHETIC RESPONSES — fight/flight

SEXUAL BEHAVIOR

DIURNAL RHYTHMS — ACTH/cortisol — am pm am

FEEDING BEHAVIOR — hunger/satiety

DRINKING BEHAVIOR — water regulation

BODY TEMPERATURE — 37°C

HORMONAL REGULATION

EMOTIONS, INSTINCT, & THE LIMBIC BRAIN

The brain may be viewed as a hierarchy of three "separate brains": a lower *vegetative/reflexive* brain, a higher *adaptive/skilled* brain, and an intermediate brain concerned with *emotions* and *instincts*. The vegetative brain corresponds roughly to the *brain stem* and is concerned with controlling vital bodily functions (respiration, digestion, circulation) as well as with integrating brain reflexes. The adaptive and skilled brain corresponds to the *cerebral cortex* (the neocortex). It has sensory, motor, and association areas that serve in complex perception and execution of skilled motor functions (e.g., hand movements, speech) as well as higher mental functions (e.g., learning, thoughts, introspection, planning).

Limbic system (LS) structures are concerned with central (neural) control over the expression of *emotions, instinctive behaviors, drives, motivation, and feelings*. In lower vertebrates, the LS is called the *rhinencephalon* (smell brain) due to its intimate connection to the central *olfactory* structures. In these animals, many instinctive behaviors are guided by the sense of smell. Both the cerebral cortex and the LS have ample access to the brain stem motor areas, permitting them to exercise their respective adaptive and instinctive (stereotyped) controls over behavior.

LS STRUCTURES AND CONNECTIVITY. The main LS structures are the *amygdala* (almond), *septum* (wall), *hippocampus* (sea horse), *cingulate gyrus, anterior thalamus*, and *hypothalamus*. The relative mass of the LS remains fairly uniform among the different vertebrates. The LS structures are connected by a number of complicated but not necessarily reciprocal pathways, some of which constitute a loop. The *loop of Papez* (hypothalamus → anterior thalamus → cingulate gyrus → hippocampus → hypothalamus) has been considered, for a variety of reasons including the fact that patients with lesions in this loop exhibit abnormalities of emotional expression, as one of the neural circuits serving in emotional and instinctive behavior.

The LS was once thought to be involved only in emotional/instinctive behavior and that the cerebral cortex and LS had few connections and little communication. *This view is changing.* Thus, the hippocampus, a prominent LS structure, thanks to its numerous connections to both the older (e.g., olfactory and limbic) and the newer (e.g., the cortex) parts of the brain, plays an essential role in *learning* and *memory* (see plate 103). Similarly, the cingulate gyrus (located on the medial hemispheric surface and possibly regulating such *social behaviors* as *parental care*) is a part of both the LS and the cerebral cortex; the cingulate gyrus, like the hippocampus, provides an important link between the adaptive and the instinctive brains. LS structures like the amygdala and septum can also communicate, via their reciprocal connections to the cingulate gyrus, with the *higher cortical association areas*. The LS structures also receive abundant *sensory input* and send *motor output* to both *voluntary* and *involuntary* motor centers via the cingulate gyrus to the *motor cortex* and

via the hypothalamus to the brain stem.

EFFECTS OF LS STIMULATION/LESIONS. In the 1940s, Walter Rudolph Hess (Swiss physiologist and Nobel laureate) discovered that electrical stimulation of certain areas in the hypothalamus and neighboring structures in cats can evoke the behavior patterns of *fear* or *aggression* similar to those observed during a cat fight (hissing, spitting, hair standing, back hunching, and slapping). Lesion studies indicated that disconnecting the forebrain from these lower limbic areas leaves the expression of these emotions intact while removing purpose and directedness from them.

Additionally, stimulation of certain areas in the amygdala often activates aggressive responses. Bulls can be forced to charge under these conditions. Stimulation of the amygdala in a normally submissive monkey causes the animal to exhibit aggressive gestures more often. As a result, the monkey temporarily moves up in the group's dominance order. Violent attacks of *rage* are occasionally seen in humans with *epileptic seizure discharges* (causing excessive local electrical activity) in the amygdala; surgical removal of the amygdala eliminates the rage attacks. Removal of the amygdala in monkeys also results in timidity and passivity. The hippocampus is one place where stimulation does not elicit any emotional/instinctive behaviors.

In the 1960s, James Olds (American psychobiologist) found that certain areas in the LS may function in *pleasure* or *reward*. When rats with electrodes in the septum or associated pathways (e.g., medial forebrain bundle) are taught to self-stimulate at will, they do so for a long time, preferring this electrical stimulation of the brain to food rewards (hence the label "pleasure centers"). Humans also report pleasure when stimulated in similar locations.

Little is known about how exactly LS circuits function in experience of feelings and expression of emotions/instincts. A person's responses to the *smell* and *sight* of a rose may help explain. These stimuli bring *pleasure, fond visual associations*, a *smile*, and *autonomic responses* (e.g., changes in heartbeat). The rose scents find access to the LS via the *olfactory system*, which feeds into the amygdala and the hypothalamus (see plate 99). The rose's visual sensations activate the LS either via the reticular formation pathways to the hypothalamus (LS) or via the visual thalamic nuclei to the anterior thalamus (LS). Activation of the LS circuits at this time presumably stimulates the subjective experience of pleasure/good feelings. The motor responses of smiling can be activated via the hypothalamic outflow to the brain stem nuclei serving in control of *muscles of expression*. The hypothalamus also serves as the LS center/output for such autonomic motor responses as changes in heartbeat. To touch/pick the rose, *voluntary motor areas* of the cortex are accessed via the cingulate gyrus or hippocampus. Similar connections to *frontal lobes* and *temporal lobes* evoke the *higher aspects of feelings* (e.g., love) and fond memories, respectively.

CN: Use a dark color for C.
1. Begin in the upper left corner with the view of the limbic system (C) compared with the brain stem (A) and cerebral cortex (B). Before going on to the actual structures of the limbic system, begin coloring its connections and responses as shown in the middle diagram. Start with the sensory inputs on the left (olfactory bulb/other senses), and follow the arrows into the darkly outlined square representing the limbic system. As you work your way through the system, simultaneously color the anatomical representation of each structure at the top of the page.
2. Color titles of limbic structures in bottom panel.

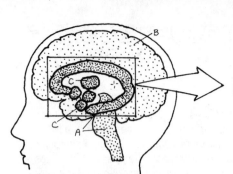

BRAIN STEM [A] CEREBRAL CORTEX [B] LIMBIC SYSTEM [C]

The limbic system (LS) consists of some forebrain structures and the hypothalamus. Positioned between the lower brain (vital functions) and the higher cerebral cortex (adaptive and skilled brain), the LS functions in motivation, emotions, and the expression of goal-directed instinctive behavior. In lower animals, the LS is intimately connected to the sense of smell.

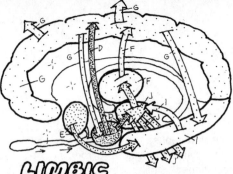

LIMBIC STRUCTURES [C]

Amygdala, septum, hippocampus, cingulate gyrus, hypothalamus, anterior thalamus, and their associated fiber pathways make up the LS. Amygdala and septum help connect the primitive senses and the cortex to the LS; vision and audition find access via the thalamus. A loop exists within the LS (the Papez circuit) such that impulses from the hypothalamus travel up to the anterior thalamus, on to the cingulate gyrus, and then via the hippocampus back to the hypothalamus. The cingulate gyrus and anterior thalamus provide connections between the LS and the cerebral cortex.

LIMBIC SYSTEM FUNCTIONS: [C] EMOTIONS [C] INSTINCTS [C] DRIVES [C] LEARNING/ MEMORY [C]

Some stimuli (aromas, strange sounds, a baby's smile) evoke emotions and bodily responses (i.e., "feelings" [pleasure], instinctive motor responses [smile], and visceral effects [heart rate]). These responses are integrated by the LS structures, including the hypothalamus, which also provides a main output for the LS. Thus, signals for somatic motor reactions (smiling) are sent to the brain stem motor centers. For visceral motor effects (heart rate), to the autonomic nervous centers. For the neurohormonal effects, to the pituitary/endrocrine system. Feelings are probably integrated at the higher cortical levels. The hippocampus is also involved in learning and memory.

RESPONSES OF THE LIMBIC SYSTEM [C]

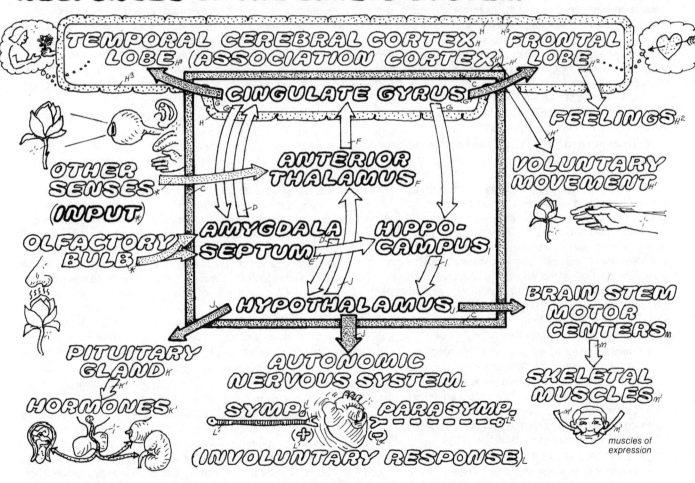

EFFECTS OF STIMULATION [C] (LESIONS MAY CAUSE OPPOSITE EFFECTS)

ANGER/RAGE * AMYGDALA [D]

PLEASURE * SEPTUM [E] HYPOTHAL. [J]

FEAR * HYPOTHAL. [J]

SEX * SEPTUM [E] HYPOTHAL. [J]

NEUROBIOLOGY OF LEARNING & MEMORY

LEARNING AND NEURAL PLASTICITY. The sensory, motor, and associative systems of the brain, particularly those involving the cortex, can show changes in their response pattern as a result of previous exposure to the same stimulus (experience). This modifiability of brain operation (plasticity) underlies the phenomenon of learning that is present in practically all animals in one form or another. Learning may be defined as the ability to change the response to a stimulus with experience. With the evolution of the brain in higher animals, resulting in the appearance of the cerebral cortex and increased complexity of the higher brain functions, the ability to learn from experience and to store learning (memory) becomes more advanced and efficient, reaching very high levels in primates and humans. Indeed, the ability to learn and transmit knowledge through teaching is the basis for the development of human culture and civilization.

ASSOCIATIVE LEARNING. The different forms and the laws of learning are discussed in psychology textbooks. Here some neural aspects of learning and memory are briefly discussed. To illustrate, consider the example of *conditioned reflexes*, a form of *associative learning* in which the subject learns to connect two different stimuli. Ivan Pavlov (Russian physiologist and Nobel laureate) studied conditioned reflexes. Dogs salivate when exposed to the smell, sight, or taste of food. This is an inborn reflexive response. The food (meat) is the *unconditioned stimulus* (US), and the salivation is the *unconditioned response* (UR). Inborn, permanent neural (synaptic) connections in the brain enable this response to occur.

Pavlov discovered that if the US was paired with a conditioned stimulus (CS) (e.g., a bell) often enough (the bell must ring a few seconds *before* the food), the animal soon shows a similar response (i.e., salivation, the UR) to the conditioned stimulus (bell) alone. The new response, called the *conditioned response* (CR), is clearly an evidence of associative learning and indicates the formation, in some way, of a *new* connection in the brain between pathways for hearing and salivation. These new connections are labile, tending to disappear gradually if the US (food) is not presented any longer. This apparent loss of the conditioned response (learning) is called *extinction*. Extinction is not the same as *forgetting*, which is an inability to retrieve learned items. The neural bases of extinction and forgetting are not known.

Instrumental conditioning (trial/error learning) is a more complex form of associative learning in which the subject takes an active part in the learning process, and reward plays a reinforcing role. Here, as in the conditioned reflexes, the ability to show and use a learned response improves with repeated exposure and practice, and the associated memories become more permanent.

NEURAL CORRELATES OF LEARNING. How does a new learning circuit form in the brain? What neural changes occur during learning? When a stimulus is presented for the first time, the associated neuronal circuits remain active for as long as the stimulus is present. A *reverberating circuit* can prolong activity in the original circuit, even when the initial stimulus ceases. In such a circuit, the original excitatory input from, say, the sensory neurons activates parallel excitatory interneurons. These make *recurrent positive feedback* connections to the original circuit, enabling the excitation to continue. Reverberating circuits can explain instantaneous (very short-term) memories.

For memories of longer duration (see below), modification in the circuitry or in the efficacy of the existing synapses is necessary. For example, changes may be brought about in the properties of the *synaptic membranes* (receptors, enzymes) as well as in intracellular synaptic compartments (presynaptic and postsynaptic). These may increase the functional abilities of a synapse (*synaptic facilitation*); so the same stimulus in a presynaptic neuron could result in a different (more or less intense) response in the postsynaptic neuron. New synapses may also be formed as a result of the exposure to new experience (*synaptic growth*). For these changes, which involve synthesis of new membranes and synapses, *protein synthesis* is necessary. Neurons are some of the most active cells in protein synthesis. The neuronal cell bodies contain large amounts of RNA and Nissl bodies (ribosomes/rough ER). When animals are given drugs that block protein synthesis, new labile learning cannot be transformed to permanent forms (see below).

STAGES OF LEARNING AND MEMORY FORMATION. Experiments with different types of associative learning indicate that learning and memory formation occur in two stages: an initial *short-term* stage, followed by a *long-term* stage. The short-term stage is further divided into a very short-term (*instantaneous*, a few seconds) and short-term (minutes to hours). Each of these stages is associated with an equivalent type of memory. Upon first exposure to a learning item (e.g., finding a number in a phone book), a very short-term memory that will dissipate rapidly if not reinforced is formed. Longer exposure or active use results in a memory that is accessible for minutes to hours. This memory is also labile and may disappear if not used. If animals are exposed to conditions that temporarily reduce brain metabolism and protein synthesis (drugs, hypothermia) or alter operation of brain electrical activity (electroshocks, blows) during short-term memory formation, the learning is lost, and memories can't be recalled.

Continuous use and exposure will transform the short-term memory into a long-term form (*memory consolidation*). Consolidated memory is in a more permanent physicochemical form, such as modified or new synapses. It cannot be erased by drugs that inhibit protein synthesis or by electroshock, blows, or hypothermia. In addition to the obvious importance of the cortex for learning and memory in general, studies on humans with brain lesions or with experimental animals show that the hippocampus and associated circuitry of the limbic system (plate 102) are essential for memory consolidation.

CN: Use dark colors for C, F, and K.
1. Color the stages of a conditioned reflex formation (top) one at a time. Note that in stage 3, neurons in the olfactory system (representing unconditioned stimulus) are not colored.
2. Note the limited number of items to be colored in the next 3 panels.

3. When coloring the material on short- and long-term memory, begin with the four neural mechanisms on the left which are believed to be involved in learning and memory formation. Color the borders of the remaining areas of this section.
4. Color the lower panel on memory loss and memory recall.

ASSOCIATIVE LEARNING: CONDITIONED REFLEX*

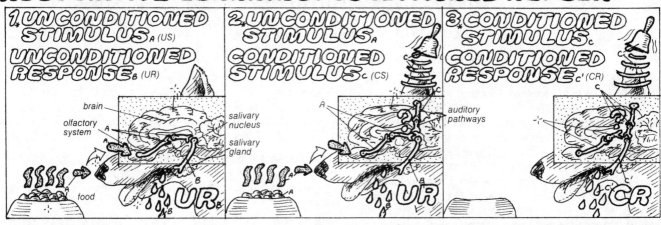

1. UNCONDITIONED STIMULUS_A (US) UNCONDITIONED RESPONSE_B (UR)
brain
olfactory system
food
UR

2. UNCONDITIONED STIMULUS_A CONDITIONED STIMULUS_C (CS)
salivary nucleus
salivary gland
auditory pathways
UR

3. CONDITIONED STIMULUS_C CONDITIONED RESPONSE_C' (CR)
CR

EXTINCTION*

If the CS is presented frequently without the US, the CR will gradually disappear (extinction).

INSTRUMENTAL CONDITIONING*

Animals given a choice between two bars learn very quickly (by trial/error) to press the right bar if rewarded for the correct choice.

TRIAL & ERROR — **TRIAL & REWARD** (food) — **LEARNED BEHAVIOR**

HABITUATION*

All animals "learn" to diminish the UR (ignore the US) if the US occurs frequently and purposelessly (habituation).

BANG — BANG — BANG — BANG

SHORT- & LONG-TERM MEMORY (NEURAL MECHANISMS)*

REVERBERATION

ORIGINAL ACTIVITY

FACILITATION

before learning — after learning

PROTEIN SYNTHESIS

SYNAPTIC GROWTH

INSTANTANEOUS LEARNING
SHORT-TERM MEMORY
REVERBERATION   FACILITATION

451-4278 → BANG

MEMORY CONSOLIDATION
limbic system — cortex
PROTEIN SYNTHESIS
HIPPOCAMPUS, TEMPORAL LOBE

451-4278

Sensory data (e.g. numbers) can be retained briefly (reverberating circuits). Increased use of this labile "short-term" memory results in permanent storage as "long-term" memory (consolidation). Consolidation requires hippocampus/temporal lobes and involves synthesis of proteins that alter synaptic function.

LONG-TERM MEMORY
PERMANENT SYNAPTIC CHANGES

451-4278 — BANG — 451-4278

MEMORY DISORDERS*

NUCLEUS BASALIS

LESIONS & INJURIES

Concussions cause retrograde amnesia. Temporal lobe damage (hippocampus/temporal cortex) interferes with consolidation.

SENILE AMNESIA

Senile amnesia and diminished consolidation may be related to aging-associated degeneration in the limbic/temporal lobe.

ALZHEIMER'S DISEASE/SENILE DEMENTIA

Cholinergic input from the nucleus basalis to the cortex and limbic system markedly influence memory functions. Senile dementia seen in Alzheimer's disease involves, in part, loss of these projections.

MEMORY RECALL*

TEMPORAL LOBE STIMULATION

Electrical stimulation of temporal lobe in conscious patients (undergoing brain operation) elicits vivid recall of past memories, particularly those with strong affective components.

BIOGENIC AMINES, BEHAVIORAL FUNCTIONS, & MENTAL DISORDERS

In the brain, as in the peripheral autonomic and neuromuscular synapses (see plates 16, 17 and 25), most synaptic communication is chemical. Brain synapses use a variety of neurotransmitter substances, such as amino acids (glutamic acid, glycine, GABA), peptides (substance P, endorphins, etc.), as well as the transmitters found in the peripheral synapses (acetylcholine and norepinephrine). *Norepinephrine* (NE), *dopamine* (DA), and *serotonin* (ST) belong to the family of *biogenic monoamines*. Brain monoamines serve as transmitters in the neural systems regulating affective states (moods, motivation, feelings) and in self-awareness, consciousness, and personality.

Reserpine (a plant alkaloid) was known as an effective agent against hypertension. This drug decreased activity of the peripheral NE synapses (hence its antihypertensive effects) by interfering with storage of synaptic vesicles. During the 1950s, it was noted that reserpine also affects the central nervous system, causing such "affective disorders" as depression and loss of appetite and interest. In animals, the effects include reduced activity and sedation. Indeed, reserpine had been used for centuries in India to relieve mental patients' mania (abnormally elevated moods). The idea that these amines (NE, ST, DA) may be involved in regulating mood and feelings (affective states) helped establish the fields of chemical psychobiology and psychopharmacology, which focused on the actions of the behaviorally important neurotransmitters and drugs that influence their action and metabolism.

NEUROANATOMY OF AMINE NEURONS. Mapping studies using fluorescent staining techniques have shown that the neurons releasing these amines make up nerve groups within the reticular formation (see plate 100). The cell bodies are generally located in the brain stem, and the fibers ascend to the forebrain. The *NE projections* (neurons) originate mainly in the *locus ceruleus* of the medulla and course up along the *medial forebrain bundle* to innervate the *cerebral cortex* and *limbic system*. NE fibers do not innervate the basal ganglia. The *ST neurons* originate in the *raphe nucleus* of the pons-medulla and course along the medial forebrain bundle to innervate all forebrain areas. The *DA pathways* (neurons) also begin in the midbrain. One pathway ends in the hypothalamus, another in the basal ganglia, and the third, behaviorally most important, ends mainly in the structures of the *limbic system* and the *frontal lobes* (see also plate 102).

Thus, the NE/ST systems, as may be construed from their reticular nature, regulate arousal, moods, motivation, pleasure, and well-being. The DA system in particular serves also in more complex functions related to the frontal lobe-limbic system (i.e., goal-directed behaviors, self-awareness, planning, anxiety, etc.).

MONOAMINE BIOCHEMISTRY. Monoamine neurotransmitters are derived from amino acids (NE and DA from tyrosine, which is hydroxylated by the enzyme tyrosine hydroxylase

eventually to form DOPA and then DA). DA can be metabolized to NE. DA neurons lack the enzyme for this last conversion. ST is derived from the amino acid tryptophan and is converted by the neuronal tryptophan hydroxylase to ST.

MONOAMINE-SYNAPSE PHARMACOLOGY. Drugs influence amine synpases function either presynaptically or postsynaptically. Actions on *presynaptic neurons* include interference with: 1. transmitter synthesis by inhibiting the *synthesizing enzyme*, tyrosine hydroxylase; 2. transmitter *storage* in *vesicles*; 3. transmitter *release* from vesicles; 4. *reuptake* of transmitter after release. *Postsynapic* actions include: 1. stimulation or blockage of *receptor binding* by the transmitter and 2. inhibition of the *deactivating enzyme* (plates 16, 17).

In broad functional terms, the neurotransmission drugs either enhance or suppress synaptic function. Thus, drugs that inhibit reuptake or the deactivating enzymes enhance synaptic function by increasing transmitter availability in the synapse. Drugs that block the postsynaptic receptors or inhibit transmitter synthesis or release, suppress synaptic function by reducing *impulse transmission* and transmitter availability, respectively. *Amphetamine* (one of the "upper" drugs) increases release and blocks transmitter reuptake. This increases transmitter availability in the synapse, which in turn enhances synaptic function. As a result, arousal, mood, excitability, and the ability to concentrate is increased. Of course amphetamines, like other drugs, have unpleasant side effects, which may appear later.

BIOGENIC AMINES AND THE TREATMENT OF MENTAL ILLNESSES. There are two major classes of mental disorders: *major depressions* and *schizophrenia*. Both are believed to be largely hereditary disorders caused not by real brain damage but by functional and chemical abnormalilties related to biogenic neurotransmission. Depression has been linked to *reduced activity* in the ST and NE synapses, and drugs that ameliorate this neurochemical condition also improve the behavioral symptoms of the depression. For example, amphetamine, which increases release and inhibits reuptake of the ST/NE transmitters, or substances that inhibit the deactivating enzyme monoamine oxidase (MAO inhibitors) tend to relieve both the neurochemical and behavioral deficiencies by increasing the NE/ST levels in the synapses.

Another promising advance in psychochemotherapy has been in treating the most common mental disorder, schizophrenia. Victims of this disease have delusions, deranged thoughts, and demented concepts of the self. The symptoms are sometimes accompanied by anxiety and psychosis. Drugs that decrease function in dopaminergic synapses, like the *DA receptor blockers* (e.g., haloperidol), have been very effective in ameliorating some of these symptoms. The particular hyperactive dopaminergic pathway appears to be the mesolimbic one connecting the brain stem to the limbic system and frontal lobes. The dopamine-excess hypothesis does not necessarily apply to all types of schizophrenias.

CN: Use dark colors for G, M, Q, and red for H.
1. At the top of the page color the introduction and chemistry of biogenic amine neurotransmitters.
2. Color the panel on depression and its treatment, begin with the brain diagram and the related structures. Then go to the site of drug action where an increase in levels of NE in the synapse is shown in the enlargement of the synaptic area. Conclude this section with the chemical structure of NE.
3. Do the same with the panel on dopamine.

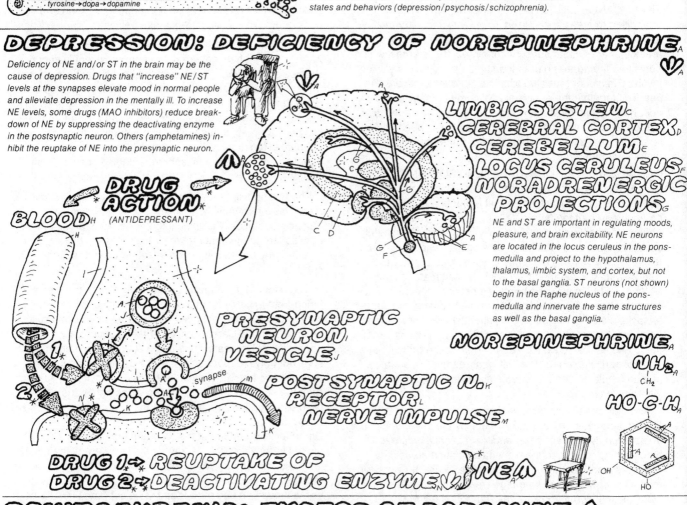

NOREPINEPHRINE — tyrosine→dopa→dopamine→norepinephrine

SEROTONIN — tryptophan→serotonin

DOPAMINE — tyrosine→dopa→dopamine

Norepinephrine (NE) and dopamine (DA) are related substances made in neurons by conversion of the amino acid tyrosine. Serotonin (ST) is derived from tryptophan. By acting as neurotransmitters, these biogenic amines regulate affective states (moods, motivation, and emotions). The aminergic neurons have their cell bodies in the brain stem and send extensive axonal branches to various higher brain areas. Altered function in these systems causes aberrant mental states and behaviors (depression/psychosis/schizophrenia).

DEPRESSION: DEFICIENCY OF NOREPINEPHRINE

Deficiency of NE and/or ST in the brain may be the cause of depression. Drugs that "increase" NE/ST levels at the synapses elevate mood in normal people and alleviate depression in the mentally ill. To increase NE levels, some drugs (MAO inhibitors) reduce breakdown of NE by suppressing the deactivating enzyme in the postsynaptic neuron. Others (amphetamines) inhibit the reuptake of NE into the presynaptic neuron.

DRUG ACTION (ANTIDEPRESSANT)

BLOOD

LIMBIC SYSTEM / CEREBRAL CORTEX / CEREBELLUM / LOCUS CERULEUS / NORADRENERGIC PROJECTIONS

NE and ST are important in regulating moods, pleasure, and brain excitability. NE neurons are located in the locus ceruleus in the pons-medulla and project to the hypothalamus, thalamus, limbic system, and cortex, but not to the basal ganglia. ST neurons (not shown) begin in the Raphe nucleus of the pons-medulla and innervate the same structures as well as the basal ganglia.

PRESYNAPTIC NEURON / VESICLE

synapse

POSTSYNAPTIC N. / RECEPTOR / NERVE IMPULSE

NOREPINEPHRINE

NH_2 — CH_2 — $HO-C-H$ — OH — HO

DRUG 1→ REUPTAKE OF
DRUG 2→ DEACTIVATING ENZYME NE

SCHIZOPHRENIA: EXCESS OF DOPAMINE

Drugs that reduce or block transmission at the DA synapses (DA receptor blockers, e.g., haloperidol) are the most effective agents in the treatment of schizophrenia (antipsychotic drugs). In contrast, drugs that markedly "increase" brain DA levels and transmission (amphetamine, cocaine in high doses) can cause paranoid/schizoid behavior, even in normal humans.

DRUG ACTION (DA RECEPTOR BLOCKER) (ANTIPSYCHOTIC DRUG)

MIDBRAIN DOPAMINERGIC PATHWAY

In the brain, three separate DA pathways are known, one in the hypothalamus, the second in the basal ganglia. A third pathway originates in the midbrain, projecting to the limbic system and frontal cortex. Excessive activity in this mesolimbic (cortical) pathway may be involved in causing schizoid psychosis. Schizophrenia is the most abundant form of mental disorder.

DOPAMINE NH_2 — CH_2 — $H-C-H$ — OH — HO

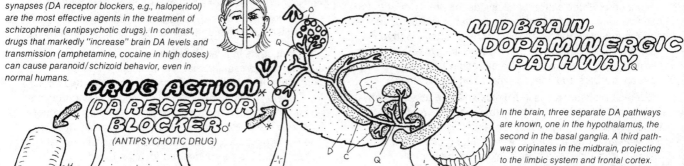

DRUG → RECEPTOR → DOPAMINE (DA) TRANSMISSION
(DA RECEPTOR BLOCKER)

LATERALITY, LANGUAGE, & CORTICAL SPECIALIZATION

SIGNIFICANCE OF CORTICAL ASSOCIATION AREAS. In addition to areas specialized for purely *sensory* or *motor* functions, the human *cerebral cortex* contains extensive areas that are neither sensory nor motor. These areas, which constitute the greater part of the human cerebral cortex, are involved in higher order *associational* and *integrative activities*. The fact that the equivalents of some of these areas are not present in animals may indicate that they serve in the higher behavioral and mental capacities (e.g., *speech* and *language*) that distinguish humans from other mammals and even other primates.

ORGANIZATION OF LANGUAGE FUNCTIONS IN THE BRAIN. In the middle of the last century, Paul Broca, the famous French scientist, noted that patients with lesions in a particular area of the *left frontal lobe* (*Broca's area*) could understand speech but had difficulty producing meaningful sentences (*motor* or *nonfluent aphasia* [aphasia = speech disorder]) without any evidence of speech paralysis. Later Karl Wernicke, a German neurologist, noted that left brain lesions, limited to a region bordering between the *parietal* and *temporal lobes* (*Wernicke's area*) caused *sensory* or *fluent aphasia*, a disorder in which the patient showed poor speech comprehension without having any hearing problems. From these and later studies, a cerebral organization for language and speech was formulated that localized this important human faculty to certain discrete association areas of the *left hemisphere*.

According to this scheme, words and sentences in the spoken language are analyzed initially by the *primary auditory areas*, then by the *secondary auditory association areas* before being relayed to the higher association areas (i.e., Wernicke's area of the left temporal lobe). Here the symbolic meanings of words and language are understood. To speak words, signal commands are relayed from the Wernicke's area via a special *association fiber pathway* (the arcuate fasciculus) to Broca's area in the frontal lobe of the same left hemisphere. Broca's area functions as the *premotor area* for speech, sending programs for the activation of appropriate speech muscles and their proper order of contraction to the *speech motor cortex* in the lower precental gyrus. Activation of upper motor neurons in this area results in contraction of speech muscles and speech production (see plate 90).

Based on observations in patients showing abnormal ability in reading (dyslexia) and writing (agraphia) of words, a similar scheme has been drawn for processing visual language *reading, writing*). Thus, images of words, after processing by the *visual association areas*, are relayed via the *angular gyrus* (a higher order *visual association area*) to the *hands premotor area*. Between the angular gyrus and hands premotor cortex, the impulses may pass through the Wernicke's area. The hands premotor area communicates to the neighboring *hand motor cortex* the necessary programs for movement of the hand muscles, resulting in writing. Sign language may involve a similar scheme. The signals between the different

association areas are sent via the intrahemispheric and inter-hemispheric association tracts (see below).

HEMISPHERIC DOMINANCE VS. HEMISPHERIC SPECIALI-ZATION. Right hemisphere damage in areas equivalent to Broca's and Wernicke's areas of the left hemisphere causes few speech defects. This and the fact that most people are right-handed (i.e., motor control areas of the left hemisphere are superior to those of the right) led to the notion that the two hemispheres, though fairly symmetrical in form, are unequal in function, with the left hemisphere being *dominant*. Functions of the right hemisphere remained obscure until recently.

The two hemispheres are connected by the *corpus callosum*. This interhemispheric association tract, massive in humans, specifically connects the association areas of one hemisphere to the exact mirror-image areas in the opposite one, thereby transferring information between the hemi-spheres.

Occasionally, the corpus callosum in human patients suffering from epileptic convulsions is sectioned to prevent the spread of seizures from one hemisphere to the other (*split brain* operation). Careful testing of these patients by Roger Sperry (Nobel laureate) revealed that each hemisphere functions not only independently but in a different manner, as though each had functional abilities and a "mind" of its own. After the surgery, if a key is placed in the right hand of a blindfolded patient, the sensory signals reach the left hemis-phere due to crossing of sensory pathways (see plate 86). Upon being asked about the nature of the object in hand, the patient verbally replies, "A key." If the key were placed in the left hand, its sensory image would be in the right hemisphere. In this case, the patient is incapable of verbally describing the key, although he can recognize the object (can point to the name or shape of a key). These results imply that (1) centers for verbal expression are in the left hemisphere, (2) the right hemisphere has access to the speech centers only via the corpus callosum, and (3) the right hemisphere has full perceptive, cognitive and non-verbal motor competence.

Further, tests indicate that the right hemisphere is actually superior in *representational* and *visuospatial* functions, in perception and discrimination of *musical tones* and *speech intonations*, in *emotional responses*, and in understanding *humor* and *metaphor*. In broad terms, the right hemisphere functions are holistic and spatial (hence labeled "artistic"). The left hemisphere, in addition to its motor and verbal super-iority, appears to be specialized for logical and analytical operations; it categorizes things and reduces them to their parts in order to understand them. The division of functions between the two hemispheres notwithstanding, under normal conditions, and especially with regard to global, cognitive, and adaptive functions (memory, learning), the brain func-tions as a whole, utilizing the capacities of its different parts in concert.

CN: Begin in the upper left corner with the list of 7 functional characteristics of the right hemisphere (A). Note that they don't refer to any particular structure. Do the same with the left hemisphere, and color the large drawing of the two hemi-spheres and the remaining material at the top.
1. Color the corpus callosum (C) in the large and smaller drawings below and to the left. The corpus callosum deals with communication between the two hemispheres. The other association tracts (D) handle signal transfer within the same hemisphere.
2. Color the material on the left hemisphere's role in speech and other symbolic communication functions. Starting with number 1 at the ear.
3. Color the two types of aphasia and the split-brain experiment.

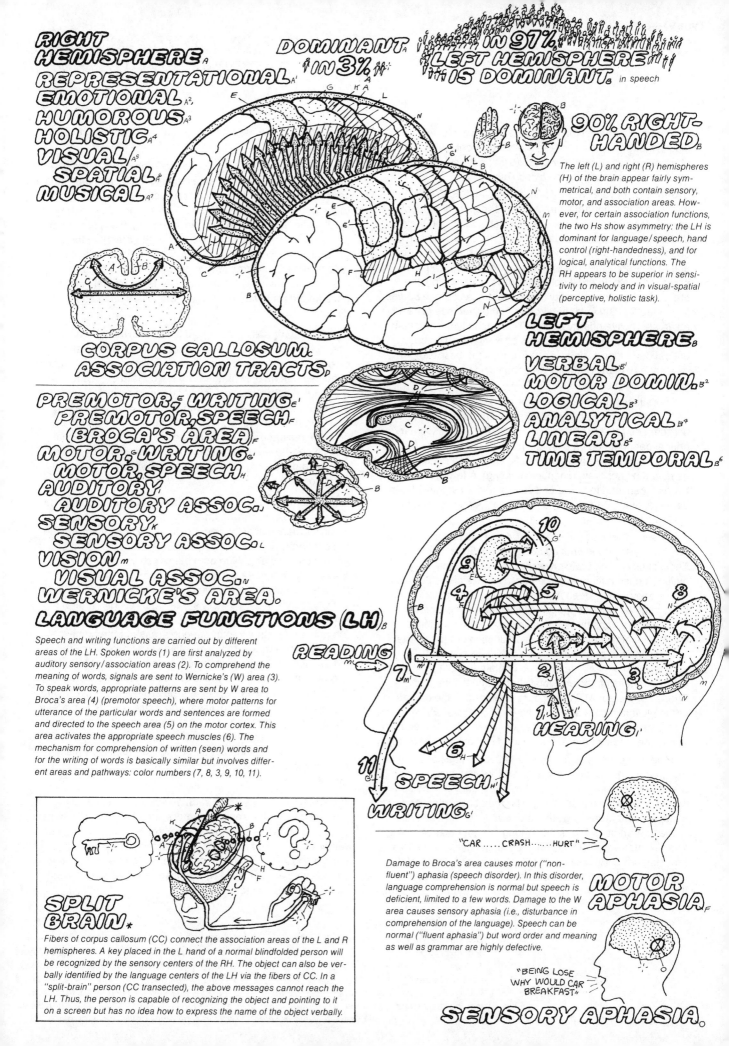

RIGHT HEMISPHERE A

REPRESENTATIONAL A1
EMOTIONAL A2
HUMOROUS A3
HOLISTIC A4
VISUAL A5
SPATIAL A6
MUSICAL A7

DOMINANT IN 3%

IN 97% LEFT HEMISPHERE IS DOMINANT in speech

90% RIGHT-HANDED B

The left (L) and right (R) hemispheres (H) of the brain appear fairly symmetrical, and both contain sensory, motor, and association areas. However, for certain association functions, the two Hs show asymmetry: the LH is dominant for language/speech, hand control (right-handedness), and for logical, analytical functions. The RH appears to be superior in sensitivity to melody and in visual-spatial (perceptive, holistic task).

CORPUS CALLOSUM. ASSOCIATION TRACTS. D

LEFT HEMISPHERE B

VERBAL B1
MOTOR DOMIN. B2
LOGICAL B3
ANALYTICAL B4
LINEAR B5
TIME TEMPORAL B6

PREMOTOR, WRITING E1
PREMOTOR, SPEECH F
(BROCA'S AREA) F
MOTOR, WRITING G1
MOTOR, SPEECH H
AUDITORY I
AUDITORY ASSOC. J
SENSORY K
SENSORY ASSOC. L
VISION M
VISUAL ASSOC. N
WERNICKE'S AREA. O

LANGUAGE FUNCTIONS (LH) B

Speech and writing functions are carried out by different areas of the LH. Spoken words (1) are first analyzed by auditory sensory/association areas (2). To comprehend the meaning of words, signals are sent to Wernicke's (W) area (3). To speak words, appropriate patterns are sent by W area to Broca's area (4) (premotor speech), where motor patterns for utterance of the particular words and sentences are formed and directed to the speech area (5) on the motor cortex. This area activates the appropriate speech muscles (6). The mechanism for comprehension of written (seen) words and for the writing of words is basically similar but involves different areas and pathways: color numbers (7, 8, 3, 9, 10, 11).

READING

HEARING

SPEECH

WRITING

SPLIT BRAIN *

Fibers of corpus callosum (CC) connect the association areas of the L and R hemispheres. A key placed in the L hand of a normal blindfolded person will be recognized by the sensory centers of the RH. The object can also be verbally identified by the language centers of the LH via the fibers of CC. In a "split-brain" person (CC transected), the above messages cannot reach the LH. Thus, the person is capable of recognizing the object and pointing to it on a screen but has no idea how to express the name of the object verbally.

"CAR ..... CRASH .... HURT"

Damage to Broca's area causes motor ("non-fluent") aphasia (speech disorder). In this disorder, language comprehension is normal but speech is deficient, limited to a few words. Damage to the W area causes sensory aphasia (i.e., disturbance in comprehension of the language). Speech can be normal ("fluent aphasia") but word order and meaning as well as grammar are highly defective.

MOTOR APHASIA F

"BEING LOSE WHY WOULD CAR BREAKFAST"

SENSORY APHASIA O

BRAIN METABOLISM & BRAIN FUNCTION

The *brain* is active all the time, not only in wakefulness but also in sleep. Therefore, it, like the heart, is critically in need of a continuous supply of *metabolic fuel substances* (energy) and *oxygen* provided by the *blood flow*.

BRAIN'S DEPENDENCY ON GLUCOSE AND OXYGEN.

In contrast to other active body organs (e.g., heart, muscle), which utilize alternative fuels like the fatty acids, the brain, under normal conditions, depends almost exclusively on *glucose* to obtain its energy needs. Marked *hypoglycemia* (e.g., due to a large insulin dose) may lead to fainting, convulsions, coma, or death. Interestingly, after days of starvation, the brain develops the capacity (enzymes) to use *ketone bodies* (a product of fatty acid metabolism in the liver, see plate 127) as an alternative energy source. This capacity is present in the newborn brain but disappears after infancy. The brain's critical dependency on glucose is one of the bases for the existence of many regulatory mechanisms for blood glucose homeostasis (see plate 125, 126).

To produce the large quantity of ATP required by brain cells, the Kreb's cycle/oxidative phosphorylation pathway is utilized (see plate 6). This accounts for the brain's high and critical dependence on oxygen. In adults, 10 seconds of *anoxia* (oxygen deprivation) is sufficient to lose consciousness and higher brain functions (fainting). A few minutes of hypoxia can lead to coma and severe and irreversible brain damage; death can occur due to the loss of function in the vital respiratory centers of the medulla.

In adults, *brain weight* is about 1.4 kg (3 lbs.), and the brain has an *oxygen consumption rate* of about 50 cc per min. Thus, although the brain's weight is only 2% of the body's, its oxygen consumption rate (*metabolic rate*) is about 20% of the whole body's. Why does the brain require such a high metabolic rate? Its work depends heavily on formation, propagation, synaptic transmission, and integration of a variety of electrochemical potentials, cellular functions requiring the maintenance of proper ionic gradients (see plates 10-12). To do this, the brain cell membranes contain one of the largest concentrations of sodium-potassium pumps in the body. These pumps are ATP dependent; they involve the operation of the plasma membrane enzyme Na-K-ATPase, which is also present in the brain in large concentrations. The sodium pump uses most of the ATP produced in the brain (plate 10).

In neurons, the *synapses* on dendrites and cell bodies use the greatest amount of energy. Therefore, the synapse-rich areas (e.g., the gray matter [cortex, basal ganglia, and subcortical nuclei]) have generally *high* metabolic rates, and synapse-poor areas (e.g., the fatty white matter [myelinated nerve fibers]) have *low* rates. Among the gray matter areas, relative rates vary. The forebrain basal ganglia and the midbrain inferior colliculi show *very high* rates; the *cortex* of the cerebrum and cerebellum have *moderately high* rates; the thalamus and the *nuclei* of the *cerebellum* and *medulla* show *medium* rates; the lowest rates are associated with spinal cord white matter.

BRAIN BLOOD FLOW. To support its high oxygen and glucose needs, the brain has an extensive vascular supply and a very efficient blood-flow regulation system. Normal blood flow to the brain is fairly high (750 mL/min.), amounting to 15% of the body's. Blood flow in different brain regions is regulated by poorly understood local (intrinsic to the brain) factors. In general, increase in *neural activity* in a particular brain area results in a rapid increase in *local blood flow* in that area, presumably to supply the excess needs for oxygen and glucose and to remove the excess metabolites (acid and carbon dioxide).

BLOOD FLOW AND LOCALIZATION OF BRAIN FUNCTION. The direct relation between an area's neural activity and blood flow has recently been utilized to map the brain's functional regions (mainly brain cortex) in conscious, responsive human subjects. A subject is injected with an inert radioactive gas such as xenon. As the blood flows through the various brain regions, detectors in a helmet worn on the head measure the radioactivity. These studies revealed that regional brain blood-flow pattern (hence neural activity) is dynamic, changing depending on physiological and psychological conditions.

For example, at rest, only the frontal lobes, particularly the premotor regions, show higher than average activity. During bodily or mental activity, depending on the task involved, the regional activity pattern changes. Thus, clenching of the right hand increases activity in both the sensory and motor hand areas, although more in the left than right hemisphere. Sensory stimulation of the hand alone, however, increases activity mostly in the sensory areas. Interestingly, whereas simple pronouncement of words increases activity mainly in the primary sensory/motor speech areas of both hemispheres (lips, tongue, face), creative speech involving thinking and ideas also increases activity in the Wernicke's and Broca's speech areas of the left hemisphere (see plate 105). Reading increases activity in a large area of the brain, including not only the expected visual and visual association areas but also the parieto-temporal areas, including the Wernicke's area, which are involved in word comprehension. Activity in the frontal lobes is increased not only during contemplation, problem solving, and planning, but also during pain and anxiety.

Certain mental diseases such as schizophrenia and depression and the senile disorders such as the dementias (reduced cognitive and memory capacities) are associated with *reduced* blood flow/metabolic activity. Brain diseases such as epilepsy, in which convulsions are observed due to excessive electrical activity, are associated with *increased* blood flow and metabolic activity.

CN: Use red for I, a dark color for E, and a very light color for H.
1. Begin in the upper left corner, and go on to the diagram on the right side of the page.
2. Color the diagrams on metabolic rate, noting that the sagittal section of the brain shown is intended to be a composite of several views in order to show the structures involved. Note that, in the diagram on the right, only the borders are colored to indicate general regions of high or low activity.
3. In the lower panel, color only the shaded areas of heightened blood flow and metabolic activity. Do not color the brain in the middle which depicts the position and names of the cortical areas.

BRAIN VS. TOTAL BODY

Compared to its weight and compared to other body organs, the brain has a very high rate of blood flow and metabolism. Even though its weight is only 2% of the body's, it receives 15% of the body's blood supply and 20% of the oxygen.

percent of body

100 — 75 — 50 — 25 — 0

1.4 kg **2%** A1

750 ml/min **15%** B

50 cc/min **20%** C

WEIGHT A1 **BLOOD FLOW** B **OXYGEN CONSUMP.** C

BRAIN FOOD

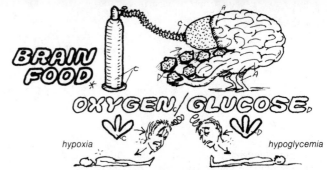

OXYGEN / GLUCOSE D

hypoxia C

hypoglycemia D

The adult brain is almost entirely dependent on glucose for fuel. Low blood sugar levels can lead to mental confusion, motor disturbances, and coma. Brain cells contain numerous mitochondria and rely heavily on oxygen to oxidize glucose. A lack of oxygen for 10 minutes leads to fainting (loss of higher brain functions). A few minutes of hypoxia leads to permanent brain damage, coma, or death.

METABOLIC RATES *
VERY HIGH E
HIGH F
MEDIUM G
LOW H

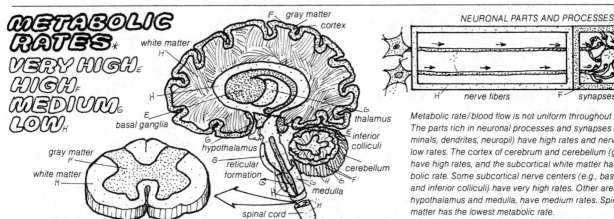

gray matter cortex F
white matter H
basal ganglia E
gray matter F
white matter H
hypothalamus G
reticular formation G
spinal cord H
thalamus G
inferior colliculi E
cerebellum F
medulla G

NEURONAL PARTS AND PROCESSES

nerve fibers H
synapses/dendrites F

Metabolic rate/blood flow is not uniform throughout brain tissue. The parts rich in neuronal processes and synapses (axon terminals, dendrites, neuropil) have high rates and nerve fibers have low rates. The cortex of cerebrum and cerebellum (gray matter) have high rates, and the subcortical white matter has a low metabolic rate. Some subcortical nerve centers (e.g., basal ganglia and inferior colliculi) have very high rates. Other areas, like the hypothalamus and medulla, have medium rates. Spinal cord white matter has the lowest metabolic rate.

CHANGES IN BRAIN BLOOD FLOW & METABOLISM

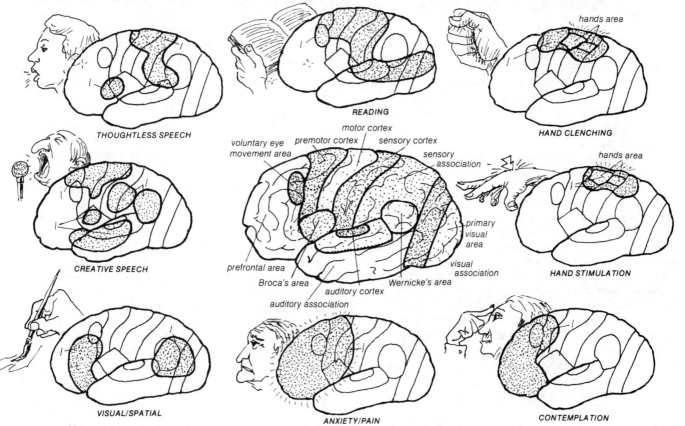

THOUGHTLESS SPEECH

READING

hands area

HAND CLENCHING

CREATIVE SPEECH

voluntary eye movement area
premotor cortex
motor cortex
sensory cortex
sensory association
prefrontal area
Broca's area
auditory cortex
Wernicke's area
auditory association
primary visual area
visual association
hands area

HAND STIMULATION

VISUAL/SPATIAL

ANXIETY/PAIN

CONTEMPLATION

Recent research using new methods for measurement of blood flow and metabolic rates has shown that different brain regions change their activity in different physiological and psychological states. When a person clenches the right hand, blood flow in the premotor cortex and hand area of the motor cortex in the left hemisphere increases. Interestingly, similar increases are observed in the sensory cortex. During resting and contemplation, activity is higher in the frontal areas than in the posterior areas. During concentration, awareness, anxiety, and pain, frontal lobe activity is markedly increased, indicating the importance of frontal areas in these mental states. During silent reading, visual association areas and the area for voluntary eye movements in the premotor cortex show increased activity. Intensive talking involving expression of ideas increases activity in auditory and speech motor cortex as well as in Wernicke's and Broca's areas.

THE ENDOCRINE SYSTEM AND FORMS OF HORMONAL COMMUNICATION

The importance of organization in the body is implicit in the concept of the body as an "organism." To be organized, the parts of the body must be regulated to work in synchrony with one another and in harmony with the external environment. This regulation is carried out by the nervous system and *endocrine system*. The nervous system, by sending nerve signals along the peripheral nerves, functions very rapidly, adjusting the activities of the internal organs within seconds. These effects (e.g., changes in blood pressure, respiration, and temperature) are relatively short lasting. In contrast, the endocrine system, which works by secreting hormones into the blood, acts slowly, its effects taking minutes to hours to days to develop, but they are longer lasting than those produced by the nerves. Hormones are chemical substances secreted in minute amounts into the bloodstream by the cells of the *endocrine glands*. Traversing through the circulation, hormones bind with appropriate receptors, which are selectively present in the cells of their target organs, inducing the desired effects on metabolism or function in those organs.

In certain instances, the nervous and endocrine systems can regulate each other's activities as well as act in concert to bring about desired changes in body functions. The special advantage of this *neuroendocrine* system of hormonal communication is that it allows the mediation of the effects of both the environment and the brain's systems on the endocrine system.

ENDOCRINE GLANDS. The endocrine cells in the body are usually found clustered in aggregates called the endocrine glands. These are the *pineal* (melatonin), *anterior pituitary* (growth hormone, tropins), *posterior pituitary* (ADH, oxytocin), *thyroid* (thyroxine), *parathyroid* (parathormone), *adrenal cortex* (corticosteroids), *adrenal medulla* (catecholamines), *pancreatic islets* (insulin and glucagon), and *testes* (male steroids) or *ovaries* (female steroids).

Another category of cells with endocrine functions is made up of those found scattered individually or in small aggregates within other organs with distinctly nonendocrine functions. These organs are the *kidney* (renin, erythropoietin, calcitriol), *liver* (somatomedin), *thymus* (thymosin), *hypothalamus* (hypothalamic hormones), *heart* (natriuretic peptide), *stomach* (gastrin), and *duodenum* (secretin, CCK). The testis and ovary can also be included in this category because the bulk of these glands is involved in producing male and female gametes, which is not an endocrine function but which is stimulated by hormones. The presence and location of endocrine cells within another organ are dictated by some special functional relationship between that organ and the endocrine cells it is hosting.

ENDOCRINE HORMONAL COMMUNICATION. The hormone was initially conceived of as a substance secreted by an endocrine gland (cell) into the blood to reach a target organ that affected the activity of that organ's cells. This form of purely hormonal communication is still applicable to many of the endocrine glands and their hormones, e.g., the pancreatic islets (insulin and glucagon). Hormonal communication may also occur between two endocrine glands. For example, the anterior pituitary secretes several tropic (trophic) hormones that stimulate other endocrine glands (*target glands*) to secrete hormones of their own (target gland hormones).

NEUROENDOCRINE COMMUNICATION. In the *neuroendocrine* form of hormonal communication, several subtypes can be recognized. In the simplest case, axons of certain nerve cells in the brain hypothalamus are extended into the posterior pituitary, secreting hormones (e.g., ADH) directly into the bloodstream to reach their targets (e.g., the kidney). In a more complicated case, hypothalamic nerve cells secrete certain regulatory hormones into a special *portal vascular system* connecting the hypothalamus with the anterior pituitary gland to control the secretion of some of the anterior pituitary hormones (e.g., growth hormone and prolactin) in the bloodstream. The anterior pituitary hormones then reach their target organ(s) (e.g., adipose tissue and mammary glands).

In a more complicated subtype of neurohormonal communication, the pituitary hormone (e.g., ACTH) secreted in response to hypothalamic hormone (e.g., CRH) will traverse the circulation to act as stimulating (tropin) hormones on some other endocrine glands (e.g., adrenal cortex). The latter will then secrete the final target hormone (e.g., cortisol) to reach, via the blood, the desired target organ (e.g., the liver).

Another type of neurohormonal communication is the secretion of a hormone from an endocrine gland directly in response to nerve signals from *autonomic nerves*. For example, the secretion of hormones of the adrenal medulla and pineal are subject to nerve signals from the sympathetic nervous system.

PARACRINE COMMUNICATION. The last major type of hormonal communication (recently discovered) is *local* or "tissue" hormonal communication. In this type, the definition of hormone is expanded to apply to substances secreted by special *paracrine* cells directly into the extracellular space of a particular tissue. These hormones diffuse across short distances within the extracellular space of the same tissue to act on nearby cells (paracrine effect) or the same cells (autocrine effect). The blood, therefore, is not involved as the medium of transport for this type of local hormones unless the paracrine cells themselves are blood cells. Paracrine hormonal communication has been observed in many tissues. Prostaglandins, known to be involved in many local regulatory functions, are the best known examples of paracrine hormones.

CN: Use red for D and dark colors for H and J.
1. Color the upper panel, starting with the endocrine glands on the left, all receiving the same color (A). Do the same with the organ column on the right (B).
2. Color the lower panel, completing each form of communication before going on to the next one.

ENDOCRINE GLANDS (A)

PINEAL (1)
PITUITARY (2)
THYROID (3)
PARATHYROID (4)
PANCREAS (5)
ADRENAL (6)
OVARY (7)
TESTIS (8)

Endocrine glands secrete hormones into the blood. The classic endocrine glands with mainly endocrine functions are the pineal, pituitary, thyroid, parathyroid, pancreatic islets, adrenal, testis, and ovary. Testis and ovary form gametes as well.

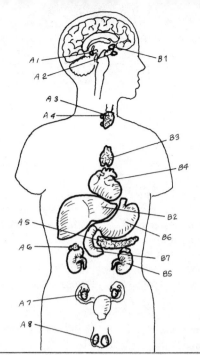

ORGANS WITH PARTLY ENDOCRINE FUNCTION (B)

HYPOTHALAMUS (1)
LIVER (2)
THYMUS (3)
HEART (4)
KIDNEY (5)
STOMACH (6)
DUODENUM (7)

Some organs contain individual or aggregates of endocrine cells releasing hormones. These hormones often relate to the functions of these organs. Among these are: hypothalamus, liver, thymus, heart, kidney, stomach and duodenum. Testis and ovary can also be included in this list.

FORMS OF HORMONAL COMMUNICATION (*)

1. ENDOCRINE (C) BLOOD CIRCULATION.

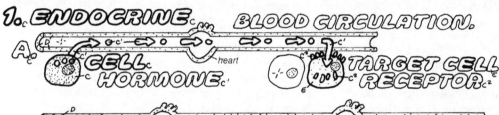

A — CELL (C) HORMONE (C') — TARGET CELL (C²) RECEPTOR (C²) — heart

B — TARGET GLAND CELL (F) HORMONE (F') RECEPTOR (F²)

Hormones are secreted into the blood to regulate the function of a distant target cell (organ). In the simplest form of hormonal communication, a hormone from an endocrine cell is transported by blood to a target cell (containing receptors for that hormone). In a more complex case, hormonal communication occurs between two endocrine glands, one serving as the target for the other. Still more complex forms involve interaction between brain and endocrines. Thus a nerve cell can secrete a hormone directly into the blood. Or a nerve cell secretes a hormone to reach the pituitary via a portal circulation. The pituitary cell then secretes a tropin which acts either on a target organ or another target endocrine gland. Nerve cells can also directly stimulate endocrine cells.

2. NEUROENDOCRINE (G)

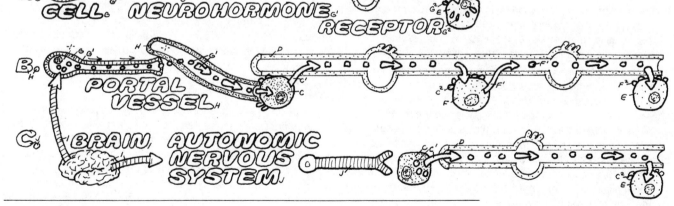

A — CELL (G) NEUROHORMONE (G') RECEPTOR (G²)

B — PORTAL VESSEL (H)

C — BRAIN, AUTONOMIC NERVOUS SYSTEM (J)

---

3. PARACRINE (LOCAL TISSUE ENVIRONMENT) (K)

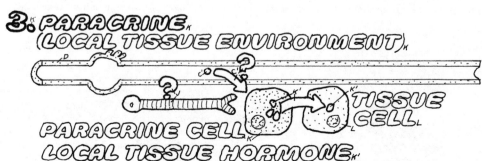

PARACRINE CELL (K) LOCAL TISSUE HORMONE (K') TISSUE CELL (L)

In local hormonal communication, paracrine cells secrete tissue hormones in the extracellular fluid to reach the neighboring target cells by diffusion, bypassing the blood entirely.

Local hormones may also act on cells which secrete them (autocrine).

AUTOCRINE (M)

CELLULAR MECHANISMS OF HORMONE ACTION

Hormones are substances secreted in small amounts by the cells of the endocrine glands. They act as blood-borne chemical messengers, regulating growth, metabolic processes, and functional activities for specific target cells, tissues, and organs. There are numerous hormones in the body, and their diversity of actions matches the diversity of bodily functions they aim to regulate, although they tend to use similar cellular mechanisms to act.

SLOW VS. FAST ACTING HORMONES. In regard to mechanisms of action, hormones may be divided into two general groups. The first, comprising the *steroid* hormones of the adrenal gland and gonads and the *thyroid* hormones, enter the target cells and influence the activity of the *nucleus* and the *synthesis of proteins*, activities that require time. For this reason, the actions of these hormones, though profound, are manifested slowly (*hours to days*). The second group, comprising the hormones of the hypothalamus, pituitary, pancreas, adrenal medulla, and gastrointestinal tract, being either *peptides* or *catecholamines*, do not enter the target cell. Instead, they primarily influence certain molecular mechanisms in the *plasma membrane* of the target cell. This mechanism in turn initiates a chain of events within the cell cytoplasm to bring about the action of the hormone. The effects of these hormones, in contrast to those of the first group, are expressed rapidly (*seconds to minutes*).

ACTIONS MEDIATED BY INTRACELLULAR RECEPTORS. The thyroid and steroid hormones, once released in the blood, will bind with specific *plasma binding protein* molecules that have a high affinity for them. In fact, the bulk of circulating hormones (>90%) is in the bound form, and only a small amount circulates as the "free" or effective form of the hormone. The carrier proteins are secreted by the liver to prevent the loss of these hormones from the kidney (proteins are not filtered in the kidney glomerulus) and to act as physiological regulators of the "free" levels of hormones.

Near the target cell, the free steroid hormone diffuses into the cell where, in the cytoplasm, the hormone binds with a specific *cytoplasmic "receptor"* to form an activated *hormone-receptor complex* that moves into the nucleus, binding with the *DNA*. Consequently, a specific *messenger RNA* is synthesized and then moves into the cytoplasm, where its code is translated into the synthesis of a *specific protein*. This protein may be an enzyme or some other functional protein. The nature of this protein differs, depending on the type of the steroid hormone and the target tissue involved. The actions of these proteins within the cell are responsible for the physiological actions associated with the hormones in the particular tissue.

The cellular mechanism action of thyroid hormones is similar to that of the steroid hormones except that thyroxine (T_4), the major thyroid hormone, is first converted in the cytoplasm to tri-iodo-thyronine (T_3), which is the cellularly active form of the hormone. T_3 then moves into the nucleus, where it binds with a *nuclear receptor*. The rest is as with the steroids. Thyroid and steroid hormones are degraded in part within the target cells or in the *liver*. Their metabolites are excreted in the *kidney*.

ACTIONS MEDIATED BY INTRACELLULAR (SECOND) MESSENGERS. The released peptide and catecholamine hormones reach the target cell, where they bind to "surface receptor" molecules in the plasma membrane. Each hormone has its own specific receptor. Depending on the hormone or tissue, the binding initiates an increase in the intracellular levels of *cyclic AMP* or the *calcium* ions (Ca^{++}). Cyclic AMP and Ca^{++} are therefore called the "second messenger" or the "intracellular messenger," the blood-borne hormone being the first messenger.

Cyclic AMP is formed from ATP by the action of *adenylate cyclase*, a plasma membrane enzyme that is activated by the hormone-receptor complex. Cyclic AMP binds to a *protein kinase*, which will in turn transform inactive proteins to active enzymes by *phosphorylating* them. The latter action also requires ATP. The phosphorylated proteins then initiate the physiologic (metabolic) events associated with the actions of these hormones. For example, the hormones glucagon, from the pancreas, and epinephrine, from the adrenal medulla, act on the liver cells, through this mechanism, to increase the release of glucose from glycogen. One of the advantages of such a complicated cascade of events is the amplification of the effects, so that a single molecule of the hormone can activate a chain of cascades resulting in the formation of millions of phosphorylated enzymes, which in turn can form billions of glucose molecules within a few seconds.

In some cells, the formation of a hormone-receptor complex leads to the release of calcium ions from cellular reserves. These combine with a regulatory protein, "calmodulin," which becomes activated. Activated calmodulin in turn activates certain enzymes called protein kinases. Kinases will catalyze the phosphorylation of some inactive proteins to turn them into active enzymes. As with cyclic AMP, these effects also result in the amplification of the original hormonal signal.

CN: Use red for A and dark colors for the hormones: C, I and J.
1. Begin by coloring the horizontal bars representing blood circulation (A) and the membranes of the two target cells (B) on the right side of the page. Color the plasma proteins (A^1) in the upper blood circulation diagram.

2. Color the steroid hormones (C) as they begin in the upper right secreting cell (C) and follow the numbered sequence 1-12. In the upper left corner for the thyroid hormones (I) do the same for 1-4.
3. Color the two lower sequences 1-6 for peptide (J) and catecholamine (J^1) hormones. Note step 7 pointing to the liver, for both hormones.

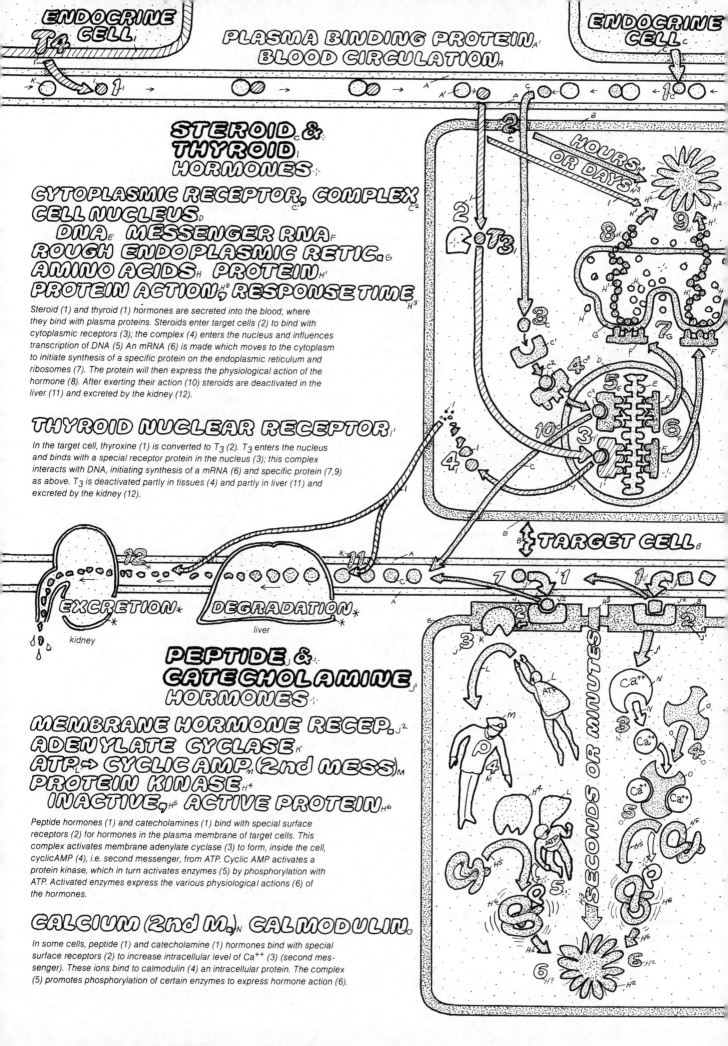

ENDOCRINE T4 CELL

ENDOCRINE CELL

PLASMA BINDING PROTEIN, BLOOD CIRCULATION.

STEROID & THYROID HORMONES

CYTOPLASMIC RECEPTOR, COMPLEX CELL NUCLEUS DNA. MESSENGER RNA ROUGH ENDOPLASMIC RETIC. AMINO ACIDS. PROTEIN PROTEIN ACTION, RESPONSE TIME

HOURS OR DAYS

Steroid (1) and thyroid (1) hormones are secreted into the blood, where they bind with plasma proteins. Steroids enter target cells (2) to bind with cytoplasmic receptors (3); the complex (4) enters the nucleus and influences transcription of DNA (5) An mRNA (6) is made which moves to the cytoplasm to initiate synthesis of a specific protein on the endoplasmic reticulum and ribosomes (7). The protein will then express the physiological action of the hormone (8). After exerting their action (10) steroids are deactivated in the liver (11) and excreted by the kidney (12).

THYROID NUCLEAR RECEPTOR

In the target cell, thyroxine (1) is converted to T_3 (2). T_3 enters the nucleus and binds with a special receptor protein in the nucleus (3); this complex interacts with DNA, initiating synthesis of a mRNA (6) and specific protein (7,9) as above. T_3 is deactivated partly in tissues (4) and partly in liver (11) and excreted by the kidney (12).

TARGET CELL

EXCRETION
kidney

DEGRADATION
liver

PEPTIDE & CATECHOLAMINE HORMONES

MEMBRANE HORMONE RECEP. ADENYLATE CYCLASE ATP → CYCLIC AMP (2nd MESS) PROTEIN KINASE INACTIVE, ACTIVE PROTEIN

Peptide hormones (1) and catecholamines (1) bind with special surface receptors (2) for hormones in the plasma membrane of target cells. This complex activates membrane adenylate cyclase (3) to form, inside the cell, cyclicAMP (4), i.e. second messenger, from ATP. Cyclic AMP activates a protein kinase, which in turn activates enzymes (5) by phosphorylation with ATP. Activated enzymes express the various physiological actions (6) of the hormones.

CALCIUM (2nd M.), CALMODULIN.

In some cells, peptide (1) and catecholamine (1) hormones bind with special surface receptors (2) to increase intracellular level of Ca^{++} (3) (second messenger). These ions bind to calmodulin (4) an intracellular protein. The complex (5) promotes phosphorylation of certain enzymes to express hormone action (6).

SECONDS OR MINUTES

MECHANISMS OF HORMONAL REGULATION

Hormones are biologically active compounds that influence many cellular and metabolic functions. To exert their effects appropriately, the hormones should be optimally secreted and finely controlled. Indeed, many diseases of the body are caused by abnormal hormonal secretion. To regulate hormonal secretion within physiological limits or in response to physiological demands, the endocrine system uses two types of control mechanisms. In one type, hormonal control is achieved by a self-regulatory system. Here, the level of a hormone in blood and the physiological parameter influenced by the hormone interact automatically to control endocrine and hormonal activity in a predetermined way and within set limits. The second type of control mechanism utilizes the influence of the nervous system over the endocrine system to override the self-regulatory operation, initiate new hormonal responses, and/or set new baselines for hormonal secretion.

FEEDBACK CONTROL. The operation of any system, be it physical or biological, involves an *input* and an *output*. If the system is intended to work by self-regulation, the output must exert some control over the input. This is referred to as *feedback control*. When the relationship between the input and the output is inverse, so that an increase in output leads to an decrease in input, the regulation is by *negative feedback*. When the relationship is direct, and an increase in output leads to a further increase in the input, the operation is by *positive feedback*.

In general, negative feedback mechanisms operate to promote stability and *equilibrium*, maintaining a system at a set point. All physiological *homeostatic* mechanisms operate by negative feedback. The regulation of numerous endocrine glands and their hormones falls into this category. Because positive feedback regulation tends to create *disequilibrium* and a *vicious cycle*, it may lead to abnormal hormonal conditions and disease. However, several physiological events depend on a positive feedback operation between hormones of the endocrine glands and the nervous system. Examples are ovulation and parturition.

SIMPLE HORMONAL REGULATION. Regulation of hormonal secretion in the body is achieved at different levels of complexity. *Simple hormonal regulation* involves only one *endocrine gland*. Here the secretion of a hormone from the endocrine gland is controlled directly, through a negative feedback mechanism, by the plasma concentration of the *parameter* the hormone is regulating. The endocrine cell usually has a receptor or a similar mechanism to detect the blood level of that parameter.

For example, a decrease in the plasma calcium level triggers an increase in secretion of parathyroid hormone from the parathyroid gland. The hormone acts on bone to release calcium. Elevated calcium levels in the plasma inhibit further release of parathyroid hormone. This negative feedback system maintains optimal calcium levels at all times. Other examples involve regulation of blood sugar by hormones of the pancreatic islets. These simple types of hormonal regulation and their automatic negative feedback operations are aimed at promoting homeostasis and equilibrium for the physiologically important variables (e.g., blood glucose and plasma Ca^{++}) within the internal environment.

PITUITARY AS THE "MASTER" GLAND. In the cases of *complex hormonal regulation*, the activity of one endocrine gland is controlled by hormones of another gland. The well-known examples are the control of thyroid, adrenal cortex, and gonads by the *pituitary gland*. If the pituitary gland is removed, these three glands will atrophy, and their hormone secretions will diminish greatly. Upon injection of extracts of pituitary, the atrophied glands will grow again, resuming their secretory function. These pituitary effects are conveyed by special *tropic hormones* that stimulate the *target glands* to grow and/or secrete their own hormones and exert negative feedback on the pituitary to inhibit the secretion of their respective tropic hormones. For this reason, the pituitary was once considered as the "master" endocrine gland, orchestrating the activities of several target glands, the hormones of which influence so many functions in the body. The importance of this mastery was considerably diminished when it was learned that the pituitary itself is subordinate to the brain.

BRAIN AND ENDOCRINE CONTROL. *Complex neurohormonal regulation* involves the interaction between the *brain* and the endocrine system. The pituitary gland is attached to the *hypothalamus*, a part of the brain involved in regulating visceral, emotional, and sexual functions. A special portal vascular system connects the hypothalamus to the anterior pituitary. Blood flowing through this system delivers the hormonal secretion from the nerve endings of certain hypothalamic neurosecretory cells directly to the pituitary cells. These *hypothalamic hormones* regulate the release of pituitary hormones. Electrical stimulation of certain areas of hypothalamus causes the release of these neurohormones, which are mostly peptides.

Such a mechanism superimposes brain control over the pituitary gland, and indirectly over the target glands of the pituitary. In this way, the effects of moods, emotions, stress, rythmical neural activity, and the *environment* (e.g., light, sound, temperature and odors), which are mediated by the nervous system, can be integrated at the level of the hypothalamus and conveyed to the endocrine system and hormones. Once in the blood, the target gland and pituitary hormones act through long and short loops, exerting negative and positive feedback effects on the hypothalamic neurosecretory cells, thereby modifying their secretion of hormones. Hypothalamic neurons, like the pituitary cells, contain receptors that can detect blood hormone levels.

The brain regulatory role over the endocrine system need not be solely conveyed through the anterior pituitary gland. The hypothalamus controls water regulation, milk flow, and parturition by releasing its hormones directly into the blood at the site of the posterior pituitary. The hypothalamus can also exert rapid and direct effects on the secretion of several endocrine glands by modifying the activity of the sympathetic and parasympathetic nerves, which innervate these glands.

CN: Use dark colors for A, E, F, and G.
1. Work from the top down. Note that the size of the output arrow determines the amount of ouput.
2. Color each of the three levels of hormonal regulation, working from top to bottom.

GENERAL FEEDBACK REGULATION.

INPUT. → SYSTEM → OUTPUT.
← FEEDBACK (CONTROL).

Regulation of hormone secretion is often based on feedback control. Feedback control operates when output of a system controls input. Feedback control enables endocrine systems to operate automatically by self-regulation.

NEGATIVE FEEDBACK → EQUILIBRIUM → (HOMEOSTASIS)

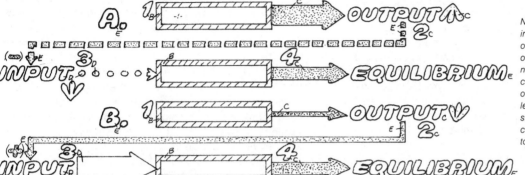

A. SYSTEM → OUTPUT ↑
INPUT. → → EQUILIBRIUM

B. → OUTPUT ↓
INPUT. → → EQUILIBRIUM

Negative feedback operates when input level is inversely related to output level. Thus in A, increased output (2) of a system (1) exerts negative feedback effect, decreasing input (3). This diminishes output (4) toward an equilibrium level. In B, a decrease in output (2) stimulates an increase in input (3), causing an increase in output (4) toward equilibrium.

POSITIVE FEEDBACK → DISEQUILIBRIUM ⇗ (VICIOUS) CYCLE. (ACTIVATION).

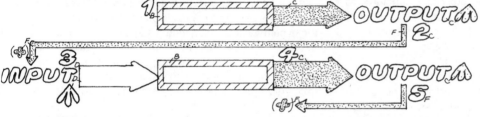

→ OUTPUT ↑
INPUT. → → OUTPUT ↑ (5)

Positive feedback regulation operates when output (2) of a system (1) stimulates further increase in input (3) resulting in still further increases in output (4) and on and on . . .(5). This type of regulation can lead to disequilibrium and vicious cycles. Some normal hormonal surges, however, are activated by positive feedback effects.

LEVELS OF HORMONAL REGULATION.

1. SIMPLE HORMONAL.
ENDOCRINE GLAND CELL.
HORMONE.
TARGET ORGAN CELL.
EFFECT OF HORMONE.

2. COMPLEX HORMONAL.
PITUITARY GLAND CELL.
TROPIC HORMONE.
TARGET ENDO. GLAND CELL.

3. COMPLEX NEUROHORMONAL.
ENVIRONMENT.
BRAIN.
HYPOTHALAMUS NEUROSECRETORY
CELL. HYPOTHALAMIC HORMONE.

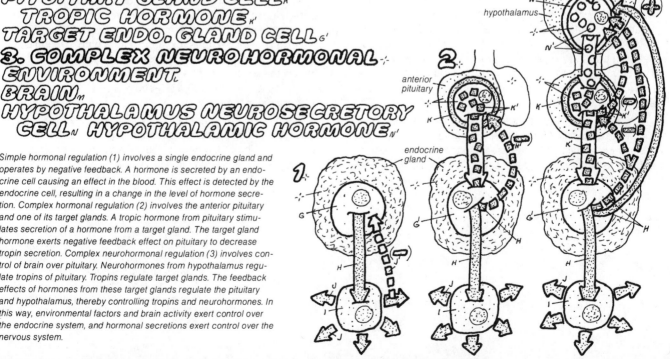

hypothalamus

anterior pituitary

endocrine gland

Simple hormonal regulation (1) involves a single endocrine gland and operates by negative feedback. A hormone is secreted by an endocrine cell causing an effect in the blood. This effect is detected by the endocrine cell, resulting in a change in the level of hormone secretion. Complex hormonal regulation (2) involves the anterior pituitary and one of its target glands. A tropic hormone from pituitary stimulates secretion of a hormone from a target gland. The target gland hormone exerts negative feedback effect on pituitary to decrease tropin secretion. Complex neurohormonal regulation (3) involves control of brain over pituitary. Neurohormones from hypothalamus regulate tropins of pituitary. Tropins regulate target glands. The feedback effects of hormones from these target glands regulate the pituitary and hypothalamus, thereby controlling tropins and neurohormones. In this way, environmental factors and brain activity exert control over the endocrine system, and hormonal secretions exert control over the nervous system.

HYPOTHALAMUS & POSTERIOR PITUITARY: NEUROSECRETION

PITUITARY STRUCTURE. The *pituitary* gland, located underneath the *hypothalamus* of the brain, is vital to body physiology because its hormones not only exert direct action on body organs (e.g., prolactin on mammary glands and antidiuretic hormone on the kidney) but also regulate the activity of several target endocrine glands (e.g., thyroid and gonads). The pituitary gland is controlled by the brain and mediates the effects of the central nervous system on hormonal activity in the body, which explains its critical anatomic position in relation to the brain.

The pituitary (*hypophysis*) is divided into an *anterior lobe* (adenohypophysis), a *posterior lobe* (neurohypophysis), and an *intermediate lobe*. In humans, the intermediate lobe either does not exist or is vestigial, consisting of a few cells with no known functions. The pituitary is connected to the brain via the *hypophyseal stalk*. This plate focuses on the structure and functions of the posterior lobe, to illustrate the concept of *neurosecretion*. Neurosecretion is also essential for understanding of anterior lobe function, and is the cornerstone of the modern science of neuroendocrinology.

POSTERIOR PITUITARY AND HYPOTHALAMUS. The posterior lobe of the pituitary secretes two hormones, *antidiuretic hormone (ADH)* and *oxytocin*. The posterior pituitary is not an endocrine gland because it does not contain true secretory cells. In fact, the gland, being an extension of the brain hypothalamus, consists mainly of nerve fibers and nerve endings of the neurons of two hypothalamic nuclei. These neurons have their cell bodies in the hypothalamus and send their axons (*hypothalamo-hypophyseal tract*) to the posterior pituitary through the hypophyseal stalk.

NEUROSECRETION. These hypothalamic nuclei are the *supraoptic* and *paraventricular*. The neurons of these nuclei are typical examples of *neurosecretory* cells. The *cell bodies* of these special neurons are the site of the synthesis of the hormones which, in the case of the posterior pituitary, are synthesized as larger prohormone molecules. These molecules contain the true hormone and a nonhormonal portion called *neurophysin*, which may function in hormone transport. The prohormone complexes are packed within the vesicles (*Herring bodies*), which flow down the axon by rapid axoplasmic transport.

Before reaching the nerve terminals in the posterior lobe, the hormone is split off the larger prohormone and stored in the *axon terminals*, to be released into the blood capillaries and carried out to the target tissues. The stimulus for hormone release is the nerve impulse arriving from the cell body down the axon membrane to the terminal, which causes calcium ions to flow into the terminal. This leads the secretory vesicles to fuse with the terminal membrane and the hormone to be released into the extracellular fluid and blood capillary.

ANTIDIURETIC HORMONES (VASOPRESSIN). The cells of the supraoptic nucleus make and secrete principally the antidiuretic hormone (ADH, also called vasopressin). Involved in regulating body water, ADH is secreted whenever the amount of water in the blood is decreased, as in dehydration due to excessive sweating or osmotic diuresis (caused by an increase in glucose or ketone bodies or sodium loss in the urine), as well as during hemorrhage and blood loss.

The signal for ADH release is believed to be an increase in the osmolarity of the blood mediated by an increase in the concentration of sodium ions in the plasma. The sodium elevation is sensed by specific *osmoreceptor* neurons in the hypothalamus, which in turn stimulate the supraoptic neurons to release ADH from the posterior pituitary. ADH acts principally on the collecting ducts in the kidney, by increasing their permeability to water. Water moves by osmosis from the kidney ducts to the plasma, decreasing plasma osmolarity. (See plate 62.)

ADH is also secreted when mechanoreceptors (volume receptors) in the heart, and pressure receptors in the vasculature, are stimulated after hemorrhage and blood loss. After a hemorrhage, ADH causes vasoconstriction, leading to an increase in blood pressure (vasopressive action).

OXYTOCIN HORMONE. Oxytocin is secreted principally by the cells of paraventricular nuclei, stimulated by sensory mechanoreceptors in the nipples of the breasts and cervix of the uterus, as part of neurohormonal reflex arcs. Sensory nerves convey the signals from the sensory receptors to the hypothalamus, leading to the secretion of oxytocin from the posterior pituitary. During labor, oxytocin acts on the myometrium of the uterus to cause massive contractions, eliciting the expulsion of the fetus (oxytocin = swift birth). During lactation, oxytocin acts on the myoepithelium of the mammary glands to elicit their contraction and cause the ejection of milk. (See also plates 150, 151.) There are no known functions for oxytocin in the male.

Oxytocin and ADH are both polypeptides containing nine amino acids. Their structures are identical except for the substitution, in ADH, of *phenylalanine* and *arginine* in place of one of the *tyrosines* and the *leucine* found in oxytocin.

CN: Use red for J, purple for K, and blue for L. Use dark colors for G.
1. Begin in the upper right corner and note that the profile of the head is left completely uncolored. Proceed to the large illustration on the left.
2. Color the diagrammatic neurosecretory cell on the right, noting its relationship to cell bodies and tracts in the large drawing.
3. Color the two peptide chains below.

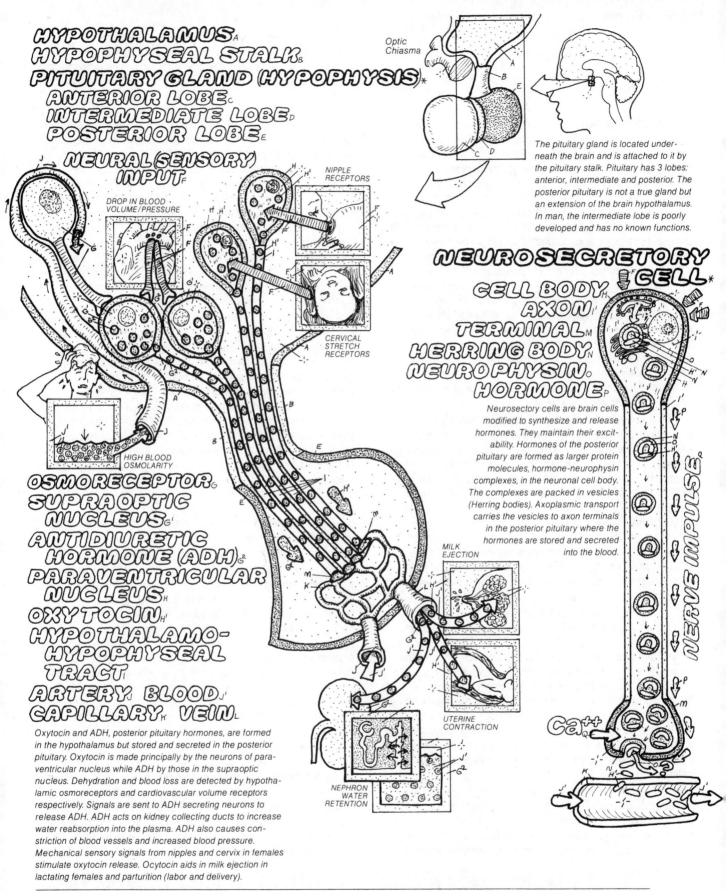

HYPOTHALAMUS A
HYPOPHYSEAL STALK B
PITUITARY GLAND (HYPOPHYSIS) *
ANTERIOR LOBE C
INTERMEDIATE LOBE D
POSTERIOR LOBE E
NEURAL (SENSORY) INPUT F

DROP IN BLOOD VOLUME/PRESSURE

Optic Chiasma

NIPPLE RECEPTORS

CERVICAL STRETCH RECEPTORS

HIGH BLOOD OSMOLARITY

The pituitary gland is located underneath the brain and is attached to it by the pituitary stalk. Pituitary has 3 lobes: anterior, intermediate and posterior. The posterior pituitary is not a true gland but an extension of the brain hypothalamus. In man, the intermediate lobe is poorly developed and has no known functions.

NEUROSECRETORY CELL *
CELL BODY H
AXON L
TERMINAL M
HERRING BODY N
NEUROPHYSIN O
HORMONE P

Neurosectory cells are brain cells modified to synthesize and release hormones. They maintain their excitability. Hormones of the posterior pituitary are formed as larger protein molecules, hormone-neurophysin complexes, in the neuronal cell body. The complexes are packed in vesicles (Herring bodies). Axoplasmic transport carries the vesicles to axon terminals in the posterior pituitary where the hormones are stored and secreted into the blood.

OSMORECEPTOR G
SUPRAOPTIC NUCLEUS G'
ANTIDIURETIC HORMONE (ADH) G²
PARAVENTRICULAR NUCLEUS H
OXYTOCIN H'
HYPOTHALAMO-HYPOPHYSEAL TRACT I
ARTERY J BLOOD J'
CAPILLARY K VEIN L

MILK EJECTION

NERVE IMPULSE

Oxytocin and ADH, posterior pituitary hormones, are formed in the hypothalamus but stored and secreted in the posterior pituitary. Oxytocin is made principally by the neurons of paraventricular nucleus while ADH by those in the supraoptic nucleus. Dehydration and blood loss are detected by hypothalamic osmoreceptors and cardiovascular volume receptors respectively. Signals are sent to ADH secreting neurons to release ADH. ADH acts on kidney collecting ducts to increase water reabsorption into the plasma. ADH also causes constriction of blood vessels and increased blood pressure. Mechanical sensory signals from nipples and cervix in females stimulate oxytocin release. Ocytocin aids in milk ejection in lactating females and parturition (labor and delivery).

UTERINE CONTRACTION

NEPHRON WATER RETENTION

Ca^{++}

STRUCTURE OF POSTERIOR PITUITARY HORMONES *

OXYTOCIN H'
CYS * TYR * TYR * GLN * ASN * CYS * PRO * LEU H' GLY * NH₂ *

ADH G
CYS * TYR * PHE G * GLN * ASN * CYS * PRO * ARG G * GLY * NH₂ *

Oxytocin and ADH are similar peptides, each with 9 amino acids. In ADH, one of the two tyrosines and the only leucine found in the oxytocin are replaced by phenylalanine and arginine respectively.

HYPOTHALAMUS & ANTERIOR PITUITARY

ANTERIOR PITUITARY HORMONES. The *anterior pituitary* gland secretes several protein hormones that either regulate the activities of other endocrine glands or directly control the activity of particular *target* organs. Of these hormones, *thyroid-stimulating hormone* (TSH) and *adrenocorticotropin hormone* (ACTH) regulate the activity of the *thyroid* and *adrenal cortex* respectively; *follicle-stimulating hormone* (FSH) and *luteinizing hormone* (LH) regulate the activities of the *gonads* (*testes* and *ovaries*). Two other hormones, *prolactin* and *growth hormone* (GH) (also known as *somatotropin* [STH]), can also act directly on non-endocrine target organs. Prolactin acts on the mammary gland, an exocrine gland, to promote milk secretion. Growth hormone promotes the breakdown of fat in adipose tissue as well as anabolic effects in muscle and bones. Anabolic and growth-promoting actions of GH are mediated by *somatomedins*, hormones released by the liver and other tissues in response to GH. Recent research in experimental animals has implicated the secretion of two other hormones, beta-lipotropin and beta-endorphin, whose functions are under investigation.

All the hormones of the anterior pituitary have growth-promoting effects on the cells of their *target glands* and generally increase their activity. For these reasons, they are collectively called pituitary *tropin* (trophin, trophic) hormones (hence thyrotropin, corticotropin, gonadotropin, and somatotropin). The tropins that regulate other endocrines generally increase the synthesis and release of the hormones of their target glands as well. Thus, TSH promotes the secretion of *thyroxine*, ACTH of *cortisol*, and FSH and LH of sex steroids (estrogen, progesterone, testosterone). Removal of the anterior pituitary (hypophysectomy) leads to atrophy of these target glands and cessation of their hormonal secretion.

PITUITARY CELLS. The anterior pituitary contains different cell types, each secreting one of these tropic hormones. Thus, *thyrotrops* secrete TSH; *corticotrops*, ACTH; *somatotrops*, GH; *mammotrops*, prolactin, and *gonadotrops* are believed to secrete both FSH and LH. The cell types of the anterior pituitary can be differentiated using various histologic stains. Classically, cells were divided into *acidophils* and *basophils* (collectively, chromophils), based on whether they reacted to acidic or basic dyes. Thyrotrops and gonadotrops are basophilic; cortictrops are lightly basophilic; somatotrops and mammotrops are acidophilic. Cells that do not stain with these dyes (chromophobes) may be immature or resting cells.

HYPOTHALAMIC CONTROL. Secretion of the hormones of the anterior pituitary is under the control of specific substances (hypophyseotropin hormones) released from certain neurons in the hypothalamus. As in the posterior pituitary, these substances (mainly peptides) are formed in the cell body of special secretory neurons, transported down their axons, and secreted in extremely small amounts at the axon terminals into a specific portal circulatory system (*hypophyseal portal capillaries*), which delivers these neurohormones directly to the pituitary cells, bypassing the general circulation. These hormones are also called hypothalamic-releasing or release-inhibiting (–RH, –IH) hormones, depending on whether their action leads to the release or inhibition of release of the pituitary hormones. For TSH, a releasing hormone (TRH), a tripeptide, has been found; for GH, there is a GRH and a GIH (somatostatin); for ACTH, a CRH; for prolactin, a PRH and a PIF (dopamine); and for gonadotropins, a GnRH, which is a decapeptide.

Hypophyseotropin hormones, from neurons in the hypothalamus, regulate the release of pituitary hormones, which in turn regulate the release of hormones of their target glands. Two factors control the release of these hypothalamic hormones. One involves signals from other brain areas, including those mediating endogenous rhythms in addition to exogenous (environmental) stimuli and stresses. The other factor is the operation of a feedback system whereby changes in the plasma level of the hormone of a target gland (e.g., cortisol) act on the hypothalamic neurons, altering the secretion of the specific hypophyseotropic hormone (e.g., CRH). This leads to a change in the secretion of the pituitary tropic hormone (e.g., ACTH), which in turn alters the secretion of the target gland hormone (e.g., cortisol). The principal site of feedback regulation for several target gland hormones (e.g., thyroid) is at the level of the pituitary gland.

IMPORTANCE OF BRAIN CONTROL. The main value of the releasing and inhibiting hormones is to permit the brain to exert a dynamic control over the endocrine system, adjusting its operation to the needs of the body. Thus, in animals, seasonal changes in light and length of the day can result in the appropriate activation and inhibition of the gonads. Long-term changes in environmental temperature can result in the appropriate adjustment in basal metabolic rate and heat production by altering thyroid secretion. Similarly, the brain's response to various stresses can increase the secretion of glucocorticoids from the adrenal cortex to bring about adaptive metabolic responses in order to increase bodily resistance and survival.

PULSATILE RELEASE. The secretion of most of the pituitary hormones has recently been shown to occur in an episodic (pulsatile) manner, i.e., there is a rhythm and a peak of secretion in regular intervals. The intervals are specific for each hormone and are in the range of one to several hours. These rhythms are believed to be caused by the episodic release of the hypothalamic-releasing hormones triggered by signals from other brain centers.

CN: Use red for E, purple for F, and blue for G. Use dark colors for A, B, N, and O.
1. Begin with the orientation drawing in the upper right, and continue with the large illustration on the left. Tiny circles and squares in the latter drawing represent hypothalamic and anterior pituitary hormones in general. Names of specific hormones appear in the list of titles. Color in the background colors of the illustration first, and then fill in the squares and circles over those colors. Note that a sixth cell type within the anterior pituitary, chromophobe, is left blank.
2. Color the chart at the bottom, noting that the target glands and the bulk of their own hormones exerting feedback on the hypothalamus and pituitary are given the color of the anterior pituitary hormone, which stimulates them, but the arrows of feedback are colored gray for additional emphasis. Note the entrance of these gray arrows in the upper illustration.

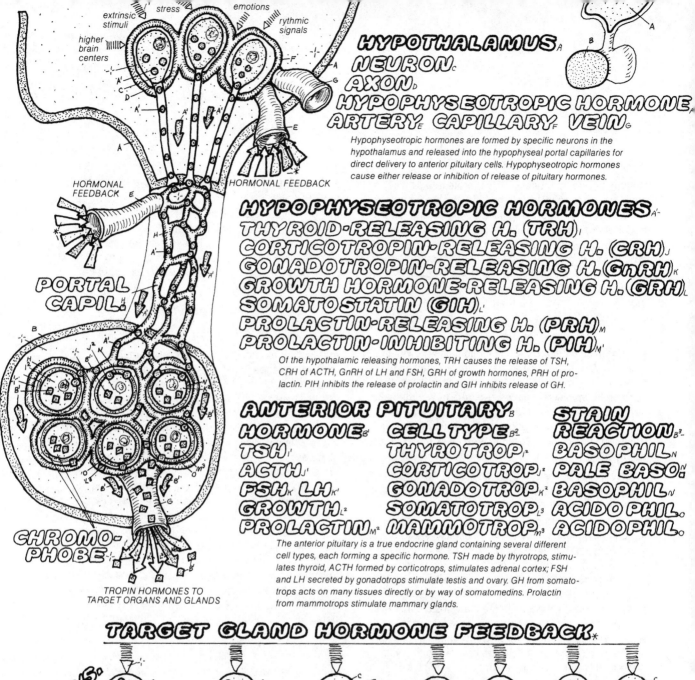

HYPOTHALAMUS
NEURON
AXON
HYPOPHYSEOTROPIC HORMONE
ARTERY CAPILLARY VEIN

extrinsic stimuli • stress • emotions • rhythmic signals

higher brain centers

Hypophyseotropic hormones are formed by specific neurons in the hypothalamus and released into the hypophyseal portal capillaries for direct delivery to anterior pituitary cells. Hypophyseotropic hormones cause either release or inhibition of release of pituitary hormones.

HORMONAL FEEDBACK

HORMONAL FEEDBACK

HYPOPHYSEOTROPIC HORMONES
THYROID-RELEASING H. (TRH)
CORTICOTROPIN-RELEASING H. (CRH)
GONADOTROPIN-RELEASING H. (GnRH)
GROWTH HORMONE-RELEASING H. (GRH)
SOMATOSTATIN (GIH)
PROLACTIN-RELEASING H. (PRH)
PROLACTIN-INHIBITING H. (PIH)

Of the hypothalamic releasing hormones, TRH causes the release of TSH, CRH of ACTH, GnRH of LH and FSH, GRH of growth hormones, PRH of pro-lactin. PIH inhibits the release of prolactin and GIH inhibits release of GH.

PORTAL CAPIL.

CHROMO-PHOBE

TROPIN HORMONES TO TARGET ORGANS AND GLANDS

ANTERIOR PITUITARY

HORMONE	CELL TYPE	STAIN REACTION
TSH	THYROTROP	BASOPHIL
ACTH	CORTICOTROP	PALE BASO:
FSH LH	GONADOTROP	BASOPHIL
GROWTH	SOMATOTROP	ACIDOPHIL
PROLACTIN	MAMMOTROP	ACIDOPHIL

The anterior pituitary is a true endocrine gland containing several different cell types, each forming a specific hormone. TSH made by thyrotrops, stimulates thyroid, ACTH formed by corticotrops, stimulates adrenal cortex; FSH and LH secreted by gonadotrops stimulate testis and ovary. GH from somato-trops acts on many tissues directly or by way of somatomedins. Prolactin from mammotrops stimulate mammary glands.

TARGET GLAND HORMONE FEEDBACK

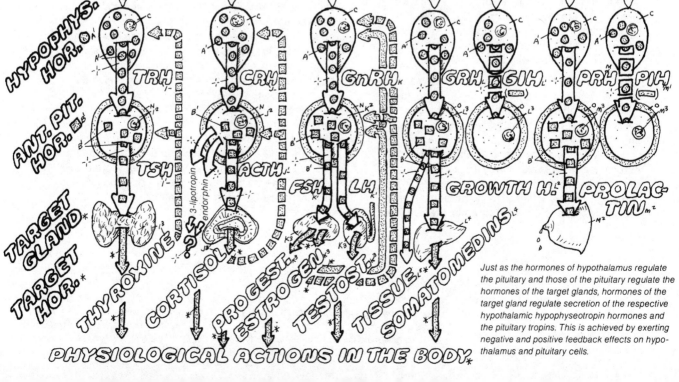

HYPOPHYS. HOR.

ANT. PIT. HOR.

TARGET GLAND

TARGET HOR.

TRH — TSH — THYROXINE

CRH — ACTH — CORTISOL — β-lipotropin endorphin ?

GnRH — FSH LH — PROGEST. ESTROGEN — TESTOST.

GRH — GROWTH H. — TISSUE SOMATOMEDINS

GIH (−)

PRH — PROLACTIN

PIH (−)

Just as the hormones of hypothalamus regulate the pituitary and those of the pituitary regulate the hormones of the target glands, hormones of the target gland regulate secretion of the respective hypothalamic hypophyseotropin hormones and the pituitary tropins. This is achieved by exerting negative and positive feedback effects on hypo-thalamus and pituitary cells.

PHYSIOLOGICAL ACTIONS IN THE BODY

GROWTH HORMONE: METABOLIC AND GROWTH EFFECTS

Human *growth hormone (GH)* is a protein (a single chain polypeptide of 191 amino acids) secreted by a specific cell type in the *anterior pituitary gland* (somatotrops). *Somatotrops* constitute the majority of the cells in the pituitary. Actions of growth hormone can be divided into two categories: first, those that promote *growth* of hard (e.g., bone) and soft (e.g., muscle) tissues in the body and, second, those that influence *metabolism*.

EFFECTS ON GROWTH. One of the classic demonstrations in endocrinology was that the removal of the anterior pituitary in a growing animal stops growth. Injection of GH in this animal leads to a resumption of growth. In the long bones of growing animals, the thickness of the *epiphyseal* plates — where new bone cells are produced and bone formation is enhanced, leading to a lengthening of the long bones — is increased by treatment with growth hormone (see plate 115). GH also promotes growth of many types of soft tissue.

GROWTH ABNORMALITIES. In growing humans with pituitary or hypothalamic tumors, hypersecretion of GH leads to *gigantism*. Pituitary giants are more than 8 ft. tall. Absence or reduced levels of growth hormone during childhood leads to *dwarfism*. A pituitary dwarf has a small body but normal head size and usually does not show mental retardation. In certain individuals who lack only the gene for GH but have an otherwise normal pituitary, stature is short, but sexual maturation and pregnancy may still occur. Some individuals may have GH but not its receptors at the tissue level; these will also have short stature. Children with unusually short stature due to low GH levels can now be treated with human GH protein which has become available through modern biotechnology methods. Dwarfism and gigantism can also be produced in growing animals by removing the pituitary gland (hypophysectomy) or appropriate treatment with GH, respectively.

In adults with excessive growth hormone secretion, bones — which can no longer grow in length due to fusion of the epiphyseal plates — grow in width. This leads to the typical picture of *acromegaly*, where the abnormal growth of bones in the digits, toes, mandible and back lead to a characteristic bodily deformity. Acromegalic people also have large visceral organs. GH does not influence embryonic growth, its action on human growth being limited to the postnatal period, especially between the ages of two and twelve years.

SOMATOMEDINS. The effects of GH on tissue growth are not direct, but are believed to be mediated by certain growth factors: *somatomedins*, secreted by the *liver* and possibly other tissues in response to stimulation by growth hormone. The somatomedins, which resemble insulin in their molecular structure, promote *cell proliferation* and *protein synthesis* in their target cells, but the specific mechanisms of their actions are not known.

METABOLIC EFFECTS. In addition to its growth-promoting (anabolic) actions, GH exerts diverse effects on metabolism of fats and carbohydrates, leading to marked changes in the levels of *fatty acids* and *glucose* in the blood. GH stimulates the *fat cells* of *adipose tissue*, stimulating lipolysis (breakdown of fat stores) and mobilization of fatty acids. The adrenal hormone, cortisol, is necessary for these actions of growth hormone. The mobilized fatty acids are released into the blood and are consumed by the heart and muscle in preference to glucose. GH also acts directly on muscle cells, promoting amino acid uptake and inhibiting glucose uptake by opposing the action of insulin (anti-insulin action). This leads to an increase in blood sugar levels. GH also acts on the liver to mobilize its glucose reserves. These actions are very important during stress, sustained exercise, and particularly during fasting, because glucose, spared and provided in this manner, can be used by the brain, which depends exclusively on glucose and cannot use fatty acids to obtain energy. After long periods of fasting (more than a week), brain tissue adapts and begins to utilize ketone bodies for energy (see plates 127, 128).

HYPOTHALAMIC CONTROL. In humans and animals, secretion of GH shows *episodic bursts* (peaks) with about 4-hr. intervals. At night, shortly after the onset of sleep, there is a big burst of GH secretion, the significance of which is not clear. The secretion of GH is regulated by two hormones from the *hypothalamus*: *growth hormone-releasing hormone (GRH)*, which stimulates the secretion of GH, and *somatostatin* (growth hormone-inhibiting hormone, *GIH*), which inhibits the release of GH. There is some evidence that GH can regulate its own secretion by exerting feedback effects on the hypothalamic neurons that secrete GRH and GIH. The plasma levels of fatty acids and glucose also influence GH secretion by affecting the hypothalamic neurons.

CN: Use the dark colors from the previous page for hypothalamus (A), and anterior pituitary (D). Use yellow for G and red for K.
1. Begin with the large illustration at the top, and color down to the two illustrations of growth hormone (F^1) action: adipose tissue (G) and liver (H).
2. Color the section in the upper right corner.
3. Color the GH effect on cell metabolism.
4. Color the secretions of somatomedins (I) from the liver and the effects on growth.

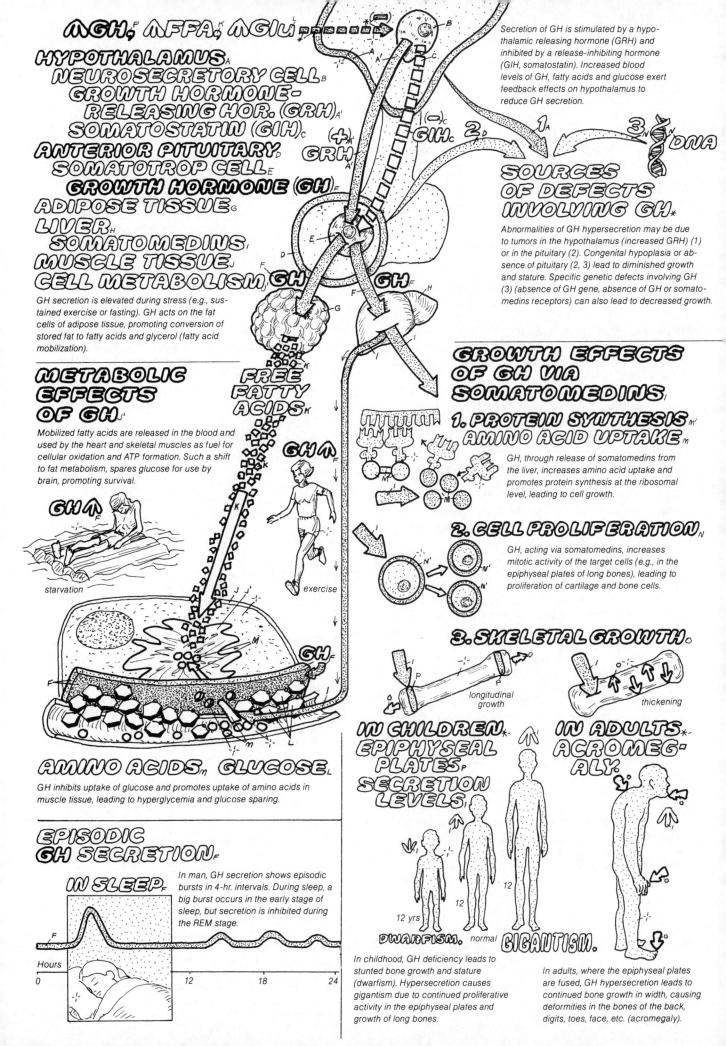

↑GH, ↑FFA, ↑GIU

HYPOTHALAMUS A
NEUROSECRETORY CELL B
GROWTH HORMONE-RELEASING HOR. (GRH) A'
SOMATOSTATIN (GIH) C
ANTERIOR PITUITARY D
SOMATOTROP CELL E
GROWTH HORMONE (GH) F
ADIPOSE TISSUE G
LIVER H
SOMATOMEDINS I
MUSCLE TISSUE J
CELL METABOLISM J'

GH secretion is elevated during stress (e.g., sustained exercise or fasting). GH acts on the fat cells of adipose tissue, promoting conversion of stored fat to fatty acids and glycerol (fatty acid mobilization).

Secretion of GH is stimulated by a hypothalamic releasing hormone (GRH) and inhibited by a release-inhibiting hormone (GIH, somatostatin). Increased blood levels of GH, fatty acids and glucose exert feedback effects on hypothalamus to reduce GH secretion.

SOURCES OF DEFECTS INVOLVING GH *

Abnormalities of GH hypersecretion may be due to tumors in the hypothalamus (increased GRH) (1) or in the pituitary (2). Congenital hypoplasia or absence of pituitary (2, 3) lead to diminished growth and stature. Specific genetic defects involving GH (3) (absence of GH gene, absence of GH or somatomedins receptors) can also lead to decreased growth.

METABOLIC EFFECTS OF GH J'

Mobilized fatty acids are released in the blood and used by the heart and skeletal muscles as fuel for cellular oxidation and ATP formation. Such a shift to fat metabolism, spares glucose for use by brain, promoting survival.

FREE FATTY ACIDS K

starvation

exercise

GROWTH EFFECTS OF GH VIA SOMATOMEDINS I

1. PROTEIN SYNTHESIS M' AMINO ACID UPTAKE M

GH, through release of somatomedins from the liver, increases amino acid uptake and promotes protein synthesis at the ribosomal level, leading to cell growth.

2. CELL PROLIFERATION N

GH, acting via somatomedins, increases mitotic activity of the target cells (e.g., in the epiphyseal plates of long bones), leading to proliferation of cartilage and bone cells.

3. SKELETAL GROWTH O

longitudinal growth

thickening

AMINO ACIDS M, GLUCOSE L

GH inhibits uptake of glucose and promotes uptake of amino acids in muscle tissue, leading to hyperglycemia and glucose sparing.

EPISODIC GH SECRETION F

IN SLEEP F

In man, GH secretion shows episodic bursts in 4-hr. intervals. During sleep, a big burst occurs in the early stage of sleep, but secretion is inhibited during the REM stage.

Hours
0   12   18   24

IN CHILDREN EPIPHYSEAL PLATES P SECRETION LEVELS

DWARFISM — 12 yrs, 12   normal, 12

In childhood, GH deficiency leads to stunted bone growth and stature (dwarfism). Hypersecretion causes gigantism due to continued proliferative activity in the epiphyseal plates and growth of long bones.

IN ADULTS ACROMEGALY

GIGANTISM

In adults, where the epiphyseal plates are fused, GH hypersecretion leads to continued bone growth in width, causing deformities in the bones of the back, digits, toes, face, etc. (acromegaly).

ACTIONS OF THYROID HORMONES

The *thyroid* is a butterfly-shaped endocrine gland located in the neck, anterior and lateral to the larynx. It receives a rich blood supply and secretes two closely related hormones, *thyroxine* (T_4, tetra-iodothyronine) and *tri-iodothyronine* (T_3). These hormones are the only iodine-containing substances of physiologic importance in the body.

ACTIONS OF T_4 AND T_3. Thyroid hormones regulate the body's metabolic rate. They increase metabolic rate (*oxygen consumption*) and heat-production in the heart, muscle, visceral tissues, but not in the brain, lymphatics, and testes. This *calorigenic action* of thyroid hormones is critical for adaptation of animals and human infants to environmental cold and heat, but it plays a lesser role in adult humans. Thyroid hormones have profound effects on body growth and development. By promoting protein synthesis in numerous tissues, including soft (muscle) and hard (bone), they ensure appropriate differentiation and growth. The most critical action in this regard is on the brain and nervous tissue (see below). Thyroid hormones act synergistically with growth hormone and may be necessary for the synthesis of GH in the pituitary.

Thyroid hormones affect heart and blood vessel functions, such as increasing heart rate and contractility and vascular responsiveness to catecholamines. These effects tend to increase blood pressure. Thryoid hormones also affect brain function and behavior, possibly by enhancing the actions of catecholamines on the nervous tissue.

CONTROL OF THYROID. The synthesis and release of thyroid hormones are under the control of a *pituitary hormone, thyrotropin* (TSH). TSH not only promotes hormonal synthesis and secretion by the thyroid, but can lead to increased cell number (hyperplasia) and size (hypertrophy) of the gland. This condition is referred to as a *goiter*. The secretion of TSH is regulated by direct negative-feedback effects of circulating thyroxine on the pituitary as well as by the stimulating effect of *TRH* (thyrotropin-releasing hormone) from the hypothalamus. Increased plasma levels of thyroid hormones can act directly on the pituitary and diminish TSH release, and vice versa. The brain also exerts control on pituitary TSH by changing the rate of release of TRH, usually in response to environmental stresses such as heat and cold. A rise in TRH increases TSH levels, leading to greater secretion of thyroid hormones.

HISTOPHYSIOLOGY, SYNTHESIS, AND SECRETION. The thyroid gland is comprised of numerous *follicles*. Each follicle consists of a single row of *follicular cells* (thyroid epithelial cells) surrounding a cavity (lumen) filled with a colloidal substance, the *colloid*. The colloid is the storehouse of a special large protein, *thyroglobulin*, which is synthesized by thyroid

cells and secreted into the lumen to participate in the synthesis of thyroid hormones. Blood capillaries run through the space between the thyroid follicles.

The *iodide* in the blood is pumped actively into thyroid cells and is then rapidly transported into the colloid. There, enzymes catalyze the oxidation of iodide into *iodine*. The iodine is attached to the *tyrosine* (amino acid) residues in the thyroglobulin. Further chemical reactions involving tyrosine residues result in synthesis of thyroxine and T_3. Pinocytotic vesicles on the apical cell surfaces (the colloid border) then reabsorb pieces of the colloid, containing the hormone, and bring them into the thyroid cells, where they are united with lysosomes. The lysosomes help release the hormone from the protein. The free hormone diffuses to the basal surface of the thyroid cell and is secreted into the blood. In the blood, thyroid hormones bind with special *blood proteins* (thyroid-binding-globulins, TBG), which carry them to the target tissues. There they are liberated, entering target cells to exert their actions.

HYPERTHYROIDISM AND HYPOTHYROIDISM. Excessive secretion of thyroid hormones (*hyperthyroidism*) is often associated with an autoimmune disease (Graves' disease), in which an antibody against TSH receptors on the thyroid cells is produced in the body, pathologically stimulating the thyroid cells. Hyperthyroid individuals have a high BMR (up to +100%). The enhanced heat production depletes energy reserves (liver glycogen and body fat) leading to wasting and thinness. These individuals are also irritable and nervous, and show increased cardiovascular and respiratory activities. Protruding eyeballs (exophthalmus) is one of the signs of hyperthyroidism. Some individuals develop goiters. The follicular cells become enlarged, and the colloid appears depleted.

In infants and children, thyroid deficiency (*hypothyroidism*) results in the syndrome of *cretinism*. Cretins are dwarfed and mentally retarded due to growth deficiency in the brain. They have potbellies, small mandibles, large tongues, and short necks. Cretinism can be due to maternal iodine deficiency or congenital absence, or abnormalities, of thyroid. Cretinism can be largely prevented (reversed) by thyroid hormone replacement therapy if the treatment is begun from birth or during the early neonatal period.

In adults, hypothyroidism results in the syndrome of *myxedema*. Myxedemic individuals have diminished BMR (down to –40%), thick skin and puffy face (edema), husky voice, and coarse hair. They are slow in physical and mental activities and may exhibit deranged mental behavior. Besides disorders of the thyroid, pituitary or hypothalamic failures may also be responsible for hypothyroidism.

CN: Use red for J. Use the same colors for hypothalamus (A) and anterior pituitary (C) as were used on the previous pages.
1. Begin with the upper panel. Include the chemical structure (the points of iodine attachment) of the two hormones.
2. Color the middle panel beginning with the thyroid gland (E). Follow the numbered sequence in the large illustration. Note that at step 5, iodide (K) becomes iodine (F^3) and it and the attached thyrosine molecule receive the new color. The arrows of movement reflect the color of the molecules involved.
3. Color the arrows in the lower illustrations.

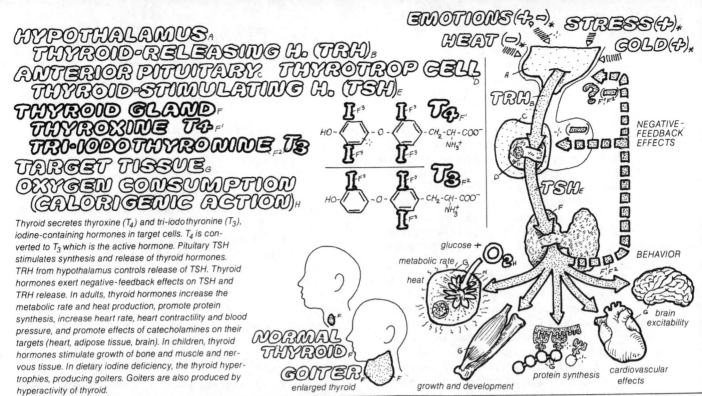

HYPOTHALAMUS A
THYROID-RELEASING H. (TRH) B
ANTERIOR PITUITARY C THYROTROP CELL D
THYROID-STIMULATING H. (TSH) E
THYROID GLAND F
THYROXINE T4 F'
TRI-IODOTHYRONINE T3 F2
TARGET TISSUE G
OXYGEN CONSUMPTION
(CALORIGENIC ACTION) H

EMOTIONS (+, −)* STRESS (+)*
HEAT (−)* COLD (+)*

TRH B

TSH E

NEGATIVE-FEEDBACK EFFECTS

BEHAVIOR

T_4 F'

T_3 F2

glucose +
metabolic rate G
heat
O_2 H

brain excitability G

protein synthesis G

cardiovascular effects

Thyroid secretes thyroxine (T_4) and tri-iodothyronine (T_3), iodine-containing hormones in target cells. T_4 is converted to T_3 which is the active hormone. Pituitary TSH stimulates synthesis and release of thyroid hormones. TRH from hypothalamus controls release of TSH. Thyroid hormones exert negative-feedback effects on TSH and TRH release. In adults, thyroid hormones increase the metabolic rate and heat production, promote protein synthesis, increase heart rate, heart contractility and blood pressure, and promote effects of catecholamines on their targets (heart, adipose tissue, brain). In children, thyroid hormones stimulate growth of bone and muscle and nervous tissue. In dietary iodine deficiency, the thyroid hypertrophies, producing goiters. Goiters are also produced by hyperactivity of thyroid.

NORMAL
THYROID F
GOITER F'
enlarged thyroid

growth and development

THYROID HORMONE: T4 T3 I
MANUFACTURE, STORAGE
AND RELEASE F'
FOLLICLE CELL E'
COLLOID (CAVITY) I
CAPILLARY J
BLOOD PROTEIN J'
IODIDE I⁻ K IODINE I F3
THYROGLOBULIN L
TYROSINE M

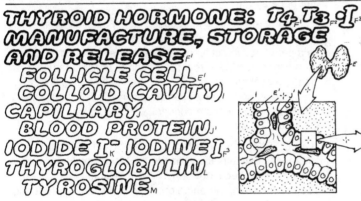

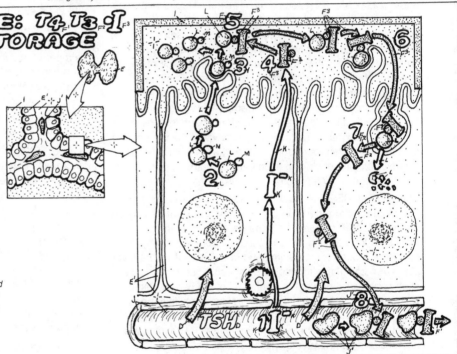

Thyroid consists of follicles: balls of cell surrounding a colloid, containing thyroglobulin. Thyroid cells pump in iodide from plasma (1), transporting it to the colloid along with their own secretions, thyroglobulins (2) and enzymes. In colloid (3) iodide is oxidized to iodine (4) and incorporated into tyrosine residues in thyroglobulin, forming thyroid hormones (5) (iodinated thryonines). When stimulated by TSH, thyroglobulin, containing T_4 and T_3, is taken in by endocytosis (6). The hormones are liberated from protein in lysosomes (7) and secreted into blood (8), where they bind with protein carriers and are transported to tissues.

TSH b   I⁻   I⁻

HYPOTHYROIDISM F4

IN CHILDREN:
CRETINISM

Thyroid deficiency in infants leads to cretinism. Cretins are dwarfed and mentally retarded, with pot bellies, small mandibles, protruded tongues, while maintaining infantile appearance and fat.

HYPOACTIVE
or resting thyroid

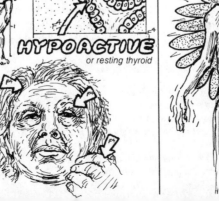

IN ADULTS:*
MYXEDEMA

Hypoactive thyroid has large colloids and small flat cells. Hypothyroid patients are sluggish, have coarse voices, hair and skin (myxedema). They show reduced BMR and poor tolerance to cold.

HYPERTHYROIDISM E' I

BMR H

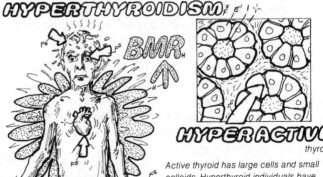

HYPERACTIVE F5
thyroid

Active thyroid has large cells and small colloids. Hyperthyroid individuals have goiter due to thyroid stimulation by TSH or a TSH-like substance (Graves' disease). There is also increased BMR, sweating, poor heat tolerance, and cardiovascular abnormalities. They are thin, nervous, show increased appetite and rapid mentation. Eyes may be protruded (exophthalmus).

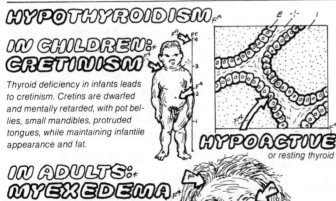

PARATHYROID GLANDS & HORMONAL REGULATION OF PLASMA CALCIUM

In humans, the *parathyroid* glands are four small bodies, each the size of a lentil, embedded in the superior and inferior poles of *thyroid* tissue, although there are no anatomic or physiologic connections between the *thyroid* and parathyroids. Two types of cells are found in the parathyroid gland: *chief* and *oxyphil*. The chief cells, in response to a decrease in the level of *calcium* ions (Ca^{++}) in plasma, secrete the *parathyroid hormone*, which acts on bone and the kidneys to elevate the plasma calcium level. The function of oxyphil cells is not well understood.

IMPORTANCE OF CALCIUM IONS. Plasma calcium ions are critically important in the physiology of excitable cells (nerve and muscle), contraction of the heart, clotting of blood and several other functions. Calcium levels are intricately regulated by complex hormonal mechanisms. The normal level of plasma calcium in humans and other mammals is *10 mg/100 cc* of plasma. A marked reduction below the critical limits (hypocalcemia) increases the excitability of nerve fibers and muscles but decreases the release of neurotransmitters at the synapses and neuromuscular junction. The net result of hypocalcemia is spasmic contractions in muscles (tetany). A characteristic clinical sign seen in humans with *hypocalcemic tetany* is the Trousseau's sign, (i.e., flexion of the wrist and thumb with extension of the fingers). Spasms of respiratory muscles interfere with respiration and can be fatal. Such problems are usually produced after surgical removal of the thyroid tissue, if the parathyroids have been inadvertently removed. In experimental animals, removal of the parathyroid gland leads, within four hours, to a marked reduction in plasma calcium levels and eventually to death unless calcium is given by infusion.

HORMONAL REGULATION OF CALCIUM. Three hormones participate in the regulation of plasma calcium: parathyroid hormone (*parathormone*) from the parathyroid gland, *calcitonin* from the thyroid and *calcitriol* from the kidney. The role of parathormone is central and most important. The level of plasma calcium is monitored by specific detectors, "calcium receptors," present in chief cells of the parathyroid. When the calcium level decreases below a set limit, these detectors signal the release of parathormone, which acts directly on bone and the kidneys and indirectly on intestinal mucosa to raise the calcium level. The elevated calcium level then exerts negative feedback action on the parathyroid to depress the secretion of parathormone. The interaction between plasma calcium and parathormone is an example of efficient negative feedback regulation by hormones without the involvement of the nervous system.

In chronic cases of depressed calcium levels, as in *rickets* or kidney diseases, or in certain physiological conditions when calcium utilization is intensive, such as pregnancy or lactation, parathyroid glands increase in size (hypertrophy) in response to the prolonged hypocalcemic stimulation. The

hypertrophied parathyroid is more sensitive to reduction in calcium levels and secretes parathormone more efficiently, so that a 1% decrease in calcium results in a 100% increase in parathormone concentration.

Parathormone acts on bone tissue, increasing resorption of calcium from the bone matrix and elevating calcium levels in the plasma. The mechanism of action of parathormone on bone tissue appears to work at two levels. The principal one is to stimulate *osteoclasts*. These bone cells digest the bone matrix, increasing the level of calcium ions in the fluid of the matrix. This calcium is then exchanged with plasma. There may even be an increase in the number of osteoclasts. A more rapid action that elevates calcium levels in minutes is the pumping of calcium from a readily available pool in the bone fluid to the plasma. This transport occurs across an extensive membrane system separating the bone fluid from the extracellular fluid (plasma). These membranes are formed by the processes of osteocytes and *osteoblasts* (see plate 115 for bone structure).

Parathormone also increases plasma calcium by increasing the renal tubular reabsorption of this ion. However, a more effective action of parathormone in this connection is the increase in the renal excretion of *phosphate* (HPO_4^{--}) ions (phosphaturia). Usually, the product of calcium and phosphate ion concentrations in plasma is constant. Thus, a decrease in phosphate concentration would lead to an increase in calcium concentration.

CALCITRIOL AND VITAMIN D. Parathyroid hormone increases calcium absorption in the small intestine by enhancing the effects of vitamin D on intestinal mucosa. *Vitamin D_3* (cholecalciferol) can be obtained in the diet or produced from cholesterol in the skin in the presence of ultraviolet radiation in sunlight. To become active, vitamin D_3 must first be converted in the liver to *calcidiol* and then in the kidney to *calcitriol*. Calcitriol is much more active than vitamin D. Parathormone is necessary for the formation of calcitriol in the kidney. Calcitriol stimulates the intestinal epithelium to synthesize more *carrier protein molecules* for calcium transport, thus elevating plasma calcium levels. Calcitriol further enhances the action of parathormone on bone cells.

CALCITONIN. Calcitonin is a hormone secreted by the *parafollicular C cells* of the thyroid. It is secreted in response to an increase in the plasma calcium level. Within minutes, calcitonin acts to reduce the calcium level, by decreasing bone resorption by inhibiting the activity of osteoclasts and stimulating the function of osteoblasts: Calcitonin is very important in growing children because bone growth requires inhibition of bone resorption and stimulation of bone deposition. Calcitonin is also important during pregnancy and lactation, when it helps protect maternal bones from the excess calcium loss initiated by parathormone.

CN: Use red for E, yellow for G, a very light color for bone (P), and a dark color for B.
1. Begin at the top by coloring the thyroid (A) and parathyroid (B) and the plasma calcium (E) material on the right.
2. Color the parathormone title (B¹) and follow its influence on bone (P) and then color the C cells

(O) of the thyroid on the far left, and follow the influence of calcitonin (O¹) (and PTH) down to the bottom of the page.
3. Go back to the parathormone title and follow its actions in the intestine and the material below it (including rickets (I) at the bottom).
4. Complete the kidney (M) and related structures.

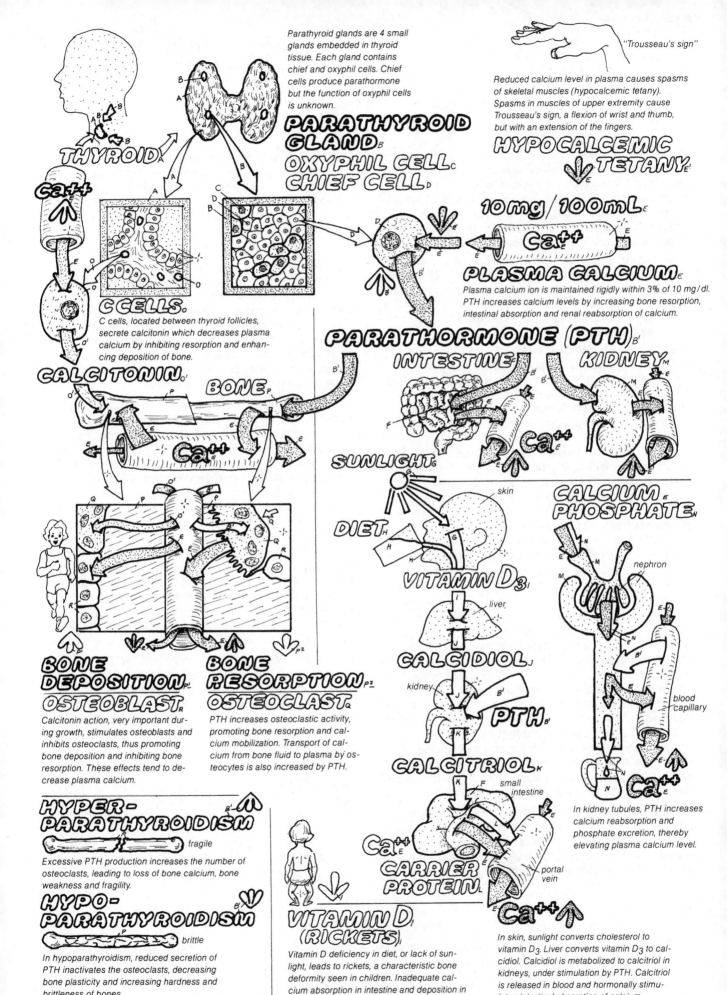

Parathyroid glands are 4 small glands embedded in thyroid tissue. Each gland contains chief and oxyphil cells. Chief cells produce parathormone but the function of oxyphil cells is unknown.

"Trousseau's sign"

Reduced calcium level in plasma causes spasms of skeletal muscles (hypocalcemic tetany). Spasms in muscles of upper extremity cause Trousseau's sign, a flexion of wrist and thumb, but with an extension of the fingers.

THYROID

PARATHYROID GLAND

OXYPHIL CELL
CHIEF CELL

HYPOCALCEMIC TETANY

Ca++

10 mg/100 mL
Ca++

PLASMA CALCIUM

Plasma calcium ion is maintained rigidly within 3% of 10 mg/dl. PTH increases calcium levels by increasing bone resorption, intestinal absorption and renal reabsorption of calcium.

C CELLS.

C cells, located between thyroid follicles, secrete calcitonin which decreases plasma calcium by inhibiting resorption and enhancing deposition of bone.

CALCITONIN

BONE

Ca++

PARATHORMONE (PTH)

INTESTINE
KIDNEY

SUNLIGHT
DIET

skin

Ca++

CALCIUM PHOSPHATE

nephron

VITAMIN D₃

liver

blood capillary

CALCIDIOL

kidney

PTH

CALCITRIOL

small intestine

BONE DEPOSITION
OSTEOBLAST

BONE RESORPTION
OSTEOCLAST

Calcitonin action, very important during growth, stimulates osteoblasts and inhibits osteoclasts, thus promoting bone deposition and inhibiting bone resorption. These effects tend to decrease plasma calcium.

PTH increases osteoclastic activity, promoting bone resorption and calcium mobilization. Transport of calcium from bone fluid to plasma by osteocytes is also increased by PTH.

Ca++
CARRIER PROTEIN

Ca++

In kidney tubules, PTH increases calcium reabsorption and phosphate excretion, thereby elevating plasma calcium level.

HYPER-PARATHYROIDISM

fragile

Excessive PTH production increases the number of osteoclasts, leading to loss of bone calcium, bone weakness and fragility.

HYPO-PARATHYROIDISM

brittle

In hypoparathyroidism, reduced secretion of PTH inactivates the osteoclasts, decreasing bone plasticity and increasing hardness and brittleness of bones.

VITAMIN D (RICKETS)

Vitamin D deficiency in diet, or lack of sunlight, leads to rickets, a characteristic bone deformity seen in children. Inadequate calcium absorption in intestine and deposition in bone weakens the growing bones. Long bones of lower extremities bend under stress.

In skin, sunlight converts cholesterol to vitamin D₃. Liver converts vitamin D₃ to calcidiol. Calcidiol is metabolized to calcitriol in kidneys, under stimulation by PTH. Calcitriol is released in blood and hormonally stimulates intestinal absorption of calcium.

STRUCTURE AND GROWTH OF BONE

Bones are the building blocks of the skeleton, which supports the body and provides leverage for muscles and movement. In addition, bones harbor the brain, spinal cord, and bone marrow and provide a storehouse for calcium. In response to hormonal stimulation, bone calcium can be readily exchanged with plasma calcium, preventing alterations in the level of this important ion in the blood. (See plate 114).

Although bone appears hard and inert, it is in fact an active tissue, supplied by nerves and blood vessels. Various bone cells (see below) are continuously active, even in the adult, building and rebuilding, repairing and remodeling the bone in response to strains, stresses, and fractures.

BONE STRUCTURE. To understand bone structure, let us examine a typical long bone such as the tibia. It consists of two heads (*epiphysis*) and a shaft (*diaphysis*). A cross-section of the long bone reveals dense and cavernous areas. Dense areas contain *compact bone*; cavernous areas consist of *spongy bone*. Diaphysis of a long bone contains mainly compact bone; epiphysis contains both compact and long bone.

Microscopic examination of the compact bone in the diaphysis reveals many cylindrical units, called the *Haversian systems* (osteon). These units, which run along the bone length, are packed tightly and held together by a special cement. Each Haversian system consists of concentric plates (lamella) surrounding a central canal through which *blood vessels* and *nerves* run. The central canal communicates with numerous smaller *lacunae* located throughout the Haversian system. The many lacunae in turn communicate via smaller passageways (canaliculi), which permit blood and nerves to reach bone cells.

Physiologically, bone tissue consists of two compartments: first, a metabolically active cellular compartment made up of *bone cells* and second, a metabolically inert extracellular compartment, the *bone matrix*, consisting of a mixture of organic and inorganic materials. The organic part is made of *collagen* fibers, extremely tough fibrous proteins, and the *ground substance* (*glycoproteins* and *mucopolysaccharides*). The inorganic part of the bone matrix consists of a mineral of calcium and phosphate ($Ca_{10}[PO_4]_6[OH]_2$) — *hydroxyapatite crystals*. To make the bone matrix, the hydroxyapatite crystals are deposited on a mesh of collagen fibers and glycoproteins, a process called *calcification*. The calcified matrix gives the bone its remarkable hardness and strength.

BONE CELLS. Bone cells are *osteoblasts*, *osteocytes*, and *osteoclasts*. Osteoblasts, usually found near the bone surfaces, are the young bone cells that secrete the organic substances of the matrix. Once totally surrounded by the secreted matrix, osteoblasts markedly diminish their bone-

making activity. At this stage, they are considered mature and referred to as osteocytes. Osteocytes are found in or near the lacunae. They develop extensive processes (filopodia) that run through the canaliculi, connecting with the other osteocytes. These membranous process facilitate the exchange of nutrients, especially calcium between the bone and blood.

The third type of bone cell is the osteoclast, which resembles the blood macrophages. Osteoclasts have important functions in repairing fractures and remodeling new bone. To accomplish their tasks, osteoclasts secrete *lysosomal enzymes* (e.g., the protease collagenase) into the bone matrix. These enzymes digest the matrix proteins, liberating calcium and phosphate. Thus, osteoclasts, in addition to their involvement in remodeling and repairing of fractures, are targets for hormones, such as parathormone, that promote bone resorption and calcium mobilization.

BONE GROWTH. The development of bone is usually preceded by the formation of cartilage, a type of connective tissue. In long bone, the growth and elongation begins during the postnatal period, continuing through adolescence. Elongation is achieved by the activity of two cartilaginous plates, called the *epiphyseal plates*, located between the shaft and the heads. Germinal cells in these plates continuously produce new cartilage cells, which migrate toward the shaft, where they form a template. Next, bone cells move into these areas, constructing new bone over these templates. In this manner, the length of the shaft increases at both ends, and the head move progressively apart. As growth proceeds, the thickness of epiphyseal plates gradually decreases. The plates are thus wide in growing children, narrow after puberty and disappearing entirely by adulthood (*epiphyseal closure*). Longitudinal growth is not possible after this stage, which occurs at different ages for different bones.

HORMONAL REGULATION OF BONE GROWTH. During childhood, *thyroid* and *growth hormones* stimulate plate growth. *Androgens* stimulate bone growth in puberty and are important in the *adolescent growth spurt*. However, in late adolescence, androgens enhance the closure of epiphyseal plates, thus terminating growth. In adults, excess GH promotes bone growth only in width, leading to the thick bones characteristic of *acromegalic* individuals (see plate 112).

FRACTURE REPAIR. During repair of a fracture, a special type of connective tissue, the *hyaline cartilage*, develops at the fracture site, forming a *callus*. The callus serves as a model for new bone growth and protects the healing bone against the deforming stress forces acting on it. When new bone replaces the callus, osteoclasts remodel the bone into its original shape by digesting the extra bone.

CN: Use red for A, a pale yellow or tan for B, and very light colors for C and D.
1. Begin with the bone structures at the top. Notice that only one group of Haversian systems (D¹) has been selected for coloring.

2. Color the three types of bone cells and two illustrations underneath demonstrating their functions.
3. Color the three steps in fracture repair.
4. Color the hormonal regulation of bone growth shown at the bottom.

BONE STRUCTURE *

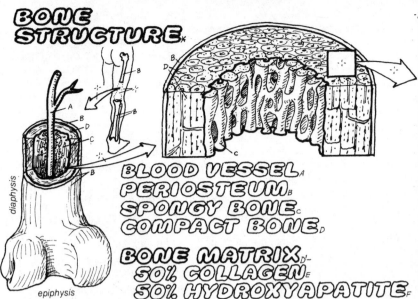

diaphysis

epiphysis

BLOOD VESSEL _A_
PERIOSTEUM _B_
SPONGY BONE _C_
COMPACT BONE _D_

BONE MATRIX _D'_
50% COLLAGEN _E_
50% HYDROXYAPATITE _F_

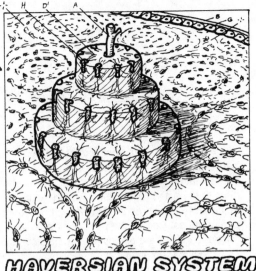

HAVERSIAN SYSTEMS _D'_

Spongy bone contains cavities, compact bone does not. Mature bone is lamellar. Several concentric lamella form a cylindrical unit, the Haversian system (osteon). A Haversian canal at its middle contains blood vessels and nerves. A long bone consists of numerous Haversian systems running parallel to bone length.

BONE CELLS *

OSTEOBLAST _G_

Osteoblasts are bone-forming cells, secreting collagen into the bone matrix.

OSTEOCYTE _H_

Osteocytes are mature osteoblasts. Their processes engage in metabolic exchange with blood.

OSTEOCLAST

Osteoclasts are polynucleated macrophage-like cells that digest bone matrix by secreting protease enzymes.

Bone matrix is a mixture of organic and mineral elements. Collagen fibers and glycoproteins form an organic net overlaid by calcium phosphate crystals, producing the hard material of bone.

 LYSOSOMAL ENZYMES _J_

(COLLAGENASE)

FRACTURE REPAIR *

HYALINE CARTILAGE _M_

callus

NEW BONE _G'_

REMODELED _G''_

At edges of bone fractures, hyaline cartilage proliferates forming a callus. This helps support fracture and serves as a model for bone formation. Infiltration by bone cells transforms callus to bone which is then remodelled and shaped by digestive actions of osteoclasts.

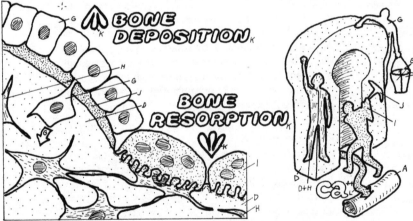

BONE DEPOSITION _K_

BONE RESORPTION

Ca!

During bone formation, osteoblasts act as builders, secreting matrix proteins. Calcification of matrix isolates osteoblasts, maturing in osteocytes with extensive processes. Osteocyts participate in exchange of nutrients and calcium with blood. Osteoclasts help shape bone by digesting extra pieces. When stimulated by PTH, calcium released by osteoclastic digestion is transported to blood to compensate for calcium deficiency.

GROWTH *

epiphysis

diaphysis

EARLY YOUTH *

new bone

GROWTH HOR.. _o_
THYROID HOR.. _p_
ANDROGENS _Q_

In long bone growth, epiphyseal plates (hyaline cartilage) expand, forming new cells. These form bone models at shaft ends. Bone is formed on this model, increasing shaft length. Growth hormone, thyroxine and androgens stimulate plate growth.

EPIPHYSEAL PLATE _N_

MATURITY *
"epiphyseal closure"

GROWTH HORMONE _o_
ANDROGENS _Q_

By maturity, epiphyseal plates fuse with bone, terminating bone growth. High androgen level in maturity enhances plate closure. Excess growth hormone can now stimulate bone growth only in width, thickening shafts and heads (acromegaly).

ACROMEGALY _o'_

ENDOCRINE PANCREAS & SYNTHESIS OF INSULIN

The *pancreas* is a major *mixed* (both *exocrine* and *endocrine*) gland. In the human body, it is located in the abdominal cavity, underneath the stomach. By bulk, most of the gland (99%) deals with the exocrine functions; (i.e., secretion of digestive enzymes and bicarbonate solution by the pancreatic *acini* and *ducts*). These functions are discussed in detail in the section on digestion (see plate 72).

PANCREAS ISLETS. The endocrine part of the pancreas, called the *islets of Langerhans*, consists of one to two million round clusters (islets) of cells, scattered throughout the gland between the exocrine acini. A rich bed of special blood capillaries with large pores surrounds the islets. Each islet is a collection of several different types of cells. Each type of cell is believed to secrete one of the pancreatic hormones. By using specific immunocytochemical staining methods, researchers have identified three types of cells: A, B, and D cells (also known as alpha, beta, and delta cells). *A cells*, less numerous and located peripherally, secrete the hormone *glucagon*. *B cells*, located centrally, are more numerous. They secrete the hormone *insulin*. *D cells* are sparse and secrete the hormone *somatostatin*.

INSULIN, GLUCAGON, AND SOMATOSTATIN. Insulin and glucagon regulate the metabolism of carbohydrates in tissues and ensure the maintenance of optimal *blood glucose levels* (*blood sugar*). Insulin, by facilitating the transport of glucose across cell membranes, enhances the availability of glucose in cells and promotes its utilization. In this regard, insulin functions as a *hypoglycemic hormone*: i.e., one that decreases blood sugar. Blood sugar usually increases after ingestion of a meal.

Glucagon also enhances carbohydrate utilization. Its function is to mobilize glucose from its major storage source, the *liver glycogen*. In doing so, glucagon functions as a *hyperglycemic hormone*: i.e., one that increases blood sugar. Low blood sugar levels are encountered between meals and during fasting.

Somatostatin, the third pancreatic hormone, may act locally as a tissue hormone (see plate 107), inhibiting secretion of both insulin and glucagon. Insulin and glucagon also exert direct influence on each other's secretion: glucagon promotes insulin secretion and insulin tends to inhibit glucagon secretion. (See plate 117 for a more thorough discussion of the actions of insulin and glucagon. Carbohydrate metabolism and its nervous and hormonal regulation are discussed on plates 124-126.)

SYNTHESIS OF INSULIN. Insulin is a protein hormone made up of two peptide chains (an *A chain* and a *B chain*) connected at two locations by disulfide bridges (bonds). This is the form in which insulin is released into the blood and acts on target cells. Insulin is synthesized within the B cells on the *endoplasmic reticulum* as a much larger peptide chain called *proinsulin*.

During later processing, this long peptide folds, as a result of formation of disulfide bridges. During packaging in the vesicles of *Golgi apparatus*, *protease enzymes* convert proinsulin to insulin by attacking the long chain at two locations, splitting the original single chain into two pieces, one being the insulin molecule (the A and B chains) and the other, a *C chain*.

Insulin and the disconnected C chain are transported together, within *secretory vesicles*, toward the *plasma membrane* of the B cells. Near a *capillary*, the contents of the vesicles are released in the blood by exocytosis of the vesicles. The function and fate of the C chain is uncertain.

REGULATION OF INSULIN SECRETION. The release of insulin by B cells is regulated by the level of glucose in the blood through the operation of a negative feedback system. An increase in glucose level, occurring usually after a meal, is detected by the B cells, resulting in increased secretion of insulin. Insulin is transported to the tissues by blood, promoting glucose uptake and utilization. This action decreases blood sugar.

The cellular mechanism within the B cells involved in the feedback system is not well understood. The change in glucose level may be detected by glucose receptors on the surface or in the cytoplasm of B cells. Alternatively, high blood glucose level may cause increased metabolic activity in the B cells. In either case, a signal is generated in the B cell, resulting in calcium ion release.

The calcium ions interact with secretory vesicles, promoting their fusion with the cell membrane. The resulting exocytosis releases insulin into the blood. The calcium ions also promote a longer-lasting response, i.e., increased synthesis of insulin at the cytoplasmic level. This response ensures availability of insulin for prolonged secretion (hours), until the hyperglycemia is eliminated.

CN: Use a yellowish color for A, red for H, purple for I, and another bright color for E.
1. Begin at the top and color down to and including the two test tubes symbolizing blood glucose levels.
2. Color the formation of insulin (E¹) from proinsulin (M) shown at the bottom left. This process occurs at step (7) in the B cell illustration.
3. Color the steps of insulin synthesis shown in the diagrammatic B cell. Note that insulin (E¹) is represented by two small parallel bars which stand for the A chain (E²) and B chain (E³) in the previous illustration. Don't color the interior of capillary (I).

PANCREAS_A

EXOCRINE: 99%_*
PANCREATIC ACINI_A'
PANCREATIC DUCT_B

ENDOCRINE: 1%_*
ISLETS OF LANGERHANS_C
A CELLS - GLUCAGON_D'
B CELLS - INSULIN_E'
D CELLS - SOMATOSTATIN_F'

SPLENIC VEIN_G
GLUCOSE LEVEL (BLOOD SUGAR)_H

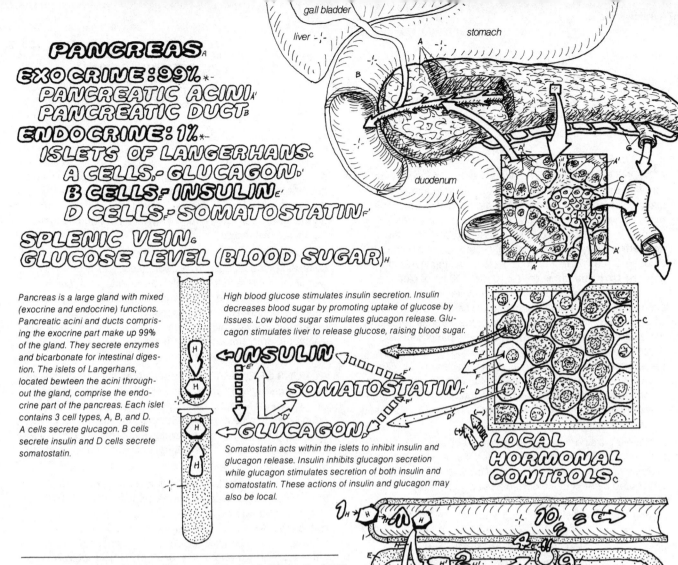

gall bladder

liver

stomach

duodenum

Pancreas is a large gland with mixed (exocrine and endocrine) functions. Pancreatic acini and ducts comprising the exocrine part make up 99% of the gland. They secrete enzymes and bicarbonate for intestinal digestion. The islets of Langerhans, located between the acini throughout the gland, comprise the endocrine part of the pancreas. Each islet contains 3 cell types, A, B, and D. A cells secrete glucagon. B cells secrete insulin and D cells secrete somatostatin.

High blood glucose stimulates insulin secretion. Insulin decreases blood sugar by promoting uptake of glucose by tissues. Low blood sugar stimulates glucagon release. Glucagon stimulates liver to release glucose, raising blood sugar.

← INSULIN_E'

SOMATOSTATIN_F'

← GLUCAGON_D'

Somatostatin acts within the islets to inhibit insulin and glucagon release. Insulin inhibits glucagon secretion while glucagon stimulates secretion of both insulin and somatostatin. These actions of insulin and glucagon may also be local.

LOCAL HORMONAL CONTROLS_C

SYNTHESIS OF INSULIN_E'
CAPILLARY_F
CELL MEMBRANE_E
GLUCOSE_H DETECTOR_H'
CALCIUM ION (Ca++)_J
MESSENGER RNA_K
ROUGH ENDOPLASMIC RETICULUM_L PROINSULIN_M
ENZYME_N
GOLGI APPARATUS_L'
SECRETORY VESICLE_O'

A rise in blood glucose (1) is detected by glucose receptors (2) or other metabolic sensors in B cells, triggering an increase in calcium ions (3) within B cells. Ca++ promotes exocytosis of secretory vesicles containing insulin and release of insulin within seconds (4). Sustained hyperglycemia stimulates insulin synthesis. A mRNA is formed in the nucleus (5), moving to cytoplasm

B (BETA) CELL_E

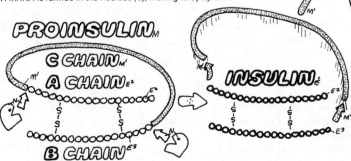

PROINSULIN_M
C CHAIN_M'
A CHAIN_E²
B CHAIN_E³

→ INSULIN_E

and promoting synthesis of a large single-chain polypeptide (proinsulin) on the rough ER (6). During packaging of Golgi apparatus, proinsulin is folded by forming 2 disulfide bridges. Also a piece of the chain (C chain) is split off at the middle (7). The remainder of polypeptide is the insulin molecule which now consists of two peptide chains (A and B) connected together by two disulfide bridges. Insulin is stored in vesicles (8) and secreted by exocytosis (9) into the blood (10).

PHYSIOLOGY OF INSULIN & GLUCAGON

ACTIONS OF INSULIN. The primary action of *insulin* is to facilitate and promote the transport of *glucose* across the plasma membranes of cells in certain special tissues, chiefly *muscle* (heart, skeletal, and smooth) and adipose (*fat*) tissues. In the absence of insulin, the membranes of these cells are impermeable to glucose, regardless of how much glucose is present in the blood. Normal fasting levels of blood glucose are in the range of 70-110 mg/100 cc of blood, a value that remains constant throughout life. When this level is exceeded, such as after a meal rich in carbohydrates (bread, potatoes, rice), the excess glucose in the blood is sensed by the *glucose detectors* in the *B* cells of the pancreatic islets, resulting in release of insulin into the blood. Insulin is then transported with the blood to its target tissues, binding with specific *insulin receptors* located in the plasma membranes of the target cells. This binding somehow increases the permeability of the target cells to glucose, resulting in increased uptake of this substance.

Muscle cells normally prefer to use glucose for oxidation and *cellular energy metabolism*. Once inside a muscle cell, glucose is either directly oxidized to provide ATP or is conserved by being incorporated into *glycogen*, a polymer of glucose (see below). The glycogen formation occurs at rest. During muscle activity, glycogen is broken down into glucose.

Insulin also promotes glucose entrance into the fat cells of adipose tissue. Here, the increased glucose supply is not utilized to provide energy for the fat cells. Instead, each glucose molecule is metabolized to form two molecules of *glycerol*, which is used along with *fatty acids* to form *triglycerides*, the storage form of fat. The fatty acids are usually obtained from the blood, though their source is ultimately the liver. Insulin also inhibits a special lipase enzyme (the *hormone-sensitive lipase*) present in fat cells. This important action prevents fat breakdown.

Insulin acts directly on the liver cells. However, this action does not promote increased transport of glucose because liver cells are normally permeable to glucose. Instead, by stimulating the synthesis or actions of specific enzymes, insulin promotes utilization of glucose for synthesis of glycogen, amino acids and *proteins*, and fats, particularly fatty acids. These fatty acids are used by the adipose tissue to form triglycerides. (See also plates 127, 128.)

Insulin does not influence glucose uptake by such tissues as the brain, kidney tubules, and intestinal mucosa, mainly because these tissues are normally permeable to glucose. However, this may be an adaptive response because the nervous tissue relies solely on glucose for its energy needs. Alterations in insulin secretion would have major consequences on brain function, as shown by the fact that a large dose of insulin (by injection) may result in coma or death by causing marked hypoglycemia and depriving the brain of fuel and energy. The intestinal mucosa and kidney tubules are involved in special transport functions of glucose, unrelated to its utilization for energy, thus precluding their regulation by insulin.

As a result of the effects of insulin on muscle, liver, and fat cells, the blood glucose level decreases. This is sensed by glucose detectors in the B cells, resulting in diminished insulin output until the next meal, when once again glucose supply is enhanced, and glucose level is increased. The interaction between blood glucose and insulin provides another example of simple hormonal regulation, by negative feedback, with no nervous system involvement.

ACTIONS OF GLUCAGON. The primary stimulus for the release of *glucagon* is a decrease in the level of blood sugar, below its normal limits. This occurs in between meals or during fasting and starvation. Special glucose detectors in the *A cells* of the pancreatic islets sense the reduction in the blood glucose level, resulting in increased secretion of glucagon.

Glucagon binds with specific *glucagon receptors* in the membranes of liver cells. This binding activates the enzyme adenylate cyclase, increasing the concentration of cyclic-AMP within the liver cells. Cyclic-AMP in turn acts as a second messenger, initiating a cascade of chemical reactions involving *activation* of *enzymes* (not their synthesis). Through an amplification mechanism, within seconds, billions of enzyme molecules are mobilized to break down glycogen, the highly branched polymer of glucose (*glycogen tree*) and release its monomers, the glucose molecules. This process is called glycogenolysis. Glucagon also stimulates synthesis of new glucose molecules from amino acids in the liver (*gluconeogenesis*). This action takes longer and is probably more important in adaptation to fasting and starvation.

The glucose molecules mobilized by the action of glucagon leak out into the blood, increasing sugar levels and supply for such constant users as the brain and heart. The increased blood glucose level acts via negative feedback on A cells, decreasing glucagon release until the glucose level falls again due to constant use, at which time glucagon release will be initiated again.

Although insulin and glucagon appear to have opposite effects on blood glucose levels (insulin being a hypoglycemic hormone and glucagon being a hyperglycemic one), their true functions in the body as a whole should be considered complementary, aimed to regulate carbohydrate metabolism and provide ample glucose supply for tissues. (See also plates 118, 124-128.)

CN: Use same colors as previous page for insulin (A) cand glucagon (H). Use red for B, yellow for G.
1. Being with the actions of insulin (A). Follow the numbered sequence. Number one begins with a rise in the blood sugar level due to ice cream ingestion. Color all glucose and insulin molecules.
2. Color glucagon action, beginning with the hungry individual with a low blood sugar level.

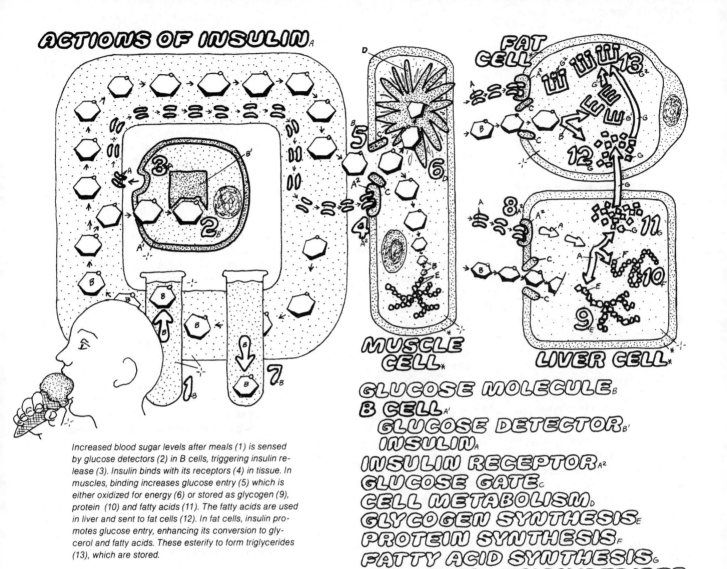

ACTIONS OF INSULIN

MUSCLE CELL *

FAT CELL

LIVER CELL *

Increased blood sugar levels after meals (1) is sensed by glucose detectors (2) in B cells, triggering insulin release (3). Insulin binds with its receptors (4) in tissue. In muscles, binding increases glucose entry (5) which is either oxidized for energy (6) or stored as glycogen (9), protein (10) and fatty acids (11). The fatty acids are used in liver and sent to fat cells (12). In fat cells, insulin promotes glucose entry, enhancing its conversion to glycerol and fatty acids. These esterify to form triglycerides (13), which are stored.

GLUCOSE MOLECULE B
B CELL A'
GLUCOSE DETECTOR B'
INSULIN A
INSULIN RECEPTOR A²
GLUCOSE GATE C
CELL METABOLISM D
GLYCOGEN SYNTHESIS E
PROTEIN SYNTHESIS F
FATTY ACID SYNTHESIS G
GLYCEROL → TRIGLYCERIDES G²

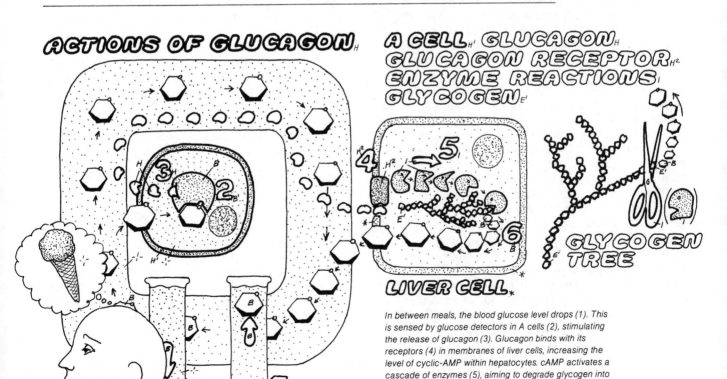

ACTIONS OF GLUCAGON H

A CELL H' GLUCAGON H
GLUCAGON RECEPTOR H²
ENZYME REACTIONS I
GLYCOGEN E'

LIVER CELL *

GLYCOGEN TREE

In between meals, the blood glucose level drops (1). This is sensed by glucose detectors in A cells (2), stimulating the release of glucagon (3). Glucagon binds with its receptors (4) in membranes of liver cells, increasing the level of cyclic-AMP within hepatocytes. cAMP activates a cascade of enzymes (5), aiming to degrade glycogen into glucose (6). Glucose enters blood, elevating blood sugar levels (7), and glucose supply to tissue.

EFFECTS OF INSULIN DEFICIENCY: DIABETES

INSULIN DEFICIENCY. *Insulin* is one of the most important hormonal regulators of *metabolism* and body physiology, as evidenced by the widespread deleterious and sometimes catastrophic consequences that follow its deficiency. Indeed, one of the best ways to understand the normal actions of insulin is to observe the effects of its lack or deficiency that occur after surgical removal of the pancreas, accidental toxic damage to B cells, or as a consequence of developing the disease *diabetes mellitus*.

Insulin-deficient individuals have high blood sugar levels (*hyperglycemia*), ranging from two times the normal (before a meal) to four times (after a meal). The hyperglycemia is caused by both decreased uptake of *glucose* by muscle and adipose tissues and increased glucose output by the liver. As a result of insulin deficiency, it also takes longer (6-8 hrs.) before postmeal glucose levels can return to premeal levels, compared to 1-2 hrs. under normal conditions. This diminished ability of the body to handle the increased glucose load is the basis for the *"glucose tolerance test,"* used clinically to diagnose diabetes (see lower illustration in the plate).

In insulin deficiency, muscle cells deprived of glucose begin to utilize alternative energy sources. Thus, *fat* and *protein* reserves of the muscle tissue are utilized for oxidation and energy production, resulting in wasting of muscles, weakness, and weight loss. Weight loss is further worsened by what happens in the fat cells of the adipose tissue. Not only can glucose not enter these cells, but the loss of insulin removes the inhibition of the enzyme "hormone-sensitive lipase," resulting in increased breakdown of stored triglycerides and mobilization of fatty acids.

The loss of body fat contributes to the characteristic thinness of the young diabetic patient or insulin deficient individuals. The malnourished state of the tissues and the individual, in the presence of high blood sugar, is why diabetes is called the disease of "starvation in the midst of plenty," and insulin is called the "hormone of abundance."

The mobilization of fatty acids provides a ready source of fuel for the energy-starved heart and muscle tissue. However, excessive production of fatty acids results in formation of keto acids (*ketone bodies*), particulary in the liver. The ketone bodies enter the blood, causing *ketosis* and *ketoacidosis*. This condition is very dangerous and if untreated results in metabolic acidosis. The increased blood acidity suppresses the higher nervous centers (*coma*). Ultimately, the depression of the brain respiratory centers leads to death. In addition, the ketone bodies are excreted in the urine, worsening the *osmotic diuresis* caused by glucose (see below).

In normal individuals, plasma glucose is filtered in the Bowman's capsule of the *kidney nephrons* but is subsequently reabsorbed completely in the proximal tubules. As a result, the urine is normally free of sugar. In hyperglycemia, above the limit of 170 mg glucose per 100 cc blood, the reabsorptive capacity of kidney tubules is exceeded. The extra glucose spills over in the urine, leading to one of the most well-known signs of diabetes mellitus and insulin deficiency: the presence of sugar in urine (*glycosuria*).

Glycosuria has two consequences: *polyuria* and *polydipsia*. The extra glucose molecules in the urine cause osmotic diuresis (excess water in urine) and polyuria (excess urine production). Polyuria results in decreased plasma volume and increased plasma osmolarity. These conditions lead to activation of hypothalamic thirst centers and excessive drinking of water (polydipsia). Diabetic individuals are characterized by frequent urination and drinking during the night. Excessive loss of water may lead to severe *dehydration* and osmotic shock, conditions that can also lead to irreversible brain damage, coma and death.

DIABETES. Diabetes mellitus as a spontaneous disease has been known since early history. In the United States, nearly 5% of the population has the disease. Two types of diabetes are now recognized: the juvenile type (Type I), seen in children and young adults, and the maturity-onset type (Type II), seen usually in obese individuals over forty. The juvenile type is associated with the lack or serious deficiency of insulin. It may be an autoimmune disease and is probably without genetic or familial traits. If untreated it is often fatal due to ketoacidosis and dehydration shock. The best treatment involves the injection of insulin.

The maturity-onset type shows strong familial association. In this type of diabetes, insulin deficiency is relative, because its absolute amounts in the blood may be even higher than in normal individuals. However, due somehow to prolonged obesity, there occurs a reduction in insulin receptors in target cells (i.e., a down-regulation of the receptors), due perhaps to high and steady insulin production. In this condition, the available insulin is ineffective, resulting in signs similar to complete insulin deficiency, such as hyperglycemia, glycosuria, polydipsia, and weight loss. Only ketosis does not occur. Although these individuals can be treated with extra insulin, simple weight reduction will frequently ameliorate the diabetic condition. For reasons not completely understood, maturity-onset diabetes, if untreated, can lead to vascular diseases that, among other things, cause blindness, atherosclerosis, heart attacks, kidney disease, and gangrene.

The chain of pathological events seen in diabetic patients can be stopped and reversed to some extent by regular treatment with exogenous insulin, which is usually obtained from the pancreas of livestock. One complication of this treatment is that the animal insulin is antigenic, becoming gradually ineffective as the body makes antibodies against this foreign protein. Recently, researchers have been attempting to use genetic engineering and new methods of molecular biology to obtain large quantities of human insulin, against which the human body does not produce antibodies. Insulin pumps, which can deliver insulin in small and repeated doses after each meal, are now becoming available.

CN: Use purple of A, red for B, yellow for F. Use a dark color for I.
1. Begin by coloring the title: Starvation . . . and the membrane of the shrinking body cell in the upper left corner. Then begin with number one to the right of the glucose level at the top of the page and follow the numbered sequence. It would be useful to color all the glucose (B), ketone body (I) and water molecules (J) in the illustrations.
2. Color the glucose tolerance test down below.

DIABETES:
STARVATION IN THE MIDST OF PLENTY

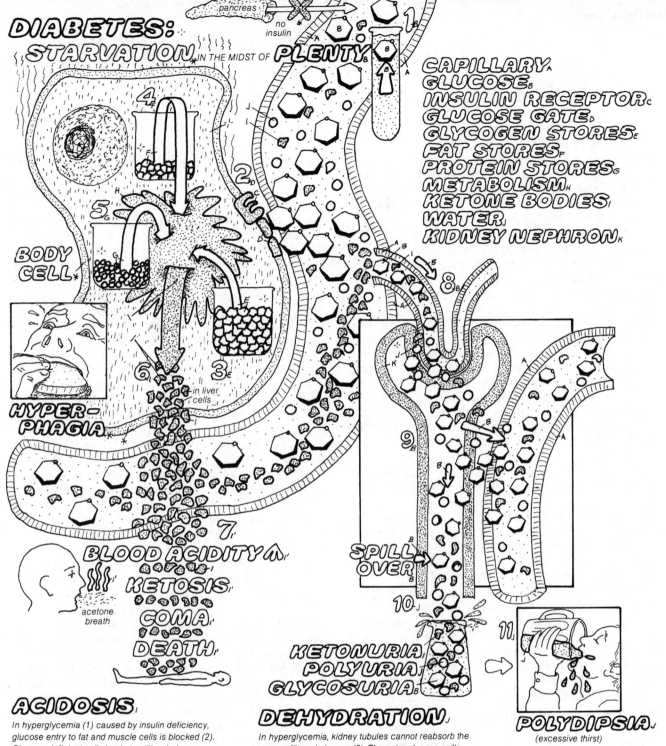

pancreas
no insulin

CAPILLARY A
GLUCOSE B
INSULIN RECEPTOR C
GLUCOSE GATE D
GLYCOGEN STORES E
FAT STORES F
PROTEIN STORES G
METABOLISM H
KETONE BODIES I
WATER J
KIDNEY NEPHRON K

BODY CELL *

HYPER-PHAGIA *

in liver cells

BLOOD ACIDITY ⬆

acetone breath

KETOSIS
COMA
DEATH

SPILL OVER

KETONURIA,
POLYURIA,
GLYCOSURIA. J

POLYDIPSIA J
(excessive thirst)

ACIDOSIS I

In hyperglycemia (1) caused by insulin deficiency, glucose entry to fat and muscle cells is blocked (2). Glucose deficient cells begin to utilize their own stores of glycogen (3), fat (4) and protein (5) to obtain energy. Reduced glucose entry in hypothalamic hunger centers leads to overeating (hyperphagia). Excessive fatty acid utilization leads to formation of ketone bodies by liver (6), causing "acetone breath", ketonemia and increased blood acidity (7) (ketosis). If untreated, ketosis causes coma and death.

DEHYDRATION J

In hyperglycemia, kidney tubules cannot reabsorb the excess filtered glucose (8). The extra glucose spills over in urine (glycosuria) (9) and causes osmotic diuresis (polyuria) (10). Polyuria reduces plasma water, leading to thirst and increased water intake (polydipsia) (11). If untreated, dehydration, osmotic shock and death will follow.

GLUCOSE-TOLERANCE TEST B

Insulin deficient or diabetic people have higher fasting blood sugar level. When given a load of glucose after fasting (glucose-tolerance test), blood glucose increases to a much higher level and decreases over a much longer duration than normal individuals.

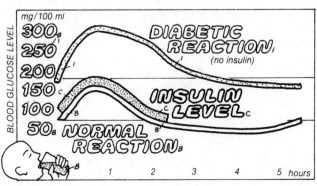

mg/100 ml

BLOOD GLUCOSE LEVEL

300
250
200
150
100
50

DIABETIC REACTION (no insulin)

INSULIN LEVEL C

NORMAL REACTION B

1   2   3   4   5 hours

THE ADRENAL MEDULLA, CATECHOLAMINES, & STRESS

Adrenal glands are paired organs located above the kidneys. Each adrenal consists of two separate glands, an outer *adrenal cortex* and an inner *adrenal medulla*. Though the two glands have different embryological origin, structure, and hormonal secretions, at least with respect to responses to stress, their functions are synergistic and aimed at a common goal.

CHROMAFFIN CELLS AND CATECHOLAMINES. The adrenal medulla is essentially part of the *sympathetic* division of the *autonomic* nervous system, being in fact a modified *sympathetic ganglion*. The secretory cells of the adrenal medulla, called the *chromaffin* cells, are equivalent to *postganglionic* sympathetic neurons that have lost their axons (see plate 25). The chromaffin cells contain vesicles filled with *epinephrine* and *norepinephrine*. These biogenic amines, collectively called *catecholamines*, are produced from the amino acid pheynlalanine via several enzymatic chemical reactions in the chromaffin cells, as follows: Phenylalanine → p-tyrosine → Dopa → Dopamine → Norepinephrine → Epinephrine. Recently, the opiate peptide *endorphin* has also been found in the chromaffin cells. This analgesic (anti-pain) peptide is co-secreted with the catecholamines.

NEURAL REGULATION. Upon stimulation of the adrenal medulla by the sympathetic nervous sytem, the vesicles release their content of epinephrine and norepinephrine into the blood to act as the hormones of the adrenal medulla. There are probably two cell types in the medulla, one secreting epinephrine and the other, norepinephrine. In humans, the secretion of epinephrine comprises 80% of the total catecholamine output, the remainder being that of norepinephrine.

Each time the sympathetic nervous system is strongly stimulated, the activity of the adrenal medulla increases. Thus, during fear and excitement or stressful muscular exercise (running, physical exertion, and struggle), stimuli from various parts of the nervous system impinge on the *hypothalamus*, which, among other things, acts as the highest center for the regulation of sympathetic responses (plate 101).

Excitatory fibers from hypothalamic neurons descend in the spinal cord, stimulating the *preganglionic* sympathetic neurons. The preganglionic fibers enter into the two chains of *sympathetic ganglia*, one on each side of the vertebral column, releasing *acetylcholine* and synaptically stimulating the *postganglionic* neurons. The latter send their fibers to the visceral organs and skin. Norepinephrine is the neurotransmitter at these nerve endings. The adrenal medulla receives a long preganglionic sympathetic fiber (via a splanchnic nerve).

CATECHOLAMINE ALPHA AND BETA RECEPTORS. Catecholamine hormones reach their target organs and bind with specific *adrenergic receptors* present on the cell membranes of their target organs. The adrenergic receptors are divided into the *alpha* and *beta* types. Norepinephrine binds mostly with the alpha receptors; epinephrine can bind with both types. The particular responses of the target organs depend on the kind and number of receptors present in the cells of the organ. Also, because sympathetic nerve fibers release only norepinephrine, they tend mainly to activate the alpha receptors. The adrenal medullary secretion, being a mixture of both catecholamines, tends to activate both types of receptors.

CATECHOLAMINES IN FIGHT-FLIGHT RESPONSES. The functions of the adrenal medulla and the effects of its hormones are best understood in terms of the preparation of the body for unexpected stressful situations such as fight-flight, or exercise. In all these responses, the intense muscular activity demands increased blood flow, nutrients, and oxygen supply.

Consider a person who is running fast. The need for increased oxygen and fuel for muscles demands increased delivery of blood by the heart. Thus, cardiac output (heart rate and cardiac contractility) must be increased (see plate 39). At the same time, the blood vessels to the heart and muscles must be dilated while those to the skin and visceral organs must be constricted, shunting the blood to where it is most needed (muscles and heart). The respiratory activity must be increased and the bronchioles dilated to supply more oxygen and remove more carbon dioxide. All these responses are brought about by various effects of epinephrine and norepinephrine acting on the target organs.

Epinephrine acts mainly on the heart, causing increased rate and contractility; norepinephrine acts on the visceral blood vessels (arterioles) to cause vasoconstriction, increasing blood pressure and shunting blood to muscles. This differential response occurs because the heart contains mainly beta receptors which bind preferentially with epinephrine, and visceral arterioles have the alpha type, which bind with norepinephrine. The smooth muscles of bronchioles and those of arterioles of the heart and muscle contain beta receptors. These receptors, when activated by epinephrine, relax the smooth muscles, causing vasodilation and bronchiolar dilation.

Metabolically, the body demands increased nutrient supply. Epinephrine increases glycogen breakdown in the liver and fat in the adipose tissue to mobilize ample fuel substances (glucose and fatty acids). Lastly, catecholamines act on the brain to increase arousal, alertness, and excitability. They also act on the iris of the eye to dilate the pupil, thus permitting more light into the eyes and enhancing peripheral vision.

In short, the functions of the adrenal medulla should be construed as complementary and synergistic with the functions of the sympathetic nervous system.

CN: Use red for K, and bright colors for C & D.
1. Begin with the material in the upper panel.
2. Color the middle section. Note the four inputs from the brain in the upper left side. Color the titles and the bar lines gray.
3. In the lower panel, color all the responses to catecholamines and their respective number or letter in the illustrations, using the two colors denoting norepinephrine (D) and epinephrine (C).

ADRENAL GLANDS A

ADRENAL MEDULLA B

adrenal cortex

CATECHOLAMINES: *
EPINEPHRINE (E) (ADRENALINE) C

The adrenal medulla, being the inner endocrine part of the adrenal gland, is part of the sympathetic NS. Chromaffin cells of adrenal medulla secrete norepinephrine (NE) and epinephrine

80% C

NOREPINEPHRINE (NE) D

(E), (catecholamines). These hormones are derived from the amino acid phenylalanine, stored in vesicles and secreted in response to stresses which activate the sympathetic NS. In man, epinephrine comprises 80% of secretory output.

20% D

STRESS *
EMOTIONS *
EXERCISE *

Chromaffin cells of adrenal medulla are modified post-ganglionic sympathetic neurons. Sympathetic fibers secrete NE only, while chromaffin cells secrete both E and NE. Stimuli activating sympathetic NS also activate the adrenal medulla. In target cells, E and NE bind with specific alpha and beta adrenergic receptors. Some tissues have alpha type, some beta and some both. E binds more avidly with a beta type, while NE binds mostly with an alpha type. The differential distribution of receptors in tissues underlies the different actions of catecholamines.

BRAIN/HYPOTHALAMUS E
SPINAL CORD F
PREGANGLIONIC NEURON G
ACETYLCHOLINE H
SYMP. GANGLION I
POSTGANGLIONIC N. D'
NOREPINEPHRINE D
TARGET CELL J*
ALPHA RECEPTOR D²
BETA RECEPTOR C'
ADRENAL MEDULLA B
EPINEPHRINE C
NOREPINEPHRINE D
BLOOD VESSEL K

SYMPATHETIC NERVOUS SYSTEM

ENDORPHIN L

FIGHT, FLIGHT, OR EXERCISE *

NOREPINEPHRINE CAUSES: D

A. VASOCONSTRICTION IN SKIN, KIDNEY, DIGESTIVE TRACT & SPLEEN.

B. DECREASES DIGESTIVE ACTIVITY.

C. GLYCOGENOLYSIS.

D. LYPOLYSIS (FATTY ACID MOBILIZATION)

E. INCREASED HEART ACTIVITY.

F. BRAIN AROUSAL.

G. HAIR ERECTION.

H. BLOOD PRESSURE RISE.

EPINEPHRINE CAUSES: C

1. INCREASED HEART ACTIVITY.

2. VASODILATION IN MUSCLE.

3. BRONCHIOLE DILATION.

4. GLYCOGENOLYSIS.

5. LYPOLYSIS.

6. BRAIN AROUSAL.

7. PUPIL DILATION.

8. INCREASED BMR.

9. VASOCONSTRICTION IN SKIN, KIDNEY, ETC.

10. BLOOD CLOTTING.

11. BLOOD PRESSURE RISE.

THE ADRENAL CORTEX: FUNCTIONS OF ALDOSTERONE

Adrenal glands are paired organs located on top of the kidneys. Each adrenal consists of two separate glands, which have different structures, embryonic origins and hormonal secretions. The inner part is the *adrenal medulla*, secreting epinephrine and norepinephrine (plate 119). The *adrenal cortex*, the outer part of the gland, secretes a variety of *steroid* hormones (*corticosteroids*).

ADRENAL CORTEX ZONES AND HORMONES. Histologically, the adrenal cortex is divided into three distinct zones, each specialized to secrete specific corticosteroids with distinct functions. The outer part (*zona glomerulosa*) secretes the hormone *aldosterone*. Aldosterone is a *mineralocorticoid* involved in the regulation of plasma salts (*sodium* and *potassium*), *blood pressure*, and *blood volume*. The middle part of the adrenal cortex (*zona fasciculata*) secretes *glucocorticoid* hormones, chiefly *cortisol*, which regulates the metabolism of glucose, especially in times of stress. The most inner part of the cortex (*zona reticularis*) secretes *sex steroids*, chiefly *androgens*. In this plate, we shall focus on aldosterone and its actions in salt balance and blood pressure regulation.

IMPORTANCE OF SODIUM AND POTASSIUM. Sodium is the chief electrolyte of the plasma and extracellular fluid. It influences the functions of plasma membranes of all cells, especially those of excitable (nerve and muscle) tissues (see plates 10-15). Sodium levels are also crucially important in regulating total body water and blood pressure. For these reasons, reductions in sodium levels are hazardous to bodily functions.

Potassium is the chief intracellular electrolyte. An abnormal rise in plasma potassium concentration leads to disturbances in cardiac and brain functions that may be fatal. The potassium level in plasma is therefore kept low, within appropriate limits. Aldosterone is the principal hormone involved in the maintenance of appropriate levels of sodium and potassium in the blood plasma. Indeed, the absence of aldosterone as occurs with removal of the adrenals (adrenalectomy) is fatal unless followed by proper treatment (see below).

ACTIONS AND REGULATION OF ALDOSTERONE. Aldosterone acts mainly on the cells of kidney tubules, stimulating them to synthesize new protein molecules. By acting as enzymes or carriers, these molecules enhance the tubular transport of sodium from the lumen into the plasma as well as promote (indirectly) secretion of potassium from plasma into the kidney tubules (*excretion*). (See plate 61, 65.)

The stimuli that activate release of aldosterone are reduction in the plasma sodium level and elevation in the plasma potassium level. These conditions may arise through alterations in the dietary or intestinal intakes of these electrolytes. Also, loss of blood, blood volume, and blood pressure (as occurs during hemorrhage) are strong stimuli for aldosterone release. Increased levels of potassium have a marked and rapid effect on aldosterone secretion, as they act directly on the cells of the zona glomerulosa. In contrast, the mechanism by which sodium decrease stimulates aldosterone release is slow, because it involves several steps.

RENIN-ANGIOTENSIN-ALDOSTERONE SYSTEM. Let us assume that a decrease in sodium intake or a loss of blood by hemorrhage has led to a reduction in blood pressure. This is detected by sensors in the renal arterioles adjacent to the *juxtaglomerular apparatus*, resulting in the release of an enzyme called *renin* from the cells of this apparatus. Upon entering the blood, renin breaks down a large polypeptide called *angiotensinogen*, which is secreted by the *liver* and normally circulates in the blood. The resultant smaller polypeptide called *angiotensin I* is rapidly converted to a still smaller peptide called *angiotensin II* as the blood circulates through the lungs. An enzyme in the lung capillaries is responsible for this conversion.

Angiotensin II performs two functions: the first is to increase blood pressure directly, and the second is to stimulate aldosterone secretion. Aldosterone also increases blood pressure, but indirectly. To elevate blood pressure directly, angiotensin II binds with receptors on the smooth muscle of the arterioles, causing *vasoconstriction*, which in turn causes increased peripheral resistance. These conditions rapidly increase the blood pressure. (See plate 42.)

Angiotensin II also acts on the cells of the zona glomerulosa, stimulating aldosterone secretion. Aldosterone acts on the renal tubules, enhancing sodium reabsorption. Increased plasma sodium increases plasma osmolarity and blood pressure. In addition, the increased obligatory reabsorption of water that occurs after sodium reabsorption restores plasma water, blood volume, and blood pressure. Although the effects of aldosterone on blood volume and blood pressure are slow to develop, taking hours, these effects are more prolonged and stable than those caused by direct actions of angiotensin II. Aldosterone can also increase plasma sodium by promoting similar reabsorptive effects on the cells of salivary glands and sweat glands.

The role of aldosterone in regulating plasma sodium and potassium is so important that, in the absence of the adrenal cortex the experimental animal or the human patient soon dies, because the loss of body reserves of sodium lead to heart and brain abnormalities as well as dehydration and shock. Only by treatment with exogenous aldosterone, which increases salt reabsorption, or by the increase of sodium in the diet and by drinking water, can the patient be saved. Rats with adrenal insufficiency are known to spontaneously increase the amount of their salt intake. This happens, presumably, because a greater desire for salt results from an increase in their salt taste threshold. Aldosterone is one reason the adrenal cortex is so essential to life.

Elevated potassium levels, as mentioned above, directly stimulate release of aldosterone by the cells of the zona glomerulosa. Aldosterone decreases potassium levels in the plasma by increasing secretion of this ion in the urine. In the renal tubules, potassium is secreted in exchange for sodium, which is then reabsorbed. (See plate 61).

CN: Use red for E, very light colors for A, B, C.
1. Begin with the zones of the adrenal cortex.
2. Proceed to Aldosterone: Sodium Control, and follow the stages from the initial drop in blood volume and pressure (and loss of sodium) to the reabsorption of sodium at the bottom of the page.
3. Color Aldosterone: Potassium Control, and the diagram of the kidney nephron in the lower right corner summarizing the effects of aldosterone on both sodium and potassium.

ADRENAL CORTEX:*
CORTICOSTEROIDS*

ADRENAL MEDULLA

ZONA GLOMERULOSA
MINERALOCORTICOID:*
ALDOSTERONE
ZONA FASCICULATA
GLUCOCORTICOID:*
CORTISOL
ZONA RETICULARIS
SEX STEROIDS:*
ANDROGEN
ESTROGEN

hemorrhage

upright posture

excessive perspiration

ALDOSTERONE:
SODIUM CONTROL

JUXTA-GLOMERULAR APPARATUS

BLOOD VOL./PRESS.

Aldosterone increases reabsorption of sodium in renal tubules elevating plasma sodium and leading to increased blood water, volume and pressure. A decrease in sodium or blood pressure is detected by juxtaglomerular apparatus, stimulating the release of renin in blood. Renin acts as an enzyme converting angiotensinogen — secreted by liver — to a shorter polypeptide, angiotensin I. In the lung, angiotensin I is further converted to a shorter and highly active peptide, angiotensin II.

afferent arteriole

distal tubule

Na^+

ALDOSTERONE:
POTASSIUM CONTROL

Aldosterone acts to decrease the potassium level in plasma. Increase in potassium is detected by the adrenal cortex, stimulating the release of aldosterone. Aldosterone acts on the kidneys to increase the urinary excretion of potassium, by exchanging potassium with sodium, which is reabsorbed in the tubules.

RENIN

ANGIOTENSINOGEN

blood

Angiotensin II causes vasoconstriction, elevating blood pressure. Angiotensin II also stimulates the adrenal cortex to release aldosterone. Aldosterone acts on the kidneys to increase sodium reabsorption. Water follows by osmosis. Increased plasma sodium and water will elevate blood pressure and volume as well as compensating for decreased sodium.

ANGIOTENSIN I

ENZYME ACTION

lungs

K^+

EXCRETION

KIDNEY NEPHRON

afferent arteriole

capillary

collecting duct

ANGIOTENSIN II

VASOCON-STRICTION (RAPID RESPONSE)

arteriole

adrenal

kidney

ALDOS-TERONE

Na^+ X

Na^+ H_2O

REABSORPTION

blood

(SLOW RESPONSE)

proximal tubule

distal tubule

K^+

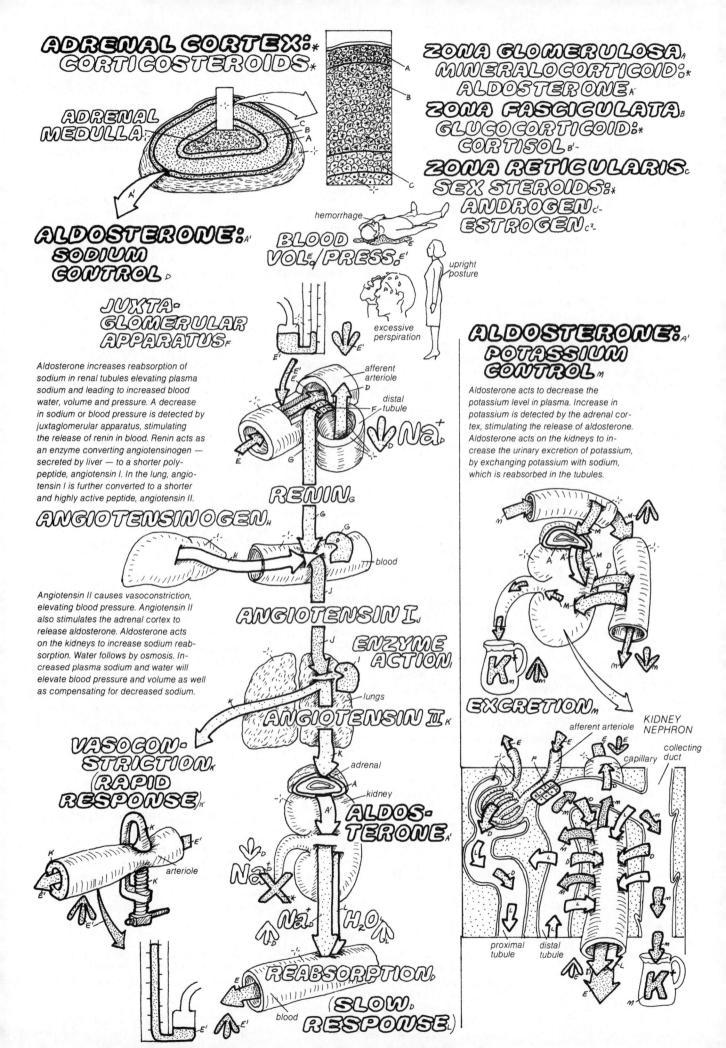

THE ADRENAL CORTEX: CORTISOL AND STRESS

Cortisol is the chief steroid hormone secreted by the cells of the middle part of the *adrenal cortex (zona fasciculata)*. The best known action of cortisol is to increase the blood *glucose* supply for tissues, mainly the brain and heart. Cortisol exerts this action by promoting *catabolism* of *proteins* and by stimulating the conversion of the resultant *amino acids* to glucose, a process known as *gluconeogenesis*. Gluconeogenesis occurs principally in the *liver*. It is for this role in carbohydrate metabolism that cortisol and similar steroids are called "gluco"-corticoids.

Cortisol has numerous other effects in the body. Many of these, along with the gluconeogenic action are intimately related to body responses in various "stress" conditions. Some of these responses are short term, exerted in conjunction with the catecholamines from the adrenal medulla. Other cortisol actions are exerted independently and are longer lasting. Because of the importance of cortisol in defense of the body against noxious and traumatic stresses, this hormone is considered essential for life. Adrenalectomized animals and humans may die if exposed to sudden unexpected stresses.

REGULATION OF SECRETION. A variety of stressful conditions (cold, fasting, starvation, loss of blood pressure [hypotension], hemorrhage, surgery, infections, pains from wounds, fractures, inflammations, severe exercise, and even emotional traumas) all act on the brain to elicit the release of *CRH* (corticotropin-releasing hormone) from the *hypothalamus*. CRH stimulates the release of *ACTH* (corticotropin), a polypeptide hormone, from the *corticotrop cells* of the anterior pituitary. ACTH acts on the adrenal cortex, stimulating the synthesis and release of cortisol.

Once the cortisol level is sufficiently elevated, CRH and ACTH secretion are decreased through the negative-feedback effect of cortisol on the hypothalamus. This reduces the cortisol level back to the normal baseline condition. When stress is chronic, the brain overrides this control. Continued stimulation of zona fasciculata by ACTH leads to hypertrophy (excess growth) of this area and enlargement of the adrenal cortex. Other zones remain unaffected.

SYNERGISM OF CORTISOL AND CATECHOLAMINES. In many instances of *short-term* responses to stress, both cortisol and catecholamines are secreted from the adrenal gland. The increased release of cortisol occurs rapidly, within a few minutes. Although the effects of catecholamines in these instances are well known, those of cortisol are not. Cortisol may promote the effects of catecholamines. For example, the vasoconstriction and fatty acid-mobilizing effects of catecholamines are markedly reduced in the absence of cortisol.

CORTISOL AND ADAPTATION TO STRESS. The effects of cortisol in promoting *long-term* metabolic adaptation are better known. This adaptation is necessary to improve defenses, promote tissue repair and wound-healing, and to provide adequate nutrients in the form of glucose and amino acids.

Consider, for example, an animal, hurt, with broken bones and immobilized, or a man stranded in the sea, overcome by starvation, fatigue, sunburn, anxiety, and despair (stress conditions). Food intake being nil, liver and muscle glycogen stores are soon exhausted, threatening the supply of glucose to the nervous system and the heart. This may have disastrous consequences, because, under normal conditions, the brain relies practically entirely on glucose for its energy needs. Adequate supplies of amino acids are also needed for tissues that must regenerate, repair, or grow. The amino acid-mobilizing and gluconeogenic actions of cortisol are essential in combating these stress related deficiencies.

Increased secretion of cortisol acts on muscle, connective tissue (bones, etc.), and lymphatic tissue, stimulating the catabolism of their labile protein reserves. The "mobilized" amino acids are taken to the liver, where, after deamination (removal of the amine group), they are converted to glucose (gluconeogenesis). Cortisol stimulates the synthesis of gluconeogenic enzymes in the liver. The newly formed glucose ensures adequate fuel supply for the brain and heart. In addition, cortisol reduces the uptake of glucose by muscle cells, sparing glucose supply for the brain and heart.

Amino acids liberated by tissue catabolism are not all utilized for gluconeogenesis; some are shunted to tissues that need them for repair and regeneration. Others are used in the liver for synthesis of blood proteins necessary for survival. Under the influence of cortisol and catecholamines, the triglycerides of the fat cells are broken down, and fatty acids are mobilized. The latter can be used by the muscle, heart, and liver for energy.

PERMISSIVE ACTIONS AND DIURNAL VARIATION. Several actions of cortisol are "permissive." Thus, cortisol must be present for glucagon and growth hormone to exert their actions on the liver (glycogenolysis) and adipose tissue (lipolysis) and for catecholamines to cause vasoconstriction. Normally, the secretion of cortisol shows a "diurnal" (daily) cycle, the secretion rate being highest in the morning and lowest in the evening. This cyclicity is regulated by centers in the hypothalamus and is independent of stress (see plate 101).

CORTISOL AS A DRUG. Treatment with large doses of cortisol (*pharmacologic doses*) has therapeutic effects against *inflammations* produced by wounds, allergies, or rheumatoid (joint) diseases. It is not known how these pharmacologic effects of cortisol are exerted or whether they occur during "physiological" defenses.

STRESS-RELATED DISEASES. In *chronic stress*, excess cortisol may have detrimental and harmful effects. Thus, stomach ulcers, atrophy of lymphatic nodes, reduction in white blood cells (decreased immunity), hypertension, and vascular disorders are often observed after severe stresses.

CN: *Use red for C, yellow for I, and the same color as the previous page for zona fasciculata (A).*
1. *Begin in the upper left corner by coloring the area of adrenal cortex which secretes cortisol (A¹). Then begin the sequence of events (upper right hand) in which stress causes a drop in blood pressure (C) and follow the numbered stages (1-10).*
2. *Color the numbered sequence (1-9) of the metabolic response to cortisol secretion (middle diagram), starting with the low blood glucose level (in the test tube symbol), resulting in this case, from starvation.*
3. *Down below, color the titles, tablet, capsule, and hypodermic needle (vehicles for delivery of cortisol).*

ZONA FASCICULATA

CORTISOL & STRESS

HYPOTHALAMUS

SLOW
CRH & ACTH
ADRENAL CORTEX
CORTISOL

PAIN

PRES.

SYMPATHETIC RESPONSE

infection
exercise
fright

8 CRH

CORTICO-TROP CELL

9 ACTH

RAPID
SYMP. NERV. SYS.
ADRENAL MEDULLA
CATECHOLAMINES

In response to stresses (1) e.g., fright, loss of blood pressure, short-term exercise (2), the hypothalamus activates sympathetic NS (3), sympathetic nerves (4), and adrenal medulla (5), stimulating the secretion of catecholamines (6). These rapidly elevate blood pressure and mobilize

glucose and fatty acids (7). Simultaneously, the hypothalamus secretes CRH (8) which stimulates ACTH release from the pituitary (9). ACTH stimulates release of cortisol (10) from the adrenal cortex. Catecholamines and cortisol help the body fight the effects of short-term stress.

METABOLIC RESPONSE
FAT & PROTEIN CATABOLISM

fat   muscle   connective   lymph

TISSUE REPAIR

TO BRAIN

starvation

GLUCAGON

GROWTH H.

hypoglycemia

FREE FATTY ACIDS
GLYCEROL
AMINO ACIDS
ENZYME SYNTHESIS
OTHER HORMONES
LIVER
GLUCONEOGENESIS
GLUCOSE
GLYCOGEN
BODY CELL

Chronic stress (illness, starvation, pain) triggers a release of ACTH and cortisol as well as adrenal hypertrophy. Cortisol helps catecholamines mobilize fatty acids and glycerol from fat cells (2). Fatty acids are used by heart and liver. Cortisol acts on muscle, bone and lymphatic tissue, to catabolize their labile proteins, mobilizing amino acids (3). These are taken to the liver or used for tissue repair and renewal (4). Cortisol stimulates the liver

to form enzymes, converting amino acids into glucose (gluconeogenesis) (5). Cortisol enhances the actions of glucagon and growth hormone as well (6). Glucose formed by gluconeogenesis (7) is secreted into the blood, elevating blood sugar (8). Cortisol also reduces the uptake of glucose by muscle and other peripheral tissues (9), sparing glucose for use by the brain (10) and heart.

PHARMACOLOGICAL EFFECTS:
ANTI-INFLAMMATORY ACTION AGAINST:

WOUNDS AND INJURIES   ALLERGIES   RHEUMATISM

In large doses, cortisol alleviates symptoms of inflammations caused by injuries, allergies, and rheumatoid disorders of joints. Except for the promotion of wound healing, cortisol probably does not contribute to the natural and anti-inflammatory responses of the body.

EFFECT OF PROLONGED STRESS: (PROLONGED SECRETION OF CORTISOL)

ULCERS   LYMPHATIC ATROPHY   HYPERTENSION VASCULAR DISORDERS

Prolonged and excessive secretions of cortisol, usually in response to chronic and severe stress, can cause ulcers by stimulating acid secretion or decreasing resistance to acid in the stomach and duodenum. Cortisol can cause a severe decline in the number of leukocytes and the atrophy of lymph nodes, thereby decreasing resistance to microbial infection. Excess cortisol also promotes hypertension and vascular disorders.

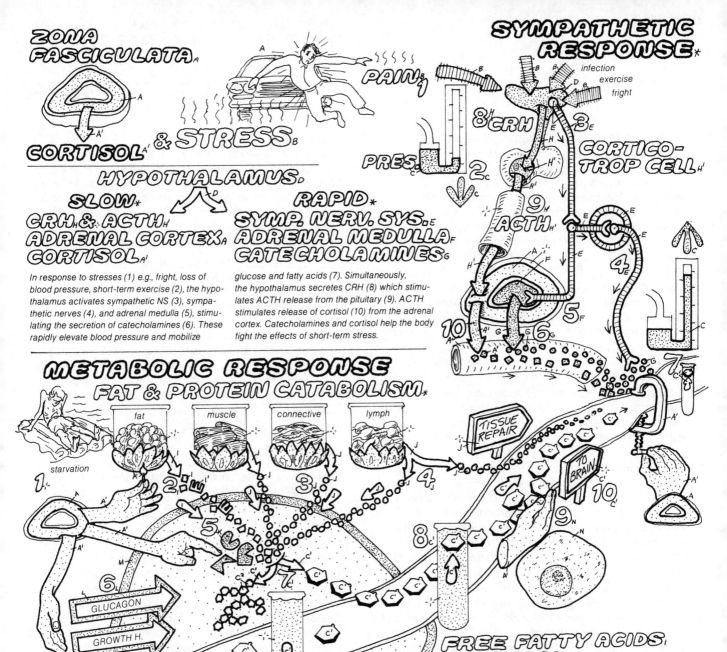

THE ADRENAL CORTEX: SEX STEROIDS & ADRENAL DISORDERS

ADRENAL ANDROGENS. Cells of the inner part of the adrenal cortex (zona reticularis) secrete sex steroids, principally androgens and small amounts of estrogen and progesterone. The major adrenal androgen is dehydroepi-androsterone (a 17-ketosteroid). Adrenal androgens have five times less potency than testosterone, the major male sex steroid secreted by the testis. The function of adrenal androgens in the adult male is probably unimportant due to the presence of large amounts of testicular androgen, testosterone (plate 144).

The adrenal androgens are, however, the main source of the male sex steroid in females. The secretion of estrogen, the female sex steroid, from the adrenal cortex is small, but some of the adrenal androgen is converted to estrogen in blood or in peripheral tissues. This accounts for the estrogen in male blood. In adults, the secretion of adrenal sex steroids is stimulated by ACTH and not by the pituitary gonatropins.

Under normal conditions, adrenal sex steroids probably exert mainly metabolic effects. It has been claimed that adrenal androgens contribute to libido (sex drive) in women. The anabolic actions of androgen would be particularly important for the female body.

In children, a marked surge in the secretion of adrenal androgens is observed between 7 and 15. This exerts significant effects on bone and muscle growth on fat accumulation and distribution. The surge in adrenal androgens (adrenarche) during pubertal growth is believed to be due either to enzymatic changes in the cells of the zona reticularis or to the secretion of a special tropic hormone from the anterior pituitary.

DISORDERS OF THE ADRENAL CORTEX. Disturbances in the secretion of adrenal steroid hormones, caused by atrophy, tumors, or enzymatic abnormalities in the cells of the adrenal cortex, bring about dramatic changes in the individual. These changes provide some of the classic demonstrations of pathological effects due to absence or excesses of the hormones.

ADRENOGENITAL SYNDROME. Normally, adrenal androgens have little masculinizing effect, as evident by the fact that eunuchs (males without testes) have femalelike appearance, even though they still have adrenal androgens. However, occasionally due to the growth of tumors or cellular (enzymatic) disorders, the adrenal cortex begins to secrete large amounts of androgens. For example, enzymes that normally convert androgens to cortisol in the adrenal cortex may become deficient. Thus, instead of cortisol, adrenal cells secrete androgens. However, absence of cortisol in the blood triggers secretion of ACTH (negative feedback), which stimulates the adrenal to secrete even more androgen. Soon a vicious cycle is set up, flooding the body with adrenal androgens.

When this occurs in mature women, secondary male sexual characteristics such as body and facial hair, muscular growth, male body configuration (due to differential fat distribution), and voice and genital changes are observed, creating the striking clinical picture of adrenogenital syndrome. Similar effects may be seen in young girls, in which case a precocious male type pseudo-puberty is observed (virilism). In young boys, this condition causes precocious development of external male characteristics in the absence of testicular development. The accelerated growth of bones and muscle in these boys often leads to stunted stature because of premature fusion of epiphyseal plates (see plate 115).

CORTISOL EXCESS: CUSHING'S SYNDROME. Excessive secretion of cortisol, which can be caused by either adrenal tumors or by ACTH secreting pituitary tumors, leads to the development of Cushing's syndrome (disease). In these conditions, the excess secretion of cortisol causes catabolism of proteins, wasting of muscles and fatigue. Decreased protein synthesis and increased protein breakdown in bones lead to weakening of the bone matrix, resulting in osteoporosis. Loss of connective tissue in the skin results in the formation of bruises and poor wound healing. Blood pressure and blood sugar levels are markedly increased.

The fat is redistributed from lower to upper parts, including the abdomen, back, neck, and face, giving rise to the "buffalo torso" appearance. In the face, loss of subcutaneous connective tissue causes edema. This condition, along with the deposited fat, gives the characteristic "moon face." The illness is often accompanied by behavioral and mental disorders ranging from simple euphoria to full-blown psychosis.

DIMINISHED ADRENOCORTICAL SECRETIONS: ADDISON'S DISEASE. Sometimes as a result of cancer or infectious diseases (tuberculosis) or in some autoimmune diseases, the adrenal cortex atrophies, resulting in diminished secretions of adrenal steroid hormones. This condition, called Addison's disease, is a very serious clinical disorder that, if untreated, can lead to death. Decreased secretion of aldosterone results in loss of sodium and water, causing loss of blood pressure, dehydration, and cardiovascular and neurologic abnormalities.

Decreased secretion of cortisol diminishes the gluconeogenic ability of the liver. Hence, blood sugar cannot be raised in fasting. Decreased cortisol diminishes resistance to stress both because of the lack of direct protective actions of cortisol in the body (e.g., reduced gluconeogenesis) and by reduced response to catecholamines. As a result, during stress the body becomes practically helpless, succumbing to shock and death in response to even such simple stresses as cold or hunger. However, most patients, if untreated, die due to inability to fight stresses caused by infectious agents (bacteria, etc.).

The decreased cortisol levels in Addison's victims leads to increased secretion of ACTH as well as MSH (melanocyte-stimulating hormone), which is co-produced by pituitary corticotrops. These hormones increase skin pigmentation, one of the classic signs of Addison's disease.

CN: Use red for G, yellow for F, and light brown for H. Use same colors for zona reticularis (A), cortisol (D) and aldosterone (I) as on the previous two plates. 1. Color the upper panel dealing with sex steroids, noting that the ovary (C) and testis (B') receive the same color as their principal hormone.

2. Color the three examples of disorders of the adrenal cortex beginning with andrenogenital syndrome. For the woman on the left, color each symbol representing an increase or decrease. In the case of Addison's disease, begin with the total adrenocortical shut down in the upper left corner.

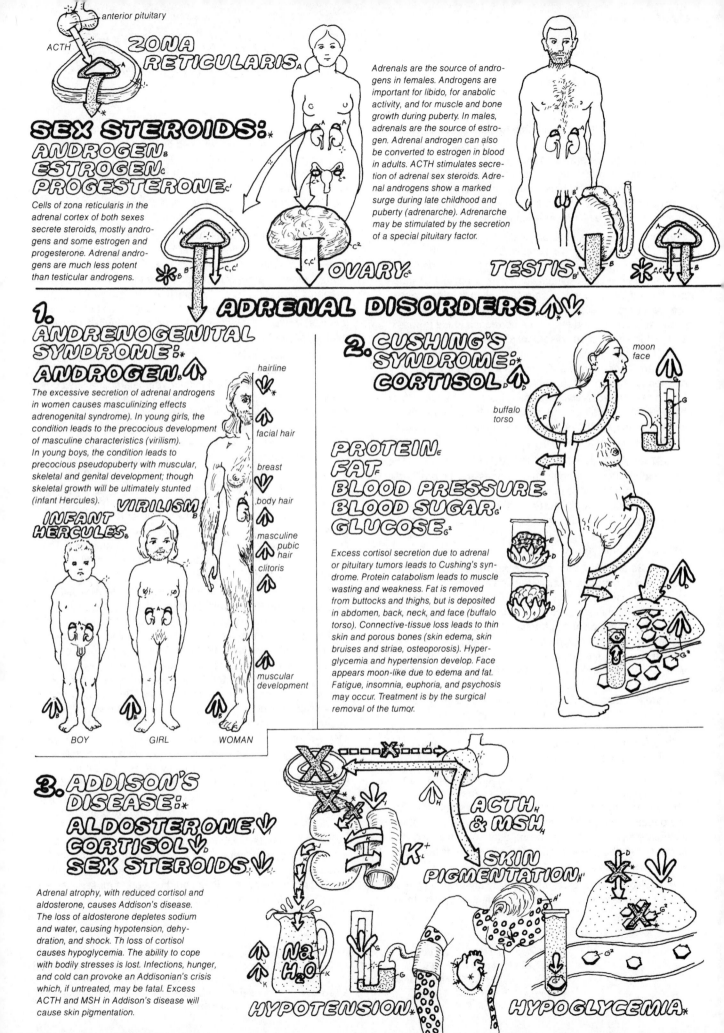

anterior pituitary

ACTH

ZONA RETICULARIS

SEX STEROIDS:
ANDROGEN
ESTROGEN
PROGESTERONE

Cells of zona reticularis in the adrenal cortex of both sexes secrete steroids, mostly androgens and some estrogen and progesterone. Adrenal androgens are much less potent than testicular androgens.

Adrenals are the source of androgens in females. Androgens are important for libido, for anabolic activity, and for muscle and bone growth during puberty. In males, adrenals are the source of estrogen. Adrenal androgen can also be converted to estrogen in blood in adults. ACTH stimulates secretion of adrenal sex steroids. Adrenal androgens show a marked surge during late childhood and puberty (adrenarche). Adrenarche may be stimulated by the secretion of a special pituitary factor.

OVARY

TESTIS

ADRENAL DISORDERS

1. ANDRENOGENITAL SYNDROME:
ANDROGEN

The excessive secretion of adrenal androgens in women causes masculinizing effects adrenogenital syndrome). In young girls, the condition leads to the precocious development of masculine characteristics (virilism). In young boys, the condition leads to precocious pseudopuberty with muscular, skeletal and genital development; though skeletal growth will be ultimately stunted (infant Hercules).

INFANT HERCULES.

VIRILISM

hairline

facial hair

breast

body hair

masculine pubic hair

clitoris

muscular development

BOY

GIRL

WOMAN

2. CUSHING'S SYNDROME:
CORTISOL

moon face

buffalo torso

PROTEIN
FAT
BLOOD PRESSURE
BLOOD SUGAR
GLUCOSE

Excess cortisol secretion due to adrenal or pituitary tumors leads to Cushing's syndrome. Protein catabolism leads to muscle wasting and weakness. Fat is removed from buttocks and thighs, but is deposited in abdomen, back, neck, and face (buffalo torso). Connective-tissue loss leads to thin skin and porous bones (skin edema, skin bruises and striae, osteoporosis). Hyperglycemia and hypertension develop. Face appears moon-like due to edema and fat. Fatigue, insomnia, euphoria, and psychosis may occur. Treatment is by the surgical removal of the tumor.

3. ADDISON'S DISEASE:
ALDOSTERONE
CORTISOL
SEX STEROIDS

Adrenal atrophy, with reduced cortisol and aldosterone, causes Addison's disease. The loss of aldosterone depletes sodium and water, causing hypotension, dehydration, and shock. Th loss of cortisol causes hypoglycemia. The ability to cope with bodily stresses is lost. Infections, hunger, and cold can provoke an Addisonian's crisis which, if untreated, may be fatal. Excess ACTH and MSH in Addison's disease will cause skin pigmentation.

K^+

ACTH & MSH

SKIN PIGMENTATION

Na
H_2O

HYPOTENSION

HYPOGLYCEMIA

LOCAL HORMONES: THE PROSTAGLANDINS

LOCAL HORMONES. Local hormones are specific, highly active, usually short-lived chemical messengers released by cells into the tissue environment (extracellular fluid) in order to act on the same or other cells in the immediate vicinity. The local hormones that act on the same cells from which they are released are called *autocrines* or *autacoids*; those acting on other cells are called *paracrines* (see plate 107).

In recent times, the physiological roles of local hormones have been extensively investigated, although much is yet to be learned, especially regarding their cellular mechanism of action. In certain cases, local hormones constitute an independent mode of hormonal communication; in other cases, the local hormones are involved in the actions of many blood-borne substances and endocrine hormones at the tissue and cellular level. Many disorders are linked to malfunctions in the local hormones. Indeed, some of the important drugs (e.g., aspirin) seem to act via mechanisms involving local hormones. Among the substances known to act as local hormones are *prostaglandins* and related substances (*thromboxanes* and *leukotriens*). Substances such as serotonin and histamine are also known to act occasionally as local hormones (e.g., in the blood)

PROSTAGLANDINS: STRUCTURE AND FORMATION. The most widely known of the local hormones are the *prostaglandins* (PGs), first isolated in human semen and the prostate gland by Swedish Noble laureate Curt von Euler in the 1930s. Research has shown that the PGs are a family of closely related substances. Many, if not all, cells can release at least some of the variety of PGs.

The PGs are 20-carbon complex fatty substances (fatty acids with a hydrocarbon ring) derived from the enzymatic modification of *arachidonic acid*, a 20-carbon unsaturated fatty acid. Arachidonic acid is a component of plasma membrane phospholipids and is formed by the action of a membrane-bound *lipase* enzyme (phospholipase A), which hydrolyzes the membrane phospholipids to release arachidonic acid. The cells, employing various enzymes, then use this product to form the different PGs. The important classes of PGs are PG-Es and PG-Fs, but other PGs (PG-A to PG-I) are also known. Arachidonic acid itself cannot be synthesized in the body and must be provided in the diet; dietary deficiency of this *essential nutrient* can lead to illness, presumably because of prostaglandin deficiency.

ACTIONS OF PGs. When the first PG, discovered in semen, was added to uterine smooth muscle, it caused contraction. This action may be important in *sperm transport* in the female reproductive tract, PGs released intrinsically by the uterine tissue are also important in uterine contractions during *labor/parturition*; in fact, certain PGs are used as drugs to induce *abortion* (see plate 150).

In addition to the uterus, the PGs also affect smooth muscles in other tissues and organs. Like the uterus, the smooth muscles of the *gut wall* contract upon stimulation by the PGs. In other cases, the effect (exerted by other PGs) may be relaxation, as in the smooth muscles of the lung bronchioles (*bronchiodilation*). Certain PGs cause *vasodilation* in the blood vessels. These *dilatory* effects are proving important in the treatment of vascular disorders like *hypertension* and respiratory disorders such as *asthma*.

In addition to the actions involving smooth muscle, the PGs play numerous roles in other body tissues. Some of these actions may be in conjunction with the *endocrine hormones*; others may be independent of them. Thus, in addition to sperm transport and parturition, PGs are imporant in several other aspects of reproductive functions such as *ovulation*, *embryonic implantation*, and *corpus luteum atrophy* (plates 145, 149).

Other PG actions include hypothalamic regulation of body temperature and platelet aggregation during formation of blood clots. Release of PGs in the hypothalamus raises *body temperature*. When excessive, this response leads to *fever* (plate 134). In fact, the antipyrogenic (fever-reducing) action of aspirin involves inhibiting the PG-forming enzyme. Similarly, PGs and leukotriens are produced during *inflammatory responses*, including those in the joints (plate 139); the analgesic (pain-reducing) effects of aspirin on these painful arthritic inflammations may also be caused by inhibiting PG formation. In the blood, certain PGs prevent *clot formation* by inhibiting platelet aggregation while other PGs (thromboxanes) tend to promote *clotting* (see plate 138). In the stomach, PGs inhibit *acid secretion* by the parietal cells. This action has important implications for the treatment of stomach ulcers (see plates 69, 77).

PG/cAMP INTERACTION AND ENDOCRINE HORMONE ACTION. Some PG actions are exerted in conjunction with those of the endocrine hormones. Thus, some PGs *mimic* the effects of *anterior pituitary hormones*, particularly those that increase *cyclic AMP* (cAMP) levels in their target cell (e.g., TSH, ACTH, and prolactin) (see plate 108). In other cases, where the pituitary hormones decrease cAMP levels, the PGs *antagonize* the action of these hormones. Thus, PGs cause *diuresis* in the kidney tubule, the opposite of the effect produced by ADH, the posterior pituitary hormone (plates 62, 110). An intimate interaction seems to exist between the PGs (released extracellularly) and the second messengers, like cAMP and cGMP (released intracellularly). Thus, certain hormones (first messengers), on reaching their target, activate receptor mechanisms, causing the release of PGs in the extracellular medium. These PGs in turn activate, in the same or nearby cells, the membrane enzyme *adenyl cyclase*. This increases cAMP levels, bringing about the action of the hormones (first messengers). This way PGs can *amplify* or *antagonize* the action of a hormone in the tissue environment, depending on the type of PGs and the second messenger (cAMP or cGMP) involved.

CN: Use red for C, a bright color for E², very light colors for A, D, G, and a dark color for B.
1. Begin in the upper left corner and complete the diagram down to the large central arrow (E²). Note that local hormones (E) refer to both paracrine (E¹) and autocrine (F).

Prostaglandins (E²) are given a paracrine color on this page
2. Complete the various actions of PGs and the related local hormones (thromboxanes [E³] and leukotriens [E⁴]). Do not color the illustrations, but do color the PG arrows and the symbols of increase and decrease.

BLOOD CIRCULATION

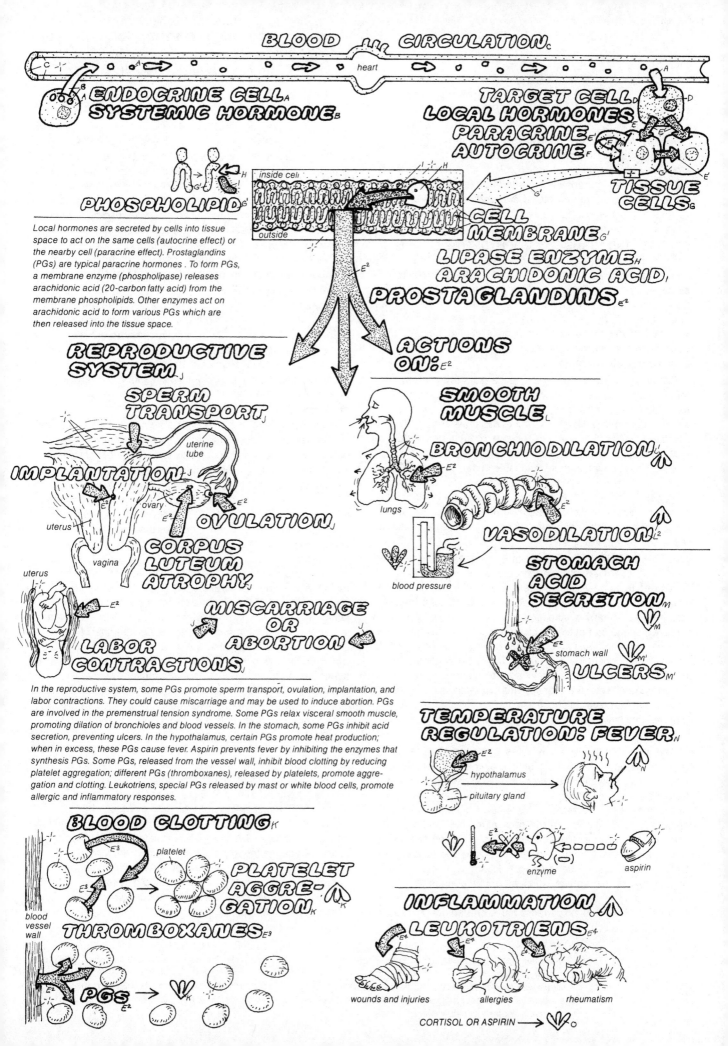

ENDOCRINE CELL
SYSTEMIC HORMONE

heart

TARGET CELL
LOCAL HORMONES
PARACRINE
AUTOCRINE

TISSUE CELLS

PHOSPHOLIPID

inside cell
outside

CELL MEMBRANE
LIPASE ENZYME
ARACHIDONIC ACID
PROSTAGLANDINS

Local hormones are secreted by cells into tissue space to act on the same cells (autocrine effect) or the nearby cell (paracrine effect). Prostaglandins (PGs) are typical paracrine hormones. To form PGs, a membrane enzyme (phospholipase) releases arachidonic acid (20-carbon fatty acid) from the membrane phospholipids. Other enzymes act on arachidonic acid to form various PGs which are then released into the tissue space.

ACTIONS ON:

REPRODUCTIVE SYSTEM

SPERM TRANSPORT

uterine tube

IMPLANTATION

uterus
ovary
vagina

OVULATION

CORPUS LUTEUM ATROPHY

uterus

LABOR CONTRACTIONS

MISCARRIAGE OR ABORTION

SMOOTH MUSCLE

BRONCHIODILATION

lungs

VASODILATION

blood pressure

STOMACH ACID SECRETION

stomach wall

ULCERS

In the reproductive system, some PGs promote sperm transport, ovulation, implantation, and labor contractions. They could cause miscarriage and may be used to induce abortion. PGs are involved in the premenstrual tension syndrome. Some PGs relax visceral smooth muscle, promoting dilation of bronchioles and blood vessels. In the stomach, some PGs inhibit acid secretion, preventing ulcers. In the hypothalamus, certain PGs promote heat production; when in excess, these PGs cause fever. Aspirin prevents fever by inhibiting the enzymes that synthesis PGs. Some PGs, released from the vessel wall, inhibit blood clotting by reducing platelet aggregation; different PGs (thromboxanes), released by platelets, promote aggregation and clotting. Leukotriens, special PGs released by mast or white blood cells, promote allergic and inflammatory responses.

TEMPERATURE REGULATION: FEVER

hypothalamus
pituitary gland

enzyme
(−)
aspirin

BLOOD CLOTTING

platelet

PLATELET AGGREGATION

blood vessel wall

THROMBOXANES

PGs →

INFLAMMATION

LEUKOTRIENS

wounds and injuries
allergies
rheumatism

CORTISOL OR ASPIRIN →

METABOLIC PHYSIOLOGY OF CARBOHYDRATES

The body uses carbohydrates as fuel to obtain energy (ATP and heat). In this plate we focus on metabolic physiology of *carbohydrates*, particularly *glucose*, in relation to *liver* function, to interconversion with other food substances, and to the preferential use of carbohydrates by various bodily tissues. (See plates 5 and 6 for the mechanisms of glucose oxidation.)

SOURCES AND FORMS OF CARBOHYDRATES. The dietary carbohydrates are provided mostly in starches, which are found in such foods as bread, rice, and potatoes. In Western societies, nearly half the body's caloric needs are derived from carbohydrates; in Eastern or developing countries, these substances are by far the major source of the calories. Milk which contains the disaccharide lactose, is also a source of carbohydrates. However, animal foods are generally poor sources of carbohydrates.

Carbohydrates are found as *simple sugars* (six-carbon *monosaccharides*, principally *glucose*, *galactose*, and *fructose*), as *oligosaccharides* (usally containing from two to ten simple sugars), or as larger polymers of simple sugars, the *polysaccharides*. Generally, the main carbohydrates of the diet are polysaccharides (e.g., starches) or *disaccharides* (e.g., the milk, sugar, lactose, and table sugar, sucrose). All carbohydrates are finally broken down by various hydrolytic enzymes (saccharidases) in the small intestine into three simple sugars: glucose, galactose, and fructose. These are absorbed across the *intestinal mucosa* and transported via the *portal vein* to the liver.

GLUCOSE AND LIVER. These simple sugars freely enter the liver cells, where galactose and fructose are enzymatically converted to glucose. This process is very efficient, so the only sugar normally found in blood is glucose. During the absorptive and early postabsorptive phases, the absorbed dietary glucose can enter the blood directly. At other times, the liver glucose constitutes the only source of blood glucose. The glucose pool in the liver can be easily exchanged with that in the blood. The tissues obtain their glucose needs from the blood pool. Only the liver can actually release glucose. Thus, when the blood glucose level is low, the liver releases glucose into the blood, when the glucose level is high, the liver cells take up and store glucose.

After a meal rich in carbohydrates, blood sugar is elevated, resulting in increased glucose uptake by the liver cells. The excess glucose within the liver cells promotes the incorporation of glucose into *glycogen*, a polymerof glucose, via a process called *glycogenesis*. This is how the excess glucose is stored in animal cells, chiefly those of the liver and muscle. The glucose residues in glycogen are bound together along branched chains, forming a treelike structure, the "glycogen tree." The excess glycogen can precipitate in the cytoplasm to form glycogen granules, which are found abundantly in liver and muscle cells. When the pool of free glucose in the liver cells diminishes, glycogen is partly broken down, by a process called *glycogenolysis*, to release free glucose.

The glucose in the liver cells can be produced from other sources as well. One such source is *proteins*, which can be broken down to form *amino acids*. *Deamination* of some amino acids (e.g., alanine) can lead to the formation of pyruvic acid, which can be converted to glucose by *reverse glycolysis*. This process is called *gluconeogenesis* and is carried out by special enzymes in the liver. Gluconeogenesis is a very important source of new glucose for the liver and eventually for the blood, particularly during fasting and starvation (see plate 121).

Another source for glucose is the *glycerol* liberated by the breakdown of *triglycerides* (*lipolysis*) in the liver and fat cells. Glycerol molecules can be recombined to form glucose through the reverse steps of glycolysis. One last source for glucose formation is *lactic acid*. Lactate, usually formed in the *muscles*, is delivered to the liver via the blood. In the liver, lactate is converted first to *pyruvate* and then through reverse steps of glycolysis to glucose.

Some tissues, such as the brain, rely principally on glucose for their energy needs. Others, such as the heart and skeletal muscle, prefer to use glucose for this purpose but are also equipped to use other fuels, such as *fatty acids*.

GLUCOSE USE IN MUSCLE. In an actively exercising muscle, glucose is taken up rapidly from the blood and converted to *glucose-6-phosphate* (*G-6-P*). G-6-P is converted to pyruvate by the enzymes of glycolysis (*aerobic glycolysis*) and, when oxygen is available, to CO_2 and water by the enzymes of the *Krebs* cycle in mitochondria. The glycolytic breakdown of glucose to pyruvate yields a small amount of *ATP*. Mitochondrial oxidation of pyruvate to CO_2 and water yields a great deal more ATP which the muscles use to do work (see plate 23). In the absence of oxygen, pyruvate is used instead to form lactic acid (lactate), a process called *anaerobic glycolysis*. This will yield more ATP, although still far less than what can be obtained in the mitochondria. If muscle activity continues, lactic acid builds up in the muscle, eventually leaking into the blood to be taken up by the liver cells. Here lactic acid is utilized for th re-formation of glucose as discussed above. The events encompassing the production of lactic acid in muscle, its conversion to glucose in the liver, and the return of the glucose to the muscle with the eventual re-formation of lactic acid constitutes the *Cori cycle*.

When the body is at rest, the glucose taken up by muscle cells is converted to G-6-P. Because the muscle is not utilizing ATP, the G-6-P is used to form glycogen, thus storing the available glucose. During activity this glycogen is converted back to G-6-P, which is shunted directly for glycolysis. Because the appropriate enzyme is lacking, the G-6-P of the muscle cannot be converted to free glucose. Therefore muscle glycogen can be used only for the muscle's own needs and cannot contribute directly to homeostasis of blood glucose.

CN: Use red for A, light blue for D, blue for G, purple for K, and a bright color for H.
1. Begin with the three types of carbohydrate molecules at the top. Color the glycogen molecule (E). Note that it receives a different color from the individual glucose (A) molecules

2. Follow the numbered sequence beginning with the entry of three monosaccharides (A, B, C) into the portal vein in the upper right. Note conversion of galactose (B) and fructose (C) into glucose (A). All stages of glucose metabolism in the active muscle cell below receive the muscle color (L).

MONOSACCHARIDE (SIMPLE SUGAR)

Glucose, galactose and fructose are the principal monosaccharides. They are six-carbon sugars and are metabolically interconvertible.

OLIGOSACCHARIDE (2-10 MONOSAC.)

Oligosaccharides are sugars containing 2 or more monosaccharides. Important disaccharides are lactose, (glu-gal), sucrose (fru-glu) and maltose (glu-glu).

POLYSACCHARIDE

Polysaccharides are polymers of monosaccharides. Polysaccharides in starches are the main dietary source of carbohydrates for man. Glucogen, a polymer of glucose with long branched chains, serves as storage for glucose in animal cells.

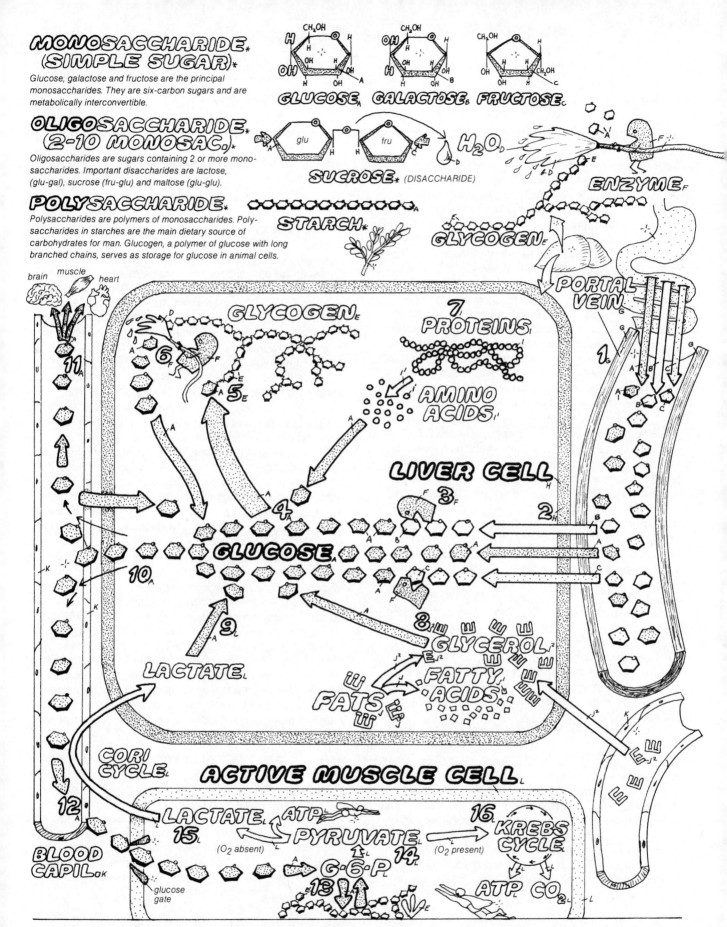

Carbohydrates are digested in the intestines to form monosaccharides, glucose (Glu), galactose, and fructose (1). These enter the liver via portal vein (2). In the liver, enzymes (3) convert fructose and galactose to Glu (4). Excess Glu is stored in the liver as glycogen (glycogenesis) (5). Glycogen can be broken down into Glu (glycogenolysis) (6). Glu may also be formed by the conversion of amino acids (gluconeogenesis) (7). Glycerol from the breakdown of triglycerides (lipolysis) can also be a source of new Glu (8). Lactate formed in anaerobic glycolysis is also convertible to Glu (9). The Glu pool in the liver freely exchanges with the pool in the blood (10). In fasting, the liver maintains a constant blood sugar level, ensuring Glu supply to tissues (11). In muscle, Glu is taken up (12) and phosphorylated to Glu-6-P. Glu-6-P is either converted to glycogen (13) or oxidized to pyruvate (glycolysis) (14) to generate ATP. In the absence of oxygen, pyruvate is converted to lactate (anaerobic glycolysis) (15) for additional ATP. Excess lactate moves into the blood and liver to form new Glu (Cori cycle). In the presence of oxygen, pyruvate is utilized by the mitochondrial Krebs cycle (16) for more efficient ATP generation.

NEURAL REGULATION OF BLOOD SUGAR

IMPORTANCE AND CONSTANCY OF BLOOD GLUCOSE. For most tissues, *glucose* is the ideal fuel substance for cellular energy production. It is the preferred fuel for some tissues, such as the heart and skeletal muscle, and the only fuel used under normal conditions by the brain. Given the central role of brain and heart in body function and body survival, an ample supply of glucose must be provided to these organs at all times. This is accomplished by regulating blood glucose content around a presumably optimal level of 1 g/l (80 to 110 mg/100 mL plasma) at all ages.

MECHANISMS OF GLUCOSE HOMEOSTASIS. The mechanisms responsible for this regulation are in part neurobehavioral and in part neurohormonal. Along with the purely hormonal mechanisms (see plate 126), they provide for a complex homeostatic system designed to restore the optimal glucose level whenever it deviates critically from the normal range. In this plate we focus on the neurobehavioral and neurohormonal mechanisms that are mainly geared to *elevated* blood sugar level when it falls below the set limits.

HYPOTHALAMIC GLUCOSTAT. Certain neurons in the *hypothalamus* that constitute a *glucostatic center* can detect changes in blood sugar levels. These neurons have a high metabolic rate (oxygen and glucose consumption) which permits them to detect changes in the glucose level within their cytoplasm and consequently in the blood (see plate 132). These neurons are the only ones in the brain where insulin is necessary for glucose entry.

HUNGER AND SATIETY. Significant reductions in the glucose level, such as occur a few hours after a meal, will result in activation of the *hypothalamic feeding* (*hunger*) center which in turn increases the appetite and other aspects of food-seeking behavior, ultimately leading to increased *food intake* (see plates 101, 132). The dietary carbohydrates absorbed in the intestine increase blood and liver glucose levels. This condition stimulates the release of insulin from the pancreatic islets; insulin in turn promotes the entry of glucose into tissues, including the neurons of the hypothalamic glucostatic center. As a result, appetite is reduced and a state of satiety prevails, at least for a few hours. Satiety and appetite suppression can also be produced by increased activity of sensory nerves from the distended stomach after food ingestion.

ROLE OF CATECHOLAMINES. In response to a relative fall in blood sugar, which may occur between meals, the glucostatic center initiates a series of actions designed to counteract the decline and elevate the blood sugar. Thus, the center initially activates the hypothalamic center for control of the *sympathetic nervous center*, leading to the release of *norepinephrine* from sympathetic nerves and *epinephrine* from the adrenal medulla. The catecholamines increase *glycogenolysis* in the liver and *lipolysis* in the adipose tissues. Glucogenolysis directly increases the glucose pool in the liver.

lipolysis provides *glycerol* for conversion to glucose in the liver. Also such tissues as muscle use the *fatty acids* mobilized from the adipose tissue, sparing glucose for the brain and heart.

GROWTH HORMONE AND CORTISOL. When food intake is delayed for a long time or in response to fasting or sustained physical exercise, blood sugar is reduced markedly. These conditions stimulate the hypothalamus to liberate *growth hormone releasing hormone* (*GRH*), which in turn stimulates the release of *growth hormone* from the pituitary gland (see plate 112). Growth hormone acts on fat cells, mobilizing fatty acids and glycerol. As mentioned above, fatty acids cause the glucose to be spared, and glycerol contributes to gluconeogenesis in the liver. As a result, the blood glucose supply increases. In addition, growth hormone acts on muscle tissues to decrease glucose utilization in exchange for an increase in the uptake of *amino acids*. This effect also spares glucose for the more essential users (e.g., the brain).

In addition to growth hormone, the hypothalamus will also stimulate the release of *cortisol* by releasing *corticotropin releasing hormone* (*CRH*), which in turn release *ACTH*. Cortisol is necessary for the action of growth hormone on fat cells. Cortisol also promotes the mobilization of amino acids from the muscle and connective tissue and stimulates their utilization for gluconeogenesis in the liver. Like growth hormone, cortisol inhibits glucose intake by nonessential user tissue, such as skeletal muscle, to spare the sugar for the brain and heart (see plate 121).

The hypothalamic glucostat senses the compensatory increase in blood sugar provided by these neurohormonal mechanisms and prevent further release of growth hormone and cortisol until the blood sugar level falls again. Meanwhile, during the immediate postabsorptive phase, *insulin* will be released to stimulate glucose entry into tissues, and *glucagon* secretion will be suppressed to decrease glycogenolysis in the liver. Later, when the glucose level drops, glucagon will be released to increase glycogenolysis, making glucose available. The nervous system does not play an important role in release of the pancretic hormones, which by themselves carry out much of the routine compensatory mechanisms to keep blood glucose constant (see plates 117, 126).

EFFECTS OF SEVERE HYPOGLYCEMIA. If starvation continues and all these restorative mechanisms fail, the blood glucose level inevitably falls below the critical limits as consumption by the heart and brain continues. Then the heart begins to weaken, and nervous and conscious activities become disturbed. Speech becomes slurred and movement uncoordinated, convulsions may occur. Further decline in blood glucose will cause loss of consciousness and coma. Eventually, the loss of activity of the brain medullary centers will stop respiration and cause death.

CN: Use red for A, and the same colors as the previous page for glycerol (C), liver (O), and glycogen (P). Use a dark color for B.
1. Begin with number one in the upper left corner, indicating a drop in the blood glucose level (A¹). Color this portion of the circulatory system (A, A¹) moving clockwise and down to the middle of the page.
2. Color the hypothalamus (B) and glucostat (A²). Color no. 2 and follow it into the liver at the bottom, but do not color the reactions within the liver just yet.
3. Do the same with 3, 4, and 5.
4. Color the conversion of substances in the liver to glucose, and the release of glucose into the blood circulation; follow this compensatory response to low blood sugar.

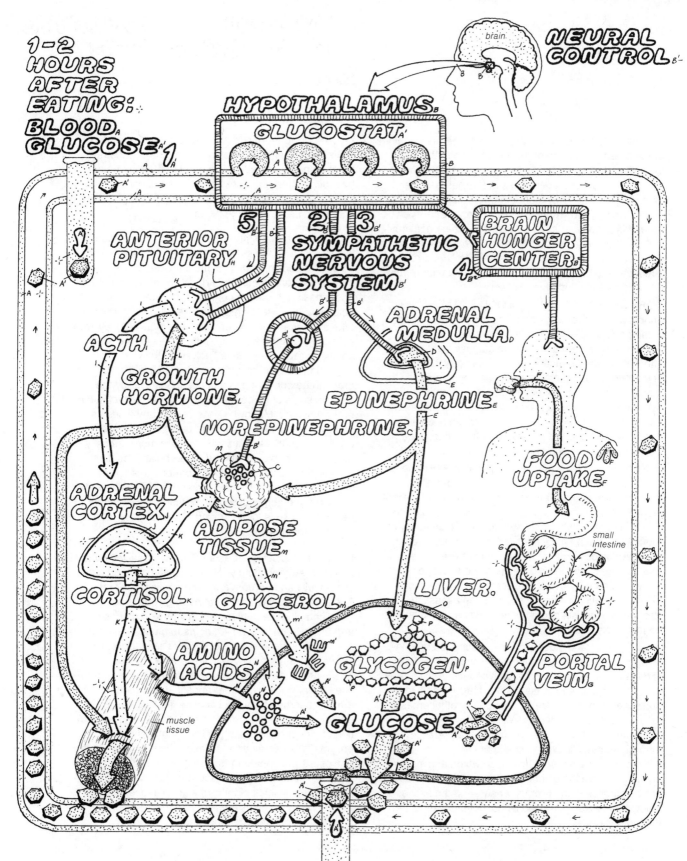

HYPOTHALAMUS

GLUCOSTAT

ANTERIOR PITUITARY

SYMPATHETIC NERVOUS SYSTEM

BRAIN HUNGER CENTER

ACTH

GROWTH HORMONE

ADRENAL MEDULLA

EPINEPHRINE

NOREPINEPHRINE

ADRENAL CORTEX

ADIPOSE TISSUE

FOOD UPTAKE

small intestine

CORTISOL

GLYCEROL

LIVER

PORTAL VEIN

AMINO ACIDS

GLYCOGEN

muscle tissue

GLUCOSE

1-2 hours after a meal, the continuous use of Glu by tissues, especially the brain, results in the relative reduction in blood Glu (1) levels. This is detected by "glucostat" neurons in the hypothalamus which initiate a series of compensatory responses. Sympathetic activation liberates norepinephrine from nerve fibers (2) and epinephrine from the adrenal medulla (3). Epinephrine stimulates glycogenolysis in the liver. Both catecholamines promote lipolysis in adipose tissue, providing fatty acids and glycerol. Mobilized fatty acids are used by heart and muscle, sparing Glu for the brain and raising blood Glu. Glyercol is converted to Glu in the liver. Later, activation of hypothalamic hunger centers (4) leads to food ingestion, increasing Glu intake, and the heptaic Glu pool. If food intake is further delayed, the hypothalamus stimulates release of GH and ACTH from pituitary (5). GH promotes lipolysis in adipose tissue and decreases Glu uptake in muscle. ACTH stimulates cortisol release. Cortisol is necessary for action of GH in fat cells. It also decreases Glu uptake by muscle and stimulates gluconeogenesis in the liver. Cortisol also increases protein catabolism in muscle to provide amino acids for gluconeogenesis in the liver. Increased Glu output by the liver and decreased Glu uptake by muscle will increase blood Glu levels, ensuring Glu availability to vital organs such as the brain and heart.

HORMONAL REGULATION OF BLOOD SUGAR

This plate decribes the integration of all the hormonal actions pertinent to the regulation of the blood sugar level. Because hypoglycemia is a potentially life threatening condition, most hormones act to increase the blood sugar level. Only one hormone, insulin, is specifically involved in lowering the blood sugar, and, even here, this action is not the hormone's primary goal but the result of its action. Being a center for storage and production of glucose, the liver serves as the target for almost all the hormones involved in blood sugar regulation and carbohydrate metabolism.

HORMONES THAT LOWER BLOOD SUGAR. The principal hypoglycemic hormone is insulin, produced in the pancreas' islets of Langerhans. Secreted in response to an increase in blood glucose level shortly after meals, insulin increases glucose entry into muscle and fat tissue and promotes glycogen synthesis and storage in the liver. The net result of these actions is a decrease in the blood sugar (see plate 117). Elevated levels of thyroid hormones can also cause hypoglycemia by increasing the metabolic rate, but they are not intended for regulation of blood sugar (see plate 113).

HORMONES THAT RAISE BLOOD SUGAR. Hormones that act to raise the blood sugar level are glucagon from the pancreatic islets, epinephrine from the adrenal medulla, growth hormone from the pituitary, and cortisol from the adrenal cortex. Epinephrine and glucagon act rapidly and are primariy intended for short-term regulation of blood sugar level, such as between meals. The actions of other hormones (e.g., cortisol and growth hormone) have longer term effects, making them important for times of stress, such as fasting, strenuous exercise, and immobility, when food intake is considerably delayed.

Epinephrine and glucagon have a common mechanism of action to increase the blood sugar level. Both stimulate liver cells to increase glycogenolysis, thereby mobilizing glucose (see plates 117, 119). These two hormones, though different chemically, both increase the concentration of cyclic-AMP in the liver cells. Acting as the "second messenger," cyclic-AMP, through an amplification cascade of effects, activates the liver enzyme phosphohydrolase which acts on the glycogen, liberating glucose molecules. Soon the pool of free glucose in the liver increases, and the excess glucose is secreted into the blood, compensating for the hypoglycemia.

All other hormones that increase blood sugar do so indirectly, either by increasing the levels of substrates for gluconeogenesis (e.g., glycerol and amino acids) or by reducing the entry or utilization of glucose in certain tissues (e.g., muscle), thereby sparing blood glucose and raising its level in the blood.

Cortisol promotes protein catabolism in peripheral tissues such as the skeletal muscle, liberating amino acids. In addition, cortisol stimulates the synthesis of certain liver enzymes, those of deamination and gluconeogenesis, which can convert the liberated amino acids into glucose. Cortisol also decreases glucose uptake by tissues such as muscle (see plate 121). Growth hormone from the anterior pituitary acts on fat cells of adipose tissue to increase lipolysis of triglycerides, mobilizing glycerol and fatty acids (see plate 112). Glycerol can be converted to glucose in the liver by the reverse steps of glycolysis, raising glucose level in the liver and blood. Meanwhile, the use of fatty acids as fuel by muscle, heart and liver spares the glucose for consumption by those tissues that are more critically dependent on this substance. Catecholamines also have actions similar to growth hormone in adipose tissue and on carbohydrate metabolism (see plate 119). However, the action of growth hormone takes longer to develop and lasts longer, being in the long run more effective for survival.

LIVER IN GLUCOSE HOMEOSTASIS. The liver is the major organ regulating the homeostasis of blood sugar specifically and carbohydrate metabolism generally. The liver has enzymes that convert glycogen to glucose and glucose to glycogen, glycerol to glucose and the reverse, and amino acids to glucose and the reverse. But the liver cannot synthesize glucose from fatty acids, a deficiency shared by all animal cells. The liver, via its special connection to the small intestine (portal vein), has direct access to the carbohydrates absorbed from the intestine, making it at once a center for the synthesis, delivery, storage, and production of glucose. Because the liver contains a special enzyme, glucose-6-phosphatase, which hydrolyzes the glucose-6-phosphate to free glucose, it is the only organ in the body that can secrete glucose into the blood when its level of this substance exceeds the blood level. This gives the liver the unique roles of glucose exchanger and glucostat.

CN: All the titles to be colored appear on the preceding page and should receive the same colors, though the letter labels may differ.
1. Begin in the upper left corner with the long arrows representing various hormones, glycerol (F), and amino acids (G).

2. Color the three primary users of blood glucose in the upper right.
3. Color the two hormones that lower glucose levels.
4. Color the hormonal influences on the liver, in the lower panel.

HORMONES THAT RAISE BLOOD GLUCOSE LEVELS*

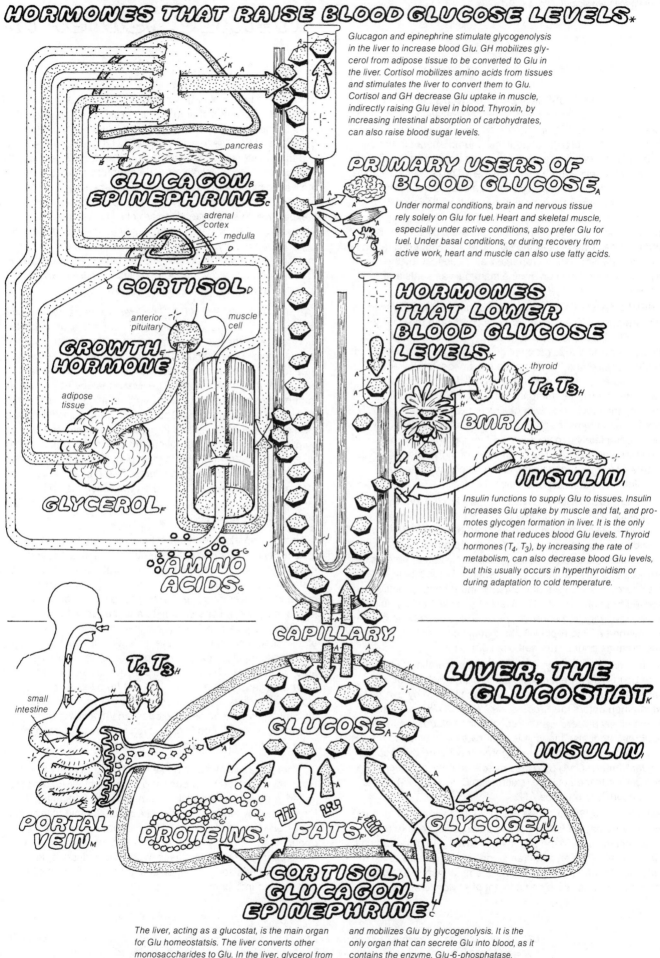

Glucagon and epinephrine stimulate glycogenolysis in the liver to increase blood Glu. GH mobilizes glycerol from adipose tissue to be converted to Glu in the liver. Cortisol mobilizes amino acids from tissues and stimulates the liver to convert them to Glu. Cortisol and GH decrease Glu uptake in muscle, indirectly raising Glu level in blood. Thyroxin, by increasing intestinal absorption of carbohydrates, can also raise blood sugar levels.

PRIMARY USERS OF BLOOD GLUCOSE

Under normal conditions, brain and nervous tissue rely solely on Glu for fuel. Heart and skeletal muscle, especially under active conditions, also prefer Glu for fuel. Under basal conditions, or during recovery from active work, heart and muscle can also use fatty acids.

HORMONES THAT LOWER BLOOD GLUCOSE LEVELS*

Insulin functions to supply Glu to tissues. Insulin increases Glu uptake by muscle and fat, and promotes glycogen formation in liver. It is the only hormone that reduces blood Glu levels. Thyroid hormones (T_4, T_3), by increasing the rate of metabolism, can also decrease blood Glu levels, but this usually occurs in hyperthyroidism or during adaptation to cold temperature.

pancreas

GLUCAGON
EPINEPHRINE

adrenal cortex
medulla

CORTISOL

anterior pituitary
muscle cell

GROWTH HORMONE

adipose tissue

GLYCEROL

AMINO ACIDS

T_4 T_3 thyroid

BMR

INSULIN

CAPILLARY

small intestine

T_4 T_3

GLUCOSE

LIVER, THE GLUCOSTAT

INSULIN

PORTAL VEIN

PROTEINS   FATS   GLYCOGEN

CORTISOL
GLUCAGON
EPINEPHRINE

The liver, acting as a glucostat, is the main organ for Glu homeostatsis. The liver converts other monosaccharides to Glu. In the liver, glycerol from fat catabolism, amino acids from protein catabolism, lactate from anaerobic glycolysis, can be converted to Glu. Liver stores Glu as glycogen and mobilizes Glu by glycogenolysis. It is the only organ that can secrete Glu into blood, as it contains the enzyme, Glu-6-phosphatase. Numerous hormones, including insulin, glucagon, cortisol, epinephrine, and thyroid act on the liver to influence these metabolic processes.

METABOLISM OF FAT

ADIPOSE TISSUE; USES OF FATS. Body *fats* serve as fuels, storage for fuels, thermal and electrical insulations, and structural components of cellular membranes. The structural fats (e.g., the phospholipids found in cell membranes) cannot be utilized for energy. Storage or depot fats are stored in the cytoplasm of *fat cells*, which comprise the *adipose tissue*. Despite the common misconception, adipose tissue is not inert; it is in fact an active and dynamic tissue that continuously forms and degrades fats (see plate 132). Adipose tissue is found in the abdominal cavity, within or around some organs (muscle and the heart), and under the skin. The subcutaneous fat has great thermal insulation properties, as evidenced by the thick layers of it found in such marine mammals as seals and whales.

CHEMISTRY OF FATS. *Triglycerides* (*neutral fats*) are the fats that are stored. They are esters of *glycerol* and three *fatty acids* (FA). FA are long hydrocarbon chains, each containing a single carboxylic acid group at one end. The longer the chain and the smaller the number of double bonds, the lower the fluidity state of the FA and the associated triglycerides. In the body, the most commonly occurring FA are palmitic, stearic, and oleic acids with chains between fourteen and sixteen carbon atoms long. Triglycerides can be broken down either completely to glycerol and FA or incompletely to FA and mono-or diglycerides. The breakdown of triglycerides (*lipolysis*) is catalyzed by various *lipase* enzymes occurring in the intestine, liver, and adipose tissue.

FATS AS A SOURCE OF ENERGY. Fats are the ideal substance for storing fuel energy because per unit weight they occupy less volume and produce more energy (ATP) than carbohydrates or proteins. In fact, 1 g of fat produces two and a half times more energy than 1 g of carbohydrate. Some tissues are well equipped to utilize FA for energy. For example, 60% of the energy requirement of the *heart*, under basal conditions, is derived from fats, chiefly FA. Skeletal muscle, too, especially during recovery from strenuous exercise, utilizes FA to obtain the needed ATP to replenish the exhausted supply of glycogen and creatine phosphate (see plate 23).

The products of triglyceride lipolysis, glycerol and FA, can both be utilized for energy production. Glycerol can be converted to intermediates of glycolysis and then to pyruvate, which then enters the *Krebs cycle* to form *ATP* (see plate 6). Alternatively glycerol can be converted to *glucose* in the liver (gluconeogenesis); glucose is used by tissues such as the brain for fuel. To liberate their energy, the FA are degraded to *acetate* (*acetyl CoA*) by a process called *beta-oxidation*. The acetyl CoA is then utilized (oxidized to CO_2 and H_2O) in the mitochondria to produce ATP.

FAT METABOLISM IN ADIPOSE TISSUE. After a meal rich in carbohydrates, the fat cells of the adipose tissue take up the abundant glucose in the blood and convert it to glycerol and FA (lipogenesis). The glycerol (an *alcohol*) and FA (*acids*) are then *esterified* to form triglycerides. After a meal rich in fats but low in carbohydrates, the blood content of *chylomicrons* (particles

transporting fats in blood) containing triglycerides increases. Within the capillaries of adipose tissue and liver, an enzyme called *lipoprotein lipase* hydrolyzes the glycerides, freeing glycerol and FA. These are taken up by fat cells and re-esterified to form storage triglycerides. Triglycerides with sufficiently long chains tend to solidify and are therefore easily stored. Increased storage of solid fats increases the size of fat cells, resulting in the formation of thick fat pads of the adipose tissue. If excessive, this condition leads to obesity (see plate 132). When stimulated by catecholamines and other hormones (see plate 128), the triglycerides are lipolyzed by lipase enzymes, mobilizing the glycerol and FA into the blood. The mobilized FA are then consumed by the heart, muscle and liver. Glycerol is usually taken up by the liver to make new glucose.

FAT METABOLISM IN LIVER. The *liver*, like the adipose tissue, is capable of forming, degrading and storing fats, although the fat granules in the liver hepatocytes are not intended for long-term storage. The particular importance of the liver lies in its great ability for metabolic interconversion between the fats, carbohydrates, and proteins. The liver hepatocytes contain all the enzymes required for these transformations. For example, excess glucose can be metabolized into fatty acids which are then either incorporated into triglycerides or mobilized for consumption by tissues. Glycerol can be converted to glucose by *reverse glycolysis* and then to *glycogen*. FA can be converted to some *amino acids* and vice versa. These amino acids are then used to make *proteins*. The only reaction that the liver (and in fact animal cells in general) cannot do in this regard is convert FA into glucose.

Another important function of the liver in fat metabolism is in its ability to make *cholesterol* (see plate 129) and to form *ketone bodies*. In response to carbohydrate deficiency in the diet or within the cells (as occurs in uncontrolled diabetes mellitus), the liver degrades its own supply of FA and that provided by the adipose tissue to acetate (acetyl CoA). When the available pool of acetyl CoA exceeds the loading capacity of the mitochondria, the acetate molecules are instead condensed together to form compounds such as acetoacetic acid, acetone, and other keto acids, collectively called the ketone bodies. Ketone bodies leak out from the liver into the blood, where they are *excreted* in the *kidneys*. Excessive amounts of ketone bodies in blood leads to *ketosis* and metabolic acidosis, conditions which may be fatal, as in the case of untreated insulin deficiency diabetes (Type 1).

In normal adults, ketone bodies are little utilized for energy. However, in newborns, pregnant women and individuals subjected to prolonged starvation, many tissues, particularly the brain, adapt by increasing their rate of uptake and utilization of the ketone bodies for energy. This recently discovered phenomenon not only accounts for the continued function of the brain (an organ that usually uses only glucose) in starvation, but also for the lack of ketone toxicity in children and starved individuals.

CN: Use red for I and yellow for C. Use light colors for glycerol (A) and fatty acids (B). Use light blue for D.
1. Color the upper materials on the structure of fat. Note that in the cube representing triglycerides in solid form, all letters receive the triglyceride color (C).

2. Color the metabolism of fat following the numbered sequence, starting with the consumption of the hamburger. Note that step (7) ends with the formation of glucose in the liver. Then proceed to (8) in the lower cell. After completing (13) return to the glucose molecule for (14).

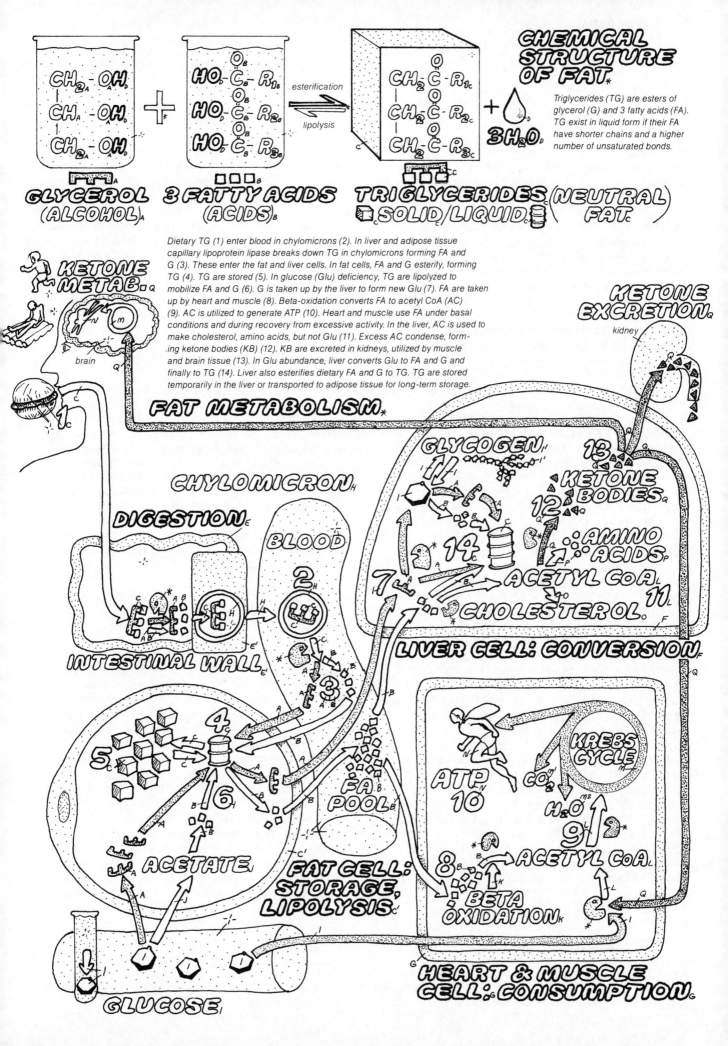

CHEMICAL STRUCTURE OF FAT*

Triglycerides (TG) are esters of glycerol (G) and 3 fatty acids (FA). TG exist in liquid form if their FA have shorter chains and a higher number of unsaturated bonds.

CH_2-OH + $HO-C-R_1$ → esterification / lipolysis → CH_2-C-R_1 + $3H_2O$

GLYCEROL (ALCOHOL) + **3 FATTY ACIDS (ACIDS)** → **TRIGLYCERIDES** SOLID/LIQUID (NEUTRAL FAT)

FAT METABOLISM*

Dietary TG (1) enter blood in chylomicrons (2). In liver and adipose tissue capillary lipoprotein lipase breaks down TG in chylomicrons forming FA and G (3). These enter the fat and liver cells. In fat cells, FA and G esterify, forming TG (4). TG are stored (5). In glucose (Glu) deficiency, TG are lipolyzed to mobilize FA and G (6). G is taken up by the liver to form new Glu (7). FA are taken up by heart and muscle (8). Beta-oxidation converts FA to acetyl CoA (AC) (9). AC is utilized to generate ATP (10). Heart and muscle use FA under basal conditions and during recovery from excessive activity. In the liver, AC is used to make cholesterol, amino acids, but not Glu (11). Excess AC condense, forming ketone bodies (KB) (12). KB are excreted in kidneys, utilized by muscle and brain tissue (13). In Glu abundance, liver converts Glu to FA and G and finally to TG (14). Liver also esterifies dietary FA and G to TG. TG are stored temporarily in the liver or transported to adipose tissue for long-term storage.

KETONE METAB. — brain

KETONE EXCRETION — kidney

CHYLOMICRON

DIGESTION

BLOOD

INTESTINAL WALL

GLYCOGEN

13 KETONE BODIES

12 AMINO ACIDS

14

7

ACETYL CoA

CHOLESTEROL 11

LIVER CELL: CONVERSION

5

4

6

ACETATE

FA POOL

FAT CELL: STORAGE, LIPOLYSIS

3

2

KREBS CYCLE

ATP 10

CO_2

H_2O

9

8 BETA OXIDATION

ACETYL CoA

GLUCOSE

HEART & MUSCLE CELL: CONSUMPTION

REGULATION OF FAT METABOLISM

INTERRELATION OF FATS AND CARBOHYDRATES. Like carbohydrates, the principal function of the body *fats* is to provide fuel (ATP) for cells. In fact, the metabolism of fats and carbohydrates is intimately linked, and many of the neural and hormonal factors that help regulate *carbohydrate metabolism* also help regulate *fat metabolism*.

REGULATION OF LIPOLYSIS. To illustrate the regulation of fat metabolism, let us consider an athlete involved in strenuous exercise or a person in a state of starvation. In these conditions, food intake is delayed, and the blood sugar level is threatened. A reduction in blood sugar (*glucose*) triggers the responses of the *hypothalamic glucostat* (see plates 125, 132). As a result, the *sympathetic nervous system* is activated, causing the sympathetic nerves and adrenal medulla to release the *catecholamines*: epinephrine and *norepinephrine*, that stimulate the *fat cells* of the adipose tissue to increase *lipolysis* (see plate 119). This action is exerted by stimulation of a special enzyme, called *hormone sensitive lipase*. The stimulation is not direct, but through increase in the level of cyclic-AMP (second messenger) in the fat cells; cyclic-AMP, in turn, activates the hormone-sensitive lipase. The lipase catalyzes the conversion of stored *triglycerides* to *fatty acids* (FA) and *glycerol*. The mobilized FA enter the blood, furnishing fuel for the *heart* and *muscle*. Gycerol is taken up by the *liver* and converted to glucose (gluconeogenesis). These events will improve the hypoglycemia, increasing the glucose supply to the brain.

GROWTH HORMONE. If exercise or starvation continues beyond a few hours, the hypothalamic glucostat will activate the release of *growth hormone* (GH) from the pituitary by increasing the secretion of *growth hormone releasing hormone* (GRH) from the *hypothalamus*. GH has a very potent and prolonged lipolytic action on the fat cells, resulting in mobilization of FA and glycerol (see plate 112). Although this effect of GH is similar to that of catecholamines, the exact cellular mechanism of this particular action of GH is not known.

CORTISOL. Another hormone that enhances lipolysis is *cortisol*, produced in the adrenal cortex. Cortisol is released in response to stimulation by the pituitary hormone, *ACTH*, which is in turn released in response to the release of the hypophysiotropin hormone, *CRH*, from the hypothalamus (see plates 111, 121). Cortisol's lipolytic actions are exerted in several ways. A very important way is the "permissive" action. Cortisol must be present before the lipolytic actions of both catecholamines and GH on the fat cells can be expressed. The nature of this permissive action of cortisol is not known. A second way is that cortisol imparts direct effects on tissues, increasing fat catabolism in adipose tissue and FA oxidation and *ketone body* formation in the liver. The second category of actions is probably part of the generalized action of cortisol in response to prolonged stress conditions (see plate 121).

INSULIN AND FAT DEPOSITION. In contrast to the above hormones, which tend to mobilize fat reserves, the pancreatic hormone insulin tends to stimulate the deposition of fat (see plate 117). Thus, after a meal rich in carbohydrates, insulin is secreted. One important insulin target is the fat cells, where it exerts two actions. One is to increase the entry of glucose into the cells, which increases the supply of a nutrient (glucose) that the fat cells are well equipped to utilize for lipogenesis. As a result, FA and glycerol are formed and further esterified to triglycerides. The second action of insulin is to inhibit the hormone-sensitive lipase, resulting in the suppression of lipolysis. Thus, the fat cells under stimulation by insulin tend to deposit fat. This is an economic and adaptive reaction to increase availability of food because, as previously mentioned, the stored fat is mobilized in times of need to provide fuel for the body.

OBESITY. One abnormality of fat metabolism occurs in certain individuals who are healthy but overeat and do not exercise or fast. Their lipolytic processes are inhibited, and fat reserves build up above and beyond the normal range, leading to *obesity* (see also plate 132). Interestingly, obesity does not develop in all individuals. In those who are genetically predisposed to it, there is presumably a higher metabolic tendency in the adipose tissue to form and/or deposit fat and less of a tendency to lipolyze it during normal stresses. The exact mechanisms involved in the pathogenesis of obesity are not understood. Nevertheless, obesity is associated with increased risk of *high blood pressure* (hypertension) and *diabetes*, so it is a serious health risk that reduces life expectancy.

WEIGHT LOSS. Physiologically, the soundest way to lose the excess fat is to *exercise*: i.e., to allow the body to do what it does naturally during strenuous muscular activity, to activate the lipolytic hormones which mobilize fat reserves. This approach is particularly effective when *intake of fats and simple sugars* in the diet is simultaneously reduced, avoiding insulin activation. Under these conditions, fat mobilization is encouraged while fat deposition is discouraged.

Exercise may not be appropriate in extremely obese subjects because it will put undue stress on the heart. These subjects may attempt to reduce by eliminating all sources of fat and carbohydrates from the diet (fasting or starvation). This should activate the hormonal mechanisms for fat loss. Because starvation in obese subjects may lead to ketosis, which is a dangerous condition (metabolic acidosis), the treatment should be carried out under a physician's care.

Administration of *thyroid hormones* can also lead to fat loss because these hormones increase the metabolic rate, resulting in the depletion and loss of carbohydrate and fat reserves as heat. In humans, thyroid hormones are not involved in regulation of lipolysis under normal conditions, although they may play a permissive role.

CN: Use red for F, purple for E, and the same colors for fatty acids (B), glycerol (C), triglycerides (D), and liver (L) as were used on the previous page.
1. begin at the top and follow numbers 1-5 to both the fat

(D¹) and liver (L) cells without coloring the actions in those cells. Then color 6 and 7 and follow the elements floating in the capillary to the liver cell (L) on the right.
2. Color the bottom panel, beginning with number one.

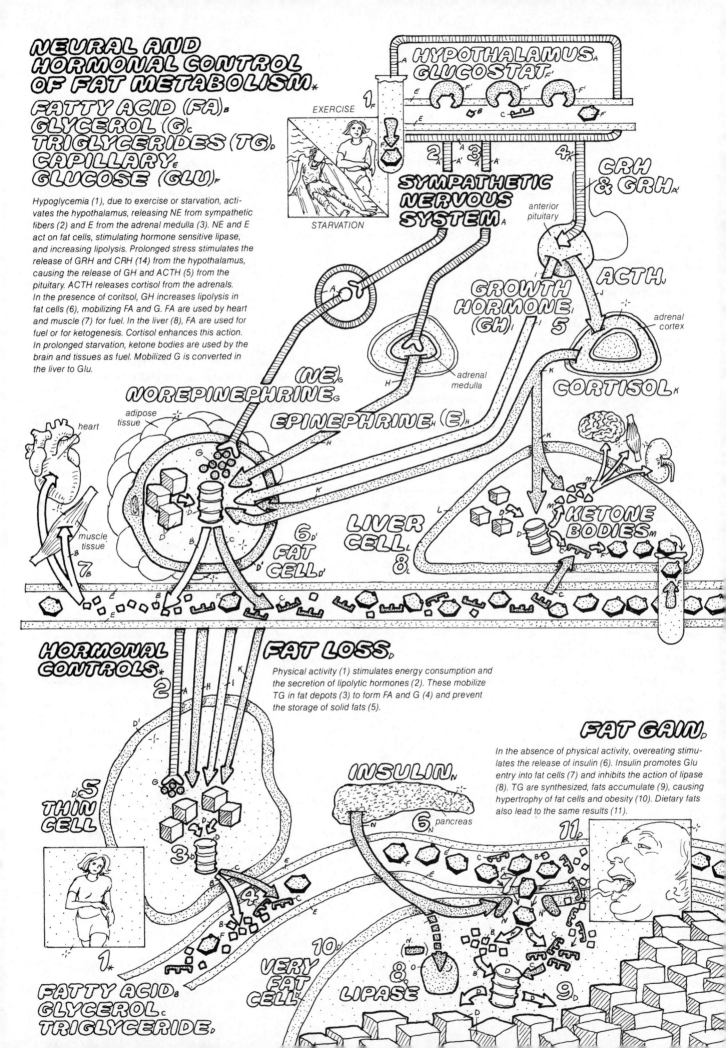

NEURAL AND HORMONAL CONTROL OF FAT METABOLISM*

FATTY ACID (FA)ʙ
GLYCEROL (G)c
TRIGLYCERIDES (TG)ᴅ
CAPILLARYᴇ
GLUCOSE (GLU)ꜰ

Hypoglycemia (1), due to exercise or starvation, activates the hypothalamus, releasing NE from sympathetic fibers (2) and E from the adrenal medulla (3). NE and E act on fat cells, stimulating hormone sensitive lipase, and increasing lipolysis. Prolonged stress stimulates the release of GRH and CRH (14) from the hypothalamus, causing the release of GH and ACTH (5) from the pituitary. ACTH releases cortisol from the adrenals. In the presence of coritsol, GH increases lipolysis in fat cells (6), mobilizing FA and G. FA are used by heart and muscle (7) for fuel. In the liver (8), FA are used for fuel or for ketogenesis. Cortisol enhances this action. In prolonged starvation, ketone bodies are used by the brain and tissues as fuel. Mobilized G is converted in the liver to Glu.

HYPOTHALAMUS GLUCOSTATA

EXERCISE

STARVATION

SYMPATHETIC NERVOUS SYSTEMA

anterior pituitary

CRH & GRHA'

ACTHⱼ

GROWTH HORMONE (GH) 5

adrenal cortex

NOREPINEPHRINE (NE)ɢ

EPINEPHRINE (E)ₕ

adrenal medulla

CORTISOLₖ

heart

adipose tissue

muscle tissue 7ʙ

6ᴅ' FAT CELLᴅ'

LIVER CELL 8ʟ

KETONE BODIESₘ

HORMONAL CONTROLS*

FAT LOSSᴅ

Physical activity (1) stimulates energy consumption and the secretion of lipolytic hormones (2). These mobilize TG in fat depots (3) to form FA and G (4) and prevent the storage of solid fats (5).

THIN CELL

FAT GAINᴅ

In the absence of physical activity, overeating stimulates the release of insulin (6). Insulin promotes Glu entry into fat cells (7) and inhibits the action of lipase (8). TG are synthesized, fats accumulate (9), causing hypertrophy of fat cells and obesity (10). Dietary fats also lead to the same results (11).

INSULINₙ

pancreas

VERY FAT CELL

LIPASE

FATTY ACIDʙ
GLYCEROLc
TRIGLYCERIDEᴅ

PHYSIOLOGY OF CHOLESTEROL AND LIPOPROTEINS

CHEMISTRY, SOURCES, AND USES OF CHOLESTEROL. *Cholesterol* is a fatty substance present in animal tissues and foods of animal origin. It is a *sterol* with a complex ring structure that is synthesized from *acetate*. In the body, the synthesis occurs chiefly in the *liver*, but other tissues, particularly the adrenal cortex and gonads, are also capable of making cholesterol. Cholesterol is not used in the body for energy. Instead, it participates in several other important functions. It serves as the precursor for *bile acids* that are formed in the liver and secreted in the *bile* to facilitate intestinal fat digestion. In the gonad and adrenal cortex, cholesterol is used to synthesize all *steroid* hormones. In the *skin*, it is used to form vitamin D_3, a reaction requiring ultraviolet sun radiation.

Cholesterol is found in abundance in the *nerve tissue*, where it is a component of the *myelin sheath* that electrically insulates the axons. In the *corneum* of skin (the outer keratinized layer), cholesterol helps minimize evaporation of body water as well as making the skin waterproof. Last but not least, cholesterol is a component of the *membranes* of cells and their organelles, helping to stabilize the *phospholipids*.

Cholesterol in the body is of two possible origins: exogenous (i.e., derived from diet) and endogenous (i.e., derived from synthesis in the tissue, chiefly the liver). Cholesterol in the diet is derived solely from foods of animal origin (egg yolk, liver and fatty meats, cheese). Because cholesterol is highly fat soluble, it is absorbed, in the intestine, with other fats as part of the *chylomicrons* that enter the circulation via the *lymphatic vessels* (see plate 75). Once in the capillaries of liver and adipose tissue, chylomicrons, being large lipoprotein particles with a core of fat wrapped in a coat of protein, are digested by the enzyme *lipoprotein lipase*. The *triglycerides* are delivered to the adipose tissue while the remnants, containing mainly cholesterol and phospholipids, are delivered to the liver cells (see plate 127).

LIVER AND CHOLESTEROL. In the liver, the dietary cholesterol acts on liver enzymes, inhibiting the synthesis of cholesterol in the hepatocytes. In this manner, the liver regulates the plasma cholesterol level. Most of the *labile* pool of cholesterol in the body is in the liver. Of this pool, most is used to form the bile salts (e.g., cholate) that facilitate fat digestion by *emulsification* of fats in the chyme (see plate 75). Only a small portion of the bile salts is excreted in the feces. Most is reabsorbed by the *enterohepatic circulation*. In the absence of dietary cholesterol, the liver makes its own cholesterol, starting with acetate (acetyl-CoA). *Saturated* fatty acids in the diet promote this synthesis; *polyunsaturated* fatty acids reduce it. This is the principal reason for recommending polyunsaturated fats in one's diet.

CHOLESTEROL AND LIPOPROTEINS. The liver also supplies cholesterol to most tissues. Cholesterol is packed in lipoprotein particles that resemble chylomicrons in basic structure but have different composition of fats and proteins and are of different size and density. Cholesterol made by the liver for tissues is transported in the largest of lipoprotein particles (*very low density lipoproteins, VLDL*). In the *plasma*, these are transformed to smaller lipoproteins (*intermediate density lipoproteins, IDLP*, and *low density lipoproteins, LDL*) by the actions of enzymes. Cholesterol delivered directly to tissues is in the LDL form. Once inside the tissue cells, cholesterol is utilized for the variety of functions previously outlined. The excess cholesterol is packed in the smallest of lipoprotein particles (*high density lipoprotein, HDL*) and transported back to the liver for processing.

HORMONES AND CHOLESTEROL. Hormones influence the cholesterol level in plasma. Thyroid hormones decrease plasma cholesterol by increasing its uptake by the liver and tissues. The female sex steroids, *estrogens*, decrease cholesterol level; the male steroids, *androgens*, increase it. The mechanisms of these hormonal actions that may relate to the higher incidence of arteriosclerosis in men are not known.

CHOLESTEROL AND ARTERIOSCLEROSIS. Cholesterol plays an important role in causing *atherosclerosis*, a specific type of *arteriosclerosis* (*hardening* of the arteries). These diseases are responsible for nearly half of all deaths, mostly in *men* and in the *elderly*. Cholesterol is deposited in large amounts in the victim's arterial wall. When the inner wall of an *artery* is damaged, *platelets* adhere to the site of damage, stimulating *fibrosis*. Plasma cholesterol is deposited on these lesions, along with *calcium ions*, forming hard, calcified *cholesterol plaques* (atherosclerosis). These plaques lead to hardening of arterial walls and loss of elasticity and responsiveness to changes in blood pressure. Plaques in the *kidney* may lead to chronic high blood pressure (*hypertension*). The plaques can protrude in the arterial lumen, reducing blood flow to a region (*ischemia*) or facilitating formation of *blood clots*. These clots can block blood flow to a region (thrombosis). Most heart attacks and strokes are due either to atherosclerosis directly or to thrombosis caused by it (see plate 138).

Previously, it was widely believed that high cholesterol in the *diet* is the main determinant of this vascular disease. It is now believed that high levels of cholesterol in the *plasma*, particularly the cholesterol associated with the LDL, favor the pathogenesis of atherosclerosis. Presumably, the cholesterol in the LDL, coming from the liver to the tissues, is more likely to be deposited at the sites of arterial wall lesions, and cholesterol in the HDL, traveling to the liver, is less likely to contribute to the lesion. Dietary cholesterol may or may not contribute to the disease, depending on how the individual's liver is able to regulate the plasma cholesterol level and the production of LDL. Women, probably because of their higher estrogen levels, have lower LDL and higher HDL contents than men, accounting for their lower incidence of the disease.

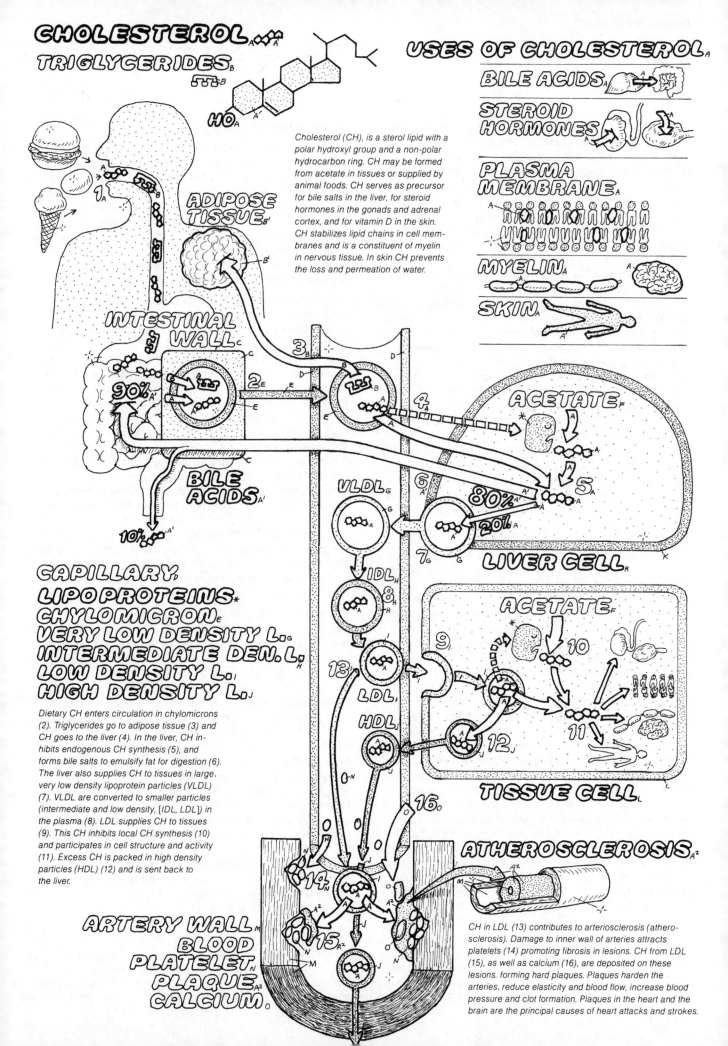

CHOLESTEROL
TRIGLYCERIDES
HO

USES OF CHOLESTEROL
BILE ACIDS
STEROID HORMONES
PLASMA MEMBRANE
MYELIN
SKIN

Cholesterol (CH), is a sterol lipid with a polar hydroxyl group and a non-polar hydrocarbon ring. CH may be formed from acetate in tissues or supplied by animal foods. CH serves as precursor for bile salts in the liver, for steroid hormones in the gonads and adrenal cortex, and for vitamin D in the skin. CH stabilizes lipid chains in cell membranes and is a constituent of myelin in nervous tissue. In skin CH prevents the loss and permeation of water.

ADIPOSE TISSUE

INTESTINAL WALL

ACETATE

90%

2

3

4

5

80%

20%

BILE ACIDS

10%

VLDL
6
7

LIVER CELL

CAPILLARY
LIPOPROTEINS
CHYLOMICRON
VERY LOW DENSITY L.
INTERMEDIATE DEN. L.
LOW DENSITY L.
HIGH DENSITY L.

IDL
8

ACETATE

9

10

11

LDL

13

HDL

12

Dietary CH enters circulation in chylomicrons (2). Triglycerides go to adipose tissue (3) and CH goes to the liver (4). In the liver, CH inhibits endogenous CH synthesis (5), and forms bile salts to emulsify fat for digestion (6). The liver also supplies CH to tissues in large, very low density lipoprotein particles (VLDL) (7). VLDL are converted to smaller particles (intermediate and low density, [IDL, LDL]) in the plasma (8). LDL supplies CH to tissues (9). This CH inhibits local CH synthesis (10) and participates in cell structure and activity (11). Excess CH is packed in high density particles (HDL) (12) and is sent back to the liver.

TISSUE CELL

16

ATHEROSCLEROSIS

ARTERY WALL
BLOOD
PLATELET
PLAQUE
CALCIUM

14

15

CH in LDL (13) contributes to arteriosclerosis (atherosclerosis). Damage to inner wall of arteries attracts platelets (14) promoting fibrosis in lesions. CH from LDL (15), as well as calcium (16), are deposited on these lesions, forming hard plaques. Plaques harden the arteries, reduce elasticity and blood flow, increase blood pressure and clot formation. Plaques in the heart and the brain are the principal causes of heart attacks and strokes.

METABOLISM OF PROTEINS

STRUCTURE, VARIETY AND SIGNIFICANCE. *Proteins* are of *primary* importance in cellular, tissue, and bodily structures and functions. All proteins are made of different combinations of about twenty naturally occurring *amino acids*, which vary in structure but share a common feature: the presence of a *carboxyl acid* and *amino* group. Amino acids can be joined by *peptide bonds*, forming *peptide chains*, hence *dipeptides*, *tripeptides*, *oligopeptides* ("oligo" = few), and *polypeptides* ("poly" = many). Proteins are basically large polypeptides of one or more chains. All body proteins are synthesized within cells from amino acids, using elaborate cellular and molecular machinery and codes (see plate 4). Of the twenty amino acids making up the proteins, the body can synthesize twelve, starting with glucose or fatty acids; the remaining eight must be provided in the diet (the dietary *essential amino acids*, e.g., leucine and tryptophan). Proteins are found in abundance in animal and plant foods. The sources particulary rich in essential amino acids are meats, egg white, milk, plant seeds, and nuts. Dietary proteins are broken down in the *digestive system*, yielding individual amino acids that are absorbed across the *intestinal mucosa*, reaching the *liver* via the *portal vein* (see plate 75).

ROLE OF THE LIVER IN PROTEIN METABOLISM. Within the liver cells, the amino acids form a pool that can be utilized to make the various proteins of the liver and blood as well as glucose, fats, and energy (ATP). The amino acids of the liver pool can be exchanged with a *second* similar pool in the *blood*, which in turn exchanges amino acids with a *third* pool within the *tissue cells*. The liver is a major center for the synthesis and degradation of amino acids and proteins. In addition to making proteins for its own needs (e.g., enzymes), the liver forms and secretes most of the *blood proteins*: *albumins* (transport of hormones and fatty acids, plasma osmotic pressure), *globulins* (enzymes, transport of hormones), and *fibrinogen* (blood clotting). The human liver is capable of forming up to 50 g of these proteins every day. In addition to their use as building blocks for proteins, the amino acids in the liver can also be utilized as fuel to obtain ATP. For this purpose, they are first converted by *deamination* to various *keto acids* such as *pyruvic* and alpha-keto-glutaric acids. These substances are then oxidized in the *Krebs cycle* to release energy and form ATP. Amino acids are approximately equal to carbohydrates in their capacity to release metabolic energy and form ATP. However, consistent with their function as the building blocks of proteins, amino acids are normally spared from metabolic oxidation, carbohydrates and fats being used preferentially. Only during starvation are the amino acids of the liver and tissues catabolized as fuels. If the diet is rich in protein, the excess amino acids in the hepatic pool will be converted to glucose (*gluconeogenesis*) or to *fatty acids* and *glycerol* (*lipogenesis*), from which glycogen and fats, respectively, can be formed for storage (see plates 121, 127). Here the amino acids are first converted to *keto acids* (e.g., pyruvate), which are then utilized to form glucose, fatty acids, and glycerol.

During deamination of amino acids, *ammonia* is formed. Ammonia can be toxic for the liver and, if it diffuses to the blood, or other tissues. The liver detoxifies ammonia by converting it into *urea*, a far less toxic, water soluble substance. To form urea, two molecules of ammonia react with one molecule of carbon dioxide. The actual formation of urea occurs in a complicated series of enzyme catalyzed reactions called the *Urea cycle*, involving the amino acids *ornithine*, *citruline*, and *arginine*, which act as intermediates. Urea diffuses into the blood and is excreted by the kidney.

METABOLISM OF PROTEINS IN TISSUE. For their growth, repair, or normal turnover of cellular proteins, tissues require a continuous supply of amino acids. They obtain these from the pool of amino acids in the blood, which in turn in equilibrium with that in the liver. The cells of each tissue make their own specific proteins (i.e., those characteristic of each cell). These and other general cellular proteins are continuously formed on *ribosomes* and broken down in *lysosomes*, the *turnover rate* depending on tissue and protein type. The liver enzymes show a high rate (a few hours); structural proteins show a slow rate (e.g., for bone collagen, a few months). Besides the general metabolic enzymes common to all cells, different tissues contain special proteins which perform unique functions. Thus, *antibodies*, the defense proteins, are secreted by the white blood cells (leukocytes). *Hemoglobin*, the oxygen transporting protein, is found in the red cells. *Collagen*, the most abundant protein in the body, is secreted by the bone and cartilage cells and fibroblasts. *Actin* and *myosin* are the contractile protein of muscle tissue. Only during extreme starvation are the tissue proteins utilized for fuel, and even then heart and brain proteins are spared.

HORMONAL REGULATION OF PROTEIN METABOLISM. Hormones profoundly influence protein metabolism. Thus, *growth hormone* and *insulin* increase the uptake of amino acids and protein synthesis in certain tissues such as muscle. *Thyroid hormones* also regulate protein metabolism. In the heart, *thyroxine* increases protein synthesis by increasing the number of ribosomes. In the liver and kidney, thyroid hormones induce the formation of specific proteins (e.g., Na-K-ATPase of the membrane pump). Thyroid and growth hormones are not only critical, but are synergistic in their anabolic effects on protein synthesis. Indeed, in the absence of thyroid and growth hormones, protein synthesis ceases, and growth is retarded or ceases entirely, resulting in various forms of *dwarfism* (see plates 112, 113). *Androgens* from the testes and adrenal glands stimulate protein synthesis, especially in bone and muscle. This action is very important for the growth spurt that occurs during puberty (see plate 144). Cortisol from the adrenal cortex increases protein catabolism in many tissues, but in the liver, it increases amino acid uptake and synthesis of some special enzymes for gluconeogenesis (see plate 121).

CN: Use a bright color for A, red for H, and very light colors for I and J.

1. Color the chemical formulas in the upper panel.
2. Color the four degrees of protein structure.
3. In the protein metabolism panel, follow the numbered sequence after coloring the liver (I) and tissue (J) cells, and the blood vessel (H).
4. In hormonal controls, note that growth hormone, somatomedins, and insulin operate together and are given a single color (N).

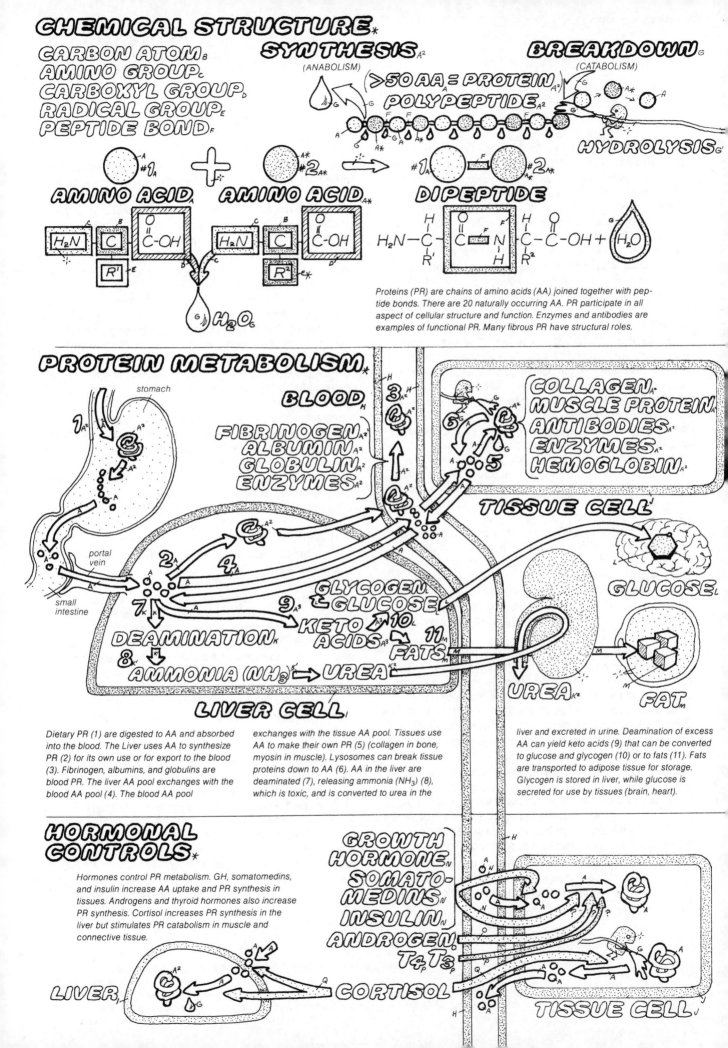

CHEMICAL STRUCTURE *

CARBON ATOM B
AMINO GROUP C
CARBOXYL GROUP D
RADICAL GROUP E
PEPTIDE BOND F

SYNTHESIS A²
(ANABOLISM)

>50 AA = PROTEIN
POLYPEPTIDE

BREAKDOWN G
(CATABOLISM)

HYDROLYSIS G'

AMINO ACID A + AMINO ACID A* → DIPEPTIDE

H_2O G

Proteins (PR) are chains of amino acids (AA) joined together with peptide bonds. There are 20 naturally occurring AA. PR participate in all aspect of cellular structure and function. Enzymes and antibodies are examples of functional PR. Many fibrous PR have structural roles.

PROTEIN METABOLISM

BLOOD

FIBRINOGEN
ALBUMIN
GLOBULIN
ENZYMES

COLLAGEN
MUSCLE PROTEIN
ANTIBODIES
ENZYMES
HEMOGLOBIN

TISSUE CELL

GLYCOGEN & GLUCOSE
KETO ACIDS
FATS

GLUCOSE

LIVER CELL

DEAMINATION
AMMONIA (NH_3) → UREA

UREA

FAT

Dietary PR (1) are digested to AA and absorbed into the blood. The Liver uses AA to synthesize PR (2) for its own use or for export to the blood (3). Fibrinogen, albumins, and globulins are blood PR. The liver AA pool exchanges with the blood AA pool (4). The blood AA pool exchanges with the tissue AA pool. Tissues use AA to make their own PR (5) (collagen in bone, myosin in muscle). Lysosomes can break tissue proteins down to AA (6). AA in the liver are deaminated (7), releasing ammonia (NH_3) (8), which is toxic, and is converted to urea in the liver and excreted in urine. Deamination of excess AA can yield keto acids (9) that can be converted to glucose and glycogen (10) or to fats (11). Fats are transported to adipose tissue for storage. Glycogen is stored in liver, while glucose is secreted for use by tissues (brain, heart).

HORMONAL CONTROLS *

Hormones control PR metabolism. GH, somatomedins, and insulin increase AA uptake and PR synthesis in tissues. Androgens and thyroid hormones also increase PR synthesis. Cortisol increases PR synthesis in the liver but stimulates PR catabolism in muscle and connective tissue.

GROWTH HORMONE
SOMATO-MEDINS
INSULIN
ANDROGEN
T_4, T_3
CORTISOL

LIVER

TISSUE CELL

METABOLISM, HEAT, AND METABOLIC RATE

FUELS, HEAT, AND WORK. The burning of most organic *fuel* substances in the air (i.e., combining them with *oxygen*: oxidation, oxygenation) causes oxidative catabolism of the substances, producing carbon dioxide and water. However, in this process all the *energy* stored in the chemical bonds of the fuel substance is released as *heat*, with no *work* being produced. Thus, the *efficiency* (i.e., ability to generate work from a form of energy) of this process is zero. In a powerhouse, a fuel (e.g., coal) is burned to generate heat; heat drives the generator turbines; turbines produce electricity, which is put at the service of homes and shops to be utilized for a variety of work. Here heat, as a form of energy, is converted into usable energy (i.e., work).

METABOLIC HEAT AND BODY WORK. The human body is also a machine. To carry out its vital functions, it has to perform work. For this, the body requires energy. Like other heat engines, the body obtains this energy by consuming and burning fuels (e.g., carbohydrates and fats). The oxidation of *foodstuffs* releases heat; however, in contrast to the powerhouse in the above example, the body is not able to convert the liberated heat directly into work. Instead, as explained in plates 5 and 6, body cells can couple the oxidation of foodstuff with the generation of ATP, the energy-rich chemical intermediate. ATP is then used for the variety of chemical (e.g., synthesis), mechanical (e.g., muscle contraction), and electrical (e.g., nerve activity) *body functions* (*work*) that the body cells need to carry out to survive and thrive. Like all machines, the body is not totally efficient. Some of the energy liberated during the oxidation of fuel substances is released as heat (metabolic heat), which is not wasted entirely because it can be utilized to keep the body *warm*. This is very useful in cold-blooded animals (*poikilotherms*) and an absolutely essential requirement in warm-blooded animals (*homeotherms*).

Even the energy utilized to do cellular work is ultimately converted to heat because not only the hydrolysis of ATP that occurs during work generates heat, but some of the energy used to do the actual work is also converted to heat (e.g., muscle contraction creates *friction*, friction creates heat). The energy values of food substances (their usefulness for the body energy needs) are also best measured in terms of their fuel heat value.

CALORIE AND CALORIMETRY. Based on the above-mentioned universal utility of heat, it is a common practice to measure all bodily energy processes in terms of *heat units* (i.e., the calorie). One *Calorie* (with a capital C = 1 kilocalorie) is defined as the amount of heat required to raise the temperature of 1 g of water by one degree centigrade.

The most accurate method for measuring body energy needs is by *direct calorimetry* (i.e., measuring the exact amount of heat the body produces during a particular time interval). A subject, naked and at rest, is placed in a *room calorimeter*. The calorimeter is completely insulated to minimize heat exchange with the outside. The heat released from the body is used to warm a stream of water running within a tube in the room. The increase in the temperature of water (outflow minus inflow) after conversion to calories is equivalent to the amount of calories generated by the body.

Another method to measure caloric production is by *indirect calorimetry* (*spirometry*). Here, the amount of oxygen consumed is measured using a *spirometer*. Oxygen is inhaled from the spirometer tank through a mouthpiece. The decline in oxygen content of the tank is registered by a device (e.g., a kymograph). The CO_2 produced is absorbed by a soda lime tank. One liter of oxygen gas utilized during the burning of any foodstuff, inside or outside the body, generates 4.82 Cal. By knowing the total volume of oxygen utilized per unit time, we can determine the subject's total caloric production (or requirement).

METABOLIC RATE AND FACTORS INFLUENCING IT. Using the above calorimetric methods, we can calculate the *Basal Metabolic Rate* (*BMR*) of the body. The BMR is the amount of energy required to sustain the body at rest in a supine position. In the average adult human, BMR is about 2000 Cal/day. Thus, based on the caloric values of foods, it takes nearly thirty apples or two pounds of bread or 520 g (1.2 lbs.) of table sugar or 800 g (1.8 lbs.) of meat or nine cups of beans to sustain such a person at rest during one day.

Many factors influence BMR. Thus, in *sleep*, the BMR decreases; during *activity*, it increases. Metabolic rate during walking is twice that during sitting; running involves a rate three times higher than walking; climbing stairs produces a rate twice that of running. BMR, when calculated per unit mass, is higher in *small animals*, which have high *surface area to mass ratio*. The small mass does not permit heat storage within the core, and the *relatively* large surface area allows for heat loss. Thus, a mouse, which consumes less than 4 Cal/day compared to 5000 Cal/day for a horse, has a BMR of 200 Cal/day/kg, twenty times that of the horse. This is also why *children* have a higher BMR than adults.

Hormones and other physiological factors, such as the *sympathetic nervous system*, regulate the level of metabolic rate. Thus, *androgenic* sex steroid hormones, such as testosterone in the male, may be responsible for the higher BMR in men, and elevated levels of *thyroid hormones* and progesterone may be the cause of increased BMR in pregnant women. Lowered sympathetic activity and *catecholamine* secretion may be the cause of low BMR in the *elderly*. The mere *ingestion* of foods also increases the metabolic rate, and the *absorption* of foods (independent of food utilization) has an even greater effect. These effects, called the "specific dynamic action of foods," are most marked for *proteins*.

CN: Use red for B, a bright color for C, and light blue for F.
1. Begin with the ingestion of food and follow the process throughout the diagram. Do the same with the test tube reaction to the right.
2. Color the 2 methods of measuring metabolic rate.
3. Color the factors influencing metabolic rate. Each title receives a different color.

METABOLISM & HEAT PRODUCTION *

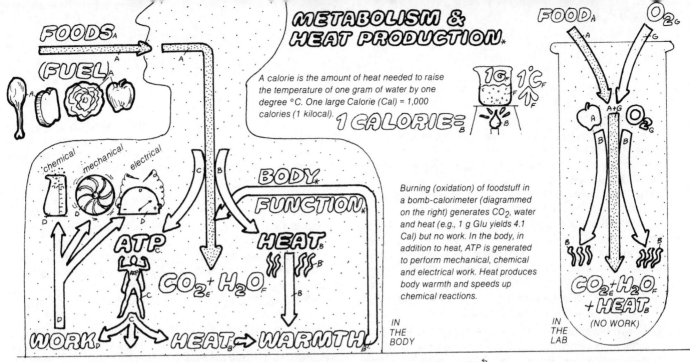

FOODS A

(FUEL) A

A calorie is the amount of heat needed to raise the temperature of one gram of water by one degree °C. One large Calorie (Cal) = 1,000 calories (1 kilocal).

1 CALORIE = B

chemical mechanical electrical

ATP C

BODY FUNCTION *

HEAT B

$CO_2 + H_2O$ E

WORK D **HEAT** B → **WARMTH** B

IN THE BODY

Burning (oxidation) of foodstuff in a bomb-calorimeter (diagrammed on the right) generates CO_2, water and heat (e.g., 1 g Glu yields 4.1 Cal) but no work. In the body, in addition to heat, ATP is generated to perform mechanical, chemical and electrical work. Heat produces body warmth and speeds up chemical reactions.

FOOD A **O_2** G

$CO_2 + H_2O$ E F + **HEAT** B

IN THE LAB

(NO WORK)

MEASUREMENT OF METABOLIC RATE *

DIRECT METHOD: CALORIMETRY H

(heat production)

calorimeter H

Metabolic rate (MR) is the rate of oxidation of foodstuff per unit of time. MR under basal conditions (supine, resting, fasting) is called Basal Metabolic Rate (BMR). MR is measured in 2 ways: In the direct method, calorimetry, total heat dissipated from the body is measured by determining changes in water temperature passing through a chamber in which the subject is sitting.

THERMOMETER I

INDIRECT: SPIROMETRY J

(O_2 consumption)

In the indirect method, spirometry, the volume of O_2 consumed is measured and converted to calories (4.82 Cal/L O_2).

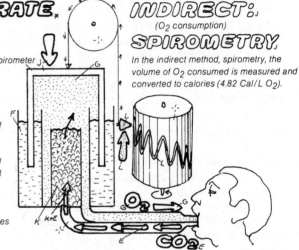

spirometer J

SODA LIME K **KYMOGRAPH** L

FACTORS INFLUENCING METABOLIC RATE *

MASS VS. SURFACE AREA M

Cal/day	4	750	2000	5000
Cal/day/kg	200	50	30	10

Absolute MR (Cal/day) increases with body mass. When expressed per unit mass (Cal/day/kg), MR is higher in smaller animals because they have larger surface area to mass ratios.

ACTIVITY N

Cal/hr 65 N 100 N 200 N 600 N 1200 N

Muscular activity increases MR, since muscles are the biggest consumers of energy. The lowest MR occurs during sleep, while the highest MR results from heavy exercise.

SEX & AGE O

Cal/hr/m^2 50 O 38 O 36 O 39 O 34 O

MR (Cal/hr/m^2) shows a decline with age. Though MR in adults is higher in men than women (androgen effect), it is the highest in pregnant women (thyroid/progesterone effect).

SYMPATHETIC NERVES & HORMONES P

Sympathetic activity releases epinephrine and norepinephrine from the adrenal medulla, and norepinephrine from nerve fibers. These increase MR by stimulating cellular metabolism. Thyroid hormones have the greatest such effect on MR. Sex hormones (androgens) and growth hormone also increase MR.

EATING Q

Food ingestion increases MR (specific dynamic action of food). Mere ingestion increases MR by 5-10%. Absorption of proteins (AA) has a longer-lasting and greater effect on MR (+ 30%). Malnutrition and starvation decrease MR by 30%.

BODY TEMPERATURE R

MR increases by 15% for each °C increase in body temperature. This may become a serious problem in the case of fever.

ENVIRONMENTAL TEMP. CLIMATIC ADAPTATION S

Increase and decrease of room temperature above and below optimal level (20°C) increases MR. In long-term adaptation, BMR is lower in tropical climate and higher in cold climates.

FOOD INTAKE, ENERGY BALANCE, OBESITY, AND STARVATION

BODY COMPOSITION AND BODY FUELS. Only organic constituents can be utilized as fuels to obtain energy. Of these, most *fats* (except cholesterol and structural fats like phospholipids), most *proteins*, and all pure *carbohydrates* can be utilized as *fuels*. Not all body fuels have similar *calorigenic* value. Caloric value of fuels is the amount of calories released upon complete oxidation to CO_2 and water. Thus, 1 g of carbohydrate or protein contains the equivalent of 4 Cal of fuel energy, but the same amount of fat contains more than twice as much. In addition, fats occupy less space. Because of these properties, fats are the ideal substances for long-term fuel energy storage. Carbohydrates are of course very efficiently utilized for fuel, but to store them, the body requires a lot of water and therefore a lot of space.

CALORIC VALUES OF FUELS AND FOODS. In the body of an adult man weighing about 70 kg (154 lbs.), the carbohydrate fuel sources, totaling about 210 g, are glucose (20 g in blood and liver) and glycogen (120 g in muscles and 70 g in liver). Proteins (about 10 kg) make up 14% of the adult male body; of this, only 6 kg can be utilized as fuel (usually during fasting and starvation). Neutral fats in the *depots* (about 15 kg) make up about 22% of the body weight, all of which can be utilized as fuel. The total fuel value of these substances is the product of the total fuel weight of each and the respective caloric value per unit weight. Thus, in the average-sized man (70 kg), fats comprise up to 84%, proteins 15%, and carbohydrates less than 1% of the total fuel value.

The fuel substances are all present in the foods we eat. Indeed, in the healthy person with an adequate daily diet, the body's fuel stores are usually not utilized to any great extent. The caloric value of the fuel foods is the same as for fuel substances within the body. Thus, per unit weight, fatty foods will provide more calories than proteins or carbohydrates.

The healthy body operates on a *balance* between *energy input* (fuel food intake) and *energy output* (expenditure of energy, i.e., work, heat). In the adult, in whom growth is over and *weight* is *stabilized*, if *food intake* equals energy needs, the body operates normally and efficiently, and the weight remains stable. Indeed, in physically fit individuals, food intake and energy utilization are very narrowly regulated, with little change in weight over many years. Thus, a construction worker who performs heavy exercise consumes much more food than a postal letter carrier of the same body weight.

FAT FUEL STORAGE AND OBESITY. If food intake is greater than the energy demands, the body promotes the storage of excess energy, usually as fat. This is a useful adaptation in expectation of times of shortage, during which the stored fuels will be mobilized (see below), resulting in the reduction of fat stores and body weight. A fat content equal to 12 to 18% of body weight in men and, 18 to 24% in women is considered normal and possibly necessary for health. For example, a marked reduction in fat content, particularly in

women, may be detrimental to a normal menstrual cycle and reproductive functions.

Excessive accumulation of fat (*obesity*), however, may also be disadvantageous, at least for individuals who are genetically predisposed to diabetes and hypertension. Clearly, obesity is a matter of degree, but when body weight exceeds 30% of the norm (due to excess fat), obesity is present. In adults, fat storage is accompanied by increase in the size of *fat cells* (i.e., *hypertrophic obesity*). This condition can be reversed by weight reduction measures. In children, excess food intake, in addition to causing increase in the size of fat cells, may also lead to increase in their *number*, a condition called *hyperplastic-hypertrophic obesity*, which is probably not reversible with weight reduction. Moreover, as adults, these children may have a higher propensity toward excessive obesity and its possible pathological consequences (see plates 118 and 128 for more on obesity).

FUEL MOBILIZATION DURING STARVATION. During prolonged *fasting* or *starvation*, all fuel reserves are *mobilized* to ensure survival of the body, particularly the *brain* and *heart*. The few hundred grams of carbohydrates (glucose and glycogen) are quickly used up during the *first day*, along with some fatty acids and labile amino acids of the *liver*. During the *next days and weeks*, all fat depots and the remaining reserves of the labile proteins are mobilized. Finally, after six weeks, structural proteins of muscle and bone are catabolized to mobilize their amino acids for gluconeogenesis. During starvation, *BMR* is reduced, and *gluconeogenesis* and *ketone body metabolism* are promoted. The massive utilization of the body fuel stores affects all body tissues, resulting in weight loss and thinness, but the brain and heart are spared. The maximum time the body can tolerate complete fuel-food starvation (but with water, minerals, and vitamins provided) is probably two months (plates 121, 125).

REGULATION OF FOOD INTAKE. Two centers in the brain *hypothalamus* regulate food intake, thus indirectly controlling weight gain and loss. Increased activity in the *feeding center* promotes appetite, feeding behavior, and food intake. The feeding center is inhibited by another hypothalamic center, the *satiety center*. The satiety center neurons are sensitive to *blood glucose* levels: high levels increase and low levels decrease their activity. Consequently, low blood sugar reduces the inhibition of the feeding center by the satiety center, resulting in the sensation of hunger and activation of feeding. Stimulation of sensors in the mouth and distension of the stomach during ingestion reduce the activity in the feeding centers. After food is absorbed, increased blood sugar activates the satiety center, which in turn inhibits the feeding center, resulting in cessation of appetite and eating behavior. Certain types of obesity as well as the loss of appetite and extreme thinness seen in *anorexia nervosa* may be of nervous and emotional origin, involving disturbances of these brain regulators of food intake (see also plates 101, 125, and 128).

CN: Use yellow for A and red for C. Use dark colors for F, G, and J.
1. Color the upper two panels. Continue down the right side of the page through the starvation section.
2. Color the food intake regulation diagram. Note the dotted line of step 2 which indicates that the normal inhibiting action (step 9) has been turned off.

FUEL VALUES OF FOODSTUFF.

FATS, PROTEINS, CARBOHYDRATES.

Caloric value (Cal/g) of fats is 9 compared to 4 for proteins (PR) and carbohydrates (CH). In a 70 Kg person, fats constitute 22%, PR 14% and CH 0.7% of body weight. All of fat but only 6 Kg of protein can be used for fuel. Hence the total fuel value of fat is 84% compared to 15% for PR and 1% for CH.

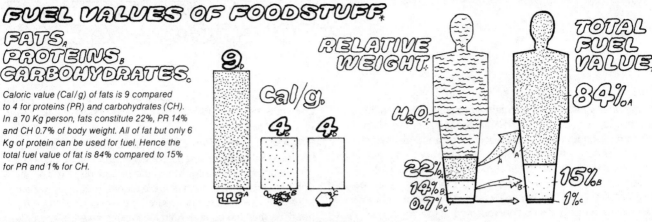

RELATIVE WEIGHT

TOTAL FUEL VALUE.

9 Cal/g.

4   4

H₂O

22%   14%   0.7%

84%   15%   1%

EFFECT OF ENERGY EXCHANGE.

To maintain optimal and constant body weight, energy exchange must be balanced: Energy input (food intake) must be equal to energy output (work + heat). If input exceeds output, excess energy is stored primarily as fat, ultimately resulting in obesity. If input is less than output, bodily stores are used, resulting in wasting and thinness.

ENERGY INPUT.

WORK.   HEAT.

ENERGY IN, > ENERGY OUT.

ENERGY IN, = ENERGY OUT.

ENERGY IN, < ENERGY OUT.

FOOD INTAKE REGULATORS.

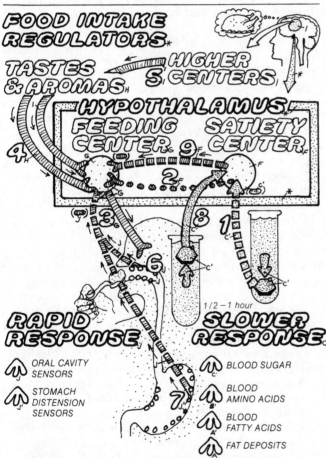

TASTES & AROMAS.   HIGHER CENTERS.

HYPOTHALAMUS.

FEEDING CENTER.   SATIETY CENTER.

1/2 – 1 hour

RAPID RESPONSE.

⋀ ORAL CAVITY SENSORS

⋀ STOMACH DISTENSION SENSORS

SLOWER RESPONSE.

⋀ BLOOD SUGAR

⋀ BLOOD AMINO ACIDS

⋀ BLOOD FATTY ACIDS

⋀ FAT DEPOSITS

To maintain energy balance, food intake must be regulated. A feeding (hunger) center and a satiety center in the hypothalamus regulate food intake. Neurons in the satiety center (glucostat) respond to changes in blood Glu (1). Reduction in blood Glu decreases neuron activity, releasing the feeding center from inhibition (2). This stimulates the appetite and food seeking (3). Odors, tastes (4) and thoughts of food also stimulate the feeding center (5). Ingestion of food activates sensory nerves in the oral cavity (6) and causes distension of stomach (7), inhibiting the feeding center. Absorption of food increases blood Glu (8), increasing the activity of the satiety center, which in turn inhibits the feeding center (9). Increased size of fat deposits also inhibit the feeding center (10).

NORMAL.   OBESITY.   FAT CELLS.

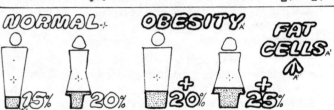

15%   20%   20%   25%

In normal adult men, fat comprises 15% of body weight, in women, 20%. Values in excess of 20% for men and 25% for women are signs of obesity. Obesity is harmful as it is associated with high blood pressure (hypertension) and metabolic (diabetes) disorders.

HYPERTROPHIC OBESITY.

Obesity may be caused by the increase in the size of the fat cells (hypertrophic obesity). This occurs usually in maturity and is easier to reverse by restricting the number of calories consumed.

HYPERBLASTIC-HYPERTROPHIC OBESITY.

Hyperplastic-hypertrophic obesity is due to the increase in the number of fat cells as well as their size. It is believed to occur mainly in growing children and is more difficult to reverse.

STARVATION.

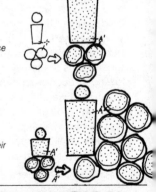

1ST DAY.   WEEK 1-6.   WEEK 7,8, ?.

Starvation leads to the loss of fuel reserves
The meager CH reserves are the first to be used.
Then fats and the labile PR are consumed. The structural PR are the last to be used. The brain and heart are usually spared.

BODY TEMPERATURE, HEAT PRODUCTION, & HEAT LOSS

HEAT GAIN AND HEAT LOSS. Mammals and birds are *homeotherms* (they can maintain a constant body temperature, usually at 37°C). To maintain the temperature of any system (or body) at a constant point, a balance must exist between *heat gain* and *heat loss*. The *metabolic heat* generated by oxidation of foodstuff in the visceral organs and tissues (*body core*) is a constant source of heat (see plates 5, 6, and 131). This heat can be increased by *muscular activity* (shivering, running) and by nervous and hormonal factors such as the *sympathetic nervous* activity, *catecholamines*, and *thyroid hormones*. *Food ingestion*, especially protein foods, by itself also increases metabolic heat.

HEAT EXCHANGE MECHANISMS. The body also gains heat from external sources (sun rays, heaters). This kind of passive heat exchange is achieved by *radiation* (from the sun), by *convection* from near sources (e.g., heat from a heater in a room), and by *conduction* involving direct contact of warm objects (a heating blanket with the skin). Conduction, convection, and radiation can also operate in the opposite direction, to increase heat loss from the body. Heat loss occurs passively if the body is exposed to outside temperatures below its temperature. Thus, sitting on a cold chair warms the seat (conduction); the presence of a crowd in a cold room increases the room temperature (convection).

SKIN IN THERMAL REGULATION. The body is also equipped with physiological mechanisms that actively increase or decrease heat loss. The *skin* is the organ that plays a central role here. Heat can be lost through the skin in two ways: by direct exchange of heat between the blood circulating in the skin and the outside or by *evaporation* of water from the skin surface. The skin contains special thoroughfarelike blood *capillaries* that do not exchange nutrients with skin cells but function solely in exchanging heat with the outside. Blood flows in these vessels whenever they are open. In cold weather, these vessels close up (cutaneous *vasoconstriction*), markedly reducing skin circulation and minimizing heat loss. In a hot environment, these special thermoregulatory vessels are totally open (cutaneous *vasodilation*), allowing through blood flow and increasing heat loss to the outside. Because of these mechanisms, blood flow in the skin, which at its height is nearly 10% of the total cardiac output, can vary more than a hundredfold, making heat exchange in the skin via blood circulation a very efficient and effective mechanism for heat loss and heat conservation.

EVAPORATION. The second way heat can be lost from the skin or other exposed surfaces such as the respiratory ducts is by *evaporation* of water. Water has a very high heat capacity (0.6 Cal/g), which means that the loss of 1 g of water from the body is accompanied by the loss of 600 cal of heat. This water loss occurs in two ways: by insensible perspiration and by active *sweating*. Loss of water from the skin at lower temperatures is called insensible perspiration because water diffuses through skin cells and pores and quickly evaporates; no sweat drops are formed. Similarly, considerable water and heat are lost in the respiratory passageways every day.

Insensible perspiration accounts for the loss of more than 0.5 L of water per day (360 Cal of heat, about 20% of daily basal caloric production)!

SWEATING. When internal body temperatures increase to above 37°C, active secretion of water and salt by the sweat glands begins, markedly raising the rate of water evaporation and heat loss. *Sweat glands* are exocrine (eccrine) glands abundantly located in many parts of the skin (e.g., the forehead, palms, and soles). Animals lacking sweat glands, like the dog, *pant*, thereby markedly increasing air flow in the respiratory passages, resulting in similar increases in evaporation and heat loss rates.

USES OF BODY HAIR. A third way by which skin is able to decrease heat loss is by the use of *body hair* (*fur*), a mechanism of little value in humans but of great value in the furry animals (bear, sheep, etc.), particularly those living in cold climates. In cold weather, skin hairs stand up (*piloerection*), which causes *entrapment* of the air in the hair web. The trapped air forms an insulating layer because blood now exchanges heat not with the flowing cold air on the outside, but with a stationary air layer in the hair web. Warm clothes in humans, particularly wool, perform the same role.

FATS IN THERMAL RESPONSES. The skin by itself is only a weak insulation. However, in many animals, including humans, the fat under the skin (*subcutaneous fat pads*) has the dual function of acting as both a very effective insulation and a source of metabolic energy. In fetuses, newborns, and infants of humans as well as in many other animals, a special type of fatty tissue, the *brown fat*, is present. The numerous *mitochondria* of these fat cells oxidize the fat in such a way as to produce a great deal more heat than ATP. This heat acts as a furnace, generating heat to protect against the cold. This heat may be one reason why newborns do not shiver when exposed to cold. Adult humans do not have brown fat (see also plate 127).

VARIATION IN BODY TEMPERATURE. Although it would be ideal for all parts of the body to operate at 37°C, only the body core (i.e., the brain and visceral organs and tissues in the trunk) operate at this optimum temperature. The tissues of the extremities and the skin, being far from the core heat source and in direct contact with the outside, have much lower temperatures. For example, in a room temperature of about 21°C, hand and foot skin temperatures are about 28 and 21°C, respectively. These values are about 34 to 35 degrees in a room temperature of about 35°C because, in the absence of limb movement, the only heat source is the arterial blood flowing from the viscera. Presumably, this heat is not sufficient to keep the limb tissues warm enough. The *frostbite* and *gangrene* (tissue death) in foot and hand that occurs at extremely cold temperatures is due to failure of an adequate blood and heat supply to these regions. Even the core temperature is not constant at all times. A circadian (diurnal, daily) cycle exists, the temperature being lowest in morning (36.7°C) and highest in the evening (37.2°C) (see plate 101).

CN: Use light blue for F, brown for J, and red for L. Use dark colors for A and G.
1. Color titles A-F, and their corresponding structures in the upper right diagram. Then color the one below it. Complete the material on perspiration (F¹) to the left.
2. Color the panels on piloerection and brown fat.
3. Color the bottom panel, beginning with the central portion, then doing the outer temperature extremes.

CORE TEMPERATURE (T)

The metabolism of visceral organs generates heat in the body core. By balancing heat loss and gain, core temperature (T) is kept constant at 37°C (98.6°F).

SKIN AS INSULATOR

Skin mediates the heat exchange between body core and environment. Special thermoregulatory mechanisms enable the skin to change its insulating capacity.

RADIATION

The body can exchange heat with distant objects of different T via radiation. Heat from the sun or from heaters is conveyed by radiation.

CONDUCTION

Direct contact with hot or cold objects allows a heat exchange by conduction. A cold chair will thus be warmed by a seated individual.

CONVECTION

Movement of air in a room (convection) increases the heat exchange between body and environment. A fan blowing air will help the body to cool off.

EVAPORATION

The evaporation of water from body surfaces is an efficient way of losing heat because of the great deal of heat needed to evaporate water (0.6 Cal/g of water).

INSENSIBLE PERSPIRATION

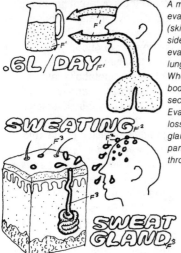

.6 L/DAY

A major source of heat loss is evaporation from body surfaces (skin, respiratory tract). When outside T is cool (below 20°C), water evaporates from exposed skin and lungs by insensible perspiration. When environmental T approaches body T, sweating (water and salt secretion by sweat glands) increases. Evaporation of sweat increases heat loss. Dogs, which have no sweat glands, lose water and heat by panting, which increases perspiration through respiratory surfaces.

SWEATING

SWEAT GLAND

BODY TEMP.

Body T is not uniform throughout. While body core is homeothermic, the extremities vary in T. In a cool room (21°C), core T is near 37°C, while hand and foot T are 28° and 21°C respectively. In a hot room (35°C), the T of the extremities approach room T.

HEAT LOSS   HEAT GAIN

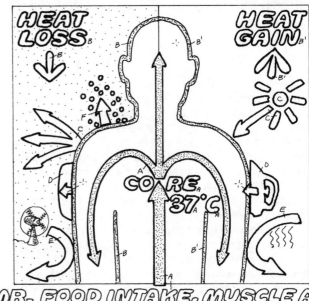

CORE 37°C

BMR, FOOD INTAKE, MUSCLE ACT.

Heat gain is increased by food intake, and muscular activity, as well as by radiation from sunlight, etc. Heat may be gained or lost by contact with warm or cold objects. Hair and clothing reduce heat loss while sweating and evaporation increase heat loss.

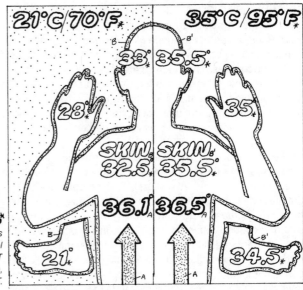

21°C/70°F   35°C/95°F
33°   35.5°
28°   35°
SKIN 32.5°   SKIN 35.5°
36.1° 36.5°
21°   34.5°

PILOERECTION
HAIR SHAFT
MUSCLE
TRAPPED AIR

In furry animals, cold T causes hair to stand up (piloerection), increasing the depth of a trapped layer of air, and minimizing heat loss. In humans, clothing performs similar functions.

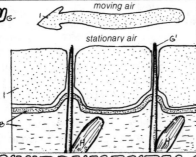

moving air

stationary air

BROWN FAT
FAT CELL
MITOCHONDRION
FAT GRANULE

A special type of fatty tissue (brown fat), rich in mitochondria, can generate much heat through lipolysis. Brown fat is found primarily in infants and animals and is located in the back and around the scapulea.

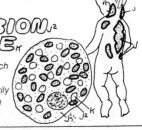

ENVIRONMENTAL TEMPERATURE

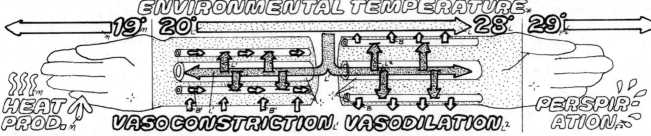

19° 20°   28° 29°

HEAT PROD.   VASOCONSTRICTION   VASODILATION   PERSPIRATION

blood flow prevents heat loss

Skin contains special blood vessels designed specifically for heat exchange. In a cold environment, these vessels constrict, decreasing skin blood flow and heat loss. In hot environments, the vessels dilate, increasing blood flow, which increases heat loss by direct contact with air, and by providing fluids for sweat glands. These vessels are supplied by sympathetic nerve fibers.

blood flow promotes heat loss

REGULATION OF BODY TEMPERATURE

How does the body keep its core temperature constant at 37°C? In response to a drop in temperature, the body activates certain mechanisms that increase *heat gain* and some that reduce *heat loss*. When the temperature increases, heat gain is decreased while heat loss is enhanced. These physiological adjustments are controlled by a "*thermostat*" center in the *hypothalamus*, the neurons of which are sensitive to changes in both skin and blood temperature. This thermostat has a normal *setpoint* at 37°C. Deviation of hypothalamic temperature away from this setpoint activates appropriate responses in the opposite direction intended to return the body temperature to the desired normal level. The hypothalamic thermostat works in conjunction with other hypothalamic, autonomic, and higher nervous thermoregulatory centers (see plate 101). Some of these thermoregulatory responses are involuntary, mediated by the autonomic nervous system, some are neurohormonal, and others are semi-voluntary or voluntary behavioral responses.

RESPONSES TO COLD. Consider a person taking a cold shower. Skin temperature quickly drops, stimulating the *skin cold receptors* and cooling the *blood* flowing in the skin. The impulse activity of these receptors increases with decreasing skin temperatures. Their signals are received by both the hypothalamic thermostat and the higher *cortical centers*. The hypothalamic thermostat is also activated by the change in *blood temperature*. The thermostat center initiates responses that promote heat gain while inhibiting centers that promote heat loss.

The activation of *sympathetic centers* results in several responses: (1) skin vessels constrict (due to norepinephrine released from sympathetic fibers), causing decreased cutaneous blood flow and decreased heat loss; (2) the *metabolic rate* increases, causing thermogenesis due to increased adrenal medullary *epinephrine* secretion; (3) body hair muscles contract, resulting in *piloerection* (hair standing), which traps the air next to the skin, decreasing heat loss (piloerection is particularly effective in furry animals and of little value in humans); and (4) *brown fat* oxidation increases, causing thermogenesis (a response only important in infants and in some animals). In addition to the above responses mediated by the sympathetic system, a *shivering center* in the hypothalamus is activated which in turn activates *brainstem motor centers* to initiate involuntary contraction of skeletal muscles, causing shivering and generating a lot of heat (see plates 101 and 102).

Cold also activates some compensatory *behavioral responses* directed at increasing heat production or decreasing heat loss. For example, *curling up* decreases surface area and heat loss. *Huddling* and cuddling seen in animals and humans, *voluntary physical activity* (rubbing the hands, pacing), and sheltering next to a heat source and wearing warm clothing are other examples of voluntary cold fighting responses. Voluntary or semivoluntary behaviors are activated by the responses of the higher brain centers (cortex and limbic system) to the uncomfortable sensation of the cold. In many animals and in children, prolonged exposure to cold climate increases the basal secretory rate of *thyroid hormones*, which by their potent calorigenic actions increase heat production (see plate 113). As a result of these compensatory responses, the body will get warmer. The hypothalamic sensors detect the warmth and diminish the heat producing and heat loss prevention responses.

RESPONSES TO HEAT. When the body is exposed to heat (e.g., from the sun, fire, or excessive clothing), body temperature rises. Here too, both skin *warmth receptors* and blood convey the changes to the hypothalamic thermostat. But warmth receptors are less effective than blood, because there are fewer of them than there are cold receptors and because blood volume and flow in the skin are high during exposure to heat (vasodilation). The hypothalamic thermostat initiates compensatory responses, some of which increase heat loss, others that decrease heat production.

Thus, the adrenergic activity of the sympathetic nervous sytem, controlling vasoconstriction and metabolic rate, is inhibited, resulting in *cutaneous vasodilation* and reduced metabolic rate, respectively. These will increase heat loss from the skin and decrease heat production in the core. If heat is sufficiently intense, a particular division of the autonomic nervous sytem (*cholinergic sympathetic fibers*, releasing *acetylocholine*) that innervates the *sweat glands* is activated, stimulating sweating. Sweating markedly increases heat loss from the skin and is the most effective involuntary heat fighting response in humans (600 Cal heat lost per liter of sweat). Behavioral responses to heat are also very effective. A hot person becomes lethargic and tends to rest or lie down with limbs spread out. These states decrease heat production and increase heat loss. Heat loss is also enhanced by wearing loose and light clothing, fanning, drinking cool drinks, swimming, etc.

FEVER. *Fever*, an increase in core body temperature of one to several degrees, is caused during illness when infectious agents enter the body. *Toxins* liberated by these bacteria stimulate the *white cells* to release *pyrogens* (e.g., interleukin-1). These substances act on the hypothalamic thermostatic neurons, raising their setpoint (e.g., to 40°C). To reach this new point, the patient shivers, increasing heat production, and becomes pale due to skin vasoconstriction. This continues until the core temperature reaches the new setpoint. Fever may be a natural defense response. The hot state of the body may be detrimental for the bacteria or their toxin. Of course, very high temperatures (above 42°C) cause *heat shock* and may be fatal if not treated. The antifever effect of aspirin is due to its preventing the effects of pyrogen on the hypothalamus. When the infection is cured, pyrogen secretion decreases, and the thermostat is set back to its original 37°C. Now the body attempts to cool down by skin vasodilation and sweating.

CN: Use red for D and dark colors for A, B, and E.
1. At the very top of the page color the titles: heat loss and heat production, and their respective "increase and decrease" symbols. These symbols will appear throughout, summarizing the result of each factor in the thermal adaptive process.

2. Color the events that follow a drop in environmental temperature. Use the order of letter labels C-J as your guide.
3. Go to the bottom panel and color the adaptation to a rise in environmental temperature.
4. Color the elements of fever by first coloring the fever producing events in the upper portion.

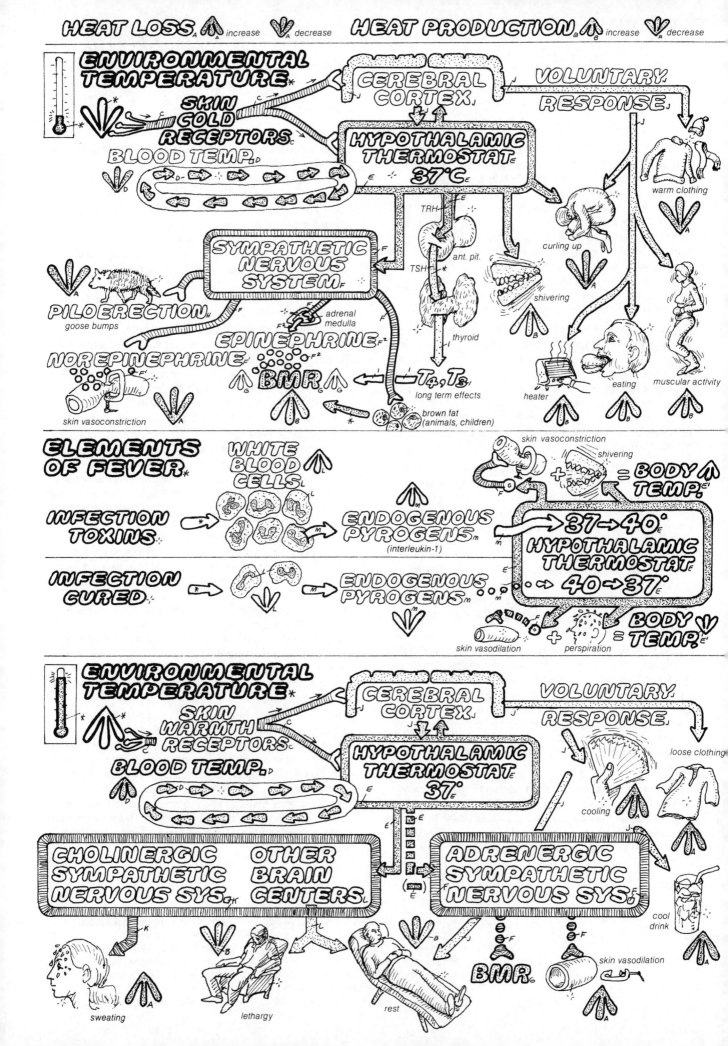

HEAT LOSS ↑ₐ increase ↓ₐ decrease   HEAT PRODUCTION ↑ᵦ increase ↓ᵦ decrease

ENVIRONMENTAL TEMPERATURE *
CEREBRAL CORTEX
VOLUNTARY RESPONSE

SKIN COLD RECEPTORS
BLOOD TEMP.
HYPOTHALAMIC THERMOSTAT 37°C

warm clothing
curling up
TRH
ant. pit.
TSH
shivering
SYMPATHETIC NERVOUS SYSTEM
thyroid
PILOERECTION
goose bumps
adrenal medulla
EPINEPHRINE
eating
muscular activity
NOREPINEPHRINE
BMR
T₄, T₃
long term effects
heater
brown fat (animals, children)
skin vasoconstriction

ELEMENTS OF FEVER *
WHITE BLOOD CELLS
skin vasoconstriction
shivering
= BODY TEMP.

INFECTION TOXINS
ENDOGENOUS PYROGENS
(interleukin-1)
37→40°
HYPOTHALAMIC THERMOSTAT
40→37°

INFECTION CURED
ENDOGENOUS PYROGENS
skin vasodilation
perspiration
= BODY TEMP.

ENVIRONMENTAL TEMPERATURE *
CEREBRAL CORTEX
VOLUNTARY RESPONSE

SKIN WARMTH RECEPTORS
BLOOD TEMP.
HYPOTHALAMIC THERMOSTAT 37°
loose clothing
cooling

CHOLINERGIC SYMPATHETIC NERVOUS SYS.
OTHER BRAIN CENTERS
ADRENERGIC SYMPATHETIC NERVOUS SYS.
cool drink

sweating
lethargy
rest
BMR
skin vasodilation

ORIGIN, STRUCTURE, AND FUNCTIONS OF BLOOD

BLOOD FUNCTIONS. *Blood* is the body's principal extra-cellular fluid. Its flow through the tissues permits its numerous *transport functions*, which ensure nutrition, respiration, physiological regulation, and defense. During its course through the tissue capillaries, blood delivers *nutrients* from the small intestine and *oxygen* from the lungs to the cells. It also removes the toxic waste products of cellular metabolism (*metabolites*), such as urea and *carbon dioxide*, from the tissue environment and eliminates them as it circulates through the kidneys and lungs respectively. In addition, blood carries the *hormones* from their sites of production in the endocrine glands to their target organs in other locations.

The *red blood cells* (RBCs), which contain the oxygen-binding protein hemoglobin, transport oxygen between the lungs and tissues and carbon dioxide between the tissues and lungs. Blood also transports the *white blood cells* (WBCs) to injury sites, where they defend the body by destroying invading microorganisms and their toxins. Another important blood function is to help maintain body temperature. This is achieved by *heat* transfer from the warmer body core to the colder periphery.

BLOOD COMPOSITION. Two compartments, a cellular compartment and a fluid medium called the *plasma*, make up blood tissue. The blood cells float freely within this medium. Separation into these two compartments is achieved by spinning (centrifuging) the blood in a small capillary tube (hematocrit tube). The centrifuged blood separates into a colorless fluid supernatant on the top and a red precipitate on the bottom. The supernatant, amounting to about 55% of the blood volume, is the *plasma*. The plasma consists mainly of *water* (90%), which helps dissolve the blood proteins (e.g., *fibrinogen*, *albumins*, and *globulins*) as well as the nutrients, hormones, and electrolytes.

The remaining 45% of the blood volume consists of the precipitate called the *hematocrit*, which is made up mainly of red blood cells (*erythrocytes*), the most abundant of the blood cells. Blood cells are also called "*formed elements*" of the blood. The white blood cells (*leukocytes*) and the *platelets* (*thrombocytes*), being smaller in number, constitute only a small fraction of the hematocrit, forming a very thin yellowish band between the red hematocrit and the plasma supernatant.

Another way to separate blood fluid and cells is to allow a drop of blood to stand for a while. It will separate into a dense red core called the *clot* surrounded by a colorless fluid called the *serum*. The clot has a composition similar to the hematocrit, and the serum resembles the plasma. However, the serum lacks the plasma protein fibrinogen, which is associated with the clot.

Blood constitutes 8% of the body weight. On the average, men have more blood (5.6 L) than women (4.5 L), although blood volume increases during pregnancy. Men's blood is also more cellular (mainly red cells), containing about 47% hematocrit, compared to 42% found in women and children. The higher blood content reflects the larger size of the males, and the higher hematocrit indicates a higher concentration of red cells. This is a response to higher metabolic rate and increased oxygen needs in males, which are compatible with higher muscle mass and work load.

FORMATION AND SOURCE OF BLOOD. The bulk of plasma proteins is manufactured by the liver; various sources in the body contribute other dissolved plasma constituents. Blood cells, however, are formed mainly in the *bone marrow*. The mass of bone marrow in a single bone may appear insignificant, but the total mass of bone marrow in the body is very large, making it one of the three largest organs of the body (liver, skin, and bone marrow).

In the adult, active bone marrow is the *red marrow* found in bones of the trunk and head (*sternum*, *ribs*, *vertebrae*, and the *skull*). The red marrow in these bones provides the *primary source* for blood cells. In growing children, however, red marrow is also found in the long bones of the lower extremity (*femur* and *tibia*). In the adult, the latter bones do not entirely lose their ability to make blood cells, but provide possible *secondary* sources for blood cell formation that are activated only when the primary sources are unable to keep up with the demand. Under such conditions, the *liver* and *spleen* can also make blood cells. Indeed, the liver is the principal source of RBCs in early embryonic and fetal periods; the spleen produces RBCs slightly later in fetal life. In extreme emergencies, such as massive blood loss due to hemorrhage or destruction of generative cells of marrow due to exposure to ionizing radiation, the adult's liver and spleen, as well as the resting *yellow marrow* in the secondary sources, can again produce new blood cells.

Blood cells are formed in the red marrow from the proliferation and differentiation of *stem cells*, which permanently reside there. One line forms the red cells, another, the white cells, and yet another, the platelets. A variety of hormonal and humoral controls adjust the production rate of various blood cells in response to physiological needs. For example, the kidney hormone erythropoietin stimulates red cell production, and a hormone called thrombopoietin stimulates platelet production. Several humoral factors are involved in regulating white cell production.

CN: Use a very light or straw color for A. Use red for D.
1. Begin with the illustrations in the upper panel. Color gray the titles of all elements that make up 2% of plasma. These substances will be given individual colors in the lower left diagram.
2. Color the formed elements. For all practical purposes, the number of red blood cells is equal to the hematocrit, and both receive the color red.
3. Color the sources of formed elements in red bone marrow. Note that the adolescent and the fetus have different primary sources and are colored accordingly.
4. Color the transport function of blood.

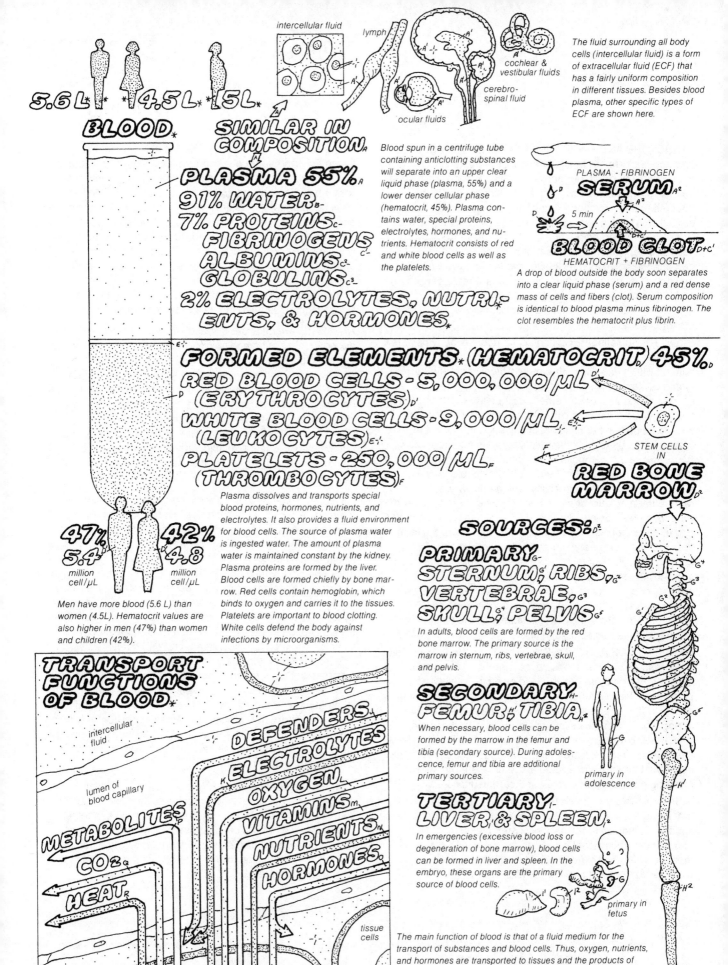

5.6 L . 4.5 L . 15 L .

BLOOD .

SIMILAR IN COMPOSITION .

PLASMA 55% .
91% WATER .
7% PROTEINS .
FIBRINOGENS .
ALBUMINS .
GLOBULINS .
2% ELECTROLYTES, NUTRI- ENTS, & HORMONES .

intercellular fluid
lymph
cochlear & vestibular fluids
cerebro-spinal fluid
ocular fluids

The fluid surrounding all body cells (intercellular fluid) is a form of extracellular fluid (ECF) that has a fairly uniform composition in different tissues. Besides blood plasma, other specific types of ECF are shown here.

Blood spun in a centrifuge tube containing anticlotting substances will separate into an upper clear liquid phase (plasma, 55%) and a lower denser cellular phase (hematocrit, 45%). Plasma contains water, special proteins, electrolytes, hormones, and nutrients. Hematocrit consists of red and white blood cells as well as the platelets.

PLASMA - FIBRINOGEN
SERUM .
5 min
BLOOD CLOT .
HEMATOCRIT + FIBRINOGEN

A drop of blood outside the body soon separates into a clear liquid phase (serum) and a red dense mass of cells and fibers (clot). Serum composition is identical to blood plasma minus fibrinogen. The clot resembles the hematocrit plus fibrin.

FORMED ELEMENTS . (HEMATOCRIT) 45% .
RED BLOOD CELLS - 5,000,000/μL .
(ERYTHROCYTES) .
WHITE BLOOD CELLS - 9,000/μL .
(LEUKOCYTES) .
PLATELETS - 250,000/μL .
(THROMBOCYTES) .

STEM CELLS IN
RED BONE MARROW .

Plasma dissolves and transports special blood proteins, hormones, nutrients, and electrolytes. It also provides a fluid environment for blood cells. The source of plasma water is ingested water. The amount of plasma water is maintained constant by the kidney. Plasma proteins are formed by the liver. Blood cells are formed chiefly by bone marrow. Red cells contain hemoglobin, which binds to oxygen and carries it to the tissues. Platelets are important to blood clotting. White cells defend the body against infections by microorganisms.

47%
5.4 million cell/μL

42%
4.8 million cell/μL

Men have more blood (5.6 L) than women (4.5 L). Hematocrit values are also higher in men (47%) than women and children (42%).

SOURCES: .
PRIMARY . STERNUM, RIBS, VERTEBRAE, SKULL, PELVIS .

In adults, blood cells are formed by the red bone marrow. The primary source is the marrow in sternum, ribs, vertebrae, skull, and pelvis.

SECONDARY - FEMUR, TIBIA .

When necessary, blood cells can be formed by the marrow in the femur and tibia (secondary source). During adolescence, femur and tibia are additional primary sources.

primary in adolescence

TERTIARY - LIVER & SPLEEN .

In emergencies (excessive blood loss or degeneration of bone marrow), blood cells can be formed in liver and spleen. In the embryo, these organs are the primary source of blood cells.

primary in fetus

TRANSPORT FUNCTIONS OF BLOOD .

intercellular fluid

lumen of blood capillary

DEFENDERS
ELECTROLYTES
OXYGEN
VITAMINS
NUTRIENTS
HORMONES

METABOLITES
CO_2
HEAT

tissue cells

The main function of blood is that of a fluid medium for the transport of substances and blood cells. Thus, oxygen, nutrients, and hormones are transported to tissues and the products of cellular metabolism (metabolites), e.g., CO_2 and urea, are transported away from the tissues. Blood also helps to exchange heat between different tissues as well as transport white blood cells (defenders) to wounds and sites of infection.

THE RED BLOOD CELLS

STRUCTURE AND FUNCTION. *Red blood cells* (RBCs, *erythrocytes*) are the most abundant cells in the blood (a total of 30 thousand billion per person). They transport the respiratory gases, particularly oxygen, and their shape is highly adapted to their function. Circulating RBCs resemble biconcave discs, having average dimensions of 7.5 by 2 microns (1 micron at the middle). As the RBCs move through blood cells and capillaries of different widths, their size and shape can change. In veins they inflate, and in the narrow capillaries they fold. The normal biconcave shape facilitates oxygen and carbon dioxide diffusion into the cells and maximizes the probability of binding with the *hemoglobin* molecules, which are stacked within the cell.

Mature circulating RBCs contain no *nucleus* or cytoplasmic organelles. Instead, all the available intracellular space is packed with hemoglobin, the oxygen-binding protein. Hemoglobin contains a protein part (*globin*) and four pigment (*heme*) molecules. Each heme is associated with one of the four *polypeptide* subunits (chains) of the globin. There are four *iron* atoms in hemoglobin, one in each heme. In the ferrous state, this iron binds reversibly with molecular oxygen (O_2). Thus, each hemoglobin can bind and transport 4 oxygen molecules. The amount of hemoglobin in the blood determines its oxygen-carrying ability, which can vary due either to reduced hemoglobin content in the RBCs or to reduced RBC production (see below). Normal blood contains about 140-160 g/L of hemoglobin in the male and 120-150 g/L in the female (nearly 2 lbs. per person). In addition to hemoglobin, RBCs contain the cytoskeletal and contractile proteins tubulin and actin, which permit shape changes; RBCs also have the necessary enzymes for glycolytic (anaerobic) glucose oxidation as well as certain regulatory chemicals such as DPG (2,3-diphosphoglycerate), which influences the binding of oxygen with hemoglobin. (See plates 48, 49.)

LIFE CYCLE OF RED BLOOD CELLS. Formation of red cells (*erythropoiesis*) occurs in the bone marrow. Special stem cells that reside there proliferate to give rise to all types of blood cells. Progenitor cells of RBCs (*erythroblasts*) contain nuclei. Within a few days, these cells differentiate into RBCs, during which time they synthesize and pack hemoglobin within their cytoplasm and then eventually lose their nuclei. At this time, the red cells are mature and ready to function. They leave the bone marrow and enter the bloodstream, where they begin to transport oxygen and carbon dioxide.

Several factors regulate RBC production, the most important being the arterial *oxygen pressure*. In low pO_2 conditions, such as at high altitude, the low oxygen pressure in the atmosphere and arterial blood causes the release of a hormone, *erythropoietin*, from the kidney. Erythropoietin acts on the bone marrow to form more RBCs.

The life span of circulating RBCs is about 4 months, after which they age and are recognized, phagocytized, and destroyed by the tissue *macrophages* (large white blood cells) residing in the *liver* and *spleen* blood capillaries. The hemoglobin content of the destroyed cells is broken down to amino acids and heme. Heme is metabolized to *iron* and *bilirubin*. The bone marrow reuses iron for hemoglobin synthesis. The liver secretes bilirubin in the bile (*bile pigments*) to be excreted in the intestine with the feces. These pigments give feces their light brown color. Some of the bile pigments are reabsorbed, recirculated, and finally excreted in the kidney. These pigments cause the yellow color of urine.

ANEMIAS. *Anemias* are diseases associated with the reduced content of hemoglobin in the blood, which decreases the blood's ability to transport oxygen to the tissues. Consequences of anemias range from simple fatigue to death. There is a variety of causes for anemias, a few of the more common ones being *direct blood loss* due to severe menstruation, internal bleeding from a gastrointestinal ulcer, accidental hemorrhage, and *failure* of *bone marrow* to produce new RBCs (such as might occur due to exposure to high doses of ionizing radiation or to certain drugs, toxins, or viruses).

Certain *digestive* or *dietary deficiencies* also cause anemias. Absence of *vitamin B_{12}* (cyanocobalamine), a substance necessary for erythropoiesis, could result in severe RBC shortage. Vitamin B_{12} is plentiful in foods of animal origin (meats) but is absent in plant foods, so a strictly vegetarian diet could result in serious deficiency in this vitamin, leading to *pernicious anemia*. However, pernicious anemia is rarely caused by dietary deficiency of vitamin B_{12}; it is often caused by a diminished ability to absorb this substance. To facilitate the absorption of vitamin B_{12}, which is the largest of the vitamins, the stomach secretes a protein (*intrinsic factor*). The absence of intrinsic factor, which can occur because of diseases of the stomach or after its surgical removal, means that dietary vitamin B_{12} is not absorbed, resulting in diminished hemoglobin synthesis and RBC production (pernicious anemia). Dietary deficiencies of *folic acid* and *iron* may also cause anemia. There is an increased need for all the dietary stimulants of erythropoiesis (such as iron and vitamins) during pregnancy and growth.

Increased destruction of RBCs occurs in individuals afflicted with *sickle cell anemia*, a hereditary disease particularly prevalent among blacks. Sickled cells stick together, hemolyze, and are rapidly destroyed by the macrophages. Kidney disease or loss results in decreased production of erythropoietin, thus diminishing the stimulation of the bone marrow to produce RBCs and resulting in anemia.

CN: Use red for A structures and dark colors for B, D, and J.
1. Color the upper panel, noting the four hemes (B) that carry the oxygen molecule (H), here represented by an arrow, with the oxygen title being found in the section below it.
2. Color the process of erythropoiesis, and follow the numbered sequence.
3. Color the regulation of red blood cell production.
4. Color the various anemias due to reduced red blood cells, noting the letter labels of the titles carefully. They are not all colored red.

RED BLOOD CELL (ERYTHROCYTE)

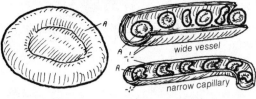

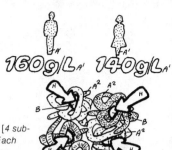

wide vessel

narrow capillary

7.5 μm

2 μm

Red blood cells (RBCs) have no organelles; instead they are packed with hemoglobin (Hb), which carries oxygen. The biconcave disc shape of RBCs allows for rapid diffusion of oxygen. This shape changes as RBCs squeeze through the narrow capillaries.

HEMOGLOBIN
4 PEPTIDE CHAINS
4 HEMES

160g/L   140g/L

Hemoglobin (Hb) has a protein part, globin: [4 subunits (2 alpha chains and 2 beta chains)]. Each subunit has a heme. Each heme has one iron, which in the ferrous state (Fe^{+2}) binds with one O_2.

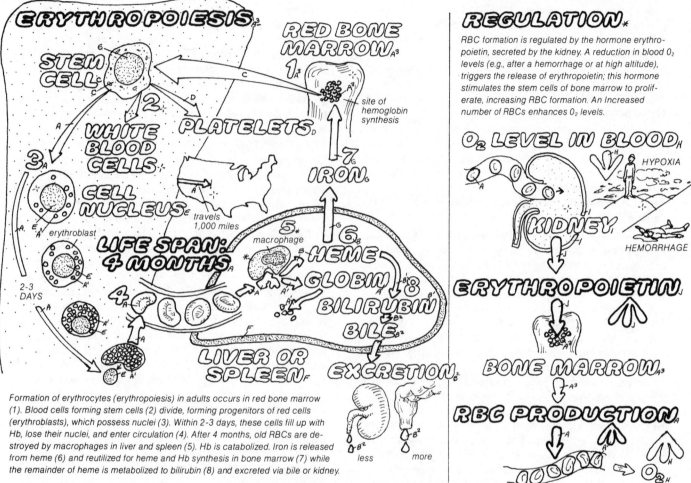

ERYTHROPOIESIS

STEM CELL

RED BONE MARROW

site of hemoglobin synthesis

WHITE BLOOD CELLS

PLATELETS

CELL NUCLEUS

travels 1,000 miles

erythroblast

2-3 DAYS

LIFE SPAN: 4 MONTHS

macrophage

IRON

HEME
GLOBIN
BILIRUBIN
BILE

LIVER OR SPLEEN

EXCRETION

less   more

Formation of erythrocytes (erythropoiesis) in adults occurs in red bone marrow (1). Blood cells forming stem cells (2) divide, forming progenitors of red cells (erythroblasts), which possess nuclei (3). Within 2-3 days, these cells fill up with Hb, lose their nuclei, and enter circulation (4). After 4 months, old RBCs are destroyed by macrophages in liver and spleen (5). Hb is catabolized. Iron is released from heme (6) and reutilized for heme and Hb synthesis in bone marrow (7) while the remainder of heme is metabolized to bilirubin (8) and excreted via bile or kidney.

REGULATION

RBC formation is regulated by the hormone erythropoietin, secreted by the kidney. A reduction in blood O_2 levels (e.g., after a hemorrhage or at high altitude), triggers the release of erythropoietin; this hormone stimulates the stem cells of bone marrow to proliferate, increasing RBC formation. An Increased number of RBCs enhances O_2 levels.

O_2 LEVEL IN BLOOD

HYPOXIA

KIDNEY

HEMORRHAGE

ERYTHROPOIETIN

BONE MARROW

RBC PRODUCTION

O_2

CAUSES OF ANEMIA

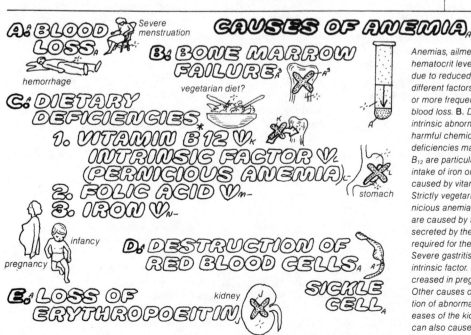

A. BLOOD LOSS

Severe menstruation

hemorrhage

B. BONE MARROW FAILURE

vegetarian diet?

C. DIETARY DEFICIENCIES
1. VITAMIN B 12
INTRINSIC FACTOR
(PERNICIOUS ANEMIA)
2. FOLIC ACID
3. IRON

stomach

pregnancy   infancy

D. DESTRUCTION OF RED BLOOD CELLS

SICKLE CELL

E. LOSS OF ERYTHROPOIETIN

kidney

Anemias, ailments or diseases associated with reduced Hb hematocrit levels and RBC numbers, cause many abnormalities due to reduced oxygen supply to tissues. Anemias are caused by different factors and conditions: A. Hemorrhage, internal bleeding, or more frequently, severe menstruation involving excessive blood loss. B. Diseases of bone marrow (aplasia) caused by intrinsic abnormalities or by exposure to ionizing radiation or harmful chemicals are important causes of anemias. C. Dietary deficiencies may also lead to anemias, because iron and vitamin B_{12} are particularly important for erythropoiesis. Thus, reduced intake of iron or vitamin B_{12} may result in anemia. Anemias caused by vitamin B_{12} deficiencies are called pernicious anemias. Strictly vegetarian diets lack vitamin B_{12} and may result in pernicious anemias. However, in most cases, pernicious anemias are caused by the absence of the intrinsic factor, a substance secreted by the parietal cells of the stomach glands that is required for the absorption of vitamin B_{12} by intestinal mucosa. Severe gastritis or a total loss of the stomach eliminates the intrinsic factor. Dietary intake of iron and vitamin B_{12} must be increased in pregnancy and during childhood development. D. Other causes of anemias are associated with increased destruction of abnormal red cells, as in sickle cell anemia. E. Lastly, diseases of the kidney resulting in the reduced level of erythropoietin can also cause anemias.

PHYSIOLOGY OF BLOOD AGGLUTINATION & GROUPING

BLOOD AGGLUTINATION. When blood of two different individuals is mixed outside the body, the *red blood cells* (RBCs) may clump together, separate from the plasma, and precipitate as solid masses. Occurrence of this *agglutination* reaction in the body, as might happen following a blood transfusion, may create a potentially lethal condition. Because of the obvious importance of agglutination to the clinical practice of blood transfusion, research has focused on expanding knowledge of the physiology and genetics of this phenomenon.

Individuals are classified into genetically determined blood groups. Blood from members of certain groups can be mixed without any undesirable consequences (agglutination), but that of members of certain other groups cannot be mixed. The basis of these differences rests on the genetically determined immunological disparities between the blood in various individuals.

CELL PHYSIOLOGY OF AGGLUTINATION. RBC surfaces contain several different *glycoprotein* substances called *agglutinogens* which have *antigenic* properties. The types of agglutinogens are unique to individuals from a common genetic pool. Thus, identical twins have the same sets of agglutinogens, but fraternal twins may have different sets. The agglutinogens may react with antibodylike substances called *agglutinins* present in the *plasma* of other individuals. (See plate 140 for more details on antigen-antibody reaction.) If the blood of an individual with a specific agglutinogen is mixed with blood containing the agglutinin against that agglutinogen, the active sites of the agglutinins will combine with agglutinogens of several RBCs, resulting in the affected RBCs clumping together or "agglutinating." Agglutination of blood may result in anemia and other serious blood and vascular disorders. The antigenic substances present on the RBC are found in some other tissues as well. However, agglutination occurs only in the blood, due to the presence of both plasma agglutinins and agglutinogens of RBCs.

BLOOD GROUPS. Based on the various agglutinogens and agglutinins present in individuals' blood and the miscibility of blood between them, several blood groups have been identified. The *ABO system* and *Rhesus (Rh) system* are the best known. In the ABO system, humans are divided into 4 blood groups, *A, B, AB,* and *O,* on the basis of two agglutinogens, A and B, and their corresponding agglutinins. Members of blood group A have agglutinogen A on their red cells and agglutinin B in their plasma. Members of group B carry agglutinogen B and agglutinin A. Group AB members have both agglutinogens but none of the agglutinins. Members of group O carry neither of the two agglutinogens but have both agglutinins in their plasma.

The blood of group A should not be mixed with that of group B because the latter contains the agglutinin A. Type A blood can be mixed with A, B blood with B, and AB with AB. Members of the O group are called *universal donors* because the absence of the agglutinogens A and B eliminates the chance of agglutination in the recipient. Members of the AB group are called *universal recipients* because the absence of the agglutinins A and B permits them to accept transfusions from the other three types.

THE RH SYSTEM. Another important blood group system is the Rh system, which is based on the presence of the *Rh factor* (*antigen-D, agglutinogen-D*) on the surface of RBCs. Those possessing this factor are called *Rh positive* (Rh+); those lacking it are called *Rh negative* (Rh-). Rh+ people markedly outnumber the Rh- ones (about 6 to 1). In contrast to the ABO system, the agglutinin-D against the Rh factor is not normally circulating but is present within several weeks of exposure to the agglutinogen (the Rh factor). Upon second exposure to the Rh+ blood, the Rh- recipient experiences a severe agglutination reaction. The most serious cases of agglutination due to Rh incompatibility are observed in fetuses and newborns. The offspring of a Rh+ male and a Rh- female will usually (but not always) be Rh+. Thus, the fetus will carry the Rh factor, which is antigenic to the mother's blood. During delivery of the first fetus, some fetal blood is mixed with the Rh- maternal blood. Within a few weeks, the mother produces an antibody against the Rh-agglutinin. During a second pregnancy with an Rh+ fetus, these antibodies may enter the fetal blood, causing agglutination and lysis of the fetal RBCs (*erythroblastosis fetalis* or the *hemolytic disease of the newborn*). These fetuses and newborns are at risk because of severe anemia.

The incidence of this disorder increases with each subsequent Rh+ pregnancy. To prevent the consequences of erythroblastosis fetalis, the newborn's blood can be replaced with Rh- blood, enabling the infant to survive for a few months. By the time the infant's own Rh+ red blood cells are produced, all traces of the maternal Rh-agglutinin will have disappeared. To prevent erythroblastosis fetalis from ever occurring, the Rh- mother can be injected (*vaccinated*) after the first Rh+ pregnancy with some *Rh-agglutinin*. In time, the treated mother produces high titers of *antibodies* against the Rh-agglutinin (itself an antibody). These anti-antibodies deactivate all maternal Rh-agglutinins, preventing their transfer to the next fetus.

CN: Use red for D.
1. Color the four blood groups at the top, noting that the plasma protein (E) in the test tube diagrams remains uncolored. The arrows at the bottom of the tubes represent acceptable transfusion possibilities.
2. Color the blood typing process on the right. Note that the cluster of clumping red cells in the upper right corner are to be colored.
3. Color the samples of blood group percentages in the box on the left. Note that the individual's head (circle) remains uncolored.

BLOOD AGGLUTINATION

RED BLOOD CELL ANTIGENS:*
AGGLUTINOGEN, A
AGGLUTINOGEN, B

BLOOD PLASMA ANTIBODIES:*
AGGLUTININ, A
AGGLUTININ, B

Agglutination is caused by the reaction between glycoprotein substances (agglutinogens) on the surface of red cells from one person, and specific proteins (agglutinins) present in the plasma of another individual. The simultaneous reaction of agglutinins with agglutinogens on several different RBC results in clumping. Agglutination reactions are like immune reactions, agglutinogens acting as antigens and agglutinins as antibodies.

A, B, O BLOOD GROUPS

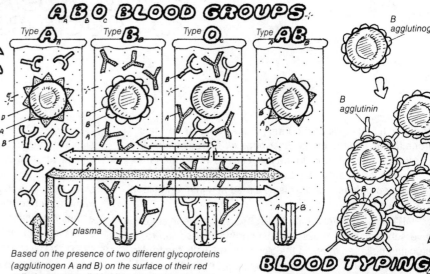

Type A   Type B   Type O   Type AB

plasma

Based on the presence of two different glycoproteins (agglutinogen A and B) on the surface of their red cells, humans are divided into 4 major blood groups. Thus, individuals possessing agglutinogen A are type A, those with agglutinogen B, type B, and those having both agglutinogens, are type AB. Type O people have neither A nor B agglutinogens. The plasma of type A blood contains agglutinin B, plasma of type B blood, agglutinin A. Plasma of type AB has neither agglutinin; plasma of type O, has both agglutinins. Thus, a person with type O blood can donate blood to all other types (universal donor) but can receive only from another O; one with type AB can receive blood from all types (universal recipient) but can give only to another AB. Type A blood must not be mixed with type B and vice versa, or agglutination will result.

Blood types are inherited. Identical twins have the same blood types. Within people of different races, the incidence of each blood type varies markedly.

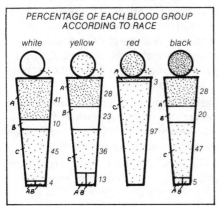

PERCENTAGE OF EACH BLOOD GROUP ACCORDING TO RACE

	white	yellow	red	black
A	41	28	3	28
B	10	23		20
O	45	36	97	47
AB	4	13		5

BLOOD TYPING AGGLUTINATION (CLUMPING)

SERUM SERUM

type B agglutinin A   type A agglutinin B

Blood types of individuals can be determined by mixing samples of their blood (each of the 4 groups is shown here on a separate slide) with a drop of a known serum (plasma) containing either anti-A or anti-B agglutinin (antibody), to see which combination causes agglutination.

RHESUS BLOOD GROUP SYSTEM*

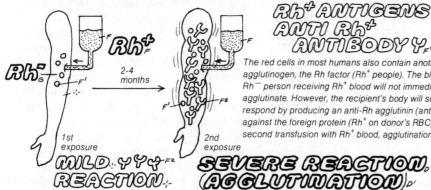

Rh⁻   Rh⁺   2-4 months

1st exposure   2nd exposure

MILD REACTION   SEVERE REACTION (AGGLUTINATION)

Rh⁺ ANTIGENS
ANTI Rh⁺ ANTIBODY Y

The red cells in most humans also contain another agglutinogen, the Rh factor (Rh⁺ people). The blood of a Rh⁻ person receiving Rh⁺ blood will not immediately agglutinate. However, the recipient's body will soon respond by producing an anti-Rh agglutinin (antibody) against the foreign protein (Rh⁺ on donor's RBC). Upon a second transfusion with Rh⁺ blood, agglutinations occurs.

HEMOLYTIC DISEASE OF THE NEWBORN*

1. 1ST PREGNANCY
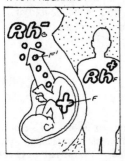
Rh⁻   Rh⁺

2. AFTER DELIVERY

Rh⁻

3. 2ND PREGNANCY
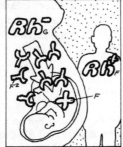
Rh⁻

4. EXCHANGE OF BLOOD

Rh⁻

5. AVOIDING FUTURE PROBLEMS IN STEP 2.

Rh⁻   Rh⁺

(1) Rh⁺ fathers and Rh⁻ mothers may have an Rh⁺ fetus. Fetal cells containing the Rh factor may enter the maternal blood during pregnancy or at birth. (2) Within months, the mother produces anti-Rh antibodies (agglutinin). (3) With a second pregnancy, these antibodies may enter the fetal blood, causing agglutination (erythroblastosis fetalis; hemolytic disease of the newborn). (4) The newborn's blood has to be replaced by Rh⁻ blood to eliminate the maternal anti-Rh. When the infant's own Rh⁺ blood is reproduced again, no maternal anti-Rh will be present to cause agglutination. (5) One preventive approach is to inject the postpartum mother once with anti-Rh agglutinin. Mother will produce antibodies that will deactivate all anti-Rh agglutinins, eliminating any future effect.

HEMOSTASIS & PHYSIOLOGY OF BLOOD CLOTTING

Because blood flows continuously in the vascular bed, it is prone to leave the body quickly whenever there is either an external or internal injury to the tissues. The vital importance of blood to tissue survival has produced a variety of preventive and defense mechanisms aimed at minimizing blood loss during injury.

IMPORTANCE OF VASOCONSTRICTION. Tissue injury often severs the connective tissue and a portion of vasculature, exposing collagen fibers in the blood vessel wall. The fragile blood platelets flowing by these rough surfaces adhere and rupture, releasing their serotonin, a potent local vasoconstrictor agent that immediately stimulates contraction of smooth muscle cells in the wall of injured arterioles and even the smaller arteries. This constriction effectively reduces and/or blocks blood flow in these vessels.

PLATELET PLUG AND BLOOD CLOT FORMATION. The vasoconstriction is a highly effective but temporary hemostatic (blood stopping) measure. This initial defense mechanism is followed by a longer lasting response consisting of formation of a plug to fill the site of injury with a temporary protective tissue until tissue regeneration repairs the wall. Thus, platelet rupture releases another substance, ADP (adenosine diphosphate), at the injury site. ADP, like serotonin, is normally stored in the platelet vesicles. ADP causes the neighboring platelets to adhere to those already bound to the injured wall, causing a clumping of the platelets (platelet aggregation). The aggregate gradually grows, finally forming a temporary hemostatic plug to prevent blood leakage. This plug resembles a blood clot when blood is allowed to stand outside the body. Next, this platelet plug is reinforced by deposition of a meshwork of fibrin fibers. This fibrin net traps the RBCs and platelets, forming a fairly rigid and strong barrier against further blood loss. Initially loose, the fibrin net becomes gradually tight, at which point it is called a blood clot.

BIOCHEMISTRY OF CLOT FORMATION. Fibrin is a fibrous protein formed by the action of the protease enzyme thrombin on fibrinogen (profibrin), a circulating protein made by the liver. Thrombin is normally present in the blood as its inactive form, prothrombin. The activation of prothrombin, the key step in the clotting mechanism, requires the presence of calcium ions and a protein factor called factor X (ten). Activation of factor X can occur by either of two pathways: the intrinsic (blood) pathway involves the activation of factor XII, which originates from blood-related sources. The extrinsic (tissue) pathway involves the production from the injured tissue of

another enzyme called thromboplastin (factor III). Thromboplastin can directly activate factor X, but factor XII must activate several other factors, which in turn activate factor X. The precipitated fibrin is initially loose; in the presence of another blood factor (factor XIII), it becomes tight, rigidifying the clot. In the absence of injury, circulating anticlotting factors such as antithrombin or possibly heparin prevent thrombin activation and clot formation.

CLOT CONTRACTION AND DISSOLUTION. Once a clot forms, it begins to contract. Contraction is an active process involving utilization of ATP and contraction of actin filaments in the platelet pseudopods. Clot contraction causes extrusion of the plasma trapped within the clot and shortening of the pseudopods. Because the edges of the clot are attached to the edges of the injured tissue, clot contraction is believed to bring the injured edges closer together, improving hemostasis and facilitating wound closure and repair.

The final stage in the life of a blood clot is its dissolution, brought about by the action of the enzyme plasmin (fibrinolysin), which digests the fibrin net, resulting in clot breakdown. Plasmin is formed from a precursor called plasminogen.

ABNORMALITIES OF CLOT FORMATION. Several disease conditions or certain nutritional deficiencies interfere with proper clotting and pose serious hazards to the individual. Hemophilia (bleeding sickness) is a series of hereditary diseases characterized by deficient hemostasis and continued blood loss after injury. The causes of these hereditary diseases are the lack of one of the blood clotting factors. In type A hemophilia, which is most frequently (75%) observed, the individual is deficient in factor VIII. The disease mainly affects males. The most famous case is the family of Queen Victoria of England, in which many of the male children fell victim to hemophilia. To prevent hemophilia, the missing clotting protein must be provided externally. Large scale production of such proteins by the application of modern bioengineering methods promises to prevent hemophilic bleeding.

Reduced platelet production (thrombocytopenia) by the bone marrow caused by ionizing radiation damage, disease, or toxic exposure of the bone marrow to drugs is another cause of deficient clotting. A third cause is dietary deficiency of vitamin K. This vitamin, normally provided in the food or by the intestinal bacteria, does not take part in clotting directly but is required for the synthesis of prothrombin in the liver. Newborn infants in whom the digestive tract is still devoid of bacteria are deficient in vitamin K and are therefore more susceptible to bleeding if injured.

CN: Use red for A and dark colors for I and L.
1. Completely color the illustrations 1-5 depicting the formation of a blood clot. Do not color the material including the two pathways (under illustrations 3 and 4) until you have completed illustration 5. Note that in 1 and 2 the blood (A) is colored as a solid band, whereas in 3-5 only the red cells (A¹) and the platelets (E) are colored.
2. Color the diagram of the two pathways, leading to the dissolution of a blood clot (under 5).
3. Color the three conditions that prevent blood clotting.

FORMATION OF THE BLOOD CLOT.

1. INJURY TO WALL OF BLOOD VESSEL *

Injury (a cut) to the blood vessel is normally followed by a series of reactions that result in the formation of a blood clot, which seals the injured opening and prevents the loss of blood (hemostasis).

2. VASOCON-STRICTION *

Adhesion of blood platelets to the exposed collagen fibers (in the wall of the injured vessel) causes the release of serotonin from platelets, resulting in strong vasoconstriction.

3. & 4. PLATELET PLUG FORMATION *

Contact of the platelets to collagen in the injured wall releases ADP, which attracts more platelets and stimulates the formation of pseudopods in the platelets. The pseudopods enable the platelets to bind together, forming a temporary plug to stop the blood loss.

5. CLOT FORMATION *

Soon, fibrinogen, a blood protein, is converted to fibrin; fibrin forms a net over the platelets. Red cells in the exterior of the plug adhere to this net. The combination of platelets and red cells entangled within a tight fibrin net forms a blood clot, a stronger and more permanent plug to stop blood loss.

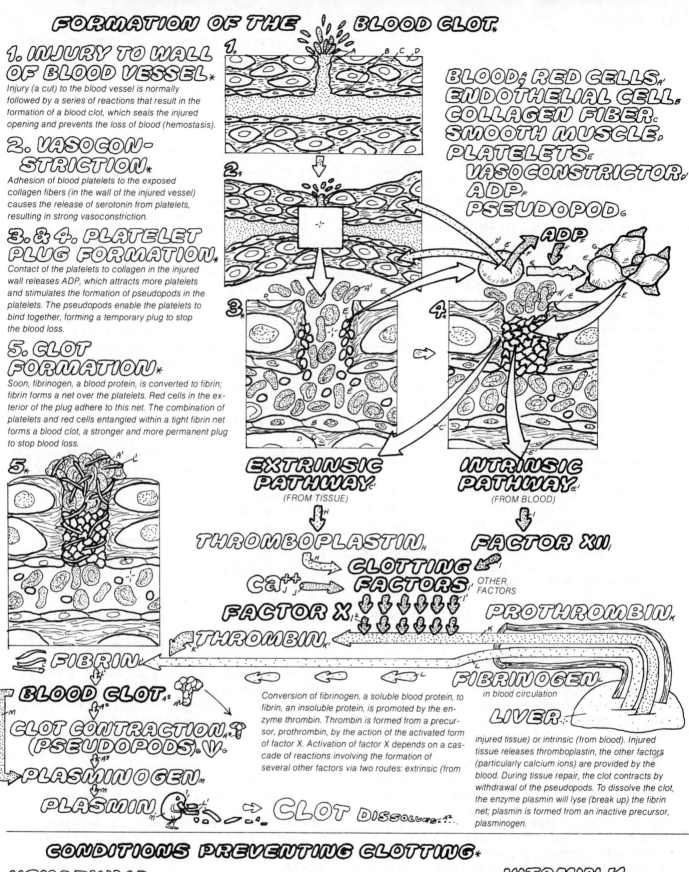

BLOOD; RED CELLS, A'
ENDOTHELIAL CELL, B
COLLAGEN FIBER, C
SMOOTH MUSCLE, D
PLATELETS, E
VASOCONSTRICTOR, D'
ADP, F
PSEUDOPOD, G

ADP, F

EXTRINSIC PATHWAY, C' (FROM TISSUE)

INTRINSIC PATHWAY, E' (FROM BLOOD)

THROMBOPLASTIN, H

FACTOR XII, I

Ca^{++}, J

CLOTTING FACTORS

OTHER FACTORS

FACTOR X, J2

PROTHROMBIN, K

THROMBIN, K'

FIBRIN, L

FIBRINOGEN, L

in blood circulation

LIVER, K

Conversion of fibrinogen, a soluble blood protein, to fibrin, an insoluble protein, is promoted by the enzyme thrombin. Thrombin is formed from a precursor, prothrombin, by the action of the activated form of factor X. Activation of factor X depends on a cascade of reactions involving the formation of several other factors via two routes: extrinsic (from injured tissue) or intrinsic (from blood). Injured tissue releases thromboplastin, the other factors (particularly calcium ions) are provided by the blood. During tissue repair, the clot contracts by withdrawal of the pseudopods. To dissolve the clot, the enzyme plasmin will lyse (break up) the fibrin net; plasmin is formed from an inactive precursor, plasminogen.

BLOOD CLOT, A2

CLOT CONTRACTION, G2 (PSEUDOPODS), V

PLASMINOGEN, M

PLASMIN, M

CLOT DISSOLVES, A2

CONDITIONS PREVENTING CLOTTING *

HEMOPHILIA *

Hemophilia is a hereditary disorder in which one or more of the blood clotting factors are absent. As a result, blood clots slowly.

CLOTTING FACTORS

FIBRIN, V

THROMBOCYTOPENIA *

In the disease thrombocytopenia, platelet formation by the red bone marrow is defective. Platelet deficiency will prevent clotting.

PLATELET PRODUCTION

RED MARROW, A4

VITAMIN K DEFICIENCY *

Vitamin K, provided in the diet or by intestinal bacteria, is one of the factors necessary for clotting, as it is needed for prothrombin synthesis in the liver.

PROTHROMBIN, K

CLOTTING FACTORS

WHITE BLOOD CELLS & DEFENSE OF THE BODY

TYPES AND GENERAL FUNCTIONS OF WHITE BLOOD CELLS. Although there are several types of *white blood cells* (WBCs, leukocytes) and they vary in morphology, they all share a common function: helping to defend the body against foreign microbial infections. On the basis of the presence or absence of *specific granules* in their cytoplasm, white blood cells are divided into *granulocytes* (those with granules, i.e., *neutrophils, eosinophils,* and *basophils*) and *agranulocytes* (those without granules, i.e., *monocytes, macrophages,* and *lymphocytes*). Functionally, white blood cells may be divided into two broad categories: (1) those that participate in *nonspecific inborn immune responses* to infections and *inflammations* caused by tissue injury; and (2) those that take part in the *acquired immune responses.* Lymphocytes participate mainly in the second category; other white cells take part in the first.

Members of the family of granulocytes and agranulocytes originate in the *bone marrow,* where they are formed by the proliferative division of committed *stem cells.* Upon entry in the circulation, most of the WBCs participate in the inborn and nonspecific defensive reactions to invading infectious agents as well as in response to tissue injury and inflammation. The less numerous lymphocytes (they have no granules) originate from another line of stem cells that reside either in the bone marrow or in parts of the *lymphatic system.* Upon formation, the immature lymphocytes temporarily migrate into certain lymphatic organs (*lymph nodes, thymus*), where they differentiate and mature, becoming specialized to carry out their major function: defending the body against invading microorganisms through acquired immune reactions.

Various types of white blood cells are present in different proportions in the blood. Granulocytes are more numerous than agranulocytes. Among granulocytes, neutrophils are the most abundant cells; among the agranulocytes, lymphocytes outnumber the others.

NATURAL (NONSPECIFIC) IMMUNITY. To understand the functions of granulocytes and the phagocytic agranulocytes, we will consider their responses to tissue injury. Upon injury to the protective epithelial tissue covering the body, *microbes* (e.g., bacteria) enter the body, release their toxins, and create local infection. This stimulates the *mast cells* (which resemble the basophils but reside in tissues) to release their granules containing *heparin* and *histamine* within the tissue spaces. Nearby *basophils* may do the same in the blood. Heparin may prevent blood coagulation; histamine causes *vasodilation* and *increased permeability* of the local blood vessels to *blood proteins* and blood cells. Blood proteins and fluids leak into the injured site, causing *edema* or *swelling.*

Gradually, the fluid in the swelling clots, trapping the bacteria and preventing their further penetration into the body.

At this time, the *tissue macrophages,* found permanently residing in many tissues like skin and lungs, attack the microbes and destroy them by *phagocytosis.* For this reason, the tissue macrophages are called the *first line of defense.* Phagocytosis consists of engulfing the microbes via the formation of the *pseudopods* followed by *endocytosis* of the *phagocytic vesicle.* Next, the endocytotic vesicle is incorporated into the *lysosomes* of the phagocytes, where the microbe is digested by *lysosomal enzymes.* If infection persists, the *neutrophils* are attracted to the injury site. Indeed, a few hours after injury, the number of neutrophils increases by several fold in the blood and particularly near the infection site. The neutrophils squeeze through the spaces between the *capillary endothelial cells* by forming *filopodia* and displacing themselves (*diapedesis*). Once inside the injured site, the neutrophils begin to phagocytize the microbes in the same manner as the tissue macrophages. Neutrophils make up the *second line of defense.*

If the tissue macrophages and neutrophils do not adequately counter the infection, then the *agranular monocytes* move into the injury site in the same manner as the neutrophils. Monocytes are initially small and incapable of phagocytosis. Within an hour after leaving the blood, they enlarge, attaining a shape like the tissue macrophages. Then they begin to phagocytize the microbes and the dead neutrophils. Monocytes may in fact be the source of new tissue macrophages, which die after phagocytosis. The monocytes are called the *third line of defense.* Usually, these three lines of defense are sufficient to eliminate the source of infection.

During the course of these anti-infectious and inflammatory responses, the number of white cells (particularly the phagocytes) in the blood increases. This is caused by humoral factors released from the injured tissue and/or certain white cells. As a result, permeability of blood *sinusoids* in the bone marrow increases, releasing fresh neutrophils and monocytes into the blood. The phagocytes find their way to the site of injury by *chemotaxis* or similar guiding mechanisms.

Gradually, the *fibroblast* cells of the connective tissue proliferate, sealing off the injured tissue to begin repair. A *pus sac,* containg fluid, dead cells, and dead microbes, forms. This pus is either extruded or gradually cleared off by the macrophages. If these nonspecific rapid natural defense reactions are not sufficient to eliminate the infection, the toxin intrusion in the blood activates other defensive responses such as the fever reaction and, more effectively, lymphocyte reactions, which lead to acquired immune responses (see plate 140).

CN: Use red for A, purple for J. Lightest colors for structures C-H. Dark colors for I, K, and N.
1. Color the various white blood cells at the top of the page, beginning with their origin in red bone marrow (A).
2. Color the nonspecific response to a microbe invasion, following the numbered titles. When coloring the second and third boxes, color in the background or larger structures before coloring the smaller ones, such as proteins (K) or microbes (I). Note that for number 3, color the tiny histamine molecules as well as the mast cell (E¹). Color the numeral 6, but not the arrow representing the movement of fluid into the tissues.
3. Color the enlargement of phagocytosis and the macrophage action below it.

WHITE BLOOD CELLS

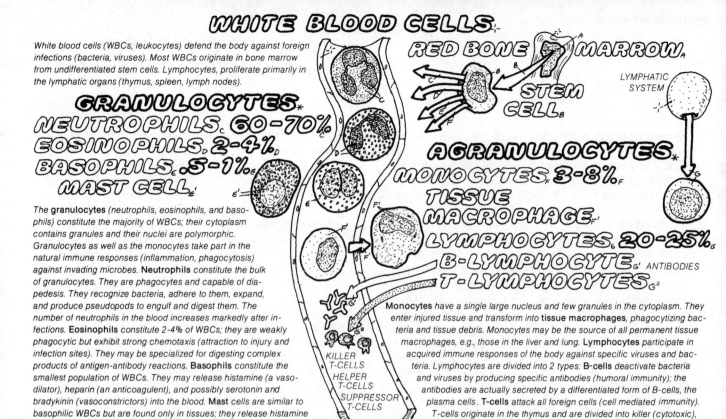

White blood cells (WBCs, leukocytes) defend the body against foreign infections (bacteria, viruses). Most WBCs originate in bone marrow from undifferentiated stem cells. Lymphocytes, proliferate primarily in the lymphatic organs (thymus, spleen, lymph nodes).

GRANULOCYTES *
NEUTROPHILS 60-70%
EOSINOPHILS 2-4%
BASOPHILS .5-1%
MAST CELL

The **granulocytes** (neutrophils, eosinophils, and basophils) constitute the majority of WBCs; their cytoplasm contains granules and their nuclei are polymorphic. Granulocytes as well as the monocytes take part in the natural immune responses (inflammation, phagocytosis) against invading microbes. **Neutrophils** constitute the bulk of granulocytes. They are phagocytes and capable of diapedesis. They recognize bacteria, adhere to them, expand, and produce pseudopods to engulf and digest them. The number of neutrophils in the blood increases markedly after infections. **Eosinophils** constitute 2-4% of WBCs; they are weakly phagocytic but exhibit strong chemotaxis (attraction to injury and infection sites). They may be specialized for digesting complex products of antigen-antibody reactions. **Basophils** constitute the smallest population of WBCs. They may release histamine (a vasodilator), heparin (an anticoagulent), and possibly serotonin and bradykinin (vasoconstrictors) into the blood. **Mast cells** are similar to basophilic WBCs but are found only in tissues; they release histamine and heparin from their granules.

RED BONE MARROW
STEM CELL

LYMPHATIC SYSTEM

AGRANULOCYTES *
MONOCYTES 3-8%
TISSUE MACROPHAGE
LYMPHOCYTES 20-25%
B-LYMPHOCYTE ANTIBODIES
T-LYMPHOCYTES

Monocytes have a single large nucleus and few granules in the cytoplasm. They enter injured tissue and transform into **tissue macrophages**, phagocytizing bacteria and tissue debris. Monocytes may be the source of all permanent tissue macrophages, e.g., those in the liver and lung. **Lymphocytes** participate in acquired immune responses of the body against specific viruses and bacteria. Lymphocytes are divided into 2 types: **B-cells** deactivate bacteria and viruses by producing specific antibodies (humoral immunity); the antibodies are actually secreted by a differentiated form of B-cells, the plasma cells. **T-cells** attack all foreign cells (cell mediated immunity). T-cells originate in the thymus and are divided into killer (cytotoxic), helper and suppressor subtypes.

KILLER T-CELLS
HELPER T-CELLS
SUPPRESSOR T-CELLS

NATURAL IMMUNITY / NON SPECIFIC RESPONSE: *
INFLAMMATION & PHAGOCYTOSIS

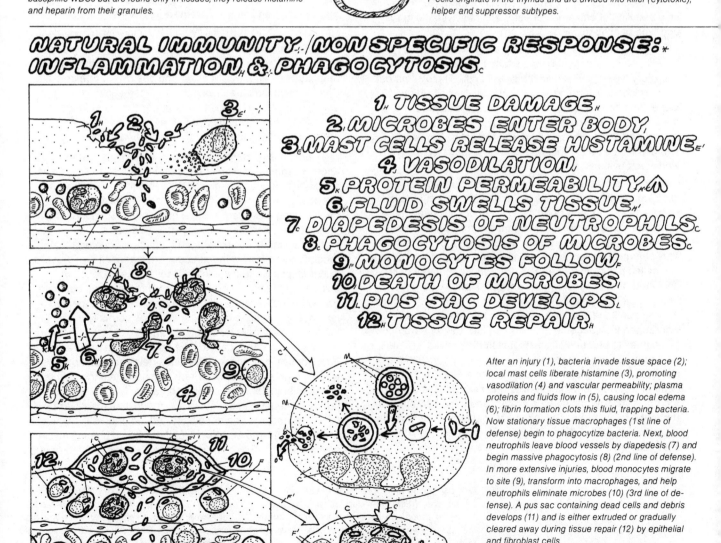

1. TISSUE DAMAGE
2. MICROBES ENTER BODY
3. MAST CELLS RELEASE HISTAMINE
4. VASODILATION
5. PROTEIN PERMEABILITY ↑
6. FLUID SWELLS TISSUE
7. DIAPEDESIS OF NEUTROPHILS
8. PHAGOCYTOSIS OF MICROBES
9. MONOCYTES FOLLOW
10. DEATH OF MICROBES
11. PUS SAC DEVELOPS
12. TISSUE REPAIR

After an injury (1), bacteria invade tissue space (2); local mast cells liberate histamine (3), promoting vasodilation (4) and vascular permeability; plasma proteins and fluids flow in (5), causing local edema (6); fibrin formation clots this fluid, trapping bacteria. Now stationary tissue macrophages (1st line of defense) begin to phagocytize bacteria. Next, blood neutrophils leave blood vessels by diapedesis (7) and begin massive phagocytosis (8) (2nd line of defense). In more extensive injuries, blood monocytes migrate to site (9), transform into macrophages, and help neutrophils eliminate microbes (10) (3rd line of defense). A pus sac containing dead cells and debris develops (11) and is either extruded or gradually cleared away during tissue repair (12) by epithelial and fibroblast cells.

PHAGOCYTOSIS
LYSOSOME

Phagocytes engulf bacteria and digest them within their lysosomes.

LYMPHOCYTES AND ACQUIRED IMMUNITY

In contrast to phagocytes, the *lymphocytes* participate in a more complicated type of immune response that develops *slowly* and *specifically* against particular foreign substances (*antigens*). This response is expressed only *after* exposure to the antigen (hence the term *acquired immunity*), although the ability to respond to specific types of antigen is genetically programmed. There are two categories of acquired immune responses, each mediated by a different family of lymphocytes: *humoral-* or *antibody-mediated* and *cell-mediated*.

ANTIBODY-MEDIATED RESPONSE: FUNCTIONS OF B-LYMPHOCYTES. Let us assume that a certain bacteria penetrates the blood after an injury. The bacterial wall contains *proteins* or *polysaccharides*, which are foreign to the body and considered harmful. These substances are called antigens, and their presence is sensed by special receptor molecules located on the surface of a certain types of circulating lymphocytes called the *B-lymphocytes*. (The "B-" comes from Bursa of Fabricus, an avian lymph organ generating these cells; the source of B-cells in the human body is probably the *bone marrow*.) Each type of B-lymphocyte contains only one kind of antigenic receptor and can respond to only one type of antigen.

In the *lymph nodes*, the intruding antigen is sensed by the B-cells, which become sensitized and transform to a larger secretory type of cell called the *plasma cell*. The plasma cell then proliferates, forming a *clone*, and all the cells in the clone synthesize a specific protein molecule called the *antibody*, which is secreted to the plasma. Upon encountering the antigen, the antibodies bind with the antigen molecules and deactivate them. This whole process takes from *days to weeks* to develop.

After the antigens are deactivated, the antibodies usually diminish in number. However, upon second exposure to the same antigen, the body's antibody production is often more *rapid* and more *intense*, as though the immune system has "learned" to deal with this particular antigen more efficiently. This enhanced response is due to a particular type of plasma cell called the *memory cell*. B-cells produce memory cells upon their first exposure to the antigen. Memory cells learn how to produce the antibody but do not do so at first. Instead, they rest until the second exposure to the same antigen, whereupon they become activated rapidly and form numerous clones to produce large amounts of the antibody. The memory cells are involved in immunization by *vaccination*. Here the body is intentionally exposed to a small amount of dead or transformed antigen (e.g., dead smallpox virus) in order to sensitize the immune system and form memory cells. When the body is exposed to the same antigen later (e.g., during a real smallpox infection), antibody production will be quick and intense.

BIOCHEMISTRY OF THE ANTIBODY-ANTIGEN REACTION. All the antibodies produced against the many different antigens are *protein molecules* (*immunoglobulins, Ig*) possessing both common and diverse features. Each antibody is roughly Y-shaped, consisting of *heavy chains* and two *light chains*. The heavy chains provide the *constant* part of the antibody, which is the same in all antibodies; the light chains, located in the arms of the Y (attached to the heavy chains), constitute the *variable* and functionally significant part of the molecule. Thus, each antibody has two sites, one on each of the variable arm, for interaction with the antigen. Antibodies can deactivate antigens by *direct* combination, causing *precipitation* (agglutination) or by *masking* the active sites of the antigens. Antibodies can also achieve the same goals *indirectly* by activating the *complement system*, which consists of a series of enzymes arranged to catalyze a cascade of chemical events. The combination of a single antibody molecule with the antigen activates this cascade, which rapidly mobilizes millions of enzymes that quickly lyse the microorganism to which the antigen is attached or cause agglutination and similar defensive reactions.

CELL-MEDIATED IMMUNITY: FUNCTIONS OF T-LYMPHOCYTES. Another family of lympocytes known as *T-lymphocytes* ("T-" for thymus) participates in acquired immune responses by directly attacking and destroying foreign cells. T-cells responses are involved in defense against the slow-acting bacteria such as tuberculosis and against fungal infections. T-cells are also involved in rejecting transplanted organs and eliminating cancer cells in the body.

When a tissue from one organism is transplanted into another organism (even of the same species), the antigenic substances in the transplant *sensitize* certain T-cells within the host's lymph nodes. The *sensitized* T-cells proliferate, transforming into a family of T-cells. The most important of these is the *cytotoxic* or *killer* T-cell. This cell contains on its surface antibody like substances (*antigen receptors*) that recognize and bind with the surface antigens of foreign cells (the transplant). Next the killer T-cell infuses *lysosomal enzymes* into the foreign cell, causing its lysis and death. In general, T-cell-mediated immunity is based upon the differentiation (recognition) of *self* antigens normally present in the host's body cells from *nonself* antigens present in the foreign cells and cancer cells. This ability to differentiate the self from nonself is acquired *early* in life (fetal-neonatal periods) when the precursor cells of T-cells migrate into the *thymus* gland and inhabit this lymphatic organ for a while. Thymus removal in early life, but not in adulthood, causes severe T-cell-mediated immune deficiency. Indeed, the adult thymus becomes fatty and atrophic. T-cell number is greatly deficient in victims of AIDS (Acquired Immune Deficiency Syndrome).

CN: Use the same colors as on the previous page for microbe/antigen (A). Use red for blood circulation (H), though the portion shown in the box in the upper left corner is actually a capillary site that normally receives purple colors (as on the previous page). The lymph node (C) receives the same color as lymphocyte on the previous page, but the B- and T-lymphocyte cells on this page will receive two different colors.

1. Start with the upper panel and follow the numbered sequence, beginning in the upper left rectangle. Color the lymph production and storage sites in the body. Color the diagrammatic material on antibodies and the complement system. When coloring the immunization chart, note that the memory cell (G), shown in the large lymph node at the top of the plate, is responsible for the increase in antibodies.

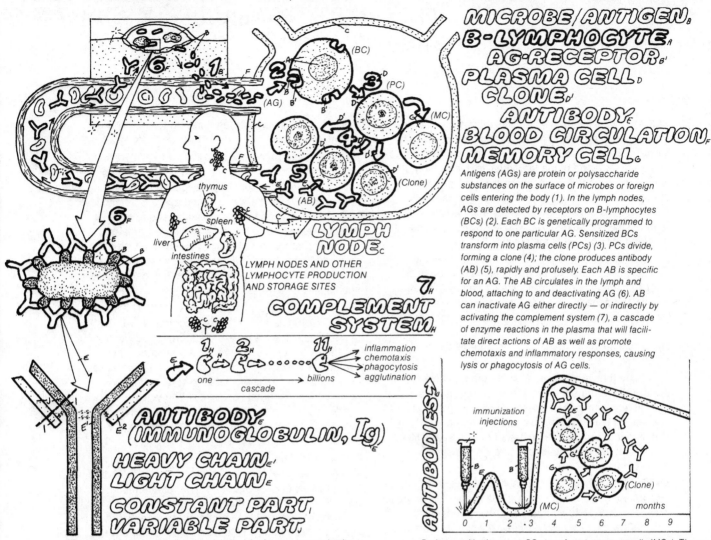

MICROBE/ANTIGEN.
B-LYMPHOCYTE.
AG-RECEPTOR.
PLASMA CELL.
CLONE.
ANTIBODY.
BLOOD CIRCULATION.
MEMORY CELL.

Antigens (AGs) are protein or polysaccharide substances on the surface of microbes or foreign cells entering the body (1). In the lymph nodes, AGs are detected by receptors on B-lymphocytes (BCs) (2). Each BC is genetically programmed to respond to one particular AG. Sensitized BCs transform into plasma cells (PCs) (3). PCs divide, forming a clone (4); the clone produces antibody (AB) (5), rapidly and profusely. Each AB is specific for an AG. The AB circulates in the lymph and blood, attaching to and deactivating AG (6). AB can inactivate AG either directly — or indirectly by activating the complement system (7), a cascade of enzyme reactions in the plasma that will facilitate direct actions of AB as well as promote chemotaxis and inflammatory responses, causing lysis or phagocytosis of AG cells.

LYMPH NODE.

LYMPH NODES AND OTHER LYMPHOCYTE PRODUCTION AND STORAGE SITES

COMPLEMENT SYSTEM.

inflammation
chemotaxis
phagocytosis
agglutination

one → billions
cascade

ANTIBODY (IMMUNOGLOBULIN, Ig)
HEAVY CHAIN.
LIGHT CHAIN.
CONSTANT PART.
VARIABLE PART.

ABs are protein molecules (immunoglobulin, Ig), of two or more subunits. Each subunit consists of a heavy and a light polypeptide chain. Each chain has a constant part (same in all ABs) and a variable part (different in each AB). The variable part endows ABs with the ability to recognize the various AGs (i.e., selectivity and specificity).

immunization injections

ANTIBODIES

months
0 1 2 .3 4 5 6 7 8 9

During sensitization, some PCs transform to memory cells (MCs). These remain dormant in the lymph nodes. Upon further exposure to the AG, MCs will evoke a pronounced, exaggerated response (AB production) that will rapidly deactivate AGs. The MC response is the basis of immunization and vaccination practices.

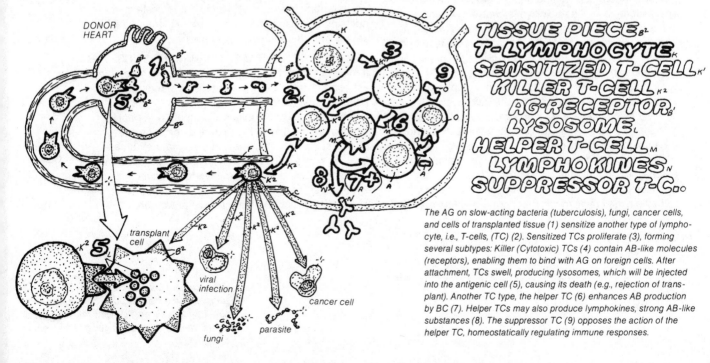

TISSUE PIECE.
T-LYMPHOCYTE.
SENSITIZED T-CELL.
KILLER T-CELL.
AG-RECEPTOR.
LYSOSOME.
HELPER T-CELL.
LYMPHOKINES.
SUPPRESSOR T-C.

DONOR HEART

transplant cell

viral infection

cancer cell

fungi

parasite

The AG on slow-acting bacteria (tuberculosis), fungi, cancer cells, and cells of transplanted tissue (1) sensitize another type of lymphocyte, i.e., T-cells, (TC) (2). Sensitized TCs proliferate (3), forming several subtypes: Killer (Cytotoxic) TCs (4) contain AB-like molecules (receptors), enabling them to bind with AG on foreign cells. After attachment, TCs swell, producing lysosomes, which will be injected into the antigenic cell (5), causing its death (e.g., rejection of transplant). Another TC type, the helper TC (6) enhances AB production by BC (7). Helper TCs may also produce lymphokines, strong AB-like substances (8). The suppressor TC (9) opposes the action of the helper TC, homeostatically regulating immune responses.

THE REPRODUCTIVE SYSTEM: AN INTRODUCTION

SEX AND REPRODUCTION. Until this point, we have studied those physiologic systems that are essentially identical in both males and females and that function to ensure the survival of the individual. In this section, we will focus on the *reproductive system* (genital system, genitalia), whose parts and organs are sexually dimorphic (i.e., they are structurally and functionally different in the two sexes) and whose function is aimed at ensuring the survival of the species.

The organs of the reproductive system grow and function in response to the stimulation provided by the male and female sex hormones secreted by the sex glands, the gonads. The gonads are in turn stimulated by the gonadotropic hormones released by the anterior pituitary gland. In the absence of these hormonal stimuli, these target glands and organs will cease to function and will atrophy.

Though the various organs of the reproductive system are formed during the embryonic period, the normal functions of this system begin during puberty and last for thirty to thirty-five years in women, terminating in "menopause" that occurs in the early fifties. In men, reproductive functions decline slowly with advancing age.

MALE REPRODUCTIVE ORGANS AND THEIR FUNCTIONS. As shown in the illustrations (lower left), in the male, the main sexual organs are *testes, prostate, seminal vesicles, vas deferens, epidiymis, bulbourethral glands, penis,* and *scrotum*. The latter two are external organs, and the rest are internal. Of the organs mentioned, the two testes (testicles) are the only ones with endocrine functions, secreting the hormone testosterone, which is the most potent of the family of androgenic hormones. The testes also produce the male gametes, spermatozoa (sperm), in a process called spermatogenesis. The epididymis consists of convoluted tubules that act to store and mature the sperm. The vas deferens is a conduit for sperm delivery during emission and ejaculation, events occurring during sexual excitation in the male. The prostate and seminal vesicles are exocrine glands producing the plasma of the semen, which is essential for the activity and survival of the sperm within the female reproductive system. The penis with its inflatable tissue acts as the organ of intromission, delivering sperm through its urethral canal and depositing them in the vagina of the female, near the uterine cervix. The scrotum is a sac containing the testicles which, through extension and retraction, maintains the the temperature of the testes a few degrees below body temperature to ensure spermatogenesis.

MALE SECONDARY SEX CHARACTERISTICS. In the human male, the secondary sexual characteristics (which appear after puberty in response to increasing testosterone levels) are active and aggressive attitudes, larger body size, enhanced muscular and skeletal growth, wide shoulders and narrow pelvis, enlarged larynx and vocal cords leading to a lower pitched voice, facial and body hair, pubic and axillary (armpit) hair, receding scalp hairlines, and baldness (if genetically susceptible).

FEMALE REPRODUCTIVE ORGANS AND THEIR FUNCTIONS. As shown in the diagrams (lower right), in the female, the main sexual and reproductive organs are the *ovary, uterus, uterine tube* (Fallopian tube, oviduct), and *vagina,* which constitute the internal sex organs. The labia majora, labia minora, and clitoris constitute the external sex organs (the vulva). The two ovaries act in part as the main endocrine glands of the system, secreting estrogen and progesterone, the female sex hormones.

In addition, the ovaries are the site of formation and release of the female gametes, the ova or eggs, by a process called oogenesis. The uterine tubes transport the unfertilized egg, as well as the young embryo. The uterus is the organ of pregnancy, providing a nest for implantation and growth of the young embyro. The uterus is also involved in labor contractions during delivery (parturition). The vagina is adapted to receive the penis and sperm during intromission and ejaculation. It also acts during delivery as the birth canal.

The female external genitalia, particularly the clitoris, are important in sexual excitation. The female breasts contain fatty tissue and the mammary glands, which secrete milk for nourishment of the newborn.

FEMALE SECONDARY SEX CHARACTERISTICS. The female secondary sexual characteristics, promoted by estrogen or absence of androgens, are enhanced subcutaneous fat deposits (providing for the shape of breasts, buttocks, and thighs in women), wide pelvis and narrow shoulders, high-pitched voice, non-receding scalp hairlines, and soft skin. Mature human females possess, like the male, axillary and pubic hair; pubic hair has the form of an inverted triangle, the opposite of its form in the male. The absence of facial and body hair is also characteristic of women.

CN: Use dark colors for C and H.
Begin with the male system, coloring the same structure in both the side view and the smaller frontal view above, before going on to the next structure.

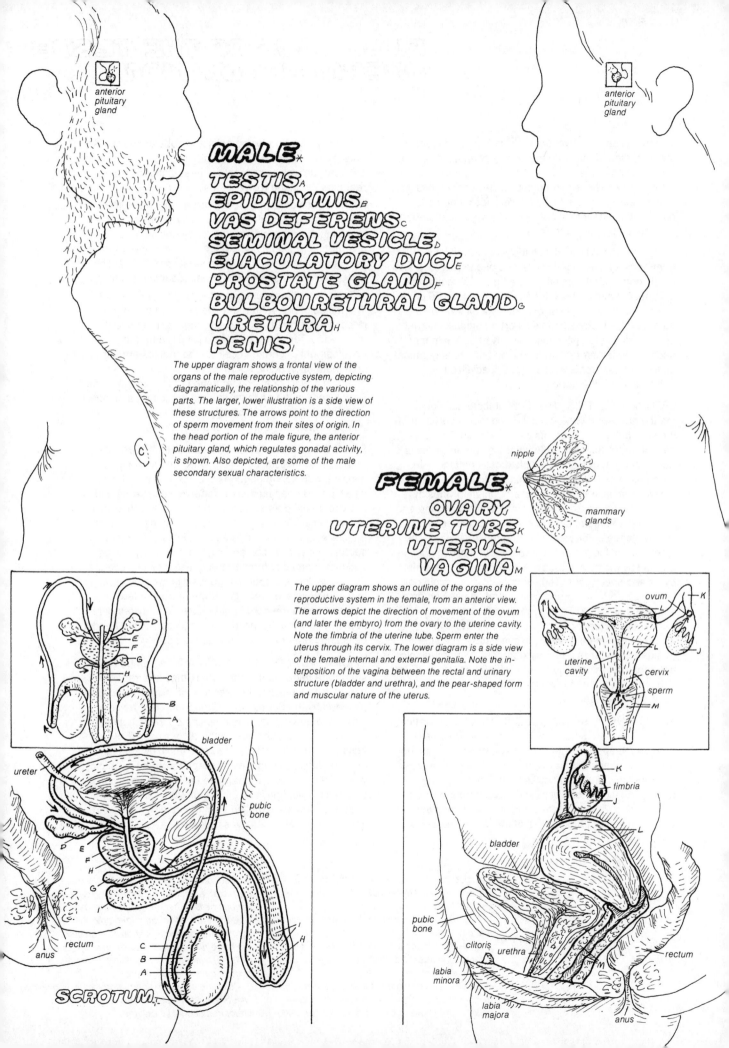

anterior
pituitary
gland

MALE*

TESTIS A
EPIDIDYMIS B
VAS DEFERENS C
SEMINAL VESICLE D
EJACULATORY DUCT E
PROSTATE GLAND F
BULBOURETHRAL GLAND G
URETHRA H
PENIS I

The upper diagram shows a frontal view of the organs of the male reproductive system, depicting diagramatically, the relationship of the various parts. The larger, lower illustration is a side view of these structures. The arrows point to the direction of sperm movement from their sites of origin. In the head portion of the male figure, the anterior pituitary gland, which regulates gonadal activity, is shown. Also depicted, are some of the male secondary sexual characteristics.

anterior
pituitary
gland

nipple

mammary
glands

FEMALE*

OVARY J
UTERINE TUBE K
UTERUS L
VAGINA M

The upper diagram shows an outline of the organs of the reproductive system in the female, from an anterior view. The arrows depict the direction of movement of the ovum (and later the embyro) from the ovary to the uterine cavity. Note the fimbria of the uterine tube. Sperm enter the uterus through its cervix. The lower diagram is a side view of the female internal and external genitalia. Note the interposition of the vagina between the rectal and urinary structure (bladder and urethra), and the pear-shaped form and muscular nature of the uterus.

ovum
K
L
J
uterine
cavity
L
cervix
sperm
M

bladder
ureter
pubic
bone
D
E
F
H
G
rectum
anus
C
B
A
I
H
SCROTUM

K
fimbria
J
L
bladder
pubic
bone
clitoris
urethra
labia
minora
M
rectum
labia
majora
anus

FUNCTIONS OF THE TESTES: SPERM FORMATION

The testes perform two functions: (1) *spermatogenesis*, the formation of the male gametes (*spermatozoa*), and (2) secretion of the male sex hormone, *testosterone*. Spermatogenesis is carried out by the *seminiferous tubules* packed in the *testis lobules*. Testosterone is produced by the *interstitial cells* of Leydig, scattered in the spaces between tubules.

TESTIS STRUCTURE. Each testis is divided into numerous lobules, each containing one to four long and highly convoluted seminiferous tubules. Each of these is about 0.2 mm in diameter and may be up to 70 cm in length. Each tubule is surrounded by a *basement membrane*, which helps to support a complex internal epithelium. In the epithelium, directly attached to the basement membrane, are found the primordial germinal cells, *spermatogonia*, and a highly specialized type of epithelial cells, the *Sertoli cells*.

SPERMATOGENESIS. This is an intricate developmental process involving mitotic and meiotic division of the spermatogonium and its daughter cells. This process requires the support functions of the Sertoli cells. Each spermatogonium is a diploid cell containing 46 chromosomes (22 pairs of somatic and one pair of sex chromosomes, XY). The spermatogonium divides by *mitosis*, giving rise to two cells, one of which adheres to the basement membrane and maintains the germinal line, while the other separates and moves inward. The latter cell, called the *primary spermatocyte*, begins division by *meiosis* to produce *secondary spermatocytes* and *spermatids*. The spermatids are haploid cells, having 22 somatic chromosomes and one sex chromosome, either X or Y. The spermatogenic cells are frequently interconnected by *cytoplasmic bridges*, which enable them to divide in synchrony. The spermatids will undergo a complex process of morphologic differentiation (*spermiogenesis*), which transforms them into unique, structurally and functionally specialized cells, the *spermatozoa*.

FUNCTIONS OF SERTOLI CELLS. These cells participate in spermatogenesis in several ways. They help support the spermatogenic cells and move them along their inward migration within the epithelium. They provide nutrients and metabolites to the spermatocytes and spermatids; these cells are isolated from the vascular supply. They secrete a fluid into the lumen that assists in sperm transport out of the testis. The Sertoli cells prevent blood-borne cells and substances (antibodies) from reaching the sperm, which contain foreign antigens, as well as preventing these antigens from reaching the blood (blood-testis barrier).

The Sertoli cells play an important role in the spermiogenesis stage of spermatogenesis by engulfing and digesting the remaining pieces of cytoplasm and cellular debris (*residual bodies*) left over from the transformation of spermatids into spermatozoa. This action also enables the sperm to be released into the lumen.

To perform these functions, the Sertoli cells require that *testosterone*, a hormone, be supplied to them directly from the Leydig cells through the basement membrane. Testosterone is also required for the development of germinal cells in the tubules and for the final maturation of the sperm in the *epididymis*. To maintain a high local concentration of testosterone, the Sertoli cells make and secrete into the lumen a special protein, the *androgen-binding protein (ABP)*, which acts as a receptor and reservoir for testosterone.

FACTORS INFLUENCING SPERM FORMATION. Malnutrition, alcoholism, cadmium salts, and some drugs interfere with spermatogenesis. Gossypol, a cottonseed oil which may be taken orally, attacks spermatids and is likely to be used as a specific male contraceptive. In some animals (e.g., the rat), but not in humans, vitamin E is essential for spermatogenesis. Physical factors, such as X-ray radiation and high temperatures (over and including body temperature), also diminish spermatogenesis. Indeed, spermatogenesis is the only process in the body whose optimum temperature is a few degrees below the core temperature. At body temperature, the germinal epithelium will regress, but the Leydig cells and hormone production are intact.

EPIDIDYMIS AND SPERM MATURATION. Once the sperm are released into the lumen, they are transported, with the aid of the pressure of the testicular fluid provided by the Sertoli cells, through the *rete testis*, which is a network of anastomosing conduits between the seminiferous tubules and the *epididymis*. The epididymis is connected to the rete testis by special efferent ducts. During their passage through the epididymis, the sperm will undergo the final stages of their biochemical and physiologic maturation. Completely mature sperm are expelled from the epididymis during sexual excitation via the *vas deferens*.

CN: Use the same colors as on the previous page for testis (A), epididymis (D), and vas deferens (E). Use red for G.
1. Begin with the diagram in the upper right corner, and then color the large illustration to its left. Note that the three tubes that extend outward (the seminiferous tubules B¹) are portions of the lobule (B) sections and receive the same color.
2. Color the enlargement of a tubule section, beginning at the bottom of the square. Interstitial cells (F) are shown outside the tubule, secreting testosterone (F') into the Sertoli cells (O), as well as into an adjacent blood capillary (G). The various cells within the tubule have their identifying labels placed within them, and both their cytoplasm and nucleus should be colored. Note the huge nucleus and cytoplasm of the Sertoli cells (O), which form a backdrop to the much smaller cells adjoining them.
3. Color the stages of spermatogenesis, beginning with the dividing spermatogonium (K). All structures should be colored.

TESTIS ᴀ
LOBULE ʙ
RETE TESTIS ᴄ
EPIDIDYMIS ᴅ
VAS DEFERENS ᴇ

Each testis has numerous lobules which contain the highly convoluted seminiferous tubules. These tubules form the spermatozoa, the male gametes, which are released into the lumen of the tubules and are transported to the epididymis through the rete testis (see H arrows). Upon final maturation in the epididymis, sperm are ejaculated via the vas deferens (H arrows) during orgasm.

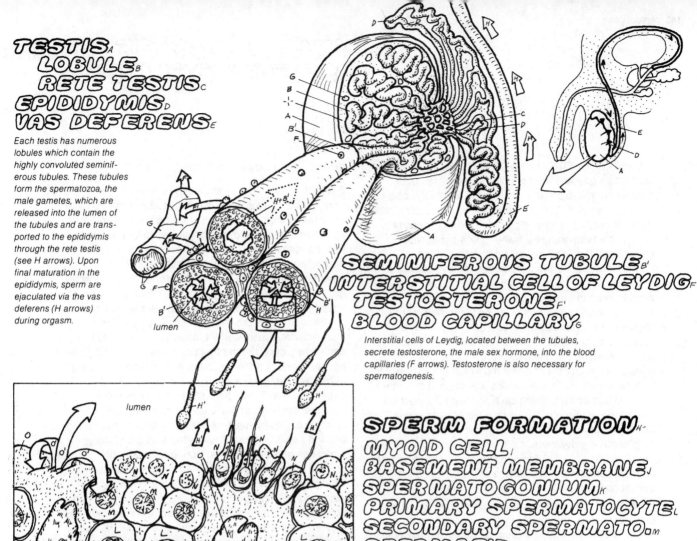

lumen

SEMINIFEROUS TUBULE ʙ'
INTERSTITIAL CELL OF LEYDIG ꜰ
TESTOSTERONE ꜰ'
BLOOD CAPILLARY ɢ

Interstitial cells of Leydig, located between the tubules, secrete testosterone, the male sex hormone, into the blood capillaries (F arrows). Testosterone is also necessary for spermatogenesis.

SPERM FORMATION ʜ
MYOID CELL ɪ
BASEMENT MEMBRANE ᴊ
SPERMATOGONIUM ᴋ
PRIMARY SPERMATOCYTE ʟ
SECONDARY SPERMATO. ᴍ
SPERMATID ɴ
SPERMATOZOON ʜ'
SERTOLI CELL ₒ/ABP ₒ'

Spermatogenesis occurs in the seminiferous tubules. Spermatogonia, attached to the basement membrane which surrounds each tubule, go through successive stages of division, forming first the primary and then the secondary spermatocytes, followed by the spermatids. The spermatids undergo morphologic changes, forming spermatozoa, which are highly differentiated, specialized cells, possessing a flagellum (tail) for motility. The Sertoli cells form androgen-binding protein and play crucial roles in support of spermatogenesis.

STAGES OF SPERMATOGENESIS ʜ
46 ᴋ' CHROMOSOMES ᴋ'
23 ɴ' CHROMOSOMES ɴ'

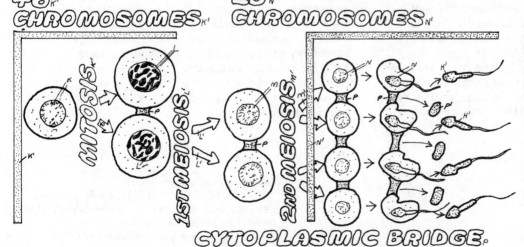

Spermatozoa contain only one half the number of chromosomes found in spermatogonia. This is accomplished through division by meiosis: spermatogonia (diploid) divide first by mitosis to produce primary spermatocytes and preserve their own line. Each diploid (2n) spermatocyte divides by meiosis, forming tetraploid spermatocytes (4n chromosomes) which then go through two meiotic divisions to form 4 haploid (n chromosomes) spermatids/spermatozoa. Cytoplasmic bridges between several spermatogenic cells (a clone) allow synchronous divisions. Spermatozoa are released into the lumen by the shedding of their extra cytoplasmic remnants, the residual bodies.

CYTOPLASMIC BRIDGE ᴘ
RESIDUAL BODY ᴘ'

PHYSIOLOGY OF SEMEN & SPERM DELIVERY

SPERM PRODUCTION AND TRANSPORT. Sperm production goes on uninterrupted in the *testes* from adolescence to old age. From the division of the spermatogonium to the release of sperm into the lumen, it takes about 2. 5 months. Sperm formation proceeds optimally at a temperature a few degrees centigrade below body temperature. In fact, the seminiferous tubules will degenerate if the testes remain within the body cavity. The *scrotum* is a sac that not only functions to keep the testes outside the body, but is equipped with nerves and muscles that reflexively regulate the distance of the testes from the body in response to changes in environmental temperature. Sperm released into the lumen are morphologically mature but incapable of either motility or the capacity to fertilize the ovum. These abilities are attained by further biochemical and physiological *maturation* during the 2 week-long passage in the *epididymis*. Mature sperm aggregate at the tail of the epididymis and the beginning of the *vas deferens*. It is from here that they are mobilized, joined with seminal plasma, and finally ejaculated as semen from the *urethra* of the *penis* during sexual orgasm.

SEMEN. This is a milky fluid containing the sperm and seminal plasma, the watery contributions of the *prostate* and *seminal vesicles*. Seminal vesicles provide nutrients (*fructose, vitamins*) needed for sperm activity and *prostaglandins*, which may aid in sperm transport. The prostate provides *alkaline substances* (bicarbonate), *enzymes* (phosphatase, fibrinolysin), *proteins* (fibrinogen), and *minerals* (zinc). These substances are important in neutralizing vaginal acids and in providing an adequate physical environment for the survival of sperm in the female reproductive tract.

ERECTION. During sexual excitation, the penis undergoes *erection*, a necessary event for intromission, the penetration of the penis into the vagina. This is brought about by the dilation of arterioles, which provide large amounts of blood to the *cavernous* and *spongiosus bodies*, specialized *erectile vascular tissue* in the penis. Turgidity and inflation of these tissues prevent blood outflow by closing the veins; this leads to hardening and erection of the penis. Erection is brought about by a *parasympathetic response* stimulated by tactile stimulation of the penis or other erogenous zones as well as by descending stimuli from the brain (sight, sound, smell, thinking, imagining). The center for control of the erection reflex is in the sacral parts of the spinal cord. The strength of the erection response

diminishes during old age.

EJACULATION. The orgasmic expulsion of semen from the penis, is divided into two stages, emission and ejaculation proper. Emission refers to the movement of sperm from the epididymis up along the vas deferens into the ejaculatory duct. This movement is accomplished by rhythmic contractions of the smooth muscles in the wall of the vas deferens. These muscles are controlled by *sympathetic nerves* from the lumbar spinal cord centers. Similar sympathetic signals cause contraction of the prostate and seminal vesicles, so that, simultaneously with the arrival of sperm into the ejaculatory duct, the contents of the prostate and seminal vesicles are added to them. At this time, a new set of reflexes for ejaculation proper is activated, causing — via signals from the pudendal nerve — rhythmic contraction of a skeletal muscle (*bulbospongiosus*) at the base of the penis, thus expelling the emitted semen through the urethra and out of the glans penis in a pulsatile manner. During emission, the urethra is lubricated and washed by mucoid and alkaline secretions of the *bulbourethral* (Cowper's) *glands* to neutralize acid from urine passage. The sensory receptors for emission and ejaculation reflexes are located mainly at the glans penis, and the spinal center for these reflexes is comparatively less subject to influences from the brain. Indeed, whereas the erection response can be interrupted at any moment by will or from fear, the ejaculation reflex, once activated, cannot be interrupted by other stimuli.

SPERM NUMBER AND MALE FERTILITY. In healthy males, the ejaculate is about 3 mL and contains about *100 million* sperm per mL. Reduction below a quarter of this amount leads to sterility. Frequent ejaculation also leads to a reduction in the volume and number of sperm in the ejaculate, thereby diminishing the chances of fertility. Approximately three to four ejaculations per week are in accord with the normal delivery of sperm from the epididymis. This frequency will provide an adequate number of sperm in the ejaculate to ensure fertility. Usually about 20% of ejaculated sperm are abnormal, having no tail, two tails, coiled tails, no heads, two heads, or small heads. Higher incidence of abnormality also causes sterility. In abstinence, sperm are either released during nocturnal emission coincident with sexual dreams or are simply stored in the epididymis, where they age and die and are destroyed by macrophages.

CN: Use dark colors for A and C.
1. Color the titles in order as you follow the numbered sequence, beginning with a rise in external temperature (A') lowering the scrotum (A) and testes (B).
2. Color the neural regulation diagrams with the title: "inputs."

SCROTUM A
TESTIS B
    SEMINIFEROUS TUBULE B'
    SPERM C
EPIDIDYMIS D
VAS DEFERENS E
SEMINAL VESICLE F
PROSTATE GLAND G
BULBOURETHRAL GLAND H
URETHRA I
ERECTION OF PENIS *
    SENSORY NERVE J
    PARASYMPATHETIC NERVE K
    ERECTILE VASCULAR TISSUE L
EJACULATION OF SEMEN *
    SYMPATHETIC NERVE M
    BULBOSPONGIOSUS M. □N

(1) The optimal temperature for sperm formation is a few degrees below body temperature. The scrotum maintains the testes at an appropriate distance from the body. (2) Sperm formation occurs in the testes and takes two months. (3) Formed sperm are continuously released into the lumen of the seminiferous tubules, and are then transported to the epididymis where they stay for two weeks and mature completely. (4) Contractions of the vas deferens transport sperm into the ejaculatory ducts, where they are joined with the secretions of the prostate and seminal vesicles and are expelled out through the urethra. (5) The seminal vesicles secrete nutrients. (6) The prostate secretes proteins, enzymes and alkalines for sperm survival. (7) The bulbourethral glands secrete alkalines and lubricants to facilitate sperm transport. (8) Erection of the penis is a vasocongestive response of special erectile tissues: the corpus cavernosum and the corpus spongiosum. (9) The bulbospongiosus muscle expels semen through the urethra.

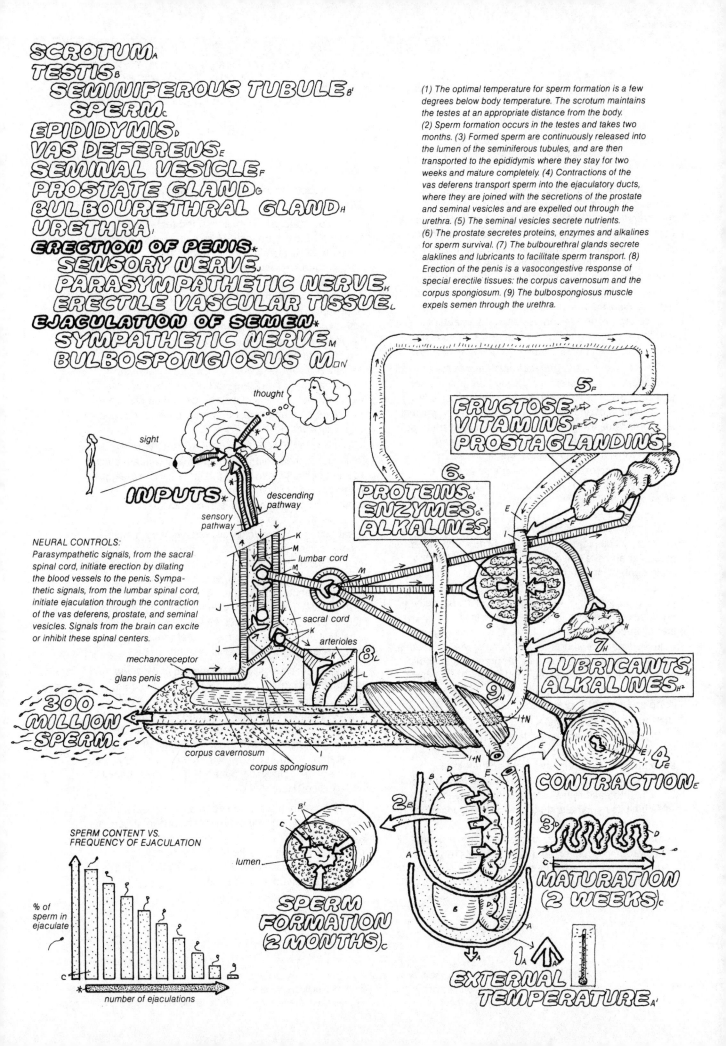

thought

sight

INPUTS *

descending pathway

sensory pathway

NEURAL CONTROLS:
Parasympathetic signals, from the sacral spinal cord, initiate erection by dilating the blood vessels to the penis. Sympathetic signals, from the lumbar spinal cord, initiate ejaculation through the contraction of the vas deferens, prostate, and seminal vesicles. Signals from the brain can excite or inhibit these spinal centers.

lumbar cord
sacral cord
arterioles
mechanoreceptor
glans penis

300 MILLION SPERM C

corpus cavernosum
corpus spongiosum

5 F
FRUCTOSE F'
VITAMINS F²
PROSTAGLANDINS

6 G
PROTEINS G'
ENZYMES G²
ALKALINES

7 H
LUBRICANTS H'
ALKALINES H²

8 L

9 N

4 E
CONTRACTION E

2 B

3 D
MATURATION (2 WEEKS) C

SPERM CONTENT VS. FREQUENCY OF EJACULATION

% of sperm in ejaculate

number of ejaculations

lumen

SPERM FORMATION (2 MONTHS) C

1 A

EXTERNAL TEMPERATURE A'

ACTIONS OF TESTOSTERONE & HORMONAL REGULATION OF TESTES FUNCTIONS

ACTIONS OF TESTOSTERONE: The male sex hormone is *testosterone*, a steroid compound formed in the *interstitial cells of Leydig* from cholesterol, a precursor substance for all steroid hormones. The *hydroxyl* (*OH*) and *ketone* (*C=O*) functional groups in testosterone are extremely important, because their chemical modification can alter the potency of this hormone or transform it to another steroid hormone. Testosterone has two classes of actions: those aimed at the reproductive organs and secondary sexual characteristics and those involving the general anabolic actions of testosterone, which are manifested in many tissues.

Early in puberty, the *gonadotropic* cells of the *anterior pituitary* begin to secrete increasing amounts of *FSH* and *LH*. FSH stimulates the *Sertoli cells* (especially the formation of androgen-binding protein) and possibly also stimulates the germinal cells, promoting spermatogenesis. LH stimulates the Leydig cells to secrete testosterone, the blood level of which increases throughout adolescence, reaching a peak by the early twenties. Testosterone promotes growth and development of primary sex organs (testes, penis, etc.) and accessory sex glands (prostate, seminal vesicles, etc.), promotes the development of secondary sex characteristics (voice, face and body hair, enhanced muscular and skeletal growth), and may activate the brain centers involved in regulating sexual activity and behavior. These actions transform adolescent boys into young men, enabling them to engage in sexual activity, produce fertile sperm, and father children. In adults, the steady secretion of testosterone (1) maintains spermatogenesis and the secretory functions of the epididymis, prostate, and seminal vesicles; (2) promotes anabolic activity and vigor in muscles and bones; (3) enhances red cell production in bone marrow; and (4) promotes libido and normal aggressive attitudes.

HORMONAL REGULATION OF TESTICULAR FUNCTIONS: In mature men, testosterone is secreted continually at a steady rate of 10 mg/day. This is accomplished by a negative feedback effect of testosterone on the *hypothalamus* and *anterior pituitary*. Thus, an increase in testosterone levels beyond set limits inhibits hypothalamic secretion of a *gonadotropin of easing hormone* (GnRH), which in turn reduces the output of LH from the anterior pituitary. This reduces testosterone production, and plasma levels return to normal. Excessive reductions will activate a reverse response. Illness and stress tend to diminish testosterone production, presumably through action on the hypothalamus.

The spermatogenic function of the testes is in part under the control of FSH from the anterior pituitary; however, LH is also important, through its control of testosterone secretion. FSH stimulates the *Sertoli cells* to secrete *androgen-binding protein* (*ABP*). ABP binds with testosterone and provides an abundant local supply of this hormone, which is needed for the activity of the Sertoli cells and for the maturation of the spermatogenic cells inside the seminiferous tubules and of the sperm inside the epididymis. The Sertoli cells in turn secrete a hormone, *inhibin*, which acts on the pituitary to regulate FSH release by negative feedback. In fact, inhibin may be used in the future as a male contraceptive, because it causes reduced sperm production.

ABNORMALITIES OF TESTICULAR SECRETION. Testicular function diminishes gradually with advancing age, but sexually active and fertile men in their eighties are not rare. *Hypogonadism* refers to the undersecretion of the testes and is frequently due to reduced pituitary function. In *eunuchoidism*, the testes are absent, or there is Leydig cell deficiency from childhood, resulting in low androgen output. Absence of testosterone prevents the development of male secondary sexual characteristics; thus, eunuchs have a female-like appearance, including skeletal and muscular development. However, they tend to be tall and have long limbs, because postadolescent closure of the epiphyseal plates in the long bones (one of the actions of androgens in high amounts) is delayed.

In rare cases, usually due to tumors in the hypothalamus and pituitary, sexual development occurs early, during childhood (*precocious puberty*). The unusually high levels of testosterone lead to the early appearance of male secondary sexual characteristics, including premature but excessive growth of muscles. The stature is stunted, however, due to premature closure of the epiphyseal plates of the bones (hence the term "boy Hercules").

In some men, excessive secretion of testosterone may be associated with overexcitability and abnormal aggressiveness. In fact, in certain habitual offenders, removal of testes (castration) or treatment with antiandrogen drugs is known to induce calmness and reduce the incidence of offenses.

CN: Use red for B and a dark color for A.
1. Begin with testosterone (A) functions as shown by three arrows from an interstitial cell (G) in the right central portion of the page. Color the relevant portions of the chemical structure of testosterone and the various arrows indicating physical manifestations of its actions in the body.
2. Go to the titles at the top of the page.

HORMONAL REGULATION OF TESTIS FUNCTION*

HYPOTHALAMUS,
GONADOTROPIN RELEASING HOR.,
ANTERIOR PITUITARY,
LUTEINIZING HORMONE (LH),
INTERSTITIAL CELL (OF LEYDIG),
FOLLICLE-STIMULATING HOR. (FSH),
SERTOLI CELL,
ANDROGEN-BINDING PROTEIN (ABP),
INHIBIN,

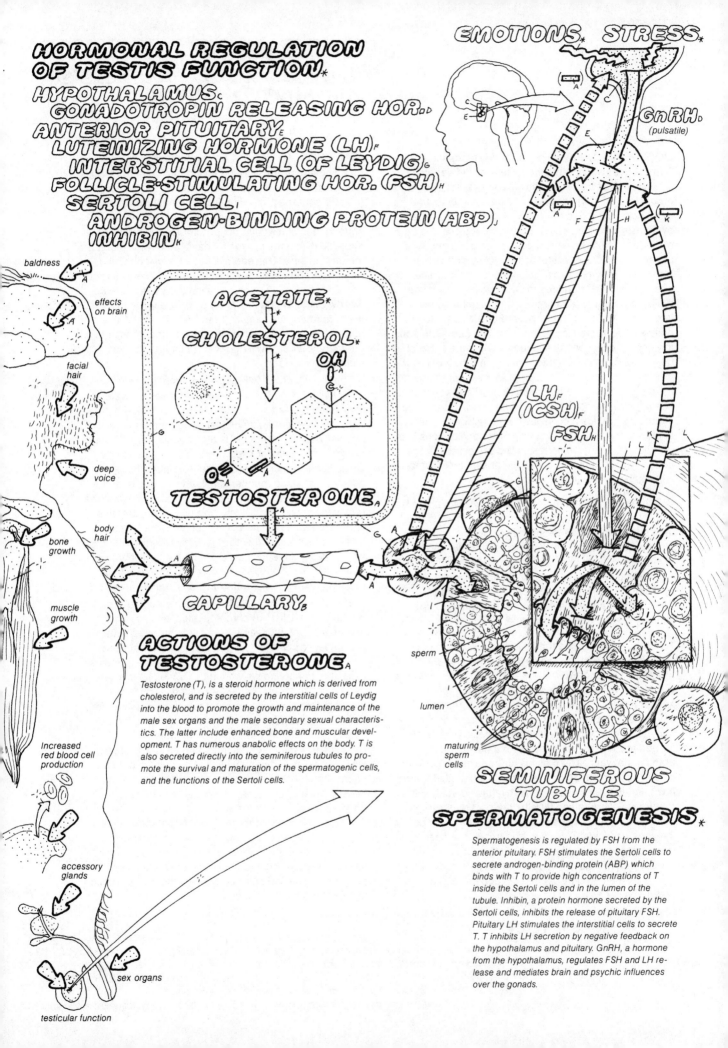

EMOTIONS* STRESS*

GnRH,
(pulsatile)

baldness
effects on brain
facial hair
deep voice
bone growth
body hair
muscle growth
Increased red blood cell production
accessory glands
sex organs
testicular function

ACETATE*
CHOLESTEROL*
OH
TESTOSTERONE,

CAPILLARY,

LH (ICSH),
FSH,

sperm
lumen
maturing sperm cells

SEMINIFEROUS TUBULE,

SPERMATOGENESIS*

ACTIONS OF TESTOSTERONE,

Testosterone (T), is a steroid hormone which is derived from cholesterol, and is secreted by the interstitial cells of Leydig into the blood to promote the growth and maintenance of the male sex organs and the male secondary sexual characteristics. The latter include enhanced bone and muscular development. T has numerous anabolic effects on the body. T is also secreted directly into the seminiferous tubules to promote the survival and maturation of the spermatogenic cells, and the functions of the Sertoli cells.

Spermatogenesis is regulated by FSH from the anterior pituitary. FSH stimulates the Sertoli cells to secrete androgen-binding protein (ABP) which binds with T to provide high concentrations of T inside the Sertoli cells and in the lumen of the tubule. Inhibin, a protein hormone secreted by the Sertoli cells, inhibits the release of pituitary FSH. Pituitary LH stimulates the interstitial cells to secrete T. T inhibits LH secretion by negative feedback on the hypothalamus and pituitary. GnRH, a hormone from the hypothalamus, regulates FSH and LH release and mediates brain and psychic influences over the gonads.

FUNCTIONS OF THE OVARY: FORMATION OF THE EGG AND OVULATION

OOGENESIS. The ovary performs two functions: (1) formation, development, and release of the eggs and (2) secretion of the female sex hormones, estrogen and progesterone. In the developing female embyro, the germ cells migrate to the ovary, where they proliferate by mitosis to form nearly a million *primary oocytes* per ovary. Beyond this prenatal period, the germ cells will cease to multiply, constituting an important difference with the male, where spermatogonia begin to divide after puberty and continue to do so until senescence. While still in the prenatal period, the primary oocytes begin their meiotic divison, but are arrested at the pro-phase stage, remaining in this condition after birth and throughout childhood until puberty, when the cycle of meiotic division is resumed. The primary oocyte is sur-rounded by a layer of follicular *granulosa cells*, forming a structure called the *primary follicle*. These are found scattered in the cortical zone of the ovary. The number of primary follicles diminishes from a million per ovary at birth to thousands by puberty. This spontaneous loss of follicles and their oocytes, called *atresia*, is not well understood but may be due to the absence of hormonal stimulation before puberty.

OVARIAN CYCLE. At puberty, the ovaries are activated by pituitary gonadotropins to develop the follicles and their eggs and to secrete sex hormones. Egg produc-tion by the ovary, in contrast to sperm production by the testis, is a cyclical process. In women, the cycle lasts an average of 28 days. The cause and regulation of cyclicity can be traced to the mode of secretion of the hypothalamic releasing hormone (GnRH) for the pituitary gonadotropins (see plate 147).

At the beginning of each 28-day cycle, which occurs randomly in one of the two ovaries, a few primary follicles begin to grow. A week later, only one continues to develop while the others regress. In the *maturing follicle*, follicular cells proliferate, initially forming several layers of the granulosa cells surrounding the oocyte. Later another layer of cells, *theca cells*, is formed around the granulosa, with a basement mem-brane in between. The masses of granulosa cells form a cavity called the *antrum*, which fills with a fluid (*antral fluid*) rich in a sticky substance, hyaluronic acid. In the *mature follicle* (Graffian follicle), the oocyte is sur-rounded by a zone of transparent jellylike substance, the *zona pellucida*. This zone is in turn surrounded by many granulosa cells, later to form the corona radiata. The corona radiata is attached to the main body of the follicle by the *cumulus oophorus* ("egg cloud"), a mass of granulosa cells.

OVULATION. By day 12-13, the oocyte with its sur-rounding structures is found floating in the antrum. By day 14, mid-cycle, the mature (Graffian) follicle, large (up to 2 cm) and protruding from the now weak surface of the ovary, ruptures, releasing the oocyte, with its appendages of follicular cells, cumulus oophorus, and antral fluid, into the peritoneal cavity near the fimbria of the uterine tube. This event is called *ovulation*. The remainder of the mature follicular cells in the ovary are transformed into a structure called the *corpus luteum* (yellow body), which grows for at least a week to form the *mature corpus luteum*. If fertilization occurs, the corpus luteum will survive and grow further to maintain pregnancy; if not, it will degenerate into the *corpus albicans* (white body).

The growth of the follicle from primary to mature, comprising the *follicular phase* of the ovarian cycle, is stimulated by the *follicle-stimulating hormone* (FSH) from the anterior pituitary. Luteinizing hormone (LH), another pituitary hormone, is also needed for follicular growth, especially for secretory activity of the follicle. Ovulation is triggered by a surge in LH. LH, through a poorly understood mechanism, leads to rupture of the follicle and expulsion of the ovum. LH also converts the remaining follicular cells into corpus luteum. LH, along with FSH, maintains the corpus luteum. The formation and maintenance of the corpus luteum comprises the *luteal phase* of the ovarian cycle. In humans, about 1% of ovarian cycles involve multiple ovulations, usually resulting in the birth of fraternal twins, triplets, etc.

DIVISIONS OF THE OVUM. The primary oocyte, containing 46 (diploid) chromosomes, begins *meiosis* during embryonic development but is arrested at the prophase. At puberty, in response to hormonal stimula-tion by FSH and LH and the secretions of the granulosa cells, the primary oocyte resumes its meiotic division and begins to grow. The first meiotic division is com-pleted before ovulation, forming the secondary oocyte and one *polar body* (cell). During this division, the cell destined to become the ovum (secondary oocyte) receives half the chromosomes and all the cytoplasm, while the polar body receives little cytoplasm but an equal share of chromosomes. While still in the ovary, the secondary oocyte begins the second meiotic division, which now stops at metaphase. At this stage, it is ovulated. It is only after fertilization that the oocyte (ovum) will attempt to complete its second meiotic division, forming the mature female pronucleus con-taining 23 (haploid) chromosomes and the second polar body. The first polar body may also divide, forming three polar bodies all together.

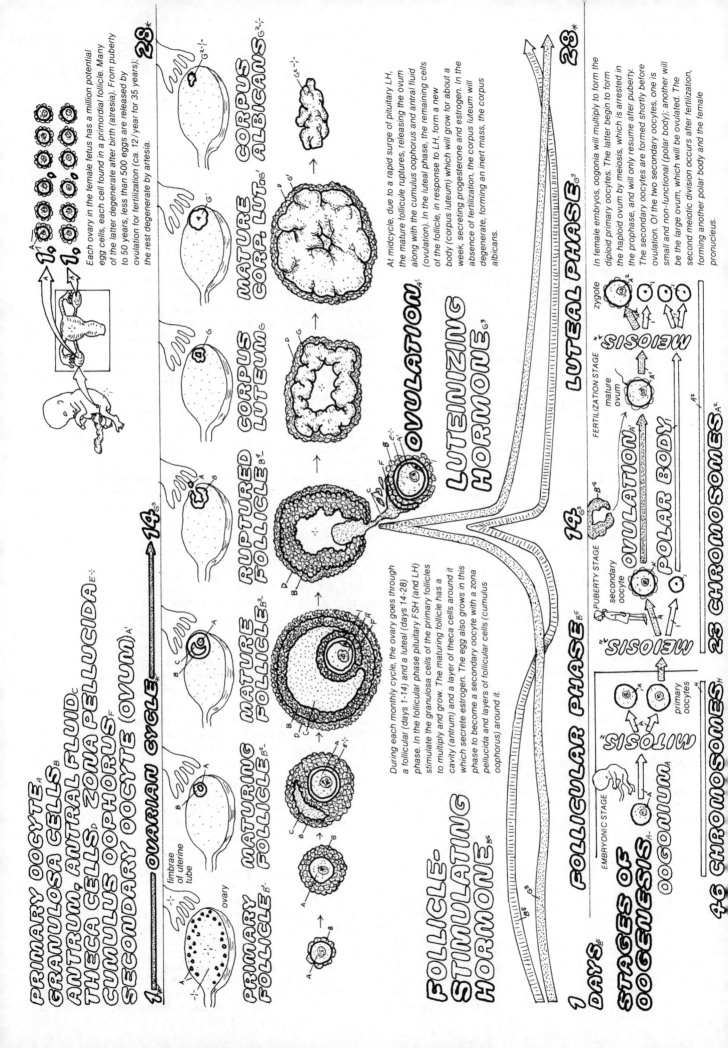

PRIMARY OOCYTE_A
GRANULOSA CELLS_B
ANTRUM, ANTRAL FLUID_C
THECA CELLS_D ZONA PELLUCIDA_E
CUMULUS OOPHORUS_F
SECONDARY OOCYTE (OVUM)_A'

OVARIAN CYCLE

Each ovary in the female fetus has a million potential egg cells, each cell found in a primordial follicle. Many of the latter degenerate after birth (atresia). From puberty to 50 years, less than 500 eggs are released by ovulation for fertilization (ca. 12/year for 35 years); the rest degenerate by artesia.

limbrae of uterine tube
ovary

PRIMARY FOLLICLE_B²   MATURING FOLLICLE_B²'   MATURE FOLLICLE_B³   RUPTURED FOLLICLE_B⁴   CORPUS LUTEUM_G   MATURE CORP. LUT._G'   CORPUS ALBICANS

OVULATION_A'

FOLLICLE-STIMULATING HORMONE_B⁵

LUTENIZING HORMONE_G³

During each monthly cycle, the ovary goes through a follicular (days 1-14) and a luteal (days 14-28) phase. In the follicular phase pituitary FSH (and LH) stimulate the granulosa cells of the primary follicles to multiply and grow. The maturing follicle has a cavity (antrum) and a layer of theca cells around it which secrete estrogen. The egg also grows in this phase to become a secondary oocyte with a zona pellucida and layers of follicular cells (cumulus oophorus) around it.

At midcycle, due to a rapid surge of pituitary LH, the mature follicle ruptures, releasing the ovum along with the cumulus oophorus and antral fluid (ovulation). In the luteal phase, the remaining cells of the follicle, in response to LH, form a new body (corpus luteum) which will grow for about a week, secreting progesterone and estrogen. In the absence of fertilization, the corpus luteum will degenerate, forming an inert mass, the corpus albicans.

FOLLICULAR PHASE_B⁵

LUTEAL PHASE_G³

DAYS_B⁶   1   14   28*

STAGES OF OOGENESIS

OOGONIUM_A⁻

MITOSIS

MEIOSIS

EMBRYONIC STAGE   PUBERTY STAGE   FERTILIZATION STAGE

primary oocytes

secondary oocyte

OVULATION_A'

POLAR BODY

mature ovum

zygote

46 CHROMOSOMES_H   23 CHROMOSOMES_A²

In female embryos, oogonia will multiply to form the diploid primary oocytes. The latter begin to form the haploid ovum by meiosis, which is arrested in the prophase, and will only resume after puberty. The secondary oocytes are formed shortly before ovulation. Of the two secondary oocytes, one is small and non-functional (polar body); another will be the large ovum, which will be ovulated. The second meiotic division occurs after fertilization, forming another polar body and the female pronucleus.

FUNCTIONS OF THE OVARY: SECRETION OF FEMALE SEX HORMONES

FEMALE SEX HORMONES. These are *estrogen* (estradiol) and *progesterone*, steroid compounds made in the ovary from cholesterol. Estradiol contains two hydroxyl groups; progesterone contains two ketone groups. Estrogen is secreted by the follicle in response to the combined stimulation by the hormones FSH and LH from the anterior pituitary during the follicular phase. During the second quarter of the ovarian cycle, the level of estrogen in the blood increases, peaking by days 12-13. After ovulation, the transformation of the follicular cells to *luteal cells* of the corpus luteum reduces estrogen output, but secretion continues into the third and fourth weeks. Progesterone is secreted by the luteal cells of the corpus luteum in response to stimulation by LH; however, FSH is also necessary. Thus, progesterone is absent in the blood during the follicular phase and appears only after ovulation, when LH begins to stimulate the corpus luteum. Progesterone secretion peaks by the middle of the luteal phase (days 20-22).

UTERINE CYCLE. During the monthly cycles, the principal actions of estrogen and progesterone in the female reproductive system are on the *endometrium* (uterine mucosa). This lining will house the young embryo after fertilization (*implantation*). To do this, the endometrium undergoes a cyclical change, building up its wall in expectation of the embryo and destroying it if fertilization does not take place.

Estrogen stimulates the epithelial cells of the *basal layer* of the endometrium to proliferate, forming a thick mucosa as well as numerous *endometrial (uterine) glands*. Extensive vascular tissue, *spiral arteries* and *veins*, also grows within the endometrium. These events constitute the *proliferative* phase of the endometrial cycle (days 6-14). At ovulation, the endometrium is fully grown (about 5 mm thick); the *myometrium*, however, does not grow extensively.

After ovulation, the cells of the growing corpus luteum begin to secrete progesterone, a hormone that acts mainly on the glands of the endometrium to promote their secretory activity. This secretion, rich in proteins and glycogen, is important for the survival and nutrition of the preimplantation and implanted embryo as well as for the adherence of the implanted embryo. Indeed, progesterone is absolutely essential for gestation. This part of the endometrial cycle promoted by the action of progesterone is termed the *secretory phase*.

MENSTRUATION. In the absence of fertilization, the hormonal signal promoting the survival of the corpus luteum, which normally comes from the embryo (see plate 147), will not arrive. The regression of the corpus luteum diminishes the output of estrogen and progesterone; this weakens the endometrial tissue, reducing blood flow to it and causing local ischemia (*ischemic phase*). The endometrium can no longer be sustained, and by day 28, it will collapse. Its debris, along with some blood, constitutes the *menstrual flow* (menstruation, menses). The menstrual phase lasts an average of five days. The growth of follicles and increasing estrogen output during the next follicular phase of the ovary will terminate the menstrual phase. Note that although the menstrual phase is the last phase of the endometrial cycle, in keeping with the events of the ovarian cycle it is represented here as the first phase.

Menstrual cycle commences at puberty (menarche), between 12 and 13 years; however, the early cycles are usually not accompanied by ovulation. The cessation of the menstrual cycle (menopause), occurring around fifty-two years of age, is related to the exhaustion of the ovarian follicles and signals the end of reproductive (but not sexual) activity. Menstrual cycles do not occur during pregnancy and in some lactating women.

OTHER EFFECTS OF ESTROGEN. In addition to its effects on the endometrium, estrogen stimulates the development of extensive mucosal folds of the oviduct as well as the formation of cilia on these epithelial cells. These cilia function in ovum transport. Estrogen also potentiates the effects of FSH on follicular growth. During puberty, estrogen (along with adrenal androgens) enhances calcium deposition in bone and stimulates bone growth. It also promotes growth of the uterus, vagina, and oviducts, as well as of the mammary glands. Estrogen allows for the development of softer skin and promotes deposition of fat in subcutaneous zones, particularly in breasts and buttocks, leading to the mature female shape and contours. Estrogen promotes the growth of wider pelvic bones as well as the closure of epiphyseal plates in long bones. However, many of the female secondary sex characteristics are due to the absence of androgens.

CN: Use the same colors for FSH (A) and LH (C) as on the preceding page. Use red for E, blue for H.
1. Begin with the bottom panel, and follow the FSH contribution to the ovarian cycle and the growth of the follicular cells into the sex hormone cycle panel. Color the estrogen elements. Then go back to the LH and luteal phase portion of the ovarian cycle. The corpus luteum structures receive the LH color, but the luteal cells therein secrete progesterone which gets a different color. Color the elements of progesterone.

2. Go to the top of the page. Starting in the left corner, color the diagram of the uterus. Note that the endometrium is shown in the menstrual phase, and it and the flow of blood are colored red. Then color the enlargement of the uterine wall section. Note that the endometrium portion is left uncolored so that the structures can stand out. Color the tiny circles representing the uterine gland secretions.
3. Color the phases of the menstrual cycle. Note the combination of colors in the secretory phase. Color the day numbers as well.

UTERUS

ovary

LH   FSH

pituitary

vagina

menstrual flow

ENDOMETRIUM
BASAL LAYER
MYOMETRIUM
SPIRAL ARTERY
VEIN
UTERINE GLAND

The uterus has the shape and size of a pear. It is connected to the uterine tubes and to the vagina through the uterine cervix. The uterine wall consists of two layers, a muscular (myometrium) and a mucosal (endometrium). The endometrium consists of a permanent basal layer, and a functional layer which is continually rebuilt and destroyed. Within the endometrium are found the uterine glands, spiral arteries, veins, and the surface epithelium.

MENSTRUAL CYCLE (ENDOMETRIAL)

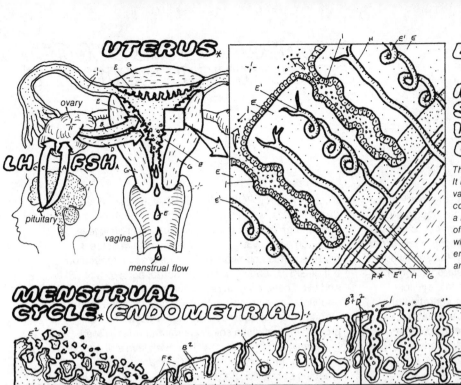

1 MENSTRUAL PHASE 6

In the first five days of the ovarian cycle, endometrium is shed, and the debris mixed with blood, constitutes the menstrual flow.

6 PROLIFERATIVE PHASE 14

Between days 6 to 14 (proliferative phase) the endometrium is rebuilt, glands are formed and the vascular supply is reestablished.

14 SECRETORY PHASE

After ovulation, in response to progesterone, endometrial glands secrete uterine fluid necessary for embryonic development.

ISCHEMIC PHASE 27 28

Without fertilization, estrogen and progesterone decline and endometrial blood flow diminishes (ischemic phase) causing the shedding of endometrium and blood.

SEX HORMONE CYCLE

ESTROGEN

OH

HO

The hormone estrogen (mostly estradiol), is one of the principal female sex steriods produced by the ovary. It is responsible for the proliferative phase of the endometrium. Estrogen is secreted by the cells of the follicle as well as by the corpus luteum.

PROGESTERONE

CH₃
C=O

O

Progesterone, produced by the luteal cells of the corpus luteum, is another female sex steroid. It appears in the blood after ovulation, and stimulates the secretion of the uterine endometrial glands (secretory phase).

OVARIAN CYCLE

FOLLIC-ULAR CELLS

FSH

Pituitary FSH promotes follicular growth, and with LH, stimulates the theca interna cells of the follicle to form estrogen.

LUTEAL CELLS

LH

Pituitary LH triggers ovulation, promotes growth of the corpus luteum, and stimulates secretion of progesterone by the luteal cells.

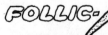

1 FOLLICULAR PHASE 14   LUTEAL PHASE 28

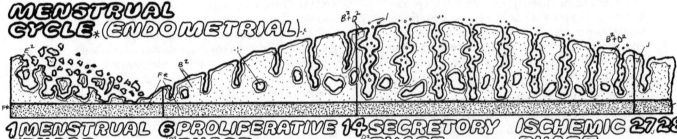

HORMONAL REGULATION OF OVARIAN ACTIVITY

In contrast to the testis, the activities of the *ovary* occur in a cycle. Thus, the formation of follicles (including the growth of the ovum), ovulation, formation of the corpus luteum, and its regression all occur in appropriate order within a single monthly cycle. Similarly, the secretion of *estrogen* at first, followed by *progesterone*, from the follicle and corpus luteum takes place in a cyclical fashion. In this plate, we study how the operation of a hypothalamic "clock," along with intricate feedback effects of the ovarian hormones on the *hypothalamus-anterior pituitary* complex, ensures the orderly operation of the ovarian cycle.

PITUITARY GONADOTROPINS. The anterior pituitary secretes two hormones that regulate the activity of the ovary. These are the *follicle-stimulating hormone* (FSH) and the *luteinizing hormone* (LH), collectively called gonadotropins. Both are glycoprotein molecules and are secreted from the basophilic gonadotropes. Both LH and FSH are necessary for all ovarian activity. In the follicular phase, FSH regulates follicular growth while LH stimulates estrogen secretion. LH, however, appears to be the predominant hormone, eliciting ovulation and growth of the corpus luteum as well as stimulation of progesterone and estrogen secretion by the corpus luteum.

HYPOTHALAMIC CONTROL. These pituitary gonadotropins are released in response to a signal from the hypothalamus in the form of a peptide neurohormone called the *gonadotropin-releasing hormone* (GnRH), released by the axon terminals of hypothalamic neurons into the portal hypophyseal capilaries, which deliver this substance rapidly and directly to the gonadotrope cells. There are receptors for GnRH on the surface of the gonadotropes. Recent research indicates that GnRH is released in pulses at hourly intervals. If increased release of gonadotropin is required, the amount of GnRH per pulse will be increased, and vice versa. It is not known how GnRH can differentially regulate LH and FSH secretions. In primates, levels of sex hormones can also influence LH and FSH secretions by direct effects on the pituitary. The pattern and amount of GnRH release is under the control of two mechanisms, a hypothalamic "clock" that sets the duration of the cycle and the timing of major events and the feedback effects of sex hormones on the hypothalamus and pituitary.

NEGATIVE AND POSITIVE FEEDBACK. Late in the ovarian cycle, when the endometrium is in the ischemic phase, preparing for menstruation, the very low level of estrogen acting via *negative feedback* stimulates the hypothalamus and pituitary. This causes increased output of GnRH, leading in turn to increased output of FSH and LH. These hormones stimulate follicular growth and increase estrogen output. By day 12 of the menstrual cycle, estrogen output is at its peak and FSH production has diminished due to the negative feedback inhibition by estrogen. The peak levels of estrogen will now act through a *positive feedback system*, increasing sensitivity of the pituitary to GnRH. This will cause a burst in the release of LH and FSH, but that of LH is many times higher and very crucial. The high levels of LH trigger the process of ovulation, which results, in several hours, in the expulsion of the ovum.

The postovulatory high levels of LH (and also of FSH) promote the secretion, mainly of progesterone and also of estrogen, by the corpus luteum cells. Gradually, the negative feedback effect will return. Thus, the increasing output of progesterone and estrogen will act on the hypothalamus and pituitary to diminish LH and FSH production. At the beginning of the fourth quarter of the ovarian cycle, progesterone and estrogen are at peak levels, and LH and FSH levels have fallen off. In the absence of fertilization, the low LH and FSH levels, as well as other factors such as locally produced prostaglandins, will cause the corpus luteum to lyse and regress, leading to diminished progesterone and estrogen output. This is the end of the cycle, and it is accompanied by menstruation. Gradually, the low levels of estrogen will relieve the inhibition over hypothalamic GnRH release, leading to increased FSH and LH output from the pituitary. This event will activate the second ovarian cycle.

Illness, malnutrition, severe stress, and emotional crises interfere with the operation of the ovarian cycle. Stress and emotional crises act on the higher brain centers and, from there, on the hypothalamus, interfering with the pattern of GnRH release. Often the release is inhibited, leading to reduction in FSH and LH levels. Depending on the timing of the stress, diminished estrogen may cause undue menstruation (spotting) or delayed menstruation (secondary amenorrhea) due to the absence of endometrial proliferation.

CN: Use the same colors as on the preceding page for FSH (D), LH (E), estrogen (G), and progesterone (H). Use light colors for A and C.
1. Color the large control illustration in the center.
2. Color the three bottom panels. Color only the bold portions of the hormone levels. Color gray that portion of the endometrium involved during the period described. The dotted ascending line in the left panel represents reduced levels of estrogen. Note that bold dotted lines in the right panel represent a cessation of FSH and LH secretion.
3. Color the diagram in the upper right.

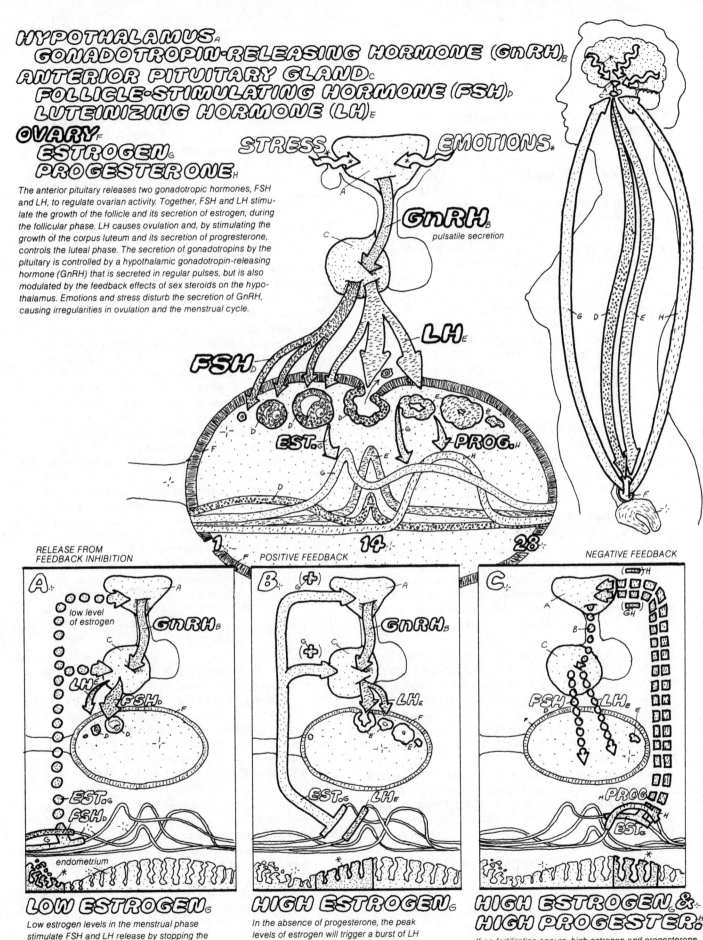

HYPOTHALAMUS[A]
GONADOTROPIN-RELEASING HORMONE (GnRH)[B]
ANTERIOR PITUITARY GLAND[C]
FOLLICLE-STIMULATING HORMONE (FSH)[D]
LUTEINIZING HORMONE (LH)[E]
OVARY[F]
ESTROGEN[G]
PROGESTERONE[H]

The anterior pituitary releases two gonadotropic hormones, FSH and LH, to regulate ovarian activity. Together, FSH and LH stimulate the growth of the follicle and its secretion of estrogen, during the follicular phase. LH causes ovulation and, by stimulating the growth of the corpus luteum and its secretion of progesterone, controls the luteal phase. The secretion of gonadotropins by the pituitary is controlled by a hypothalamic gonadotropin-releasing hormone (GnRH) that is secreted in regular pulses, but is also modulated by the feedback effects of sex steroids on the hypothalamus. Emotions and stress disturb the secretion of GnRH, causing irregularities in ovulation and the menstrual cycle.

STRESS*   EMOTIONS*

GnRH[B]
pulsatile secretion

LH[E]

FSH[D]

EST.[G]   PROG.[H]

1   14   28

RELEASE FROM FEEDBACK INHIBITION   POSITIVE FEEDBACK   NEGATIVE FEEDBACK

A
low level of estrogen

GnRH[B]

LH[E] FSH[D]

EST.[G]
FSH[D]

endometrium

LOW ESTROGEN[G]

Low estrogen levels in the menstrual phase stimulate FSH and LH release by stopping the inhibition of GnRH release (see panel C). The increase in FSH and LH will stimulate follicular growth and elevate estrogen levels to their peak by day 12 of the ovarian cycle.

B (+)

GnRH[B]

LH[E]

EST.[G] LH[E]

HIGH ESTROGEN[G]

In the absence of progesterone, the peak levels of estrogen will trigger a burst of LH (positive feedback) by day 14. High LH levels causes ovulation and growth of corpus luteum, that increase progesterone and estrogen output to a second peak by day 22.

C

FSH[D] LH[E]

PROG[H]
EST[G]

HIGH ESTROGEN &
HIGH PROGESTER.[H]

If no fertilization occurs, high estrogen and progesterone levels inhibit LH as well as FSH release (negative feedback). Reduced gonadotropin levels lead to atrophy of corpus luteum and decreased output of sex steroids, leading to menstruation. Cycle continues with panel A.

FERTILIZATION OF THE OVUM

SPERM FUNCTIONS. The fully mature human *spermatozoon* has a flattened, pearlike *head* that consists mainly of the nucleus, bearing the genetic material (chromatin, DNA). Covering the head is the *acrosome*, a large, modified lysosomal sac containing several lytic enyzmes, such as hyaluronidase and acrosin, which facilitate the penetration of egg membranes by the sperm. The *neck* contains the basal body (a pair of *centrioles*), which anchor the sperm flagellum. The flagellum (60 μ long, 1 μ thick) is divided into a short *middle piece* (5 μ), a long *principal piece* (50 μ), and a *short piece* (5 μ). The principal piece acts as the motile *tail*.

The contractile machinery of the flagellum (axoneme), which endows the sperm tail with its special wavelike swimming motion, consists of nine pairs of peripheral microtubules surrounding two central microtubules. These filaments run along the entire length of the middle piece and tail. The energy for contraction of these filaments is provided by ATP, produced by the mitochondria, which are packed tightly in a spiral around the axoneme of the middle piece region. Microtubular filaments are aggregates of tubulin, a contractile protein.

When deposited in the vagina, sperm swim through the cervical canal (cervix) and enter the uterus, where they swim in all directions, because there is no known force to attract them to the *uterine tube*, the site of fertilization. The random direction of sperm movement is the reason so many sperm are required for fertilization. Of the 300 million deposited in the vagina, only 0.1% reach the uterine tube and only a few hundred reach the egg (*ovum*).

Human sperm swim at an average speed of 3 mm/min. Thus, they can reach the uterine tube from the cervix within one hour. The fact that some sperm, in animal experiments, can reach the oviduct within a few minutes indicates that sperm transport may be facilitated by special contractions of the uterine wall believed to be induced by prostaglandins in the sperm.

THE EGG (OVUM). The human egg (ovum) is a very large cell (up to 200 μ in diameter) compared to the sperm, whose head is only 5 μ thick. This is mainly due to the large content of *cytoplasm* in the egg. There is little cytoplasm in the sperm. Each sperm functions mainly to deliver genetic material to the egg; the egg has the added function of providing the nutritive needs of the very young embryo; the nutritive substances are stored in the cytoplasmic granules (yolk).

The ovulated egg is surrounded by a layer of *follicular cells* (corona radiata), which support the egg metabolically and nutritionally. The follicular cells are small and held together by *hyaluronic acid*, a mucopolysaccharide that functions as the intercellular "cement." Just outside the egg, between the layer of follicular cells and the egg plasma membrane, is the *zone pellucida*, a membrane made of a transparent, jellylike substance about 5 μ thick. The follicular cells and the egg send fingerlike projections (microvilli) of their plasma membrane through the zona pellucida, possibly for the interchange

of substances. The zona pellucida will later help provide a mechanical support for the young embryo as well.

The egg has no motility of its own. After ovulation, the sweeping movements of the uterine tube and its *fimbriae* create suction, drawing the egg (and its associated structures, the corona radiata and cumulus oophorus) into the uterine tube. There, the contractile activity of the oviduct wall, as well as the constant oarlike beating of the numerous cilia on the epithelial cells of the *mucosal folds*, will push the egg continuously toward the uterus. Estrogen is necessary for the contraction of the uterine tube and for the formation and beating of the cilia. Within hours after ovulation, the egg will reach the ampulla of the uterine tube; at this time, it is fully ripe for fertilization.

FERTILIZATION. In order to penetrate the egg, the sperm must first be capacitated. *Capacitation* involves the removal from the acrosome of an outer glycoprotein coat that prevents premature release of the acrosomal enzymes. Substances that induce sperm capacitation may come from the oviduct, or from follicular cells of the cumulus oophorus. As the capacitated sperm prepares to penetrate the egg, the acrosome will release its *enzymes* (acrosome reaction). Hyaluronidase will lyse the hyaluronic acid, separating the follicular cells and allowing the sperm to make their way through these cells. Next, other enzymes, such as acrosin, will digest parts of the zona pellucida. Contact of sperm with the egg plasma membrane is enhanced by binding of the sperm to special sperm receptors on the egg's surface. Next, the egg plasma membrane will engulf the sperm, which will be entirely taken in, head and tail. This is the main stage of *fertilization*. Entry of the first sperm is followed immediately by the *zona reaction*, a rapid chemical modification of the zona pellucida that blocks penetration of more sperm. The cause of the zona reaction is the outflow of some substances originating from granules in the *cytoplasm* of the egg. Failure of the zona reaction leads to polyspermy, which is not compatible with normal development.

Penetration by the sperm results in activation of the egg, including triggering the last meiotic division of the egg nucleus, expelling the last polar body, and forming the *female pronucleus*. Meanwhile the sperm tail will degenerate, and the sperm nucleus will swell and enlarge, forming the male pronucleus. The last stage of fertilization is the *fusion* of male and female pronuclei, resulting in the combination of the chromosomes of the male and female gametes and the formation of the zygote nucleus.

Within the female, sperm can survive up to three to four days, especially those stored in the cervical mucosa and nourished by the cervical mucus. However, when appropriately frozen, sperm can be kept a few years and still maintain their ability to fertilize the ovum. The egg has a shorter life span after ovulation (about one day), and if not fertilized, it will age and degenerate. The optimum time for fertilization is within the first twelve hours after ovulation.

CN: Use light colors for D, G, H, and M-R.
1. Begin with the uterine tube. Color over the very tiny sperm (F) shown swimming toward and through the tube. Color the titles indicating the number. Color the two enlargements to the right.

2. Color the material on spermatozoon (F).
3. Color the ovum and the five stages of fertilization in the bottom illustrations. Note that in the last one, the dotted line represents the degenerating tail of the spermatozoon.

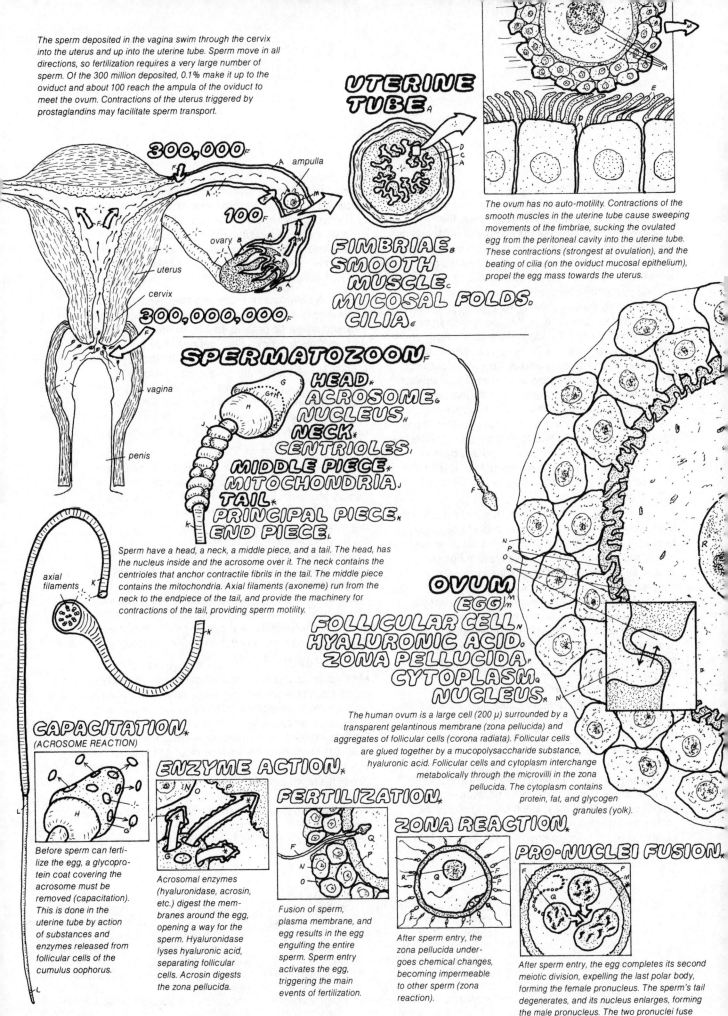

The sperm deposited in the vagina swim through the cervix into the uterus and up into the uterine tube. Sperm move in all directions, so fertilization requires a very large number of sperm. Of the 300 million deposited, 0.1% make it up to the oviduct and about 100 reach the ampula of the oviduct to meet the ovum. Contractions of the uterus triggered by prostaglandins may facilitate sperm transport.

300,000

100

300,000,000

ampulla

ovary

uterus

cervix

vagina

penis

UTERINE TUBE

FIMBRIAE
SMOOTH MUSCLE
MUCOSAL FOLDS
CILIA

The ovum has no auto-motility. Contractions of the smooth muscles in the uterine tube cause sweeping movements of the fimbriae, sucking the ovulated egg from the peritoneal cavity into the uterine tube. These contractions (strongest at ovulation), and the beating of cilia (on the oviduct mucosal epithelium), propel the egg mass towards the uterus.

SPERMATOZOON

HEAD
ACROSOME
NUCLEUS
NECK
CENTRIOLES
MIDDLE PIECE
MITOCHONDRIA
TAIL
PRINCIPAL PIECE
END PIECE

Sperm have a head, a neck, a middle piece, and a tail. The head, has the nucleus inside and the acrosome over it. The neck contains the centrioles that anchor contractile fibrils in the tail. The middle piece contains the mitochondria. Axial filaments (axoneme) run from the neck to the endpiece of the tail, and provide the machinery for contractions of the tail, providing sperm motility.

axial filaments

OVUM
(EGG)

FOLLICULAR CELL
HYALURONIC ACID
ZONA PELLUCIDA
CYTOPLASM
NUCLEUS

The human ovum is a large cell (200 μ) surrounded by a transparent gelantinous membrane (zona pellucida) and aggregates of follicular cells (corona radiata). Follicular cells are glued together by a mucopolysaccharide substance, hyaluronic acid. Follicular cells and cytoplasm interchange metabolically through the microvilli in the zona pellucida. The cytoplasm contains protein, fat, and glycogen granules (yolk).

CAPACITATION
(ACROSOME REACTION)

Before sperm can fertilize the egg, a glycoprotein coat covering the acrosome must be removed (capacitation). This is done in the uterine tube by action of substances and enzymes released from follicular cells of the cumulus oophorus.

ENZYME ACTION

Acrosomal enzymes (hyaluronidase, acrosin, etc.) digest the membranes around the egg, opening a way for the sperm. Hyaluronidase lyses hyaluronic acid, separating follicular cells. Acrosin digests the zona pellucida.

FERTILIZATION

Fusion of sperm, plasma membrane, and egg results in the egg engulfing the entire sperm. Sperm entry activates the egg, triggering the main events of fertilization.

ZONA REACTION

After sperm entry, the zona pellucida undergoes chemical changes, becoming impermeable to other sperm (zona reaction).

PRO-NUCLEI FUSION

After sperm entry, the egg completes its second meiotic division, expelling the last polar body, forming the female pronucleus. The sperm's tail degenerates, and its nucleus enlarges, forming the male pronucleus. The two pronuclei fuse to form the zygote nucleus.

EARLY STAGES OF EMBRYONIC DEVELOPMENT

YOUNG EMBRYO. The individual consists of different cells and tissues, but the *zygote*, the precursor of the individual, is a single and simple cell. Thus, the zygote must proliferate to increase the number of cells, and the cells must differentiate to form the different cell types and tissues. These events constitute the early stages of *embryonic development*. Cell proliferation is the first event observed. This is accomplished by several mitotic divisions, taking the zygote up to the stage of *morula* (berry), when the young *embryo* becomes a ball of cells. Throughout the early stages of division (cleavage), the cells maintain a uniform appearance. As the dividing cells utilize the cytoplasmic stores of the zygote, they become increasingly smaller. On the whole, no growth occurs, and the morula is as large as the zygote was. The *zona pellucida* is also maintained. During cleavage, the young embryo is propelled down the uterine tube toward the uterus by the action of the cilia of the mucosal lining and by the contractions of the oviduct. It takes about four days to traverse the uterine tube; by this time, the embryo is in the morula stage.

Upon entering the uterus, the embryonic cells, which were fairly uniform in appearance, begin to segregate, forming an internally located group of cells (*inner cell mass*) and a peripheral sheet of cells (*trophoblast*), as well as a *cavity*. At this stage (day 5), the embryo is called the *blastocyst*. The *early blastocyst* still has the zona pellucida, but this will soon degenerate, allowing growth and expansion of the embryo to form the *late blastocyst*, in which the trophoblast cells become flattened and active. The dissolution of the zona pellucida will permit the embryro to obtain nutrients and oxygen from the uterine environment. During later development, the trophoblast will give rise to the *placenta* and embryonic membranes (e.g., the amniotic sac); the inner cell mass cells will give rise to the *embryo* proper.

IMPLANTATION. By day 6-7, the embryo is ready to attach itself to the uterus. This is necessary because further growth requires enhanced nutritional and oxygen supplies, which can be obtained only through the maternal blood supply. Cells of the trophoblast will release *lysosomal enzymes*, which begin to digest the *uterine endometrium* cells, allowing the blastocyst to penetrate into the endometrial mucosa. This event is

called *implantation*. Implantation usually occurs in the dorsal wall of the uterus, but it may also occur in various *ectopic sites* in the uterine tube, in the cervix, or in the peritoneal cavity. Ectopic pregnancies are not usually viable. Tubal pregnancies are actually dangerous for the mother, because rupture of blood vessels and hemorrhage will result as the embryo enlarges.

Once the blastocyst is fully implanted within the endometrium, the damaged endometrial epithelium will heal and cover the embryo so that, in effect, the human embryo does not grow in the uterine cavity, but within the uterine endometrial wall. After implantation, the trophoblast will proliferate to form the *chorionic villi*, which will exchange nutrients, respiratory gases, and metabolites with the *maternal blood vessels* through special blood sinuses. Later on, the chorionic villi and maternal vessels form a separate, anatomically distinct organ called the placenta.

HCG FUNCTION. After implantation, trophoblast cells in the chorionic villi secrete a peptide hormone called *human chorionic gonadotropin* (HCG) into the maternal blood. HCG acts like LH and promotes continued and enhanced estrogen and progesterone secretion from the corpus luteum in the ovary. These hormones in turn maintain the endometrium in optimum condition for gestation of the embryro. During the first week after implantation, the inner cell mass transforms first into two layers and finally into three germinal layers (endoderm, ectoderm, and mesoderm). The cells of these layers will proliferate, migrate, and differentiate to give rise to the tissues and organs of the developing embryo.

MULTIPLE BIRTH. The release of more than one egg at ovulation results in the formation of two or more zygotes. Each of these will implant separately, resulting in multiple pregnancies and *fraternal twins*. These may be of the same or opposite sex. If, however, the two blastomeres from a single zygote separate at the first cleavage division, or if the single inner cell mass divides into two separate masses, then each blastomere or cell mass will proceed to form an independent embryo. Because these embryos share a common genetic origin (genotype, copies of the same sets of genes), they will be alike in sex and with respect to phenotype (*identical twins*). Identical twins may share a placenta.

CN: Use red for P. Use light colors throughout except for structures, L, N, and O, which receive dark colors. Use your lightest colors for H and M.
1. Begin with the entrance of the ovum into the uterine tube. Note that the zona pellucida (K) and polar bodies (L) titles are in the upper right corner. Color the day numbers gray. Color the ectopic implantation (N) sites (marked by large asterisks).
2. Continue with the three-dimensional drawing of the later blastocyst at day 6 and the large drawing of day 12. Color in the uterine endometrium (M) in the large drawings before dotting in the lysosomal enzymes (O) in day 6.
3. Color the three hormonal influences in the lower right corner.
4. Color the diagrams illustrating twin formation.

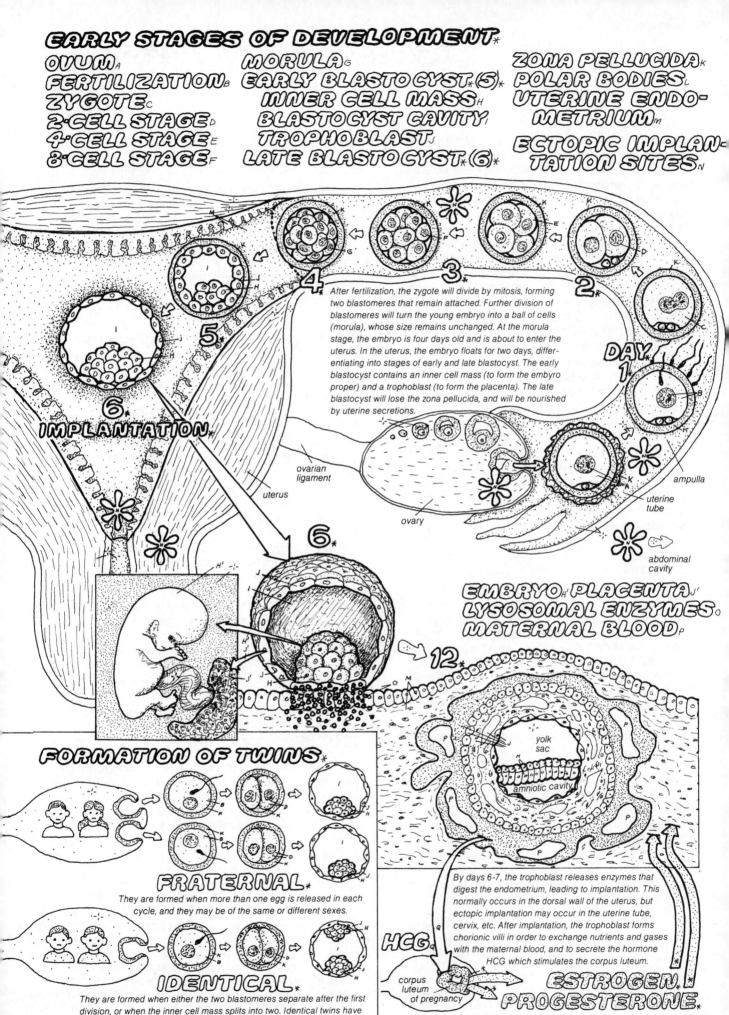

EARLY STAGES OF DEVELOPMENT*

OVUM_A
FERTILIZATION_B
ZYGOTE_C
2-CELL STAGE_D
4-CELL STAGE_E
8-CELL STAGE_F

MORULA_G
EARLY BLASTOCYST* (5)*
INNER CELL MASS_H
BLASTOCYST CAVITY_I
TROPHOBLAST_J
LATE BLASTOCYST* (6)*

ZONA PELLUCIDA_K
POLAR BODIES_L
UTERINE ENDO-
METRIUM_M
ECTOPIC IMPLAN-
TATION SITES_N

After fertilization, the zygote will divide by mitosis, forming two blastomeres that remain attached. Further division of blastomeres will turn the young embryo into a ball of cells (morula), whose size remains unchanged. At the morula stage, the embryo is four days old and is about to enter the uterus. In the uterus, the embryo floats for two days, differentiating into stages of early and late blastocyst. The early blastocyst contains an inner cell mass (to form the embryo proper) and a trophoblast (to form the placenta). The late blastocyst will lose the zona pellucida, and will be nourished by uterine secretions.

DAY 1*

2*

3*

4*

5*

6* IMPLANTATION*

ovarian ligament

uterus

ovary

ampulla

uterine tube

abdominal cavity

12*

EMBRYO_H' PLACENTA_J'
LYSOSOMAL ENZYMES_O
MATERNAL BLOOD_P

yolk sac

amniotic cavity

FORMATION OF TWINS*

FRATERNAL*

They are formed when more than one egg is released in each cycle, and they may be of the same or different sexes.

IDENTICAL*

They are formed when either the two blastomeres separate after the first division, or when the inner cell mass splits into two. Identical twins have the same sex, genotype, and similar phenotypes.

By days 6-7, the trophoblast releases enzymes that digest the endometrium, leading to implantation. This normally occurs in the dorsal wall of the uterus, but ectopic implantation may occur in the uterine tube, cervix, etc. After implantation, the trophoblast forms chorionic villi in order to exchange nutrients and gases with the maternal blood, and to secrete the hormone HCG which stimulates the corpus luteum.

HCG

corpus luteum of pregnancy

ESTROGEN*
PROGESTERONE*

REGULATION OF PREGNANCY AND PARTURITION

PREGNANCY. The successful implantation of the blastocyst will be followed by the development of special cells in the immature *placenta* that secrete an *LH*-like gonadotropic hormone (*HCG, human chorionic gonadotropin*) into the maternal blood. This hormone acts as a signal from the young embryo to the mother's *corpus luteum*. In response to this hormonal stimulation, the corpus luteum will grow further, forming the *"corpus luteum of pregnancy,"* which will secrete large amounts of *progesterone* and *estrogen* for the remainder of the *pregnancy*. Not only are estrogen and progesterone important for the maintenance and growth of the *endometrium*, but high levels of estrogen promote growth and proliferation of the *myometrium* (uterine smooth muscle wall); both hormones stimulate the growth of the breast (*mammary glands* and fat deposits).

With further development of the placenta, HCG will stimulate other cells in this versatile organ to secrete increasing amounts of estrogen and progesterone. By the second trimester of pregnancy, this secretion becomes so prodigious as to make the contribution of the ovaries insignificant. The HCG level increases profoundly during the first trimester and falls off gradually thereafter; but it continues to stimulate the ovary and placenta to secrete estrogen and progesterone.

The rising estrogen and progesterone levels from the second week of conception will inhibit, through negative feedback, the secretion of pituitary gonadotropins. In the absence of FSH and LH, further follicular development and ovulation will not occur during pregnancy. At the same time, in the presence of high levels of estrogen and progesterone, menstruation will not occur. Thus, in a healthy woman, missing a period is a well-known sign of pregnancy.

The appearance of HCG in maternal blood and urine has been the basis of most pregnancy tests. The early tests involved treatment of laboratory animals with samples of urine from the women. If the women were pregnant, the test would result in egg laying or luteal development in laboratory animals. Modern tests rely on immunochemical assays of HCG; pregnancy can be determined as early as 8-10 days after conception by blood tests and two weeks after by urine tests.

Another protein hormone with properties similar to growth hormone and prolactin has recently been discovered in pregnant women. This hormone, called *human chorionic somatomammotropin* (HCS, also human placental lactogen), is secreted in increasing amounts throughout gestation from the placenta, but only into the maternal blood. HCS antagonizes the action of maternal insulin by promoting fatty acid utilization by *maternal tissues*, so that glucose and amino acids can be spared for the fetus, which is heavily dependent on these substances for growth. In this way, HCS is indirectly involved in control of fetal growth, because in its deficiency, less glucose and amino acids may reach the fetus, resulting in decreased growth. HCS may also be involved in the growth of maternal mammary glands.

During the embryonic period (weeks 1-8), the embryo's development will consist largely of proliferation and differentiation of cells and tissues, resulting in organogenesis, the formation of the organs and systems. Major organs are formed during weeks 4-8, making this period a significant and critical one in terms of the effects of drugs and other teratological agents on embryonic development. By the third month, the embryo is called a fetus. The fetal period (3-9 months) is characterized chiefly by growth, but differentiation in several systems still continues.

PARTURITION. Throughout pregnancy, under the influence of estrogen, the smooth muscle cells of the uterine myometrium increase in number and size in order to support the fetus during pregnancy and expel it during labor and delivery. Mild contractions of the uterus are evident beginning with the fourth month, but usually about 270 days after conception, the muscular wall of the uterus will begin strong and rhythmic contractions resulting in birth (i.e., the expulsion of the fetus through the cervix and vagina — birth canal). The mechanisms of *parturition* are not fully understood, but the hormones estrogen, *oxytocin*, and *prostaglandins*, as well as *relaxin*, are believed to be involved.

A few days before birth, progesterone levels in the maternal blood drop, possibly due to metabolic changes in the placenta. Estrogen, unchecked by progesterone, increases the excitability of the uterine smooth muscles. This initiates labor. Just before contractions, prostaglandins are increased in the blood. These hormones originate from uterine glands and act on the uterine myometrium to increase its contractility. During delivery, the passage of the fetal head, being its largest part, dilates the cervix, activating the *cervical stretch receptors*. These activate a neurohormonal reflex. Sensory nerves from the cervical stretch receptors will stimulate the *hypothalamus* and *posterior pituitary* to release the hormone oxytocin. During pregnancy, the number of *receptors for oxytocin* is increased by estrogen action. Oxytocin exerts powerful contractile effects on the uterus. The release of oxytocin increases until the head passes entirely through the cervix, and the fetus is delivered. Oxytocin will also help to expel the *"afterbirth,"* or placenta, which occurs shortly after the birth of the baby. Parturition can be induced by injections of oxytocin. The hormone can also be given during labor to aid in delivery. In some animals, such as the sheep, adrenal steroids from the fetus provide the signal for the onset of labor. To facilitate birth, another peptide hormone, relaxin, is secreted during gestation by the corpus luteum of pregnancy and the placenta. Relaxin will soften the cervix as well as the ligaments and joints of the pelvic bones.

CN: Use same colors for the first four hormones (A-D) that were used in the earlier plates of this chapter. Use a dark color for F and N and a light color for O.
1. Begin with the titles of the upper half, and color all the material in the large rectangle (including the month numbers). Then color the pregnant woman, beginning with the secretion of HCG. Note the large word: PLACENTA with an HCG arrow shown secreted by and stimulating the placenta to produce greater amounts of estrogen, progesterone, and HCS, all represented by a solid line of arrows.
2. Color the parturition panel beginning at the asterisk. Complete the left illustration before going on to the next. Note that the wall of the cervix is left uncolored.

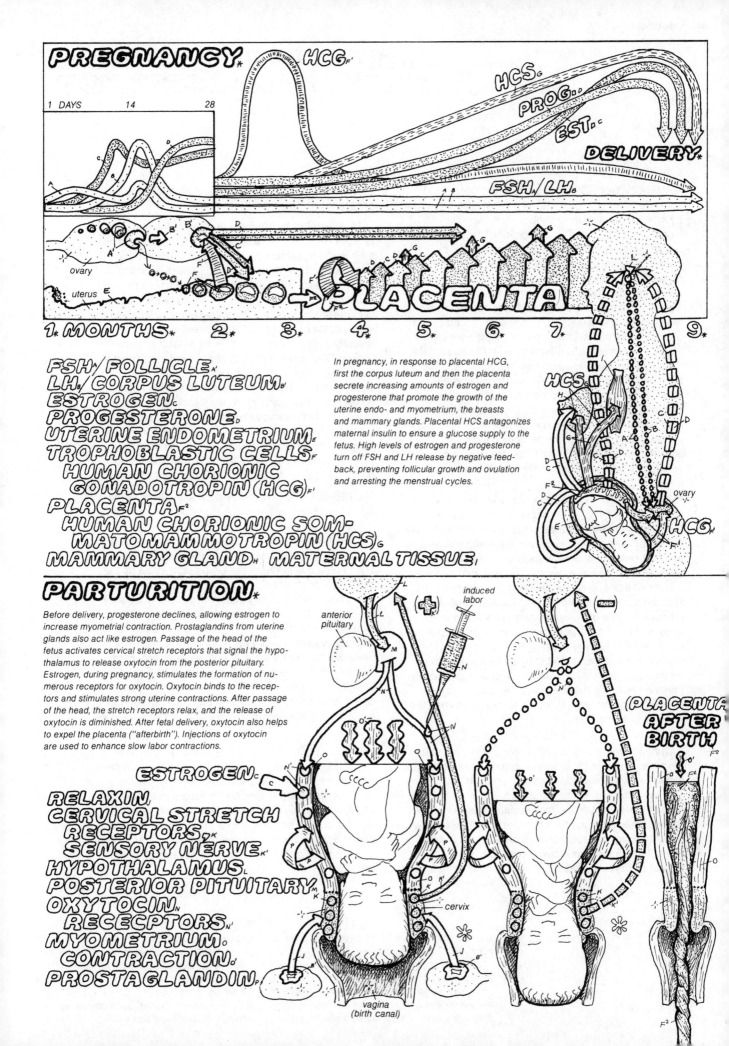

PREGNANCY*

HCG_F'
HCS_G
PROG._D
EST._C
DELIVERY*
FSH/LH_B

1 DAYS   14   28

ovary

uterus

PLACENTA

1* MONTHS*   2*   3*   4*   5*   6*   7*   9*

FSH / FOLLICLE_A'
LH_B / CORPUS LUTEUM_B'
ESTROGEN_C
PROGESTERONE_D
UTERINE ENDOMETRIUM_E
TROPHOBLASTIC CELLS_F
HUMAN CHORIONIC
GONADOTROPIN (HCG)_F'
PLACENTA_F²
HUMAN CHORIONIC SOM-
MATOMAMMOTROPIN (HCS)_G
MAMMARY GLAND_H   MATERNAL TISSUE_I

In pregnancy, in response to placental HCG, first the corpus luteum and then the placenta secrete increasing amounts of estrogen and progesterone that promote the growth of the uterine endo- and myometrium, the breasts and mammary glands. Placental HCS antagonizes maternal insulin to ensure a glucose supply to the fetus. High levels of estrogen and progesterone turn off FSH and LH release by negative feed-back, preventing follicular growth and ovulation and arresting the menstrual cycles.

HCS_G
ovary
HCG_F'

PARTURITION*

Before delivery, progesterone declines, allowing estrogen to increase myometrial contraction. Prostaglandins from uterine glands also act like estrogen. Passage of the head of the fetus activates cervical stretch receptors that signal the hypo-thalamus to release oxytocin from the posterior pituitary. Estrogen, during pregnancy, stimulates the formation of nu-merous receptors for oxytocin. Oxytocin binds to the recep-tors and stimulates strong uterine contractions. After passage of the head, the stretch receptors relax, and the release of oxytocin is diminished. After fetal delivery, oxytocin also helps to expel the placenta ("afterbirth"). Injections of oxytocin are used to enhance slow labor contractions.

anterior pituitary

induced labor
(+)   (−)

(PLACENTA AFTER BIRTH)

ESTROGEN_C

RELAXIN_J
CERVICAL STRETCH
RECEPTORS_K
SENSORY NERVE_K'
HYPOTHALAMUS_L
POSTERIOR PITUITARY_M
OXYTOCIN_N
RECEPTORS_N'
MYOMETRIUM_O
CONTRACTION_O'
PROSTAGLANDIN_P

cervix

vagina
(birth canal)

REGULATION OF MAMMARY GROWTH AND LACTATION

MAMMARY GLANDS. In mammals, the newborn is nourished for varying periods of postnatal life directly by *milk* secreted by the mother's *mammary glands*, located in the *breast*. The number of *nipples* and the size of the breast varies in different mammals. In human females, the two breasts are large, due to particularly large amounts of *fatty tissue* interspersed between the branches of the mammary glands. The mammary glands are exocrine glands containing extensive *alveoli* and *ducts*. The milk is produced by the alveolar cells, which obtain raw materials from the blood, synthesize the nutrients in the milk and secrete them into the alveolar sacs. Then the milk flows through the small ducts, which converge to form larger ducts, which finally emerge from the nipple. Around the mammary ducts, specific *myoepithelial cells* (smooth muscle) can contract to force the milk out. In the nipple, special *touch* and *pressure receptors*, stimulated by sucking actions of the infant's lips, are important in *milk ejection*, a neuroendocrine reflex (see below).

MAMMARY GROWTH. During adolescence, in response to the rising levels of sex steroids from the ovary, the mammary glands begin to develop. *Estrogen* enhances duct growth. *Progesterone* enhances alveolar development, which is sparse in adolescents. Several other hormones (*insulin, growth hormone, prolactin* and *glucocorticoids*) are also necessary for the successful actions of sex steroids at this stage. Milk is produced only if the alveolar cells are stimulated by large amounts of the anterior pituitary hormone *prolactin* and if the levels of estrogen and progesterone are low. Prolactin is regulated by a *release-inhibiting hormone* (dopamine) and also by a release-promoting hormone from the hypothalamus. In adolescent girls and nonpregnant women, low prolactin levels keep the alveolar cells inactive.

During pregnancy, the high levels of estrogen and progesterone from the placenta, as well as increasingly higher levels of prolactin, stimulate prodigious development of the mammary glands in preparation for milk production. However, the high levels of sex steriods inhibit the milk-forming effect of prolactin on the alveolar cells, avoiding unnecessary milk production. The placental hormone *chorionic somatomammotropin*, as well as cortisol, insulin, and growth hormones, also influences growth of the mammary glands during pregnancy.

MILK FORMATION. At birth, the loss of the placenta removes the major source of sex steroids, eliminating their inhibitory effect on the prolactin's action on alveolar cells. This stimulates milk production, which in women begins about one to three days after birth. Just before, and around birth, the mammary glands secrete very small amounts of a thick substance, *colostrum*, which contains no fat and little water but otherwise resembles milk. Colostrum may be a source of antibodies. Continued milk production is ensured by adequate secretion of prolactin, which is maintained by the sucking-induced sensory stimulation from the nipple tactile receptors. Each episode of sucking causes a surge in hypothalamic release hormone for prolactin (as well as a reduction in hypothalamic inhibitory hormone for prolactin "dopamine"). This causes a surge in prolactin and in milk production. Regular artificial massages of the nipples will have the same effect.

MILK EJECTION. The mechanical stimulation of the nipple also enhances milk ejection from the mammary ducts. Secreted milk accumulates in the alveoli and ducts but will not flow out unless the smooth muscle cells (myoepithelial cells) around the ducts contract. The contraction of these cells is brought about by the action of the hormone *oxytocin* from the posterior pituitary gland. Thus, sensory impulses generated by sucking stimuli will travel up the sensory nerves, reaching the brain and activating the hypothalamus-posterior pituitary system, promoting oxytocin release. The hormone will then enhance milk outflow as stated. In the absence of such regular sensory stimuli from the nipples, the secreted milk will accumulate in the glands, cause inflation of the ducts and pain, and, in the long run, diminish milk production, leading to drying up of the breast.

MILK COMPOSITION. Milk is a complete source of nutrition for the newborn during the first year. It contains carbohydrates, protein, and fat as well as vitamins, minerals, and water. Compared to cow's milk, the milk sugar, lactose, is present in higher amounts in human milk, but protein content is less, and the fat content is similar. For optimal growth, the mineral and vitamin content of human milk is nearly adequate, except for iron and vitamin D.

CN: Use same colors as on the previous page for estrogen (F) and progesterone (G).
1. Color the stages of breast development, completing each before going on to the next. Note the increased size of the prolactin (J) arrow to reflect the greater amount of flow. Note too, that in the pregnancy panel, part of the estrogen (F) output has the effect of blocking the prolactin (J) effect on the breast. In the lactation panel, the blowup of a portion of the breast development shows the secretion of milk globules from the alveoli. These are left uncolored.
2. Color the chart illustrating the comparison between mother's and cow's milk.

ADOLESCENCE *

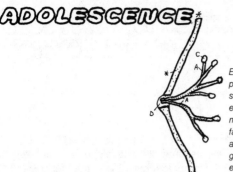

Early in puberty, the mammary glands are poorly developed. The breast contains little subcutaneous fat. During adolescence, estrogen promotes the development of the mammary ducts and the deposition of fatty tissue, while progesterone induces alveolar development. Growth hormones, glucocorticoids, and insulin are also necessary. Prolactin secretion from the anterior pituitary is low due to the strong inhibitory effect of hypothalamic-inhibiting hormone.

YOUNG ADULT *

hypothalamus

ovary

anterior pituitary

During pregnancy, increased estrogen and progesterone markedly promote mammary growth. Placental somato-corticoids and insulin are also needed for breast growth. Prolactin levels increase in pregnancy, but high estrogen and progesterone levels prevent stimulation of alveoli by prolactin, resulting in the lack of milk secretion.

PREGNANCY *

placenta

LACTATION *

posterior pituitary

sensory nerve

THE BREAST *
DUCT A
FATTY TISSUE B
ALVEOLI C
NIPPLE D, TOUCH RECEPTOR
MYOEPITHELIAL CELL E

HORMONES *
ESTROGEN F
PROGESTERONE G
PROLACTIN-INHIBITING HORMONE H
PROLACTIN-RELEASING HORMONE I
PROLACTIN J
HUMAN CHORIONIC SOMATOMAMMOTROPIN K
GLUCOCORTICOID L
INSULIN M
GROWTH HORMONE N

COMPARISON OF MILK *
MOTHER'S P
COW'S Q

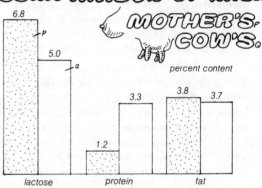

percent content

6.8 P
5.0 Q
3.3
1.2
3.8 3.7

lactose   protein   fat

Milk contains all the nutrients needed for infant growth. Human milk contains carbohydrates (lactose), protein, fat, minerals, and vitamins. Cow's milk contains the same nutrients, but in different proportions.

MILK FORMATION: J PROLACTIN J
MILK EJECTION: O OXYTOCIN O

After birth, estrogen and progesterone levels decline sharply. This permits prolactin to stimulate the alveoli, and milk is produced. The infant's sucking activity stimulates sensory receptors in the nipple and areola. Impulses are sent to the hypothalamus, increasing secretion of prolactin releasing hormone. This stimulates surges of prolactin release, ensuring continued milk formation. Sucking stimuli also causes release of oxytocin from the posterior pituitary. Oxytocin then stimulates the contractions of myoepithelial cells of the mammary ducts, forcing milk out of the nipple.

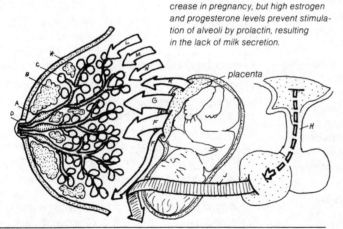

SEX DETERMINATION & SEXUAL DEVELOPMENT

SEX CHROMOSOMES. In mammals, individuals are divided into male and female. The primary determinant of the individual's gender is a genetic (chromosomal) event occurring at fertilization and regulated by two types of combinations of the two sex chromosomes, X and Y. All normal male individuals contain, in the nuclei of their cells, one X and one Y chromosome, in addition to 22 pairs of somatic chromosomes. As a result of meiosis during spermatogenesis, spermatozoa may carry either an X or a Y chromosome, along with 22 somatic chromosomes. The cells of the normal female, however, contain 22 pairs of somatic chromosomes and one *pair* of X chromosomes. Thus, meiotic division of primary oocytes can give rise only to X-chromosome-carrying eggs. At conception, fertilization of the egg by a sperm carrying an X chromosome will result in a zygote with XX combination (female); while a Y-carrying sperm will generate the XY combination (male). Thus, the genetic sex is determined by the father. Although X- and Y-carrying sperm are morphologically and physiologically somewhat different, attempts to separate them have not been successful. Also, although there should be equal numbers of these two types of sperm, for unknown reasons (possibly the lighter weight of Y-carrying sperm) more male zygotes are presumably formed, because not only are the majority of spontaneously aborted embryos male, but at birth males still outnumber females (107 to 100).

SEX CHROMATIN. The sex of the individual can also be determined by the presence of the Barr body in cells of females and the F body in those of males. The Barr body, or sex chromatin, is a piece of chromatin associated with the inactive X chromosome and easily visible near the nuclear membrane (of the two X chromosomes in the female, one will become inactive during early embryonic development). The F body is part of the Y chromosome that will fluoresce when treated with a special fluorescent dye.

DEVELOPMENT OF GONADS. Up to the sixth week, the embryo does not show signs of sexual differentiation. The primordial gonad appears identical in both sexes and is sexually bipotential. In the genetically male individual, action of a Y chromosome gene in the cells of the *medulla* of the primordial gonad leads to the formation of a specific surface antigen, *H-Y antigen*, which will cause the destruction of the *cortex* (presumptive ovary), promoting the development of the medullary zone into the embryonic testes. The *Leydig cells* in the latter secrete *testosterone* as well as a protein substance (MRF, Mullerian regression factor).

DEVELOPMENT OF SEX ORGANS. The structures forming the *internal genitalia* (Mullerian and Wolffian ducts) and the *external genitalia* are at first sexually indifferent and bipo-

tential. In the male embryo, testosterone will directly act on the embryonic structures, promoting development of the external genitalia. MRF promotes regression of the Mullerian ducts (forerunners of the female internal genitalia); MRF plus testosterone will act on the Wolffian ducts to promote the formation of the male internal genitalia. In genetic females, in the absence of Y chromosomes and consequently of H-Y antigen, the cortical zone will develop into the embryonic ovary, which is nonsecretory. In the absence of MRF and testosterone, the Wolffian ducts will degenerate, and the Mullerian ducts and other structures will spontaneously form the female internal and external genitalia.

SEXUAL DIFFERENTIATION OF THE HYPOTHALAMUS. Testosterone is also responsible for differentiation of the *male type* of *hypothalamus*, promoting a *continuous* (non-cyclical) mode of secretion of FSH and LH. In the rat, this occurs during the first days of postnatal life. The differentiation of the hypothalamus will also include induction of male sexual behavior, which will be manifested after puberty. In the absence of testosterone (as in normal females), the hypothalamus will spontaneously develop into the *female type*, showing the female regulatory mechanisms of *GnRH* and of *FSH* and *LH* (*cyclical*), as well as female sexual behavior. It is not yet known whether, how, or when similar influences occur in humans. The actions of sex hormones on sexual maturation during puberty are discussed in plates 122 and 144.

ABNORMALITIES OF SEXUAL DEVELOPMENT. Individuals with one genetic sex but with genitalia of the opposite sex are called pseudohermaphrodites. If a female embryo is exposed to abnormally high levels of androgens (e.g., from tumors of the adrenal cortex), it will develop male external genitalia and deranged internal genitalia (female pseudohermaphroditism).

Occasionally, during meiosis of gametes, sex chromosomes are disproportionately divided between gametes, resulting in a zygote lacking a sex chromosome or one having extra numbers. The X chromosome contains genes that are essential for life; embryos with no X chromosomes are aborted spontaneously. Individuals with trisomy of X chromosomes are called "superfemales" and are not abnormal. In the XO pattern (no Y chromosome, Turner syndrome), the gonads do not develop or are abnormal, but the individual will have female genitalia, which will not mature at puberty due to the absence of sex hormones. Body development is also stunted or abnormal. In the XXY pattern (Kleinfelter's syndrome), which occurs commonly, testes and male genitalia will develop, and secondary sexual characteristics may be normal, but seminiferous tubules will not develop, making the individual sterile.

CN: Use very light colors for D, E, J, and K. Use colors previously used for estrogen (Q), progesterone (G), FSH (N), LH (O).
1. Begin at the bottom, coloring in the mature testis and ovary, and color the long arrows up the sides to the top panel, where those colors become primary spermatocytes and oocytes. In the top panel,

carefully note the letter labels.
2. Color the embryonic stage, first doing the developing testis. Begin with the testis closest to the titles, and then do the outer one. Note that in the outer one, except for the Leydig cells (G), the entire structure receives the medulla color (E).
3. Color the neonatal stage and the puberty stage.

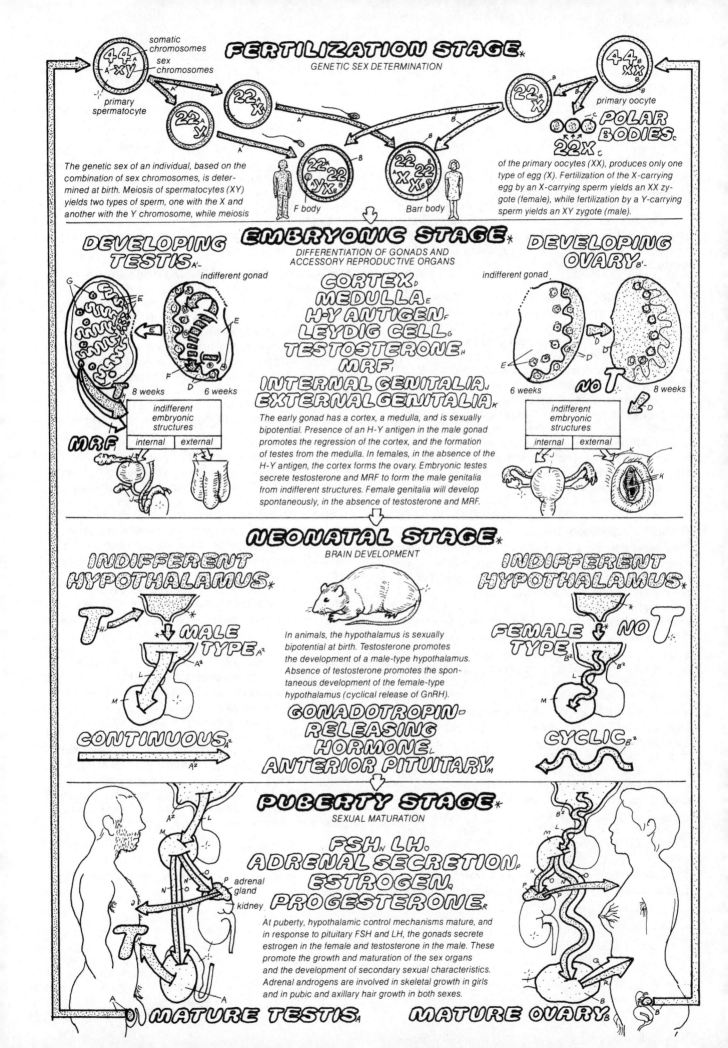

FERTILIZATION STAGE*

GENETIC SEX DETERMINATION

somatic chromosomes
sex chromosomes
primary spermatocyte

primary oocyte

POLAR BODIES

The genetic sex of an individual, based on the combination of sex chromosomes, is determined at birth. Meiosis of spermatocytes (XY) yields two types of sperm, one with the X and another with the Y chromosome, while meiosis

F body

Barr body

of the primary oocytes (XX), produces only one type of egg (X). Fertilization of the X-carrying egg by an X-carrying sperm yields an XX zygote (female), while fertilization by a Y-carrying sperm yields an XY zygote (male).

EMBRYONIC STAGE*

DIFFERENTIATION OF GONADS AND ACCESSORY REPRODUCTIVE ORGANS

DEVELOPING TESTIS A'-

indifferent gonad

8 weeks   6 weeks

indifferent embryonic structures

internal	external

MRF

CORTEX D
MEDULLA E
H-Y ANTIGEN F
LEYDIG CELL G
TESTOSTERONE H
MRF I
INTERNAL GENITALIA J
EXTERNAL GENITALIA K

The early gonad has a cortex, a medulla, and is sexually bipotential. Presence of an H-Y antigen in the male gonad promotes the regression of the cortex, and the formation of testes from the medulla. In females, in the absence of the H-Y antigen, the cortex forms the ovary. Embryonic testes secrete testosterone and MRF to form the male genitalia from indifferent structures. Female genitalia will develop spontaneously, in the absence of testosterone and MRF.

DEVELOPING OVARY B'-

indifferent gonad

6 weeks   NO T   8 weeks

indifferent embryonic structures

internal	external

NEONATAL STAGE*

BRAIN DEVELOPMENT

INDIFFERENT HYPOTHALAMUS*

T   MALE TYPE A²

CONTINUOUS A²

In animals, the hypothalamus is sexually bipotential at birth. Testosterone promotes the development of a male-type hypothalamus. Absence of testosterone promotes the spontaneous development of the female-type hypothalamus (cyclical release of GnRH).

GONADOTROPIN-RELEASING HORMONE L
ANTERIOR PITUITARY M

INDIFFERENT HYPOTHALAMUS*

FEMALE TYPE B²   NO T

CYCLIC B²

PUBERTY STAGE*

SEXUAL MATURATION

FSH N, LH O
ADRENAL SECRETION P
ESTROGEN Q
PROGESTERONE R

adrenal gland
kidney

At puberty, hypothalamic control mechanisms mature, and in response to pituitary FSH and LH, the gonads secrete estrogen in the female and testosterone in the male. These promote the growth and maturation of the sex organs and the development of secondary sexual characteristics. Adrenal androgens are involved in skeletal growth in girls and in pubic and axillary hair growth in both sexes.

MATURE TESTIS A   ## MATURE OVARY B

THE PHYSIOLOGY OF BIRTH CONTROL

The physiologically based approaches to birth control act by preventing conception (*contraception*) or implantation. The simplest contraceptive approaches are withdrawal and temporary abstinence (*rhythm method*). In the withdrawal method, widely practiced in older cultures and developing countries, the penis is withdrawn shortly before orgasm, allowing for external ejaculation. This is not a safe (i.e., effective) method.

RHYTHM METHOD. In temporary abstinence (rhythm or "symptothermal" method), vaginal intercourse is avoided during the period when the female is most fertile. The method is based on the timing of *ovulation* and the survival period of sperm and egg within the female genitalia. Time of ovulation can be estimated by measuring *basal body temperature* every morning before leaving bed. 1-2 days after ovulation, there occurs a rise of about 0.5°C (1°F), believed to be associated with the metabolic effects of progesterone from the corpus luteum. Time of ovulation can also be checked approximately by daily examination of cervical mucus. The mucus becomes increasingly thin and distensible under the influence of preovulation estrogen. It is thinnest at ovulation and dries in a fernlike arborizing pattern if spread thinly on a glass slide. After ovulation, the mucus becomes thick and indistensible in response to progesterone and no longer forms such a pattern upon drying.

Freshly ovulated eggs are mature, but within 1-2 days can age and become overripe, unable to be fertilized. Sperm may survive 3-4 days in the female genitalia, particularly those stored in the cervical mucosa. Thus, to avoid conception, sperm must not be deposited in the vagina for at least 4 days before and 3 days after ovulation, the remaining time of the monthly period constituting a relatively safe period.

MECHANICAL/CHEMICAL BARRIERS. One way to prevent sperm from reaching the egg is the use of mechanical barriers. The *condom*, a nonporous sheath made of rubber or gut, is a device used to cover the penis, preventing sperm deposition within the vagina. A *diaphragm* is a plastic dome-shaped object placed deep in the vagina to block the passage of deposited sperm into the cervix. Similar in operation to the diagragm, but more secure, is the *cervical cap*, a plastic object made to fit tightly over the cervical protrusion into the vagina. *Chemical spermicides* containing acidic or other specific antisperm substances designed to destroy sperm in the vagina can be applied in the form of creams, gels, foams, or douches usually before intercourse. Diaphragms should be used in conjunction with spermicide foams or gels for added protection.

IUDs. It is known that the presence of a foreign body in the uterus will inhibit pregnancy. IUDs (*intrauterine devices*) have been developed to exploit this response. An IUD is a thin plastic or copper wire shaped in the form of a T, a loop, or a coil placed for long periods in the uterine cavity. The mechanism by which IUDs prevent pregnancy is not completely understood, but interference with implantation of the young embryo in the endometrium is widely suspected.

VASECTOMY/TUBAL LIGATION. An effective way to prevent the meeting of sperm and egg is surgical *sterilization*, which involves cutting and tying the uterine tubes in women (*tubal ligation*) or the *vas deferens* in men (*vasectomy*). In sterilized women, the ligated portion will prevent passage of the egg as well as of sperm, but all hormonal and other aspects of sexual physiology are intact. In vasectomized men, sperm production and androgen secretion are normal, but the ejaculate contains seminal plasma only. The sperm emerging from the epididymis accumulate behind the ligated vas, where they age. After death, the sperm are phagocytized by macrophages.

THE "PILL". One *oral contraceptive* method is the *"pill,"* which is based on the negative-feedback effect of female sex steroids on the hypothalamus-pituitary axis. In a typical situation, a woman takes one pill per day for 21 days, beginning with the fifth day of menstruation. These pills contain small amounts of a synthetic estrogenlike compound and larger amounts of a synthetic progesteronelike compound. In the body, these substances mimic the effects of natural estrogen and progesterone hormones. However, because their levels in the blood are suddenly raised from the first day the pill is taken, the hypothalamus-pituitary axis, sensing high amounts of the "hormone," will shut off GnRH, FSH, and LH levels, a response similar to that occurring during early pregnancy. In the absence of FSH and LH, follicular development and ovum maturation, as well as ovulation, will not occur (as in pregnancy), making fertilization extremely unlikely. Meanwhile, the estrogenlike and progesteronelike substances in the pill promote endometrial proliferation and secretion (not a purpose but a side effect). A day or two after a woman stops taking the pill, the endometrium, losing its support, will slough off and bleed, resulting in menstruation. Women desiring pregnancy can regain their normal cycles within one to several months after discontinuing the pill. However, pregnancy within the first 1-3 months is not encouraged because of the possibility of multiple ovulation and pregnancy that may occur due to excessive rebound secretion of pituitary gonadotropins.

Gossypol, a cottonseed oil compound, is under study in China as a reversible male chemical contraceptive. It is believed to inhibit spermatogenesis reversibly by inactivating the spermatids.

CN: Use red for A and a dark color for F.
1. Begin with the three forms of abstinence.
2. Color the various methods of contraception. Note the dotted uncolored areas, which represent sperm deposits.
3. Color the two most common sterilization sites.

ABSTINENCE*

CALENDAR (RHYTHM METHOD)*

In the rhythm method, coitus is abstained from during a week-long period when the woman is fertile. This period, from a minimum of 4 days before ovulation to 3 days after, is based on the maximum survival time of sperm (4 days) and egg (2 days), in the female reproductive tract.

BASAL TEMPERATURE F

Basal body temperature, taken early in the morning, before leaving bed, shows a rise of 0.5°C (1°F) with ovulation. This lasts until the next menstruation. This elevation is a marker for ovulation.

MUCUS G

Cervical mucus is thick after menstruation, but will thin out in the next several days through the action of estrogen. It is thinnest at ovulation and forms a fern-like pattern when dried on a glass slide. After ovulation, it begins to thicken due to the effect of progesterone.

MENSTRUATION A
OVULATION B
SPERM VIABILITY C
EGG VIABILITY D
SAFE DAYS E

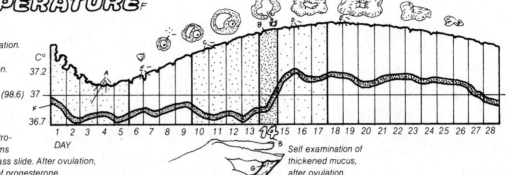

C°
37.2
(98.6) 37
36.7

1 2 3 4 5 6 7 8 9 10 11 12 13 **14** 15 16 17 18 19 20 21 22 23 24 25 26 27 28
DAY

Self examination of thickened mucus, after ovulation.

CONTRACEPTIVES*

CONDOM H   DIAPHRAGM I   CERVICAL CAP J   CHEMICAL SPERMICIDE K   INTRAUTERINE DEVICE (IUD) L

Condoms, made from rubber, plastic, or gut, are used to sheathe the penis during coitus, preventing deposition of ejaculated sperm in the vagina. They also protect against transmission of infectious agents (bacteria, viruses).

The diaphragm is a dome-shaped, plastic or rubbery barrier, placed deep in the vagina to prevent the passage of deposited sperm to the cervix.

Cervical caps, when in place, fit tightly over the protrusion of the cervix into the vagina. This too, is a mechanical barrier to the cervix.

Spermicidal foams, creams, jellies, or douches act to chemically destroy sperm cells, before they enter the uterus.

IUDs are plastic or copper wiry objects, T or coiled-shaped, that are placed permanently inside the uterus. Their presence inhibits the implantataion of the embryo into the endometrium.

ORAL CONTRACEPTIVES M

SYNTHETIC ESTROGEN AND PROGESTERONE

Oral contraceptives are pills containing synthetic estrogen and progesterone, that are taken by women for 21 days after menstruation. The rapidly elevated levels of hormone-like substances in the blood, will inhibit FSH and LH release, preventing follicular growth and ovulation.

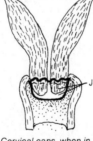

hypothalamus

GnRH P
anterior pituitary
FSH N   LH O
ovary (no follicle growth)

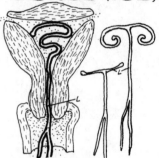

NO OVULATION
blood level of synthetic estrogen and progesterone
begin
stop
6   27

STERILIZATION*

TUBAL LIGATION Q
UTERINE TUBE Q'

By cutting and tying the uterine tubes, women can be permanently sterilized, because deposited sperm can no longer reach the ovulated egg.

VASECTOMY R
VAS DEFERENS R'

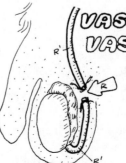

Cutting and tying the two vas deferens ducts is a simple operation which permanently obstructs the delivery of sperm through the vas deferens during ejaculation, thereby causing male sterility.

Index

Index

Index